Heterogeneous Kinetics at Elevated Temperatures

Heterogeneous Kinetics at Elevated Temperatures

Proceedings of an International Conference in Metallurgy and Materials Science held at the University of Pennsylvania September 8-10, 1969

Edited by
G. R. Belton
and
W. L. Worrell
School of Metallurgy and Materials Science
University of Pennsylvania
Philadelphia, Pennsylvania

PLENUM PRESS · NEW YORK–LONDON · 1970

Library of Congress Catalog Card Number 70-128401

ISBN 978-1-4684-8067-2 ISBN 978-1-4684-8065-8 (eBook)
DOI 10.1007/978-1-4684-8065-8

Softcover reprint of the hardcover 1st edition 1970
A Division of Plenum Publishing Corporation
227 West 17th Street, New York, N.Y. 10011

United Kingdom edition published by Plenum Press, London
A Division of Plenum Publishing Corporation, Ltd.
Donington House, 30 Norfolk Street, London W.C. 2, England.

PREFACE

This book contains nineteen of the twenty papers which were presented at the conference on "Heterogeneous Kinetics at Elevated Temperatures" held on September 8, 9 and 10, 1969 at the University of Pennsylvania.

The papers are concerned primarily with the kinetics of interfacial reactions between the gaseous phase and a condensed phase and include pertinent topics in microscopic and macroscopic transport phenomena. Several of the more recent advances in experimental techniques are covered, since progress in the interpretation of kinetic phenomena seems to be intimately connected with advances in experimental method. The discussion after each paper, as well as the general discussion at the end of the meeting, have been included in a strictly edited version; the philosophy being to include only those topics and questions which were not dealt with in the final versions of the manuscripts.

The Chairmen (editors) would like to express their thanks to Professor C. B. Alcock for vigorously leading the final discussion period and for his entertaining after-dinner remarks on the evening of the conference banquet. The Chairmen would also like to place on record their indeptedness to Dr. D. R. Gaskell who, with the assistance of Mr. P. C. Shumaker, was largely responsible for the efficient technical arrangements of the meeting; to Dr. P. Entner and the willing group of graduate students for recording the discussions and for other organizational assistance; to Miss L. Porreca for her capable handling of correspondence and registration; and to Miss G. Schindler for transcribing the discussion.

The conference was sponsored by the School of Metallurgy and Materials Science of the University of Pennsylvania through the good offices of Professor R. Maddin, Director. Initial financial support from the Board of the Pennsylvania Science and Engineering Foundation, Commonwealth of Pennsylvania, is gratefully acknowledged, as is the further support from Bethlehem Steel Corporation, Foote Mineral Company, Jones and Laughlin Steel Corporation, Leeds and Northrup Company, New Jersey Zinc Company, St. Joseph Lead Company,

Union Carbide Corporation, and United States Steel Corporation.

The editors thank Miss N. Newton for her patient re-typing of the entire manuscript and Mr. H. I. Kaplan for his willing assistance in the editing.

G. R. Belton

W. L. Worrell

CONTENTS

THE ACTIVATED COMPLEX IN SOME HIGH TEMPERATURE HETEROGENEOUS REACTIONS

Henry Eyring* and Sheng Hsien Lin**

Department of Chemistry, University of Utah*

Department of Chemistry, Arizona State University**

INTRODUCTION

Since Mendeleev's outstanding success in predicting the properties of four new elements from the periodic table, it has been clear that atomic structure largely determines thermodynamic properties. Since absolute rate theory postulates an activated complex in equilibrium with reactants, one can hope to predict successfully reaction rates by extrapolating from the kinetic behavior of neighboring atoms or molecules. The important factors in rate theory are (1) the mechanism; this fixes the structure of the activated complex, (2) the activation energy, (3) the vibrational frequencies, (4) the moments of inertia and (5) the transmission coefficient which fortunately may frequently be taken as unity.

For a reaction in which one bond is broken as another is formed, i.e., when an unbonded atom or radical attacks and breaks a bond in an exothermal reaction, the expected value of the activation energy is about 2 kcal. For the reverse endothermal reaction the heat of reaction must be added to the two kcal. In the case of very light atoms, as in the reaction $D + H_2$, the activation energy can rise to as high as 7 kcal. but for most known cases 2 kcal. is a more probable estimate.

When two diatomic molecules form from four monovalent atoms such as H_2 and X_2, where X is a halogen, the reaction involves breaking the weakest bond followed by an attack of either one or two atoms on the strong bond, whichever reaction is fastest. The estimated activation energy for a one atom attack is half the weakest bond plus 2 kcal. plus the endothermicity. The endothermicity is the difference between the bond broken and the bond formed in the three atom reaction. The competing two atom attack has an estimated activation energy of the dissociation energy of the weakest bond plus 4 kcal. These estimates predict correctly a one atom attack for the reactions involving F_2, Cl_2 and Br_2 on H_2 and a two atom attack in the case of I_2. The quantitative agreement with experiment is quite good. If one thinks of the solid as a large molecule, one is prepared to expect many parallels between homogeneous and heterogeneous reactions [1].

The periodic table can also be a guide in selecting materials for special uses. Thus when materials for the throats of missiles are needed to withstand erosion by the hot corrosive gases rushing through the throat, it is natural to try high boiling substances with suitable physical properties such as graphite and tungsten. Not only will they have less tendency to melt than low boiling substances, but other factors being equal, activation energies go up with the energy of melting or sublimation of a substance. We next consider briefly bonding and the energy of activation.

THE NATURE OF BONDING AND THE ENERGY OF ACTIVATION

Chemical bonding was inexplicable until the wave nature of particles was understood. Bohr was the first to propose a successful dynamic picture of the electrons. His three equations:

$$\frac{\mu v^2}{r} = \frac{e^2}{r^2} \tag{1}$$

$$E = 1/2\ \mu v^2 - \frac{e^2}{r} \tag{2}$$

$$nh = 2\pi rp \tag{3}$$

lead to the correct equation for the energy, E, of the hydrogen atom.

$$E = \frac{-2\pi^2 \mu e^4}{n^2 h^2} \tag{4}$$

Here μ, v, r, e and p are the reduced mass of the electron, its velocity, its radius, its charge and its momentum, respectively; n and h are the integral quantum number and Planck's constant. DeBroglie proposed later that each allowed orbit of an electron around a proton is an integral number of wave lengths. Thus:

$$n\lambda = 2\pi r \tag{5}$$

Combining this with equation (3) gives:

$$\frac{h}{\lambda} = p \tag{6}$$

which is DeBroglie's equation. Using DeBroglie's equation in calculating the kinetic energy we obtain:

$$E = \frac{h^2n^2}{8\pi^2r^2\mu} - \frac{e^2}{r} \tag{7}$$

Now according to the virial theorem twice the kinetic energy is equal to minus the potential energy. Using this equality one obtains for the allowed values of r:

$$r = \frac{n^2h^2}{4\pi^2\mu e^2} \tag{8}$$

This is also the value of r given by $\frac{\partial E}{\partial r} = 0$. Thus the allowed orbits correspond to values of r which are stationary states for the energy. The Bohr radius, a_o, corresponds to the value of r in equation (8) when n = 1.

Let us next consider how the term for the kinetic energy in equation (7) is important for bonding and in the activation energy of chemical reactions. The fact that two neutral hydrogen atoms bond with 103 kcals. while a proton is bound to a hydrogen with 60 kcals. is significant. Clearly the total charge on a molecule is less significant for bonding than the number of electrons doing the bonding. Further, two electrons bind the nuclei with nearly twice the strength of a one electron bond. Even though a pair of bonding electrons tend to correlate their motions to keep out of each others way they still repel each other enough to explain the difference between the bonding in H_2^+ and half the H_2 bond. The DeBroglie equation (6) shows us that the kinetic energy is inversely proportional to the wave length squared. Thus all particles are claustrophobic and there is a trade off between staying in a small volume to keep the average potential down and expanding this volume so as to lower the kinetic energy. Thus the two electrons on two hydrogen atoms can lower their energy by escaping from the region surrounding one hydrogen atom into the region surrounding the two, with the atoms moving to the best bonding distance simultaneously. The Pauli Principle, however, requires that the electrons sharing an orbital must have opposite spins. This provides the bases for an explanation of the activation energy in chemical reactions as well as the repulsion energy between colliding saturated molecules. In such a saturated molecule all the electrons are paired. Accordingly for the four atoms paired in the two initial bonds of a reaction to come close enough together to switch partners the bonding orbitals will necessarily overlap. This requires some of the crowded electrons to be at least partially promoted into upper states to comply with the Pauli principle. This promotional energy is thus an activation energy. Typically then an activated state involves crowding of the valence electrons which are going to switch their allegiance during the reaction. Thus if a reaction occurs on a surface capable of accommodating the electrons of the activated complex in lower lying unfilled orbitals, the promoted electrons will be partly drawn off into these unfilled orbitals and as a result the activation energy is lowered.

Ionizing solvents or ionizing surfaces circumvent the necessity of formation of a congested activated complex by breaking the initial bonds through ionization so that the ions can then regroup without passing through a state of electron congestion. In this case the ions are stabilized by solvation in which the solvent molecules orient their electron rich sides toward the positive ions and their electron deficient side toward the negative ions. In analogous fashion an ionic crystal can stabilize the intermediate ionic states of adsorbed molecules and so promote reactions with little or no energy of activation.

Catalytic cracking of hydrocarbons by clay minerals exemplifies the stabilizing of the intermediate carbonium and carbon ions by ionic surfaces, which then become involved in molecular rearrangements. Many organic reactions involving ionic intermediates are catalyzed by ionic crystals [1a].

THE ACTIVATED COMPLEX

If we have any molecular system A in equilibrium with a system B having a total energy, E, in the combined system, we can ask what is the chance that system A has the energy ε distributed in a unique way in its degrees of freedom. From equilibrium considerations we know the answer is independent of the nature of system B and depends only on the temperature. To make the calculation simple, we take system B to consist of identical oscillators which absorb quanta in multiples of $h\nu$. Thus the number of ways the n quanta, where $n = E-\varepsilon/h\nu$, can be distributed among the s oscillators is:

$$N_{\varepsilon} = \frac{(n + s - 1)!}{n!(s-1)!} \tag{9}$$

We arrive at this result by arranging the s quanta and n oscillators in all possible linear arrays, always having one particular oscillator on the extreme right and assigning the quanta lying between each pair of oscillators to the oscillator on their right. The factor n! corrects the numerator for the indistinguishability of the quanta, and the factor $(s-1)!$ results from another overcount of the numerator, since the s-1 oscillators with their assigned quanta can be permuted among themselves without changing the assignment of quanta to oscillators. We can rewrite equation (9) as:

$$N_\varepsilon = \prod_{i=1}^{s-1} \frac{(n+s-i)}{(s-1)!} = \frac{1}{(s-1)!} \prod_{i=1}^{s-1} \frac{E-\varepsilon+(s-i)h\nu}{h\nu} \qquad (10)$$

$$= \frac{1}{(s-1)!} \prod_{i=1}^{s-i} \left(\frac{E+(s-i)h\nu}{h\nu}\right) \prod_{i=1}^{s-1} \left(1-\frac{\varepsilon}{E+(s-i)h\nu}\right)$$

Now the average energy of an oscillator is $\gamma = (E-\varepsilon)/s$. Hence for $E>>\varepsilon$ and $n>>s$ then $\gamma>>h\nu$, and $E+(s-i)h\nu = E-\varepsilon+\varepsilon+(s-i)h\nu$

$$= s\gamma+\varepsilon+(s-i)h\nu \approx s\gamma,$$

since $E-\varepsilon>>\varepsilon+(s-i)h\nu$. Hence:

$$N_\varepsilon = \frac{1}{(s-1)!} \prod_{i=1}^{s-1} \left(\frac{E+(s-i)h\nu}{h\nu}\right)\left(1-\frac{\varepsilon}{s\gamma}\right)^{s-1} = C\left(1-\frac{\varepsilon}{s\gamma}\right)^{s-1}$$

$$= C\left(1-(s-1)\right)\frac{\varepsilon}{s\gamma} + \frac{(s-1)(s-2)}{2!}\left(\frac{\varepsilon}{s\gamma}\right)^2 - \cdots = Ce^{-\varepsilon/\gamma} \qquad (11)$$

Here C is independent of ε. Hence the probability, p_i, of a system having the energy, ε_i, distributed in a unique way in its degrees of freedom is

$$p_i = \frac{e^{-\frac{\varepsilon_i}{\gamma}}}{\sum_i e^{-\frac{\varepsilon_i}{\gamma}}} = \frac{e^{-\frac{\varepsilon_i}{kT}}}{\sum_i e^{-\frac{\varepsilon_i}{kT}}} \qquad (12)$$

The average energy of an oscillator in thermal equilibrium for which $h\nu<<\gamma$ is $\gamma=kT$, which accounts for the last equality in equation (12)

We now apply the considerations leading to equation (12) to a large energy rich molecule or positive ion with s internal degrees of freedom and having a total energy E and an activation energy at absolute zero, along the reaction coordinate, of ε_o. For certain simple cases such as some unimolecular decompositions, or the reverse associations, it is a useful approximation to consider that activation only involves motion along the reaction coordinate. For such cases we can treat the reaction coordinate as system A and the rest of the molecule as system B. The chance of lying at the top of the barrier in a quantum state lying between n and n+dn and moving in the forward direction is $\frac{1}{2}e^{-(\varepsilon_o+\varepsilon_n)/\gamma}(1-e^{-h\nu/\gamma})\,dn$, and the rate of reaction is obtained by multiplying this chance by the transmission coefficient, κ, and the rate, $\frac{\dot{x}}{\delta}$, of passing through the length, δ, and integrating over n. We thus obtain for the specific rate:

$$k' = \kappa \int_{\varepsilon=0}^{E-\varepsilon_0} \frac{1}{2} \frac{e^{-(\varepsilon_0+\varepsilon)/\gamma}}{(1-e^{-h\nu/\gamma})^{-1}} \frac{\dot{\chi}}{\delta} dn = \kappa \frac{e^{-\varepsilon_0/\gamma}}{h} \int_{\varepsilon=0}^{E-\varepsilon_0} \frac{e^{-\varepsilon/\gamma} d\varepsilon}{(1-e^{-h\nu/\gamma})^{-1}}$$

$$= -\frac{\kappa}{h} \frac{\gamma e^{-\varepsilon_0/\gamma}}{(1-e^{-h\nu/\gamma})^{-1}} e^{-\varepsilon/\gamma} \Big|_{\varepsilon=0}^{E-\varepsilon_0} = \kappa e^{-\varepsilon_0/\gamma} \frac{\gamma}{h} \frac{1-e^{-(E-\varepsilon_0)/\gamma}}{(1-e^{-h\nu/\gamma})^{-1}}$$

$$\simeq \frac{\kappa e^{-\varepsilon_0/\gamma}}{(1-e^{-h\nu/\nu})^{-1}} \frac{\gamma}{h} = \frac{\kappa e^{-\varepsilon_0/kT}}{(1-e^{-h\nu/kT})^{-1}} \frac{kT}{h} \tag{13}$$

In the above since $\varepsilon = (n^2h^2)/8\delta^2m^\dagger = \frac{1}{2} m \dot{\chi}^2$ it follows that

$$\frac{1}{2} \frac{\dot{\chi}}{\delta} dn = \frac{d\varepsilon}{h} \tag{14}$$

which explains the second equality in equation (13). Also we have used the fact that the average energy, γ, of an energy rich oscillator in thermal equilibrium is kT.

EVAPORATION

Simple molecules evaporating from a surface may be thought of as forming part of a dual system in which the departing molecule undergoes most of its change along the reaction coordinate. At the activated state the number of molecules per cc. at equilibrium is n/v = p/kT where p is the vapor pressure. Thus the rate of evaporation per square centimeter, R, should be:

$$R = \frac{1}{2} s \frac{n}{v} \bar{\dot{\chi}} = \frac{1}{2} s \frac{p}{kT} \sqrt{\frac{2kT}{\pi m}} = \frac{sp}{\sqrt{2\pi mkT}} \tag{15}$$

Here $\bar{\dot{\chi}} = (2kT/\pi m)^{\frac{1}{2}}$ is the mean velocity normal to the surface. This is the usual equilibrium expression for the number of molecules leaving or striking a surface multiplied by the sticking coefficient, s. In equation (15) we must include in s all the contributions to the sticking coefficient, including the contributions due to necessary adjustments in other degrees of freedom than the reaction coordinate.

To calculate sticking coefficients better we must give up our assumption that there is no change in any other degrees of freedom, during activation, than in the reaction coordinate.

Consider a compound system in which system A is a molecule which is separating from a solid B. Then the specific rate of separation is [2]:

$$k' = \kappa \frac{kT}{h} \frac{F^{\dagger}}{F_i} e^{-\varepsilon_o/kT} \tag{16}$$

where $F^{\dagger}$ is the partition function in the activated state for all degrees of freedom except the reaction coordinate in both the solid and the departing molecule, which are different than they were initially, and F_i is the partition function for the initial state for these same degrees of freedom including the reaction coordinate.

If a molecule's rotation in the condensed phase is restricted, presumably the activated complex which is about to settle into the condensed phase will be restricted in much the same way. If so, the ratio of the rotational partition function in the final state over the initial state can be treated as a significant factor in the sticking coefficient. This effect is readily estimated for liquids in terms of the excess entropy of vaporization, δS, over that given by Trouton's rule. The factor, s_r, in the sticking coefficient is:

$$s_r = e^{-\delta S/R} \tag{17}$$

In an earlier paper from this laboratory [3] it has been shown that the overall sticking coefficient on a surface takes the form:

$$s = s_r s_t s_k \tag{18}$$

Here s_r is the factor in the sticking coefficient due to rotation and has the form $s_r = F^{\dagger}/F_i$. The translational sticking coefficient is s_t which is the chance that a molecule will lose enough translational energy in collision so that it won't separate from the surface on the first outward swing. This has the form:

$$s_t = 2I\left(\frac{\theta}{T}\right) \tag{19}$$

where

$$\theta = \frac{2\alpha D}{(1-\alpha)k} \tag{20}$$

$I(\frac{\theta}{T})$ is the error function of the argument $\frac{\theta}{T}$. Here α is the fraction of the incident energy, $(D + \frac{1}{2} m\dot{\chi}^2)$, that is dissipated on the first collision with the surface. D is the dissociation energy for this particular kind of physical adsorption; m is the mass of the colliding molecule. The kinetic sticking coefficient, s_k, is the chance that a physically adsorbed molecule will shift and become chemically adsorbed before desorbing.

Thus

$$s_k = \frac{k_a}{k_a + k_d} \tag{21}$$

This is because a physically adsorbed molecule can either chemically adsorb or desorb. The specific rate constant, k_a, is of course for chemical adsorption, and k_d is for desorption, giving equation (21) for the expected probability. Many examples are available [3,4].

SMELTING AND CORROSION

Aluminum is formed, at a carbon electrode dipped in a salt bath containing aluminum ions, by reduction, using electrons fed into the carbon electrode which raise the electrode potential to the point where it can reduce the aluminum ions. In ordinary smelting, electrons are pumped by chemical reactions into conductors such as coke or a metal which form floating electrodes that reach voltages which enable the electrode to reduce many metal ions more noble than aluminum.

Thus if in a carbon container at around 1500°C [5,6], one has a metal bath covered by a slag, the oxide ions from the slag migrate to the metal surface and react with carbon dissolved in the metal bath or with the graphite container according to the equation

$$O^{=} + C = CO + 2\varepsilon \tag{22}$$

The CO escapes as a gas and leaves the two electrons to charge up the conductor. If the electrons are not carried away by other reactions, reaction (22) will finally stop when the voltage on the floating electrode reaches a value where the back reaction balances the forward one. What actually happens is that the electrons reduce metal ions in the slag, or else this reaction may run in the reverse direction depending on the electrode voltage and the activity of ions in the metal and in the slag. Elemental sulfur, when dissolved in the metal, picks up electrons from the conductor acting as an electrode and migrates into the slag. All of these reactions either draw electrons or add them to the floating electrode which acts as an electron bank. In any event a steady state is set up, and the voltage of the electrode, V, is given by the equation:

$$\frac{d(CO)}{dt} = a_1 a_2 k' e^{(\alpha Z i V F)/(RT)} \tag{23}$$

or

$$V = \frac{RT}{\alpha Z i F} \ln \left[\frac{\frac{d(CO)}{dt}}{a_1 a_2 k'} \right] \tag{24}$$

Here R, T, α, Z_i, F, a_1 a_2 and k′ are the gas constant, absolute temperature, fraction of voltage which acts to increase current, charge carried by oxide ion, the Farady, the activity of the $O^=$ in the slag and of the carbon in the melt, and the specific rate constant when V=0, respectively. Thus the voltage is indicated by the rate of evolution of CO, if the other factors are held constant. The rate of evolution of every species can be written as an equation analogous to equation (23), except that a back reaction must be included. Now since equation (24) gives the voltage, we can calculate the rate of smelting of every species from the observed rate of CO evolution. The calculated results agree with experiment [6]. Thus smelting is an electrolytic process. The specific rate constant, k′, may differ, depending on the nature of the electrode. We may think of the electrode as a catalyst which speeds up or slows the reaction as any catalyst would: Thus k′ can be studied, the mechanism of reaction can be determined, and the heat and entropy of activation can be categorized as for any other rate process.

The reverse reaction to smelting is corrosion, and it too is an electrolytic process in which the metal again acts as an electron bank. An increase of the negative charge on the metal slows down corrosion as of course it must.

THE THERMAL DECOMPOSITION OF NITRATES

If P_2O_5 is added to concentrated nitric acid, N_2O_5 may be sublimed off and collected as pure white crystals. Cold crystalline N_2O_5 can be stored indefinitely. N_2O_5 has a half life of about a half hour at room temperature in the gas phase or when dissolved in inert solvents [6a]. In solvents which complex with the N_2O_5, the dissociation free energy adds to the free energy of activation and the reaction goes correspondingly slower [6b]. The crystalline phase decomposes at its surface only, at an appreciable rate.

The following mechanism proposed by Ogg [7] has been confirmed [8].

$$N_2O_5 \underset{k_2}{\overset{k_1}{\rightleftarrows}} NO_2 + NO_3 \tag{25}$$

$$NO_2 + NO_3 \xrightarrow{k_3} NO + O_2 + NO_2 \tag{26}$$

$$NO + NO_3 \xrightarrow{k_4} 2NO_2 \tag{27}$$

where $k_2 > k_3$ and $k_4 > k_3$. Whence:

$$\frac{d\ln(N_2O_5)}{dt} = \frac{2k_3k_1}{k_2 + 2k_3} \approx \frac{2k_3k_1}{k_2} \qquad (28)$$

since in 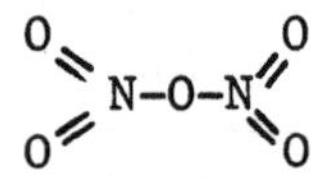 no oxygen atoms are in contact with each other, there must be a rearrangement before two of them can come together to make an O_2 molecule. That this is most easily accomplished by a preliminary dissociation corresponding to reaction(25), followed by a reassociation, reaction (26), might have been guessed. It is not surprising that the free energy for the activated complex in reaction (26)is higher than that for reaction (25), since for an oxygen molecule to be formed two nitrogen to oxygen bonds must be broken. In reaction (27) NO disappears by about the only route open to it.

A different situation arises when dry HNO_3 and sulfuric acid are added to substances like benzene and toluene. The rate determining step is the formation of the ion XNO_2^+, which then adds to a molecule like benzene or toluene displacing a proton. The XNO_2^+ reacts many times faster with the para position in toluene than it does with the position in benzene. This is convincing evidence that the added XNO_2^+ is electrophyllic and therefore positively charged. By assuming that in the activated complex: (a) the charge on XNO_2^+ was 1.6 A^o from the carbon it was adding onto, (b) the effective dielectric constant was unity and (c) the charges on the molecules being substituted obeyed certain inductive and resonance relations, Ri and Eyring [9] calculated the charge distributions in the monosubstituted benzenes, which led to calculated dipole moments in agreement with experiment.

The decomposition of N_2O_5 in molten salts is interesting. Some studies of Kust et. al. [10, 11]are instructive in this connection. In all cases the nitryl ion, NO_2^+, is generated by the reaction of the type;

$$X + NO_3^- \rightleftarrows Y + NO_2^+ \qquad (29)$$

followed by the decomposition:

$$NO_2^+ + NO_3^- \rightarrow 2NO_2 + 1/2\ O_2 \qquad (30)$$

In a series of studies X was the dichromate ion, $Cr_2O_7^=$, the pyrosulfate ion, $S_2O_7^=$, and Cu^{++}. Kust and Burke [11] studied the latter case at length assuming the following mechanism:

$$Cu^{2+} + NO_3^- \underset{k_{-1}}{\overset{k_1}{\rightleftarrows}} CuO + NO_2^+ \qquad (31)$$

$$NO_2^+ + NO_3^- \overset{k_2}{\rightarrow} 2NO_2 + 1/2\ O_2 \qquad (32)$$

The nitration of aromatic hydrocarbons in molten salt solutions containing nitrates and the strong acid, potasium pyrosulfate, indicates the presence of NO_2^+ [11a].

By assuming a steady state and that reaction (32) is catalyzed by CuO, they are led to the equation for the rate of reaction, R:

$$R = \frac{k_1 k_2 (1 + \alpha w)}{k_{-1} + k_2 \quad (1 + \alpha w)} \qquad (33)$$

which they find agrees with their experiments. Since Equation (32) has 1/2 O_2 as a product it can't represent an actual mechanism but must be replaced by two or more equations analogous to (26) and (27). Thus it will be interesting to see if equation (32) should be broken down into such steps as:

$$NO_2^+ + NO_3^- \rightleftarrows NO_2 + NO_3 \rightleftarrows N_2O_5 \qquad (34)$$

$$NO_2 + NO_3 \rightarrow NO + O_2 + NO_2 \qquad (34a)$$

$$NO + NO_3 \rightarrow 2NO_2 \qquad (35)$$

Eq. (34a) would be expected to be the slow step, and thus the step catalyzed by cupric and ferric oxide. More work will be required to establish the detailed mechanism. As a final example of high temperature activated complexes we will consider the oxidation of carbon.

THE OXIDATION OF CARBON

Diamond with its atoms each having four equivalent bonds is analogous to the alliphatic hydrocarbons, whereas graphite with its layers of carbon atoms arranged in an interconnected hexagonal pattern should behave more like the aromatic hydrocarbons. Actually since the most unstable atoms react first, the reaction rates observed involve pitting and are strongly influenced by chemical impurities and structural imperfections. We next consider the equations appropriate to a discussion of porous surfaces.

The Effect of Porosity on Reaction Rates

The thermal oxidation of graphite is an example of a heterogeneous gas-solid reaction in which most of the reacting surface may lie within the pores of the solid. The transport process of the oxidizing gas effects the concentration distribution of oxygen throughout the porous material, and this influences the rate of reaction. Consider an arbitrary volume, V, of the system. If ρ_i represents the mass density of the i-th component of gases in V, then the rate of change of the mass of this component is given by:

$$\frac{d}{dt}\int_V \rho_i \, dV = \int_V \frac{\partial \rho_i}{\partial t} dV \tag{36}$$

The surface of the volume, V, is divided into infinitesimal elements of area, dA. The vector $d\vec{A}$, of magnitude dA, is defined for each element of area. The direction of $d\vec{A}$ is taken perpendicular to the element of the surface dA and pointing outward from the volume V. If $\vec{u}_i$ represents the velocity of component i with respect to the external coordinate system, then the current density of mass of component i is the mass of component i flowing per unit area in unit time, $\rho_i\vec{u}_i$, and the flow of component i through the boundary of V is $-\int_A \rho_i\vec{u}_i \cdot d\vec{A}$, where the integration is taken over the entire surface of V. Let Φ_i be the amount of component i produced per unit time by chemical reaction. Using the law of conservation of mass we obtain:

$$\int_V \frac{\partial \rho_i}{\partial t} dV = -\int_A \rho_i \vec{u}_i \cdot d\vec{A} + \Phi_i \tag{37}$$

By means of Gauss' theorem, Eq. (35) can be written as:

$$\int_V \frac{\partial \rho_i}{\partial t} dV = -\int_V \vec{\nabla} \cdot (\rho_i \vec{u}_i) dV + \Phi_i \tag{38}$$

Now if Φ_i can be expressed as:

$$\Phi_i = \int_V \Phi_{V_i} \, dV \tag{39}$$

where ϕ_{V_i} represents the amount of component i produced per unit volume and unit time by chemical reaction, then substitution of Eq. (39) into Eq. (38) gives:

$$\frac{\partial \rho_i}{\partial t} = -\vec{\nabla} \cdot (\rho_i \vec{u}_i) + \phi_{V_i} \tag{40}$$

If Φ_i can be expressed as:

$$\Phi_i = \int_A \phi_{A_i} \, dA \tag{41}$$

where ϕ_{A_i} denotes the amount of component i produced per unit surface area and unit time by chemical reaction, then combining Eqs. (41) and (38), we obtain:

$$\int_V \frac{\partial \rho_i}{\partial t} \, dV = - \int_V \vec{\nabla} (\rho_i \vec{u}_i) dV + \int_A \phi_{A_i} \, dA \tag{42}$$

The velocity of component i, $\vec{u}_i$ can be related to the diffusion current density, $\vec{j}_i$, as follows:

$$\vec{j}_i = \rho_i \, (\vec{u}_i - \vec{u}) \tag{43}$$

where $\vec{u}$ represents the velocity of the center of mass. The diffusion current density, $\vec{j}_i$, is by Fick's first law of diffusion, related to the density, ρ_i, as follows:

$$-\vec{j}_i = D_i \, \vec{\nabla} \rho_i \tag{44}$$

Here the cross diffusion effect has been neglected. D_i is the diffusion coefficient with respect to the center of mass and is related to the diffusion coefficient with respect to solvent [12]. Using Eqs. (43) and (44), Eqs. (40) and (42) become:

$$\frac{\partial \rho_i}{\partial t} = \vec{\nabla} \cdot (D_i \, \vec{\nabla} \rho_i) - \vec{\nabla} \cdot (\rho_i \vec{u}) + \phi_{V_i} \tag{45}$$

and

$$\int_V \frac{\partial \rho_i}{\partial t} \, dV = - \int_V [-\vec{\nabla} \cdot (D_i \vec{\nabla} \rho_i) + \vec{\nabla} \cdot (\rho_i \vec{u})] \, dV + \int_A \phi_{A_i} \, dA \tag{46}$$

respectively. Whether Eq. (46) or Eq. (45) should be used depends on whether the chemical reaction involves the pore surface. If the system is in a steady state, then $\partial \rho_i / \partial t = 0$. If there is no bulk flow, then $\vec{u} = 0$.

To be more specific, we suppose that the pore is of cylindrical shape with radius r. If we assume that the system is in a steady state, and that there is no bulk flow, then Eq. (46) becomes:

$$- \pi r^2 D_i \frac{d^2 \rho_i}{dz^2} = 2 \pi r \phi_{A_i} \tag{47}$$

where z is the coordinate along the cylindrical axis. In obtaining Eq. (47), the gradient of ρ_i along the radial direction is assumed negligible, and D_i is assumed to be independent of ρ_i. If the reaction kinetics involves only a single reaction species, then in terms of concentration c_i, Eq. (46) can be written as [13-16]:

$$\pi r^2 D_i \frac{d^2 c_i}{dz^2} = 2\pi r k c_i^n \tag{48}$$

or

$$\pi r^2 D_i \frac{d^2 c_i}{dz^2} = 2\pi r R \tag{49}$$

where R represents the chemical reaction rate per unit solid surface area, and k is the rate constant which gives the number of molecules reacted per unit time and per unit area. Integration of Eq. (48) gives:

$$\frac{dc_i}{dz} = -\left[\frac{4kc_i^{n+1}}{(n+1)rD_i}\right]^{\frac{1}{2}} \tag{50}$$

Here we have assumed that c_i falls to zero at some value of z. Combining Eqs. (44) and (50), we obtain the reaction rate in the pore as:

$$\text{Rate} = 2\pi r^2 D_i \left[\frac{4kc_{io}^{n+1}}{(n+1)rD_i}\right]^{\frac{1}{2}} \tag{51}$$

where c_{io} denotes the concentration c_i at z = 0. Eq. (51) shows that the order of chemical reaction observed will not be the true order, n, but will be (n+1)/2, since the rate is proportional to $(c_{io}^{n+1})^{\frac{1}{2}}$. Here the rate constant per unit area of pore mouth is $2[4kc_{io}^{n+1}D_i]^{\frac{1}{2}}/[(n+1)r]^{\frac{1}{2}}$ instead of $2kc^n$ per unit of surface. The rate equation (51) applies if making two CO molecules for one entering CO_2 makes no appreciable back wind as when there is much inert gas such as N_2. However, the back wind will cut this rate in two, if pure CO_2 is reacting. The observed rate will be the sum of the rates thru the pore mouths and on the external surface giving a mixed effective rate constant. If the temperature dependance of the diffusion coefficient is negligible, and if the contribution of the external surface to the reaction is small, then from the Arrhenius plot one can determine the activation energy of the chemical reaction. For a very wide temperature range, the temperature dependance of D_i is not negligible. In this case a deviation from the linearity of the Arrhenius plot will be expected. It should be noticed that the temperature dependance of D_i is determined by the mechanism of diffusion, which in turn depends on the pressure of gaseous reactants used. Equation (51) is true only for a pore of cylindrical symmetry; for other cases of heterogeneous reactions and catalytic reactions, Eq. (45) or Eq. (46) can be used.

In most experimental conditions of heterogeneous reactions [16,17] involving a porous solid and a gas or a liquid with a non-equilibrium flow, the equation of continuity for component i in the fluid phase at the steady state condition is given by:

$$-\vec{\nabla}\,(D_i\vec{\nabla}P_i) + \vec{\nabla}\,(P_i\vec{u}) = 0_{Vi} \tag{52}$$

And the solid surface is given by:

$$\int_V [-\,\vec{\nabla}(D_i\vec{\nabla}P_i) + \vec{\nabla}\,(P_i\vec{u})]\,dv = \int_A \phi_{Ai}\,dA \tag{53}$$

or

$$\hat{r}\,(P_i u) - \hat{r}\,(D_i\,P_i) = \phi_{Ai} \tag{54}$$

where $\hat{r}$ represents the unit vector perpendicular to the solid surface. Eqs. (50) and (54) are general. Unless there are pressure and temperature gradients, and the viscous effect is important, we do not need the hydrodynamic equations for the equation of motion and the equation for the conservation of energy. Solutions for Eqs. (52) and (54) for practical systems will be carried out in the near future.

Oxidation of Carbon by Carbon Dioxide

Various workers [18-27] find reaction orders for the carbon-carbon dioxide reaction varying from zero to one depending on temperature, pressure, type of carbon reacted, purity of carbon, and geometric dimensions of the sample. Many of the inconsistencies seem to arise from inadequate consideration of all the pertinent variables, many of which may have decisive influence on the rate controlling processes and the ranges of their applicability. To correlate the existing experimental information, we assume that the rate of oxidation of carbon by carbon dioxide involves only one type of site on the carbon surface, according to the following scheme:

$$c_f + CO_2(g) \underset{k_{-1}}{\overset{k_1}{\rightleftarrows}} c(o) + CO(g) \tag{55}$$

$$c(o) \underset{k_{-2}}{\overset{k_2}{\rightleftarrows}} CO(g) + \Sigma_c \tag{56}$$

where c_f represents a surface carbon site and c(o) represents a chemisorbed oxygen atom. For kinetic consistency, it is necessary to include Σ_c, which represents a site on the carbon surface from which a carbon atom has been removed. The rate expressions for c(o), CO, and CO_2 can be expressed as

$$R_{c(o)} = k_1 [c_f]P_{CO_2} - k_{-1} [c(o)] P_{CO} - k_2[c(o)] + k_{-2}P_{CO} [\Sigma_c] \tag{57}$$

$$R_{CO} = k_1[c_f]P_{CO_2} - k_{-1}[c(o)]P_{CO} + k_2 [c(o)] - k_{-2}P_{CO} [\Sigma_c] \tag{58}$$

and

$$R_{CO_2} = - k_1 [c_f]P_{CO_2} + k_{-1}[c(o)]P_{CO} \tag{59}$$

respectively. Introducing the steady state approximation for c(o) we obtain

$$[c(o)] = \frac{(k_1P_{CO_2}+k_{-2}P_{CO})n_c}{k_2+k_1P_{CO_2}+(k_{-1}+k_{-2})P_{CO}} \tag{60}$$

and

$$[c_f] = \frac{(k_2+k_{-1}P_{CO})n_c}{k_2+k_1P_{CO_2}+(k_{-1}+k_{-2})P_{CO}} \tag{61}$$

where n_c is the total number of active carbon sites, i.e. $n_c = c_f + c(o)$. Substitution of Eqs. (60) and (61) into Eq. (59), after rewriting it using Equation (58) with R_{CO}=0, we have:

$$R_{CO_2} = k_2 [c(o)] \quad k_{-2}P_{CO} \quad [\Sigma_c]$$

$$= \frac{-k_2n_c(k_1P_{CO_2}+k_{-2}P_{CO}) + k_{-2}P_{CO}(k_2+k_{-1}P_{CO})n_c}{k_2+k_1P_{CO_2}+(k_{-1}+k_{-2})P_{CO}} \tag{62}$$

Here we have assumed that $[c_f] = [\Sigma_c]$. The rate constants for elementary rate processes appearing in Eqs. (55) and (56) can be calculated by using the absolute reaction rate theory [2, 28-30]. In general k_{-2} is very small and negligible. In that case, Eq. (62) can be expressed as:

$$R_{CO_2} = \frac{-k_1n_cP_{CO_2}}{1 + \frac{k_1}{k_2} P_{CO_2} + \frac{k_{-1}}{k_2} P_{CO}} \tag{63}$$

Ergun's experimental results [31] support the above mechanism. He conducted experiments in a fluidized bed using three different types of carbon - ceylon graphite, activated carbon, and activated graphite. These samples had a considerable range of mineral content (from a trace to 0.5%) and also had a wide range of specific surface area. In spite of this, Ergun finds the equilibrium constant for reaction (55) with the above mechanism to be independent of the carbon used and to have an average ΔH of 23 kcal/mole over the temperature range from 800 to 1400°C. Applying the equilibrium condition to c(o), CO, and CO_2, we can obtain the proper equilibrium constant for the following reaction:

$$CO_2(g) + C(s) = 2CO(g) \tag{64}$$

Using radioactive carbon, C^{14}, as a tracer to study the oxygen and carbon exchange reaction occurring during the carbon - carbon dioxide reaction, Bonner and Turkevich [32] find that the forward reaction of reaction (55) is rapid on charcoal at temperatures of 735 and 840°C and with initial carbon dioxide pressures of 180 and 330 torr. On the other hand, under these conditions they find the forward reaction of reaction (56) to be slow. They confirm that some carbon from the original carbon dioxide has also transferred to the surface of charcoal. It can be seen from Eq. (63) that the carbon - carbon dioxide reaction will be zero order when $P(CO)k_{-1}/k_2 << 1$ and $P(CO_2)k_1/k_2 >> 1$. At low temperatures, the rate of production of carbon monoxide is small, and the first inequality is satisfied. At high carbon dioxide pressures, the second inequality is satisfied. On the other hand, it can be seen from Eq. (63) that the carbon - carbon dioxide reaction will be first order when $P(CO)k_{-1}/k_2 <<1$ and $P(CO_2)k_1/k_2 << 1$, which will be satisfied at low temperatures and low carbon dioxide pressures. Gadsky et. al. [33]. and Lewis et. al. [26] have shown that k_1/k_2 and k_{-1}/k_2 decrease sharply with increasing temperature. Therefore, at high temperatures and increasing pressures, the inequalities, $P(CO)k_{-1}/k_2 <<1$ and $P(CO_2)k_1/k_2 <<1$ still can be found to hold and result in a first order reaction. It should be noticed that the activation energy determined for the carbon - carbon dioxide reaction need not refer to the same rate determining step in every case. From Eq. (63) it can be seen that at low temperatures and pressures the rate of reaction is proportional to k_1, the rate constant for the formation of c(o). At low temperatures and high pressures, the rate constant is proportional to k_2, the rate constant for the removal of c(o). When none of the terms can be dropped from the denominator of Eq. (63), an understanding of the exact physical meaning of an overall activation energy is difficult. If the chemisorption of CO is significant, then we have to include the following reaction in the above mechanism:

$$CO(g) + c_f \underset{k_{-3}}{\overset{k_3}{\rightleftarrows}} C(CO) \tag{65}$$

Similarly applying the steady state approximation for C(CO), we obtain:

$$[C(CO)] = \frac{k_3}{k_{-3}} [c_f] P_{CO} \tag{66}$$

The rate expression for CO_2 becomes:

$$R_{CO_2} = \frac{-k_2 n_c (k_1 P_{CO_2} + k_{-2} P_{CO}) + k_{-2} P_{CO} n_c (K_2 + k_{-2} P_{CO})}{k_2 + k_1 P_{CO_2} + (k_{-1} + k_{-2}) P_{CO} + \frac{k_3}{k_{-3}} P_{CO} (k_2 + k_{-1} P_{CO})} \tag{67}$$

If we compare Eq. (67) with Eq. (62), we can see that the introduction of the adsorbed species, C(CO) is simply to decrease the available surface carbon sites and hence to slow down the rate of reaction. The adsorption of other molecules also brings in the same effect. The effectiveness of C(CO) depends on the equilibrium constant of the adsorption of CO, k_3/k_{-3}. Gadsby and co-workers [33] find that only about 0.5% of a charcoal surface chemisorbs carbon monoxide at 850°C.

When the effect of porosity is important, we express the rate expression of CO_2 in terms of its concentration, c:

$$R_{CO_2} = - \frac{K_1 c}{1+K_2 c} \tag{68}$$

where K_1 and K_2 are constants and are related to k_1, k_{-1} and k_2. Then combining Eq. (68) with Eq. (47) of the previous section, we obtain:

$$\pi r^2 D \frac{d^2 c}{dz^2} = 2\pi r \frac{K_1 c}{1+K_2 c} \tag{69}$$

Integration of Eq. (69) gives:

$$\frac{dc}{dz} = - \left(\frac{4K_1}{rDK_2}\right)^{1/2} \left[c - \frac{1}{K_2} \log (1+K_2 c)\right]^{1/2} \tag{70}$$

Here we have assumed that c falls to zero at some value of z inside the pore. Using Eq. (70), we obtain the reaction rate in the pore as:

$$\text{Rate} + 2\pi r^2 D \left[\frac{4K_1}{rDK_2}\right]^{1/2} \left[c_o - \frac{1}{K_2} \log (1+K_2 c_o)\right]^{1/2} \tag{71}$$

where c_o is the concentration of CO_2 at z = 0. When $K_2 c_o << 1$, expanding the right hand side of Eq. (71) in a power series of $K_2 c_o$, we obtain:

$$\text{Rate} = 2\pi r \, (2rDK_1)^{\frac{1}{2}} \, c_o \, (1 - \tfrac{1}{3} K_2 c_o + \cdots) \tag{72}$$

If $K_2c_o >> 1$, we can rewrite Eq. (71) as:

$$\text{Rate} = 2\pi r \left[\frac{4K_1c_o}{K_2}\right]^{1/2} \left[1 - \frac{1}{K_2c_o} \log (1+K_2c_o)\right]^{1/2}$$

$$\approx 2\pi r \left[\frac{4K_1c_o}{K_2}\right]^{1/2} (1- \frac{1}{2K_2c_o} \log K_2c_o + \cdot \cdot \cdot) \qquad (73)$$

From the above discussion, we can see that the temperature dependence of the Arrhenius plot is very different in the two cases, and that the apparent order of reaction varies from 1/2 to 1, when the effect of porosity is important. The above results are also applicable to the carbon-steam reaction; the rate expression for the carbon-steam reaction has the same form as is given by Eq. (68).

Oxidation of Carbon by Oxygen

An extensive amount of work has been done on the oxidation of carbon and of graphite in various forms. Both CO_2 and CO are simultaneously formed as primary reaction products. Broadly speaking, the ratio CO/CO_2 is smaller than unity at high pressures and low temperatures [34-39] and is greater than unity at low pressures and high temperatures [40-43]. The experimental results under varied experimental conditions show that the order of carbon-oxygen reaction also varies from zero to unity. The mechanism for the oxidation of carbon by oxygen is assumed to be as follows:

$$2c_f + O_2(g) \underset{k_{-1}}{\overset{k_1}{\rightleftarrows}} 2c(o) \qquad (74)$$

$$c(o) \underset{k_{-2}}{\overset{k_2}{\rightleftarrows}} CO(g) + \Sigma_c \qquad (75)$$

$$c_f + O_2(g) \underset{k_{-3}}{\overset{k_3}{\rightleftarrows}} C(O_2) \qquad (76)$$

$$C(O_2) \underset{k_{-4}}{\overset{k_4}{\rightleftarrows}} CO_2(g) + \Sigma_c \qquad (77)$$

$$c_f + CO_2(g) \underset{k_{-5}}{\overset{k_5}{\rightleftarrows}} c(o) + CO(g) \qquad (78)$$

Equation (78) is included because of the fact that it is important in the carbon-carbon dioxide reaction. For the case of the carbon-oxygen reaction, Eq. (78) is a secondary reaction, so for simplicity we shall neglect this reaction in our discussion. Applying the steady state approximation for c(o) and $C(O_2)$, we obtain:

$$R_{c(o)} = k_1 [c_f]^2 P_{O_2} - k_{-1} [c(o)]^2 - k_2 [c(o)] + k_{-2} [c_f] P_{CO} = 0 \tag{79}$$

and

$$R_{C(O_2)} = k_3 [c_f] P_{O_2} - k_{-3}[C(O_2)] - k_{-4}[C(O_2)] + k_{-4} P_{CO_2} c_f = 0 \tag{80}$$

It has been shown that the chemisorbed oxygen species never can be removed from the carbon surface as such. When the surface is degassed, the oxygen is removed as oxides of carbon. In view of this, k_{-1} and k_{-3} are negligible. Thus the rate expressions for CO, CO_2 and O_2 can be written as:

$$R_{CO} = k_2 [c(o)] - k_2 [c_f] P_{co} = k_1 [c_f]^2 P_{O_2} \tag{81}$$

$$R_{CO_2} = k_4 [C(O_2)] - k_{-4} [c_f] P_{CO_2} = k_3 [c_f] P_{O_2} \tag{82}$$

and

$$R_{O_2} = -(k_3 + 1/2\, k_1 [c_f]) P_{O_2} [c_f] \tag{83}$$

Recently, Walker et. al. [38, 40] have shown that for the combustion of graphite in the temperature range 575 - 675°C, the rate of oxygen consumption is proportional to $[c_f]$ and P_{O_2}. From Eqs.(79) and (80), we can solve for $[c_f]$, $[c(o)]$ and $[C(O_2)]$. For convenience $[c_f]$ is expressed as:

$$[c_f] = \frac{n_c}{1 + \frac{k_{-2}}{k_2} P_{CO} + \frac{k_{-4}}{k_4} P_{CO_2} + \left[\frac{k_3}{k_4} + \frac{k_1}{k_2} [c_f]\right] P_{O_2}} \tag{84}$$

Substituting Eq. (84) into Eq. (83), we obtain the rate expression, R_{O_2}, for oxygen as:

$$R_{O_2} = - \frac{n_c (k_3 + 1/2\, k_1 [c_f]) P_{O_2}}{1 + \frac{k_{-2}}{k_2} P_{CO} + \frac{k_{-4}}{k_4} P_{CO_2} + \left[\frac{k_3}{k_4} + \frac{k_1}{k_2} [c_f]\right] P_{O_2}} \tag{85}$$

From Eq. (85), we can see that at low temperatures and low pressures, the denominator of Eq. (85) approaches unity and $[c_f]$ approaches n_c.

Under these conditions, the carbon - oxygen reaction becomes first order. Similarly, at low temperatures and high oxygen pressures, the order of the carbon - oxygen reaction approaches zero. The findings of Strickland - Constable [42], Day [44], Rossberg [45], Lewis et. al. [46], Armington [47], etc. [48-51] show the carbon-oxygen reaction to be first order, or close to first order. From the previous discussion, the implication of the first-order reaction is that under the experimental conditions used by the above authors, the fraction of the total active carbon surface occupied by an oxygen complex at any given instant during the reaction is very small. Gulbransen and Andrew [52], and Blyholder and Eyring [53]observed a zero - order for the carbon - oxygen reaction. Fractional order of reaction is also reported [54-57].

In the above discussion of the carbon - carbon dioxide reaction and the carbon - oxygen reaction, we have assumed that there is only one type of active site on the carbon surface. If there is more than one type of active site on the carbon surface, then the mechanisms discussed above should be modified accordingly.

REFERENCES

1. Henry Eyring and F. William Cagle, Treatise on Analytical Chemistry, Part II, Vol. II, p. 45, edited by I. M. Kolthoff and Philip J. Elving. Interscience publishers (1965).

1a. H. Eyring, H. M. Hurlbert and R. A. Harman, Ind. & Eng. Chem., 35, 511 (1943).

2. S. Glasstone, K. J. Laidler and H. Eyring, *The Theory of Rate Processes*, McGraw Hill, 1941.

3. H. Eyring, F. M. Wanlass and E. M. Eyring, *Condensation and Evaporation*, Chap. 1, edited by R. Emile, P. Goldfinger and J. P. Hutt, 1964.

4. F. M. Wanlass and H. Eyring, Advances in Chemistry Series, 33, 140 (1961).

5. T. B. King and S. Ramachandran, *The Physical Chemistry of Steelmaking*, ed. J. F. Elliott, Technology Press, Cambridge, Mass, and Wiley, New York, 1958.

6. De Hemptinne, H. Eyring and Taikyue Ree, *Physical Chemistry of Process Metallurgy*, ed. George R. St. Pierre, Part 1, Vol. VII, pp. 69-93, Interscience Publishers, N. Y., 1961.

6a. H. Eyring and F. Daniels, J. A. C. S., 52, 1472 (1930).

6b. H. Eyring and F. Daniels, J. A. C. S., 52, 1486 (1930).

7. R. A. Ogg, Jr., J. Chem. Phys., 15, 337, 613 (1947).

8. S. W. Benson, *The Foundation of Chemical Kinetics*, McGraw Hill, New York, 1960, p. 408.

9. T. Ri and H. Eyring, J. Chem Phys., 8, 443 (1940).

10. R. N. Kust and F. R. Dube, J. Am. Chem. Soc., 85, 3338 (1963).

11. R. N. Kust and J. D. Burke, Inorg. Chem., 8, 1748 (1969).

11a. R. B. Temple, C. Foy, J. Williamson, Chem. Comm., 966 (1967).
12. D. D. Fitts, *Nonequilibrium Thermodynamics*, McGraw Hill, 1962, p. 78.
13. E. W. Thiele, Ind. & Eng. Chem, 31, 916 (1939).
14. A. Wheeler, Adv. in Catalysis, 3, 249 (1952); also, Catalysis 2, 105 (1955).
15. P. B. Weisz, and C. D. Prater, Adv. in Catalysis, 6, 143 (1954).
16. P. L. Walker, Jr., F. Rusinko, Jr., and L. G. Austin, Adv. in Catalysis, 11, 133 (1959).
17. P. Hawtin, J. A. Gibson, R. Murdoch, and J. B. Lewis, Carbon, 2, 299 (1964);ibid. 4, 489 (1966); ibid. 4, 501 (1966).
18. J. N. Ong, Jr., Carbon, 2, 281 (1964).
18a. E. T. Turkdogan and J. V. Vinters, Carbon, 7, 101 (1969).
19. A. E. Reif, J. Phys. Chem., 56, 785 (1952); ibid., 56, 778 (1952).
20. A. A. Orning and E. Sterling, J. Phys. Chem. 58, 1044 (1954).
21. M. Bregazzi, E. Greenhalgh, J. W. Sutherland and D. J. Tucker, Carbon, 3, 73 (1965).
22. L. G. Overholser and J. P. Blakely, Carbon, 2, 385 (1965), ibid., 3, 269 (1965).
23. E. A. Gulbransen, K. F. Andrew, and F. A. Brassart, Carbon, 2, 421 (1965).
24. R. F. Strickland - Constable, Proc. Roy. Soc., A189, 1 (1947).
25. F. J. Long and K. W. Sykes, Proc. Roy. Soc., A193, 377 (1948).
26. W. K. Lewis, E. R. Gilland and G. T. McBride, Jr., Ind. & Eng. Chem. 41, 1213 (1949).
27. T. J. Clark, R. E. Woodley and D. R. DeHalas, *Nuclear Graphite*, edited by R. E. Nightingale, Academic Press, 1962, p. 387.
28. K. J. Laidler, S. Glasstone and H. Eyring, J. Chem. Phys., 8, 659 (1940).
29. E. M. Mortensen and H. Eyring, J. Phys. Chem 64, 846 (1960).
30. H. Eyring and S. H. Lin, *The Law of Mass-Action*, A centenary Volume, C. Oslo, 1964.
31. S. Ergun, J. Phys. Chem. 60, 480 (1956).
32. F. Bonner and J. Turkerich, J. Am. Chem. Soc., 73, 561 (1951).
33. J. O. Gadsby, F. J. Long, P. Sleightheim, and K. W. Sykes, Proc. Roy. Soc., A193, 357 (1948).
34. E. A. Gulbransen, and K. F. Andrew and F. A. Brassart, J. Electrochem. Soc., 110, 476 (1963).
35. J. R. Arthur, British Coal Utilization Res. Assoc. 13, 297 (1949); ibid., 13, 296 (1949).
36. E. S. Golovina and G. P. Khaustovitch, *Eighth Symposium on Combustion*, Williams and Wilkins, 1962, p. 1784.
37. C. W. Snow, D. R. Wallace, L. L. Lyon and G. R. Crocker, *Proceedings of the Third Conference on Carbon*, Pergamon Press, 1959, p. 279.
38. F. J. Vastola, P. J. Hart and P. L. Walker, Jr., Carbon, 2, 65 (1964).
39. J. B. Lewis, P. Connor and R. Murdock, Carbon, 2, 311 (1964).

40. N. R. Laine, F. J. Vastola and P. L. Walker, Jr., J. Phys. Chem., 67, 2030 (1963).
41. L. B. Bonnetain, X. Duval and M. Letort, *Proceedings of the Fifth Conference on Carbon*, Vol. 1, Pergamon Press, 1962, p. 107.
42. R. F. Strickland - Constable, Trans. Faraday Soc., 40, 333 (1944).
43. J. Okada and T. Ikegawa, J. Appl. Phys., 24, 1249 (1953).
44. R. J. Day, Ph.D. Thesis, The Pennsylvania State University, 1949.
45. M. Rossberg, Z. Elektrochem., 60, 952 (1956).
46. W. K. Lewis, E. R. Gilliland and R. R. Paxton, Ind. & Eng. Chem., 46, 1327 (1954).
47. A. F. Armington, Ph.D. Thesis, The Pennsylvania State University, 1960.
48. G. S. Scott and G. W. Jones, U. S. Bur. Mines. Rept. Invest., 3405 (1938).
49. M. C. Chen, C. J. Christensen and H. Eyring, J. Phys. Chem., 61, 682 (1957).
50. E. Wicke and K. Hedden, Z. Electrochem., 57 636 (1943).
51. J. R. Arthur, E. J. Newitt and M. M. Raferty, Nature., 178, 157 (1956).
52. E. A. Gulbransen and K. F. Andrew, Ind. & Eng. Chem., 44, 1034 (1952).
53. G. Blyholder and H. Eyring, J. Phys. Chem., 61, 682 (1957).
54. G. Blyholder and H. Eyring, J. Phys. Chem., 63, 1004 (1959).
55. J. B. Lewis, P. Connor and R. Murdock, Carbon, 2, 311 (1964).
56. X. Duval, J. Chem. Phys. 58, 3 (1961).
57. S. T. Gregg and R. F. S. Tyson, Carbon, 3, 39 (1965).

ADSORPTION AND KINETICS AT ELEVATED TEMPERATURES

L. S. Darken and E. T. Turkdogan

Edgar C. Bain Laboratory for Fundamental Research, United States Steel Corporation Research Center, Monroeville, Pennsylvania 15146

This work consists of a review of (1) the nature of crystalline surfaces at elevated temperature, (2) adsorption on metallic surfaces, and (3) reaction rates involving transport across surfaces. as influenced by adsorption. It is pointed out with the aid of numerous examples that the strong adsorption frequently encountered has a pronounced effect on high-temperature kinetics at such surfaces. Such pronounced effects are encountered in the following processes considered in some detail: the flux of nitrogen across an iron surface on which oxygen is adsorbed; the decarburization of Fe-C alloys; the growth of wustite or pyrrhotite on iron; the oxidation of carbon by various gases. Attention is called to the fact that the rate equations take simple form only if the extent of adsorption is small (compared to a monatomic layer) or if a single species forms a nearly complete monatomic layer.

When the older of us were graduate students, reaction rate theory was concerned principally with homogeneous solutions, occasionally liquid but mostly gaseous, and we talked about the collision of molecules. However, most reactions of metallurgical interest differ markedly in nature from such homogeneous reactions. In the first place metallic solutions are not molecular in nature. In the second place, our attention must usually center on the nucleation and growth of product phases from the reactants. Thus our interests focus to a large extent on kinetics at surfaces and interfaces and on diffusion toward or away from these, as well as the two-dimensional diffusion on such surfaces and interfaces. Hence the properties of such surfaces and interfaces become a matter of great importance in understanding the nature of reactions thereat.

Many reactions of interest are of mixed control and the activated processes at the surfaces or interfaces are difficult to separate from the diffusion processes simultaneously occurring. The isothermal reduction of hematite by hydrogen provides an excellent example of such complexity. Although earlier work including McKewan's [1] indicated that the rate was controlled dominantly by reaction at the iron-wustite interface, a detailed analysis by Spitzer, Manning and Philbrook [2] indicated that the vast complications introduced by the many reactions and diffusive phenomena could simulate the simple type of control previously postulated. Olsson and McKewan [3] confirmed the importance of transport processes in an interesting manner. The nature of the reduced shell of iron on a hematite specimen is shown by the scanning electron micrograph of a fractured surface in Fig. 1. The effective diffusivity, corresponding to the counterflux of hydrogen and water vapor through such porous iron, was measured by the equipment shown schematically in Fig. 2. A relatively large amount of iron and wustite inside the cylinder establishes the equilibrium ratio, $p(H_2O)/p(H_2)$, inside and a pure H_2 atmosphere bathes the outside. By thermogravimetric technique the effective diffusivity is thus established. As shown in Fig. 3, this effective diffusivity is much smaller than the known diffusivity of the gas mixture; the ratio, called the tortuosity factor, is considerably smaller than the ideal factor usually assumed. From the effective diffusivity thus measured one can then compute the ratio $p(H_2O)/p(H_2)$, at the iron-wustite interface during the reduction of a hematite sphere (or cylinder) [4].

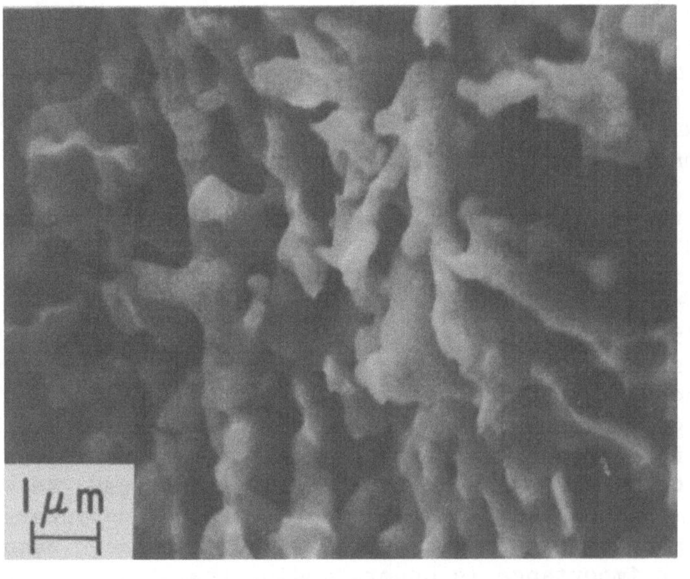

Fig. 1. Scanning electron micrograph of fractured surface of reduced iron pellet.

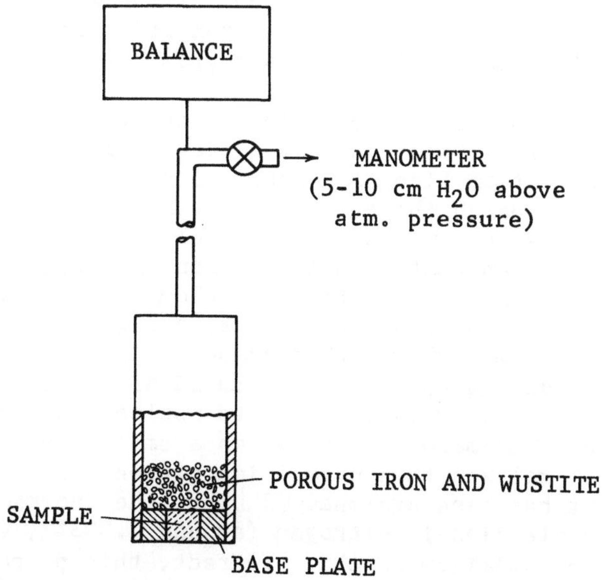

Fig. 2. Equipment used to measure H_2-H_2O diffusivity in porous iron pellet. (Olsson and McKewan [3])

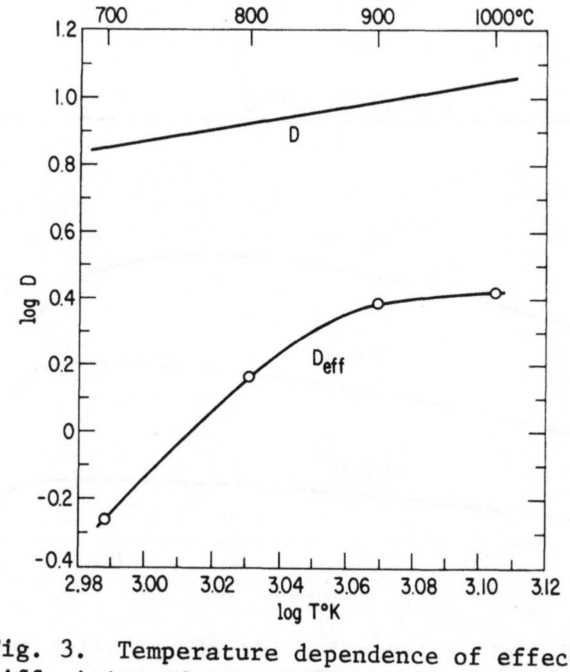

Fig. 3. Temperature dependence of effective diffusivity of H_2-H_2O in porous reduced iron pellets. (McKewan and Olsson [4])

The results obtained, Fig. 4, show quite clearly the rather complex nature of the rate control. The atomistics of the interface reaction still seems well beyond our grasp.

Another interesting example of the frustrations in interpreting metallurgical reactions had a practical origin. This was a failure of a steam line in a power plant. In the rather extensive subsequent research the faulure was traced to extensive graphitization of this carbon-molybdenum steel over a period of years. It was also shown that this phenomenon occurred only in aluminum-killed steels and the practical problem was solved by avoiding such steels for power plant boilers, and research efforts dwindled. Although it is well known that graphite is stable relative too cementite under these conditions (mild alloy steel, about 500°C), the factors normally inhibiting its nucleation are far from fully understood. The mechanism by which a small amount (perhaps 0.02%) of aluminum relaxes this normal inhibition also remains a mystery although it has been proposed [5] that the "normal" inhibitor of graphite nucleation is nitrogen (around 0.003%), which is precipitated by aluminum as AIN. In fact, this proposal is supported by the fact that a susceptible steel, held in a nitrogen atmosphere, is free of graphite only in a rim near the surface which the nitrogen has penetrated.

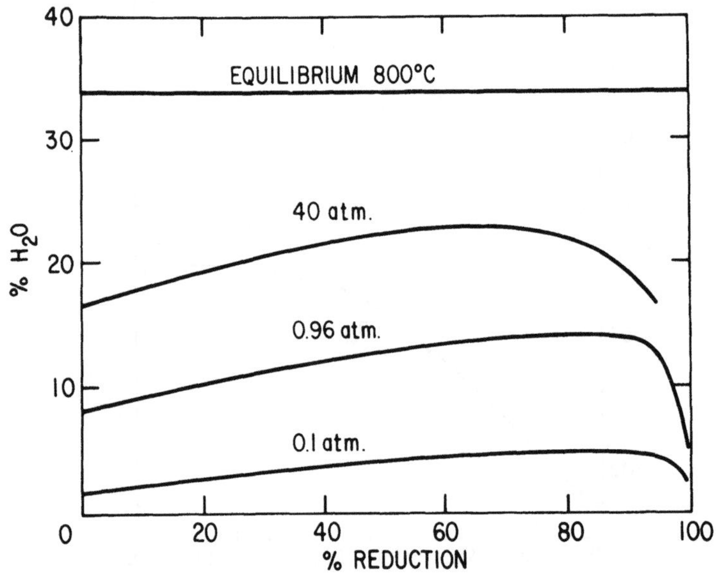

Fig. 4. $\%H_2O$ at iron-wustite interface during reduction of dense hematite spheres (7 mm diam.) in H_2 at 1 atm. at 800°C. (McKewan and Olsson [4]).

If one looks more closely it becomes clear that nucleation is not the only mystery of this graphitization problem; Figure 5 is a micrograph of graphitized aluminum-killed steel. The rate of growth of these graphite nodules is much less than anticipated on the basis that it is controlled by carbon diffusion. One explanation offered is that the volume expansion of the reaction requires the slower vacancy diffusion, i.e. diffusion of iron away from the nodules; another is that the volume expansion results in the imposition of an essentially hydrostatic pressure on the graphite thus rendering the reaction less energetically favorable. Although both these factors may play a role, it seems highly unlikely that the stresses arising from the volume change would not be substantially relieved by dislocation generation and motion at the temperature and times involved.

If either of the proposed explanations for the slowness of graphite growth is correct, clearly carbon diffusion is still required for growth and one would expect the carbon to come from the immediate vicinity of the nodule and thus give rise to a rim around each nodule completely free of cementite; a simple calculation involving the mass balance and densities discloses that this rim should have a diameter three times that of the nodule.

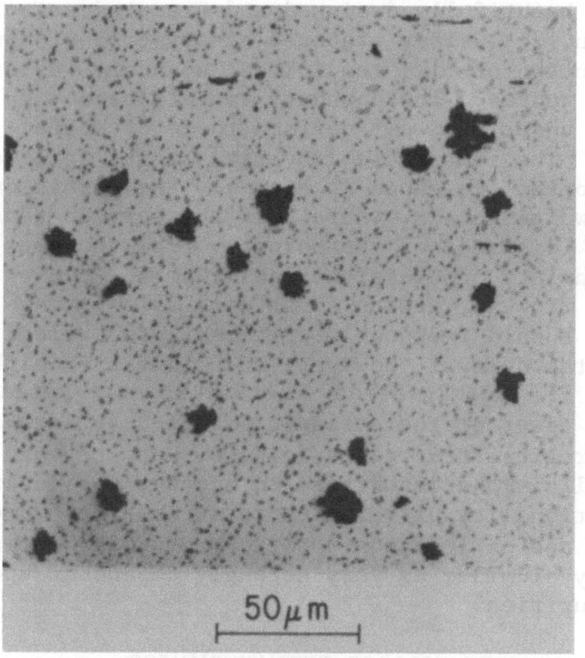

Fig. 5. Micrograph of graphitized Aℓ-killed carbon steel; picral etch (ref. 5).

However, inspection of Figure 5 discloses no such rims; in fact, the distribution of cementite appears essentially uniform. This is in contrast to the appearance of nodular iron (strongly desulphurized) where such clear rims are normally found around the graphite nodules. Further, decarburization experiments on relatively high-purity iron-carbon alloys (two-phase, -iron-cementite) always result in a cementite-free rim near the surface and (as expected) no evidence of diffusion in the interior except for Ostwald ripening. We thus seem forced to the conclusion that graphitization, such as exhibited in Fig. 5, is hindered to a substantial extent by some hindrance to the dissolution of cementite. To the best of our knowledge, the macrofactors that might be considered offer no adequate explanation and the slow dissolution is most reasonably attributed to adsorption of some element(s) at the ferrite-cementite interface. Some little light on this matter is perhaps shed by the conclusion of Podgurski [6], based on his experiments and others, that the decomposition (graphitization) of the carbides of iron, nickel and cobalt is markedly retarded by small amounts of sulphur sufficient to produce only a surface monolayer.

GRAIN BOUNDARY MIGRATION

Let us now seek relief from such complicated situations in considering a beautifully designed and executed set of experiments by Rath and Hu [7]. They prepared bi-crystals of aluminum in the shape of a section of pie; the initial boundary was essentially normal to the radius. The boundary rapidly became cylindrical and its migration under the well-defined force of surface tension only was measured. The remarkable linearity of its intersections with the top and bottom are shown in Fig. 6. The stress on the boundary is simply shown to be σ/r, where σ is the surface tension and r is the radius (distance of the boundary from the apex). By analogy with a vast variety of classical laws (Ohm's, Fourier's, Fick's, Poisueille's, Stokes', Newton's) as well as existing theories for this case, we should reasonably anticipate the velocity to be preportional to the stress. However, such is far from the case. As shown in Fig. 7, the migration velocity is more nearly proportional to the fourth power of the stress. As noted by Rath and Hu [7] existing theoretical treatments (e.g. of Cahn [8] and of Lucke and Stuwe [9]) all invoke impurities to account for departure from direct proportionality of velocity to stress for boundary migration.Thus, we are called upon to interpret this "chemical reaction" in high-purity aluminum as very strongly dependent upon unknown traces of impurities.

SITE-FILLAGE THERMODYNAMICS

These examples not only illustrate the complexities that may be involved in surface reactions, but also indicate the apparent

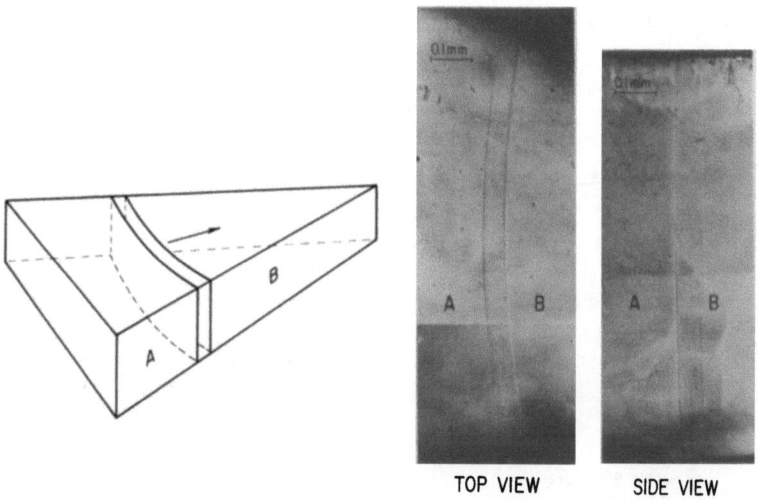

Fig. 6. Effect of driving force on migration of high angle tilt boundaries in Aℓ bicrystals [7].

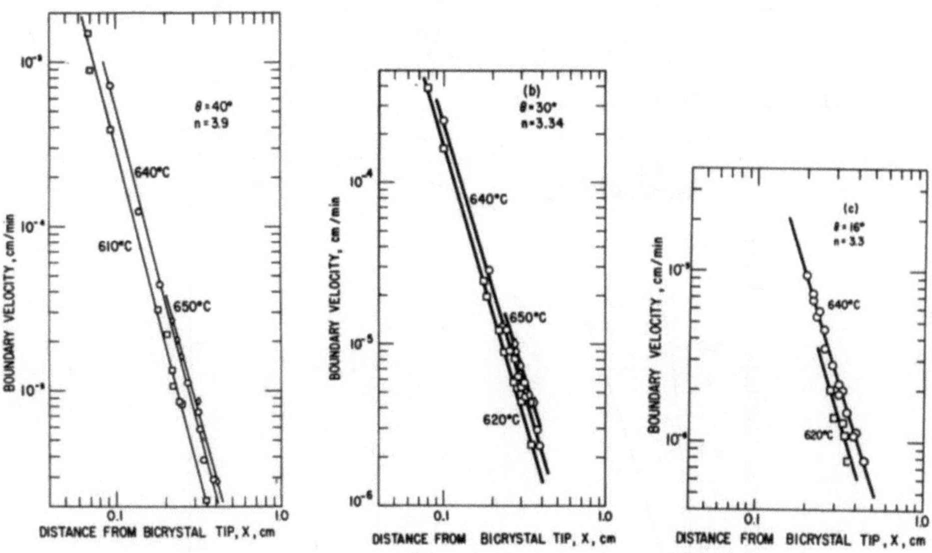

Fig. 7. Boundary velocity as a function of distance from the bicrystal tip for different misorientation angles and temperatures [7].

vast effects (strong adsorption) of trace components. Demonstration of the actuality of such strong adsorption even at steelmaking temperatures was established by the work of Kozakevitch and Urbain [10] and of Holden and Kingery [11] on the dramatic effect of small amounts of oxygen on the surface tension of liquid iron. These measurements as well as the surface concentration of oxygen as derived therefrom via the Gibbs adsorption isotherm* are shown in Figs. 8 and 9.

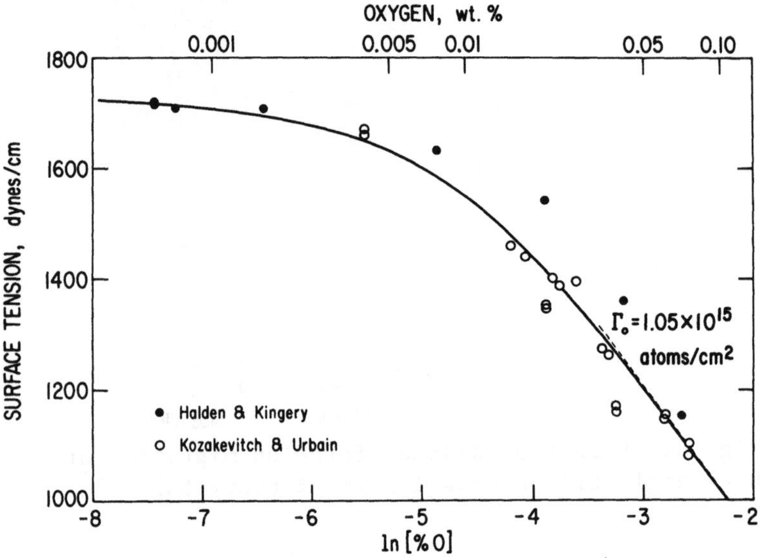

Fig. 8. Surface tension of liquid iron-oxygen alloys, 1550°C.

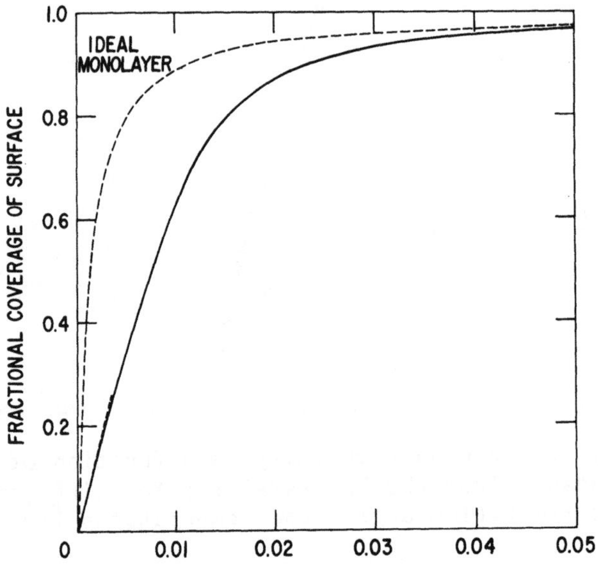

Fig. 9. Oxygen adsorption isotherm for liquid iron at 1550°C.

* The surface concentration, of oxygen in this case is $\Gamma_o = -d\sigma/d\mu_o$, where σ is the surface tension and μ_o is the chemical potential of oxygen taken as RT ln [%O].

It will be noted that at low concentrations the ratio of oxygen atoms to iron atoms is about 1,000 times greater at the surface than in the interior of the liquid phase; the surface is 90% "filled" when the bulk concentration of oxygen is only 0.02 wt. %--one-tenth the concentration in equilibrium with bulk liquid oxide. It is particularly noteworthy that "fillage" corresponds to approximately one oxygen atom per surface iron atom.

In such a case the temptation is strong to think of the surface oxygen atoms as distributed on a fixed number of sites provided by the iron atoms. Although this concept may be a bit misleading, it is worthwhile at this point to review briefly the elementary statistical mechanics leading to the thermodynamics of an ideal solution in which a solute is distributed on a fixed number of sites provided by the solvent--such as interstitial solution in a crystalline phase.

For any ideal solution (one such that the enthalpy of transfer from any component to the solution is independent of the concentration and hence of the detailed nature of the atomic distribution), the activity of the i'th component is found immediately from elementary statistical mechanics to be proportional to $g(n_i+1)/g(n_i)$, where $g(n_i)$ is the degeneracy or the possible number of distinguishable arrangements of the atoms (including n_i atoms of species i) upon the available sites and $g(n_i+1)$ is the similar quantity when the number of atoms of species i is increased by one, the number of all other atoms remaining constant.

Table I summarizes the derivation from this relation of the activity of ideal binary solution for: Case I--the more ordinary substitutional solution in which the number of sites is equal to the total number of atoms; and Case II--the case of interest here, in which the number of sites (m) available for the solute is not changed as the amount of solute is increased. Here, N_2 is the atom (mol) fraction of solute and θ is the fractional occupancy of the sites by solute.

If we have perhaps belabored the obvious here in arriving at the well-known Langmuir-type isotherm for Case II, it is because attention is usually focused at the low solute end where proportionality of activity to concentration prevails just as in Case I. Although Case II-type behavior has been known for a long time, and ramifications considerably beyond those to be discussed here are treated in many texts, those interested in high-temperature chemistry have all too frequently chosen to ignore the rise to infinity occasioned by the $1-\theta$ term in the demoninator. The high-θ region has seldom been investigated with precision. For this reason we call attention to the measurements of Wriedt [12] on the nitrogen activity in the γ' phase ("Fe_4N") of the Fe-N system; this phase is

TABLE I

Ideal Binary Solution
Activity of Solute

	Case I	Case II
	Substitutional	Fixed No. of Sites (=m)
$g(n_2)$	$= \frac{(n_1+n_2)!}{n_1!\ n_2!}$	$\frac{m!}{n_2!\ (m-n_2)!}$
$g(n_2+1)$	$= \frac{(n_1+n_2+1)!}{n_1!\ (n_2+1)!}$	$\frac{m!}{(n_2+1)!\ (m-n_2-1)!}$
$\frac{a_2}{a_2^*}$	$= \frac{g(n_2)}{g(n_2+1)} = \frac{n_2+1}{n_1+n_2+1}$	$\frac{n_2+1}{m-n_2}$
$\frac{a_2}{a_2^*}$ (for $n_2 >> 1$)	$= N_2$	$\frac{\theta}{1-\theta}$

based on fcc iron in which only the centers of the cells are occupied by nitrogen at the ideal composition, Fe_4N, and may thus be visualized as obtained by continuous addition of nitrogen to γ-iron. The activity (fugacity) of nitrogen for this system is represented in the left portion of Fig. 10. The low-concentration data (and the indicated tangent) are extrapolated from well-established data on the solubility of N_2 gas, and the high-concentration (γ') data are from Wriedt's measurements shown on an expanded scale on the right. The intermediate points were obtained by less precise extrapolation of phase diagram data. The nature of the curve clearly indicates the known miscibility gap. The departure of this curve from the idealized one is quite pronounced; clearly the intermediate region is far from ideal. The linear type of behavior at the low end is very well established and recognized. The abrupt rise at high fillage is a dominant feature.

Let us now see how well Wriedt's measurements fit the Langmuir-type isotherm previously discussed. In this case, the fillage, θ, exceeds 0.98 for nearly all the measurements. Hence the numerator of the expression (Table I) for the activity is essentially one, and we may thus anticipate the reciprocal of the activity of nitrogen to be proportional to 1- θ. Over the short range of compositions involved, the reciprocal of the nitrogen activity should then be a linear function of % N with an intercept corresponding to the composition Fe_4N (5.900% N). That such a relation is indeed valid is indicated by a plot of the data in Fig. 11.

Although these equilibrium measurements do not reveal any departures from stoichiometry, other than that occasioned by partial fillage of the cube centers by nitrogen as already discussed, kinetic measurements do reveal some nitrogen occupancy of abnormal sites*(centers of cube edges). Schwerdtfeger, Grieveson and Turkdogan [13] have measured the growth of the γ'phase on α-iron in NH_3-H_2 gas mixtures. The flux of nitrogen through this γ' phase, on the basis of the vacancy model, which satisfies the equilibrium measurements so nicely, would be expected to be proportional at any

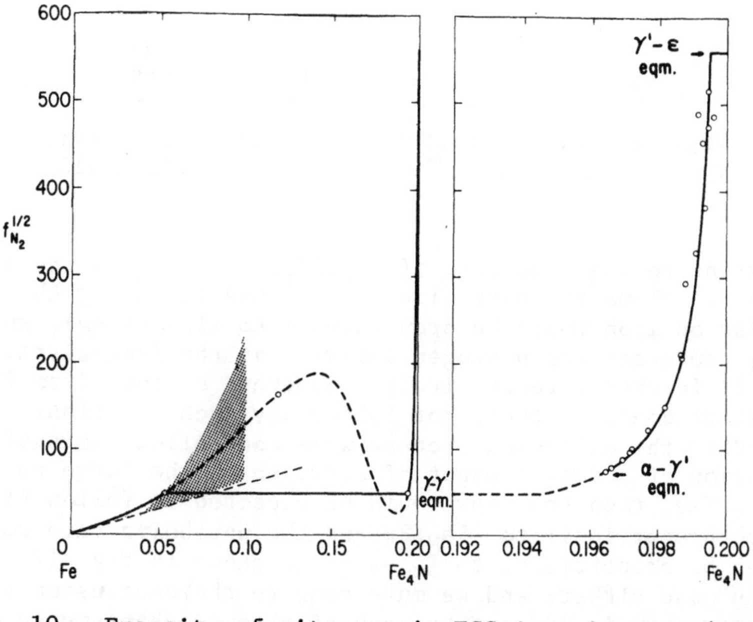

Fig. 10. Fugacity of nitrogen in FCC iron lattice (500°C).

* H. J. Grabke [Ber. Bunsenger, physik. Chem. 72, 541 (1968)] has been able to show that the concentration of nitrogen in the disorder sites (very small) is proportional to the nitrogen activity, in accord with the interpretation of the diffusion studies. Accordingly, there should be a slight departure from linearity in Fig. 11 at high nitrogen activities.

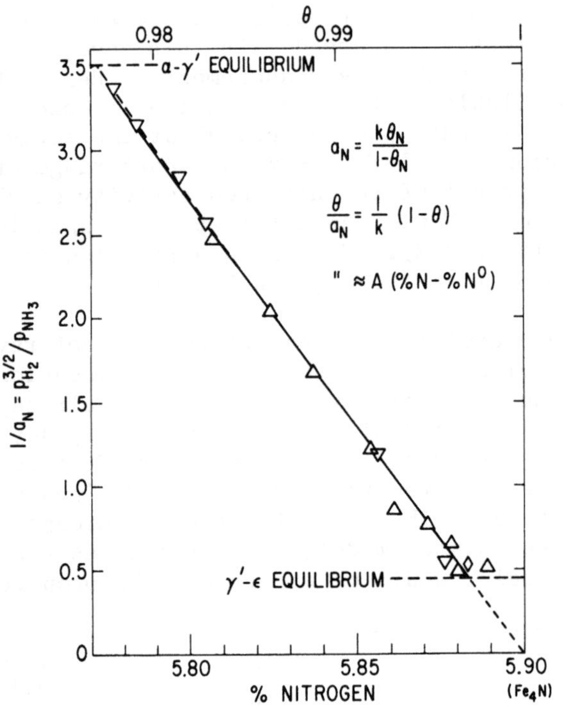

Fig. 11. Linear reciprocal relation of activity to composition γ' ("Fe_4N"), 500°C. (Wriedt [12]).

given instant to the gradient, $\partial(1/a_N)/\partial_x$, where a_N is the activity of nitrogen.* Thus the parabolic rate constant [13] for the growth of γ' phase on iron would be proportional to $(1/a_N'-1/a_N'')$, where a_N' and a_N'' represent the nitrogen activity at the iron-nitride and gas-nitride interface respectively. However, as seen from Fig. 12a, this constant does, in fact, not follow any such relation. On the other hand, if the diffusion process were controlled dominantly by the migration of a small amount of nitrogen on the large number of abnormal sites, then the flux would be expected to follow Fick's law and be proportional to $-\partial a_N/\partial x$ and the parabolic rate constant would then be proportional to $(a_N''-a_N')$; as shown in Fig. 12b, this is not the case either, and we must come to the conclusion that the diffusive process is in part by each of these mechanisms as illustrated by Fig. 12c. (The activity of nitrogen in this figure is

*) By the Einstein relation the flux is equal to the product of the mobility, concentration and chemical potential gradient. Setting the mobility proportional to the vacancy concentration, (1-θ), and the activity proportional to θ/(1-θ) gives immediately the above proportionality for θ near one.

here taken as equal to $p(NH_3)/p(H_2)^{3/2}$.) The relative fluxes by each of the two processes are shown as functions of composition in Fig. 13.

THE NATURE OF CRYSTAL SURFACES

Let us now take a look at some actual crystal surfaces. Fig. 14 shows Sundquist's [14] replicas and profiles of pure gold and of gold containing 1/4% silver equilibrated at 1000°C in vacuum. The shape of the pure gold crystallites is rather typical of pure fcc metals. Doping with a substitutional solute is seen to decrease the anisotropy of surface tension so that the {100} and {111} facets have almost disappeared and the surface becomes rounded or "rough" on an atomic scale. Increase of temperature also leads to roughening of the crystallites, i.e. disappearance of facets, as illustrated in Fig. 15. However, nickel or copper particles, which appear somewhat similar to those of gold (Fig. 14a) when annealed in dry hydrogen, exhibit definitely sharpened edges if a trace of oxygen is present, Fig. 16. For all the fcc metals investigated by Sundquist, facets (when present) were {100} and {111}, as seen more strikingly in the photograph (Fig. 17) constructed to scale from his observations.

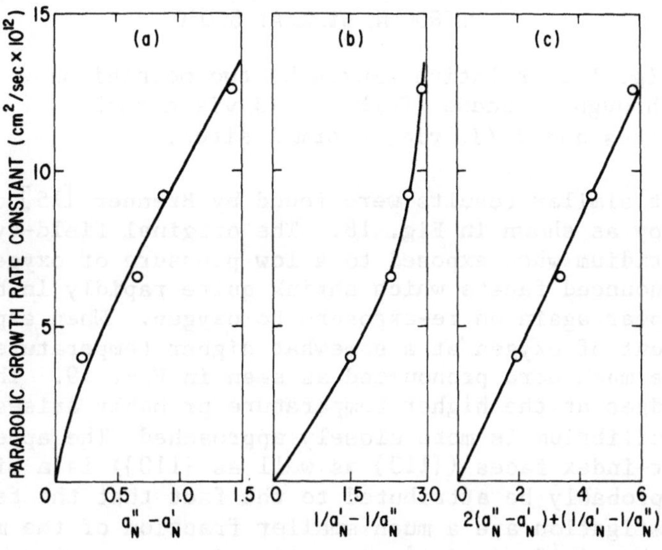

Fig. 12. Parabolic rate constant for growth of γ' phase on iron as a function of nitrogen activity, indicating diffusion is not exclusively via (a) normal sites, (b) abnormal sites, but (c) both (Schwerdtfeger, Grieveson and Turkdogan [13]).

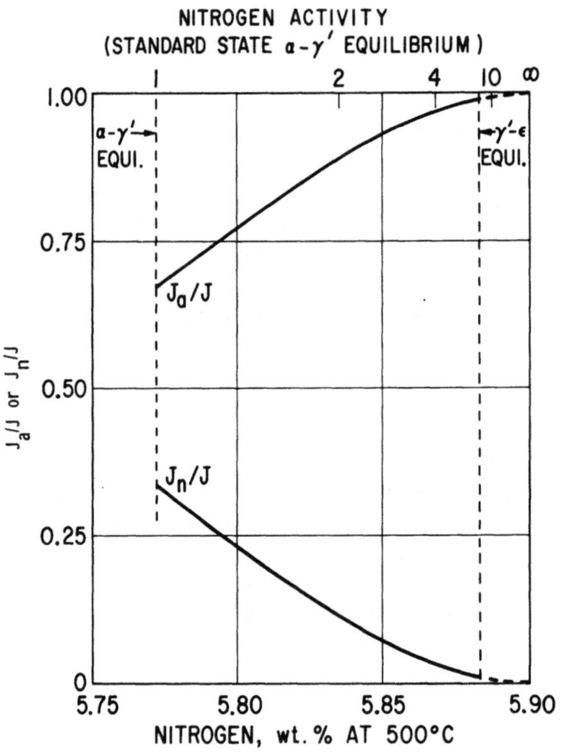

Fig. 13. Relative fluxes by two mechanisms through γ' phase (Fe_4N); J_n/J via normal sites and J_a/J, via abnormal sites.

Somewhat similar results were found by Brenner [15] using field-ion microscopy as shown in Fig. 18. The original field-evaporated surface of iridium when exposed to a low pressure of oxygen at 700°C exhibits pronounced facets which shrink quite rapidly in high vacuum but which appear again on re-exposure to oxygen. When exposed to the same amount of oxygen at a somewhat higher temperature, the facets become much more pronounced as seen in Fig. 19. The sharpening of the edges at the higher temperature probably arises from the fact that equilibrium is more closely approached The appearance of the higher-index faces ({113} as well as {110}) is a bit surprising but may probably be attributed to the fact that the temperatures of this investigation are a much smaller fraction of the melting temperature than in Sundquist's investigation.

Let us next turn our attention to the nominally flat surfaces more frequently encountered in the laboratory and in practice. There have been many observations of the development of striations on these by so-called thermal etching--i.e., by holding at elevated temperature in vacuum or controlled atmosphere for an extended period of time. An example of this is given in Fig. 20a, which is a photomicrograph in oblique light showing striations formed after annealing a polished specimen of zone-refined iron in wet hydrogen [$p(H_2O)/p(H_2) = 0.035$] for 24 hours at 1000°C. It has also been observed, for example by Moreau and Benard [16], that pronounced striations occur on the thermal etching of a polished specimen only in a certain range of oxygen pressure. This region starts at a partial pressure of oxygen just below that corresponding to equilibrium with massive oxide and terminates rather sharply at an oxygen pressure perhaps two orders of magnitude lower. At lower oxygen pressures yet (especially at high temperature), the surface is essentially smooth or slightly rumpled as viewed in the light microscope. The obvious interpretation, viewed in conjunction with the previously mentioned observations of Sundquist [14] and of Brenner [15], is that this lower limit occurs at an oxygen potential corresponding to substantial coverage of the surface sites by oxygen. The surfaces of the facets are usually found to be low-index planes, expecially {100} and {111}. It can also be observed at higher magnification that the bottoms of the valleys and tops of the hills of these striations are not sharp angles but

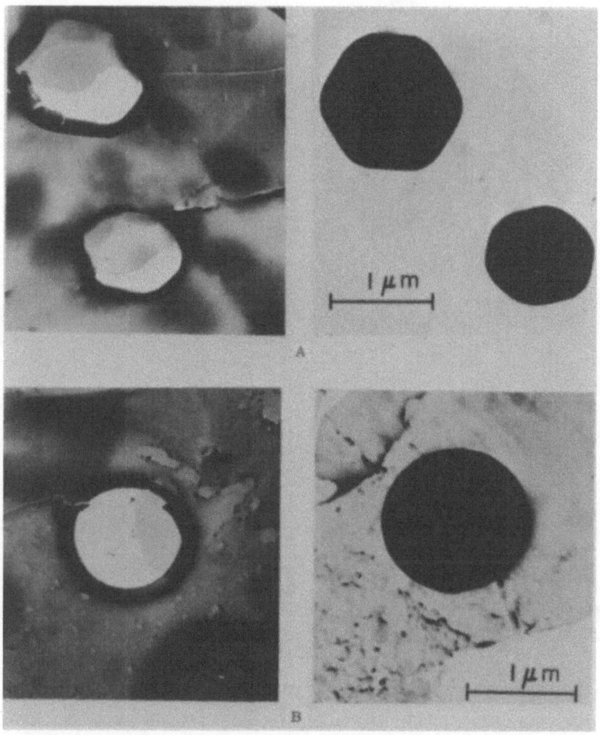

Fig. 14. Replicas and profiles of equilibrated particles: (a) Pure Au and (b) Au + 0.25%Ag, 1000°C in vacuum. (Sundquist [14]).

exhibit curvature just as did Sundquist's crystallites, as for example shown in Fig. 20b.

Limiting faceting of a 1/10 mm nickel wire annealed in hydrogen at 1100°C for 72 hours is shown in Fig. 21. Although facets and rounded edges are clearly visible in some areas, large areas are completely void of such features, appearing smooth at these magnifications and thus obviously rough on an atomic scale.

All these observations are thus quite consistent and lead to a model of a metallic surface at elevated temperatures displaying a vast array of different types of sites for an adsorbate; not only do we encounter different types of facets but we must recognize the multitude of different types of sites present on the curved (atomically rough) surfaces. Moreover, it seems clear from the effect of oxygen that even the nature of the dominant sites may change drastically with the extent of adsorption. Thus, the anticipated departures from the highly idealized type of adsorption may be visualized as arising from two sources:

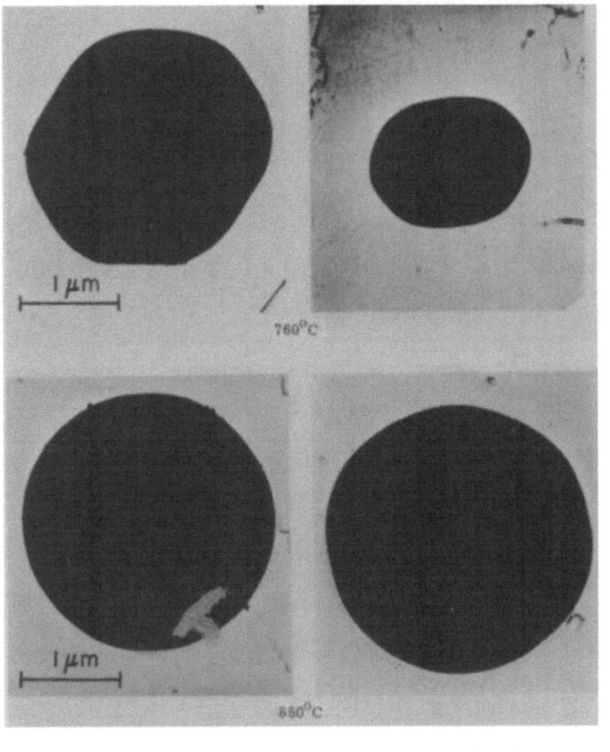

Fig. 15. Profiles of silver particles equilibrated in dry H_2. (Sundquist [14]).

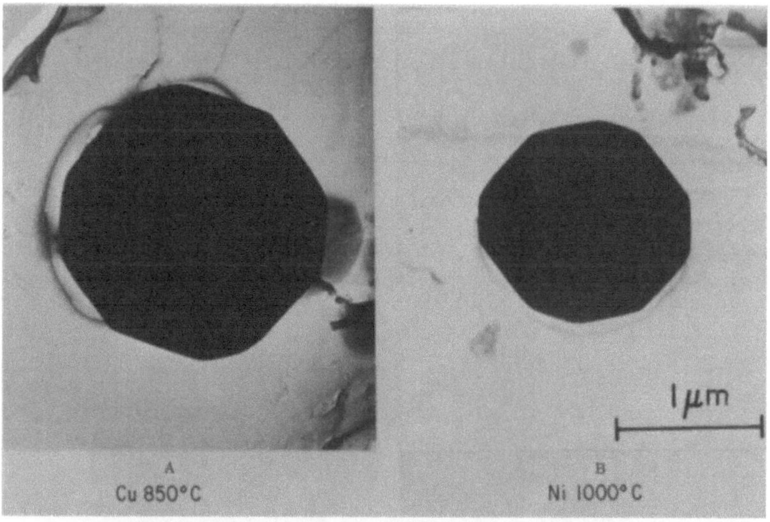

Fig. 16. Typical profiles of (a) copper and (b) nickel particles equilibrated in a vacuum of 10^{-5} to 10^{-6} torr. (Sundquist [14]).

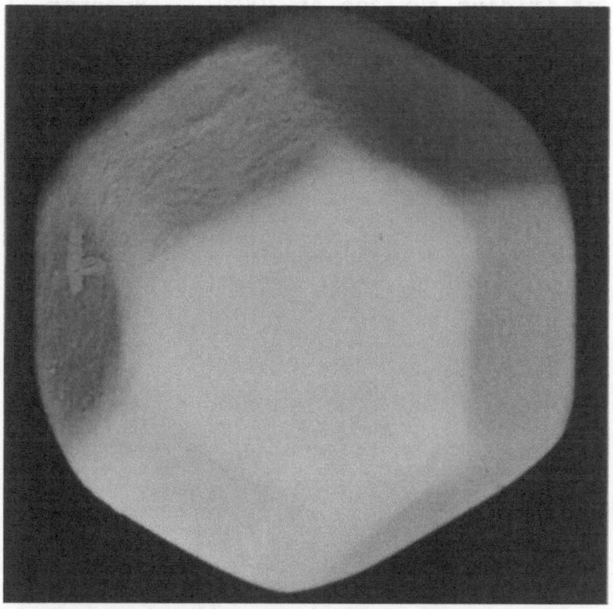

Fig. 17. Model of equilibrium shape of an FCC crystal. (Sundquist [14]).

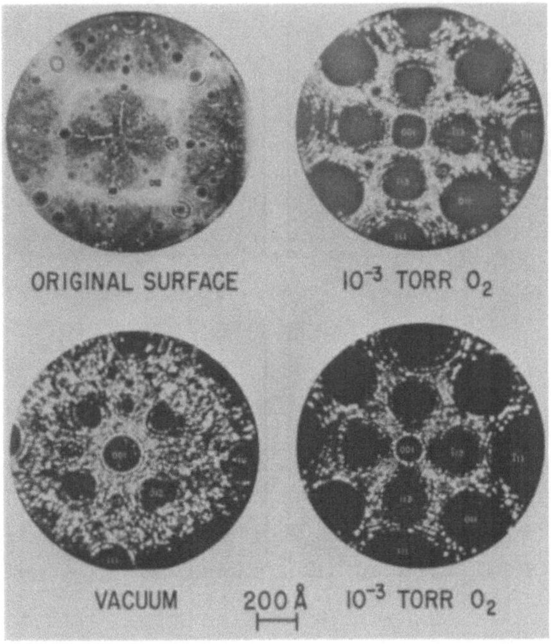

Fig. 18. Reversible thermal rearrangement of iridium surface at 700°C. (Brenner [15]).

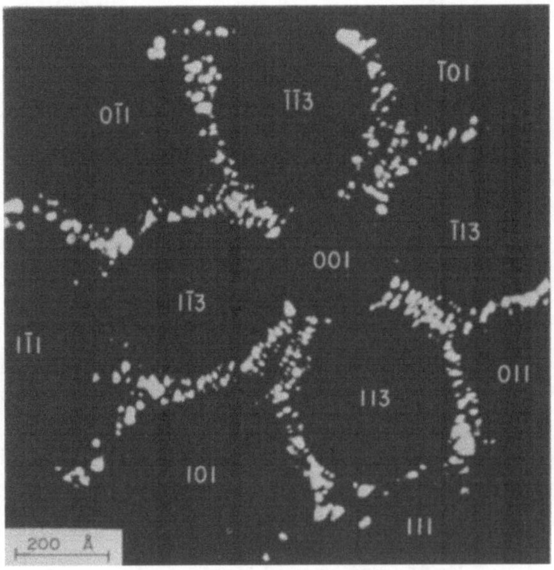

Fig. 19. Faceting of iridium tip at 880°C in 10^{-3} torr oxygen.

1. Departures from ideality of these two-dimensional solutions. Even for a well-defined set of identical sites, it would be reasonable to suppose that this type of departure would be quite analogous to that normally encountered for three-dimensional solutions including interstitial solutions as illustrated in Fig. 10, and indicated at least in part by the departure from ideality of oxygen adsorbed on liquid iron, Fig. 9.

2. The various types of sites and the changing distribution of these with increasing adsorption. In fortunate cases at least, we might anticipate that at low coverage only

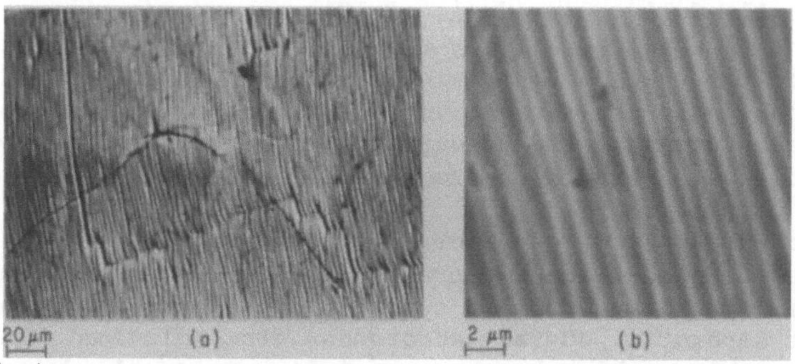

Fig. 20. Thermal faceting on zone refined high-purity iron after annealing in wet hydrogen (H_2O/H_2 = 0.035) for 24 hr. at 1000°C. (a) oblique illuminations and (b) scanning electron micrograph.

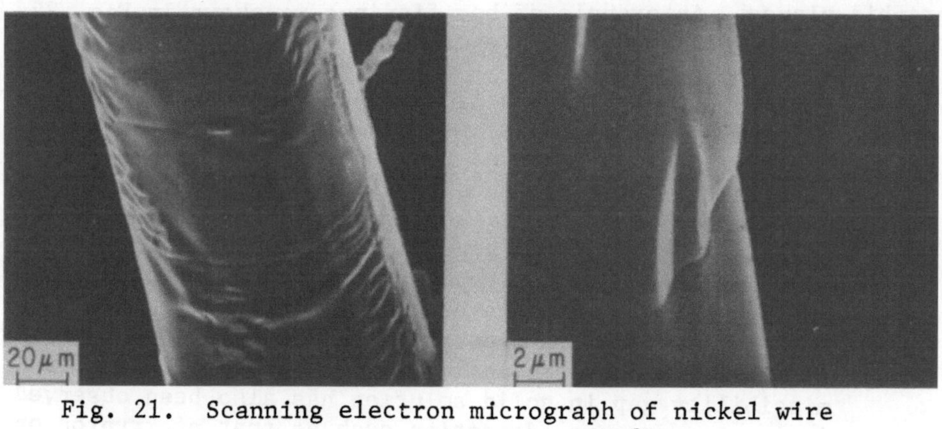

Fig. 21. Scanning electron micrograph of nickel wire annealed in hydrogen for 72 hr. at 1100°C.

one particular type of site would be occupied to appreciable extent; similarly at a high coverage only one particular type of vacancy would be available for occupancy.

Measurements of adsorption on the surfaces of solid metals at elevated temperatures are indeed difficult and very few have been reported in the literature. Mention has already been made of the pronounced adsorption of oxygen on the surface of liquid iron as deduced from surface tension measurements. By the null-creep method Hondros [17] has measured the surface tension of γ- and δ-iron and the effect of phosphorus and nitrogen thereon. From these measurements, he has calculated for δ-iron at 1450°C, the extent of adsorption of phosphorus, finding that the surface is essentially fully covered (1.4×10^{15} atoms/cm^2) when the bulk composition is about 0.1% P; by further refinement of the technique, he found that the grain boundaries were essentially "filled", but at only about half the surface concentration, at the same bulk composition. However, the adsorption isotherms cannot be obtained with any degree of precision from these measurements.

Direct measurement of adsorption isotherms at elevated temperatures present almost insurmountable difficulties on account of the solubility of the adsorbed species in the matrix and the problem of separating surface adsorption from solution. The attempt to minimize this problem by the use of fine powder leads to other difficulties such as indeterminacy and change of the surface area at elevated temperatures. However, Cabané-Brouty [18] managed to overcome these difficulties in an investigation of the adsorption of sulphur on silver by use of a radiotracer technique. The low solubility of sulphur in silver combined with a high adsorption of beta rays from the matrix permitted direct determination of the adsorption isotherms at several temperatures and on different crystallographic planes. An example of her findings is shown in Fig. 22. Two features are particularly noteworthy:

1. The isotherm approaches linearity at low activity, and a horizontal asymptote is essentially followed at a high activity over an extended range. There is no evidence of steps in the curve such as might have been anticipated if we expected an appreciable activity range between the fillage of one type of site and another. In fact, the limiting occupancy corresponds to about 2×10^{14} S atoms/cm^2, routhly to a monolayer. Her isotherms for the {100} {100} planes exhibit even greater departure from ideality. (A similar phenomenon analogous to the appearance of a miscibility gap in solid solution has also been observed in low-temperature adsorption such as that of krypton on graphite at 70°K.[19]) The intermediate range is in no case ideal. However, at higher activity all her isotherms

exhibit close approach to a horizontal asymptote in about the same activity range.

2. Fillage is essentially complete when the activity of sulphur is about 1% of that corresponding to the Ag-Ag_2S equilibrium. If we make the reasonable assumption that the solubility of Ag_2S in Ag corresponds to less than 0.01 at.%, we thus come to the conclusion that the distribution coefficient of sulphur between surface sites and lattice sites is greater than 10,000.

Cabané-Brouty's adsorption isotherms as well as the indications from the measurements of Hondros [17] and of Kozakevitch and Urbain [10] and of Holden and Kingery [11] are in strong contrast to the more familiar BET diagrams typical of low-temperature adsorption and illustrated [20] in Fig. 23. The reason for the lack of horizontal asymptote here is quite clear--the adsorption is not limited to a monolayer; the chemical bonding to the substrate is of the relatively weak van der Waal's type and the measurements are in a region relatively close to the condensation point of the gas involved. On the other hand, high-temperature adsorption, in the cases considered, involves a strong chemical bond and appears to be limited to an amount approximating at most a monolayer.

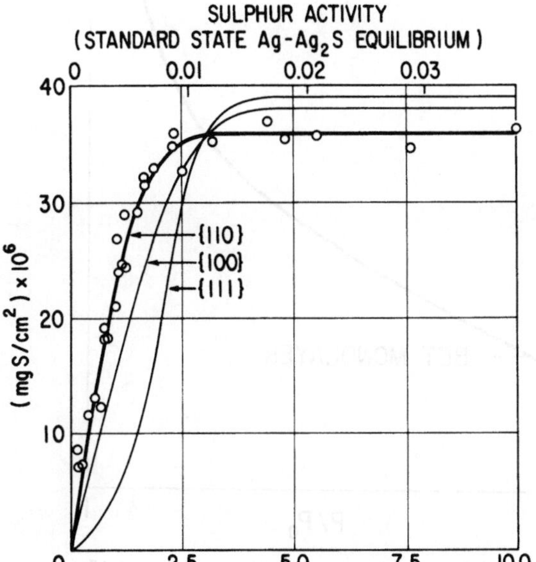

Fig. 22. Chemisorption of sulphur on silver {111},{100}, and {110} planes at 400°C. (Cabané-Brouty [18]).

REACTION RATE AT SURFACES

The evidence as to thermodynamic behavior of adsorption on surfaces and interfaces certainly leaves much to be desired. However, it seems worthwhile to proceed tentatively on the basis that the relations developed in the manner of those of Table I for a fixed number of sites may be useful at low and especially at high surface adsorption. By extension of this reasoning we may write the activity of the i'th species (including the activated complex) as

$$a_i = \frac{\theta_i}{1-\theta}\phi_i \qquad (1)^*$$

where θ_i is the fractional occupancy of the sites by the i'th species, $\theta = \sum_i \theta_i$ and ϕ_i plays the role of an activity coefficient which is now assumed to be essentially constant when θ is small or when any θ_i is near 1.

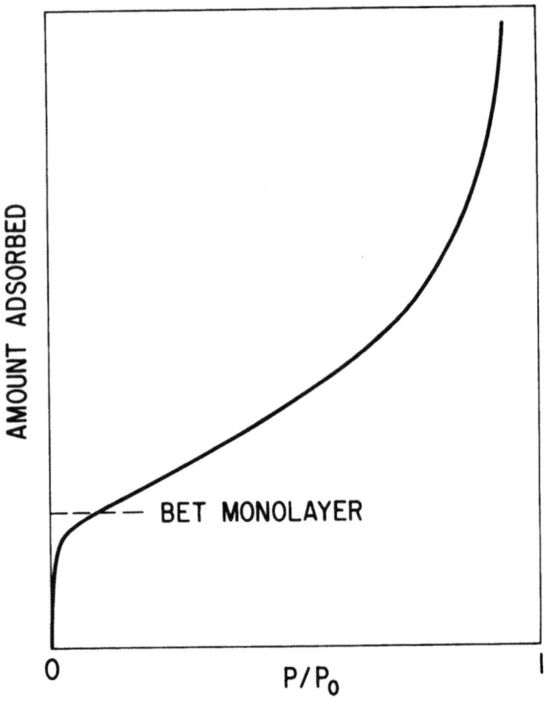

Fig. 23. Adsorption of argon on rutile (Morrison, Los, and Drain [20]).

---

*) The denominator of this expression should in general be $(1-\theta)^{x_i}$, where x_i is the number of sites occupied by a single member of the i'th species. However, in all cases to be considered $x_i = 1$.

Kinetic expressions can then be developed using this assumption in conjunction with the two basic equations of absolute reaction rate theory: The first is

$$\text{Specific rate} = \frac{kT}{h}\Gamma_{\ddagger} = \frac{kT}{h}\Gamma_{o}\theta_{\ddagger} \qquad (2)$$

where $\Gamma_{\ddagger}$ is the surface concentration of the activated complex number of activated complexes per unit area) and $_{o}$ is the number of sites per unit area. The second is the thermodynamic equilibrium constant for the equilibrium between the activated complex and the reactants which form it:

$$K_{\ddagger} = a_{\ddagger}/\Pi\, a_i^{\nu i} \qquad (3)$$

Support of the concept of a limiting rate, implied by the use of the postulate of a limited surface fillage, is provided by a simple case considered in more detail later. This is the surface-controlled decarburization of an iron-carbon alloy [21] shown in Fig. 24. We see here (1) the simple anticipated proportionality of rate to activity (or concentration) in the low concentration region and (2) a limiting rate interpreted as corresponding to essentially complete coverage of the surface by carbon at high activity. Detailed analysis of this curve reveals that ϕ_C and $\phi_{\ddagger}$ are far from constant since the initial slope computed on an ideal basis from the terminal data is nearly ten times that observed.

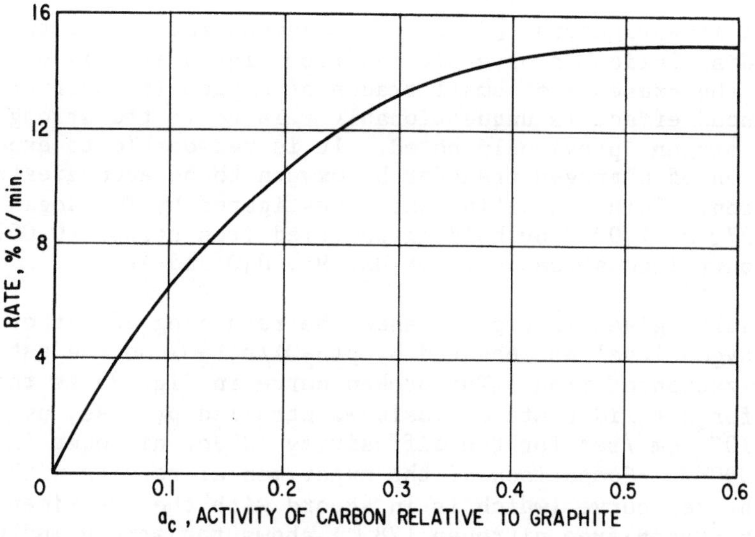

Fig. 24. Decarburization rate of Fe-1.5%C alloy (0.56 mm thick strips) in dry hydrogen at 1140°C and 0.96 atm. (Turkdogan and Martonik [21]).

The forward reaction rate, R_f, found by combining equations (1), (2) and (3) is

$$R_f = \frac{kT}{h} \Gamma_o K_{\dagger} (1-\theta) \Pi a_i^{\nu_i} / \phi_{\dagger} \quad (4)$$

In general, the reverse reaction must also be considered and, if no other activated process is involved, the net reaction rate is

$$R = \frac{kT}{h} \Gamma_o (1-\theta) \frac{K_{\dagger}}{\phi_{\dagger}} (\Pi a_i^{\nu_i} - \Pi_{eq} a_i^{\nu_i}) \quad (5)$$

where $\Pi_{eq} a_i^{\nu_i}$ is the value that would prevail at equilibrium with the products and is found from the equilibrium constant and the activities of the reaction products. In deriving equation (5), it is also assumed that $\phi_{\dagger}$ is the same for the forward and reverse reactions. At low site-fillage the term (1-θ) reduces approximately to one, $\phi_{\dagger}$ is essentially constant and if the reaction is far from equilibrium equation (5) reduces to the well-known simple proportionality of rate to the product of the activities of the reactants forming the activated complex. At high site-fillage the factor (1-θ) approaches zero and obviously plays a dominant role in decreasing the rate; it is evaluated from equation (1) with appropriate choice of i. This solution is usable only if the high site-fillage is dominantly by a single species. It is only in this case that each site has essentially the same surroundings (environment) and hence the activity coefficients are constant.

KINETICS OF NITROGEN TRANSFER THROUGH AN IRON SURFACE

Several investigators [22-26] have demonstrated that the rate of N_2 dissolution in or evolution from liquid iron is markedly retarded by the presence of small traces of oxygen (or sulphur). This pronounced effect is unquestionably related to the strong adsorption of oxygen, previously noted. It is reasonable to expect the inhibition of nitrogen transfer by oxygen to be even greater for solid iron. This inhibition was investigated by Turkdogan and Grieveson [27] at 1000°C by holding purified iron strips (0.051 cm thick) in controlled atmospheres of N_2, H_2, H_2O and Ar.

The results given in Fig. 25 show the retarding effect of increased oxygen level as measured by $p(H_2O)/p(H_2O)$ on the rate of nitrogenization of iron. The broken curve in Fig. 25 is that calculated for a solid state diffusion-controlled process, using $D = 1.12 \times 10^{-7}$ cm²/sec for the diffusivity [28] of nitrogen in γ-iron at 1000°C. Comparison of the experimental results with the calculated broken curve (which is in accord with the experimental findings for oxygen-free nitrogen [28]) shows the strong influence of oxygen, or oxygen-bearing gaseous species, on the rate of the nitrogen-iron interfacial reaction. In one series of experiments, the samples were initially annealed for 17 hours at 1000°C in a hydrogen-water vapor mixture having the ratio $p(H_2O)/p(H_2) = 0.064$ (solid triangles in Fig. 25). The rate of nitrogenization of these samples was found to be the same as those in another series (open triangles in Fig. 25) in which the samples were annealed in dry hydrogen prior to the nitrogenization experiment in an atmosphere with $p(H_2O)/p(H_2) = 0.064$. This finding indicates that the rate of saturation of the surface of iron with chemisorbed oxygen, in the particular atmosphere employed, is a relatively fast reaction. It is apparent from these experimental results that the interstitial diffusion of nitrogen in iron does not play a prominent role in the control of the rate of nitrogenization or denitrogenization of iron strips (0.051 cm thick) when oxygen is present in the reacting gas. In order to verify this indication, a thicker iron strip (0.51 cm thick) was reacted with nitrogen at 1000°C with $p_{N2} = 0.91$ atm and $p(H_2O)/p(H_2) = 0.052$. After a given reaction time, the metal was sectioned parallel to its surface and each layer analyzed for nitrogen. As seen from the results given in Fig. 26,

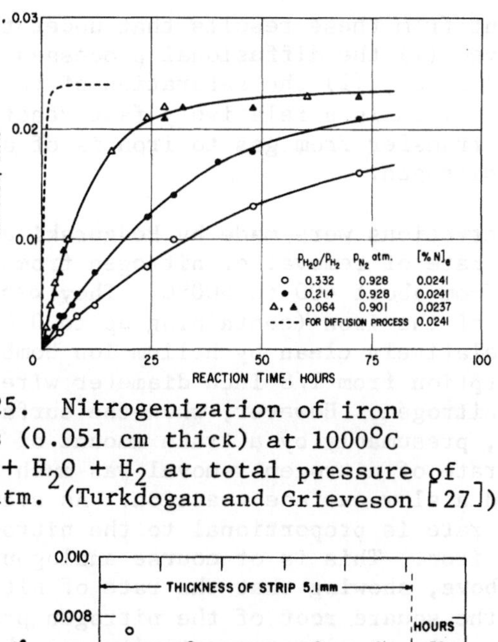

Fig. 25. Nitrogenization of iron strips (0.051 cm thick) at 1000°C in $N_2 + H_2O + H_2$ at total pressure of 0.96 atm. (Turkdogan and Grieveson [27]).

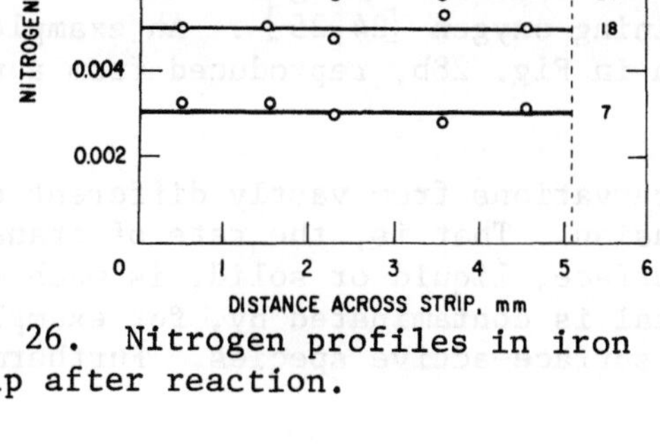

Fig. 26. Nitrogen profiles in iron strip after reaction.

the nitrogen distribution across the sample is essentially uniform. Since, in the rate studies much thinner strip was used (0.051 cm thick), the subsurface nitrogen concentration in solution can in all cases be taken to be almost the same as those determined by analyzing the entire sample.

In another series of experiments, iron was nitrogenized in atmospheres of various nitrogen partial pressures with $p(H_2O)/p(H_2)$ = 0.035. The results obtained are given in Fig. 27; as would be expected, the rate of nitrogenization decreases with decreasing partial pressure of nitrogen. It should be noted that the half-time of the reaction is independent of nitrogen partial pressure. The dotted curves given in Fig. 27 are calculated from the half-time for a first-order-type reaction. These theoretical rate curves are very close to those drawn through the experimental data. The values of $(\% \text{ N})_{eq}$ given in Figs. 25 and 27 are the equilibrium nitrogen concentrations for the nitrogen partial pressures used.

It is evident from these results that under the experimental conditions employed (1) the diffusional processes do not interfere with the measured rates, (2) the saturation of the surface of iron with chemisorbed oxygen is a relatively fast reaction, and (3) the rate of nitrogen transfer from gas to iron is of a first-order type with respect to nitrogen.

Similar observations were made by Podgurski et al. [29] who investigated the rate of removal of nitrogen from α-iron in vacuo at temperatures from about 400 to 500°C. They demonstrated that when the surface of the iron (containing up to 0.05% N in solution) was maintained relatively clean by helium ion bombardment, the rate of nitrogen desorption from 1/8 inch diameter wire was controlled by diffusion of nitrogen. However, when the surface of the sample was contaminated, presumably by a trace amount of CO in the furnace atmosphere, the rate of nitrogen removal was much retarded, and the results indicated a slow surface reaction. As seen from the data in Fig. 28a, the rate is proportional to the nitrogen content (or activity) of the iron. This is of course analogous to the other examples given above, showing that the rate of nitrogenization is proportional to the square root of the nitrogen pressure. The same dependence of the rate on nitrogen activity was observed in other investigations of the rate of nitrogen solution or removal from liquid iron containing oxygen [24,25]. An example of these measurements is given in Fig. 28b, reproduced from a paper by Pehlke and Elliott [25].

All these observations from vastly different experiments lead to the same conclusion. That is, the rate of transfer of nitrogen through an iron surface, liquid or solid, is much retarded, if the surface of the metal is contaminated by, for example, oxygen, sulphur or any other surface-active species. Furthermore, the rate for a given contaminated surface is proportional to the activity of nitrogen, i.e. the concentration of nitrogen in iron or the square root of the nitrogen pressure.

For quantitative treatment of the Turkdogan and Grieveson [27] data, let us substitute for $(1-\theta)$ in equation (5) from equation (1) taking the i'th species as oxygen--thus $1-\theta = \phi_0\theta_0/a_0 = \phi_0\theta_0\, p_{H_2}/p_{H_2O}$. Without speculating as to the detailed atomistics of the activated complex, we see from Fig. 27 and discussion thereof that it contains only one atom of nitrogen and thus that $\Pi a_i^{\nu_i}$ is simply a_N taken as (%N), the nitrogen content of the specimen, and $\Pi_{eq} a_i^{\nu_i}$ is $(\%\,N)_{eq}$, the nitrogen content of the specimen when equilibrated with the atmosphere. Equation (5) thus becomes:

$$R = k'\,[(\%N)_{eq} - (\%N)]\frac{\phi_0}{\phi_\dagger}\,\theta_0\, p_{H_2}/p_{H_2O} \qquad (6)$$

where the various isothermal constants have been lumped together as k'. Introducing now the assumption that θ_0 is near unity and hence that ϕ_0 and $\phi_\dagger$ are constant, equation (6) reduces to:

$$R = k''\,[(\%N)_{eq} - (\%N)]\,p_{H_2}/p_{H_2O}. \qquad (7)$$

As the rate, R, is proportional to d(%N)/dt (where t is time), the integrated form is

$$\ln \frac{(\%N)_{eq} - (\%N)}{(\%N)_{eq} - (\%N)_o} = -\,\Phi\,\frac{p_{H_2}}{p_{H_2}}\,\frac{t}{\ell} \qquad (8)$$

where $(\%N)_o$ is the initial nitrogen content, ℓ is the characteristic specimen dimension, and Φ is a function only of temperature.

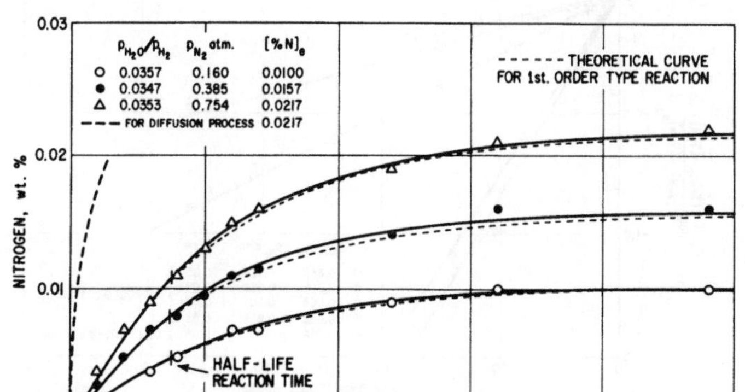

Fig. 27. Nitrogenization of iron strips (0.051 cm thick) at 1000°C in $N_2 + H_2O + H_2$ at total pressure of 0.46 atm. (ref. 27).

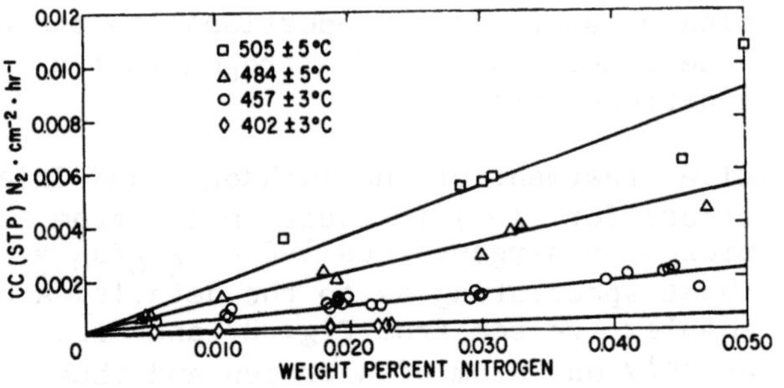

Fig. 28a. Nitrogen desorption rates from iron wire 1/8 in diam. (CO contaminated surfaces). (Podgurski, Davis and Smith [29]).

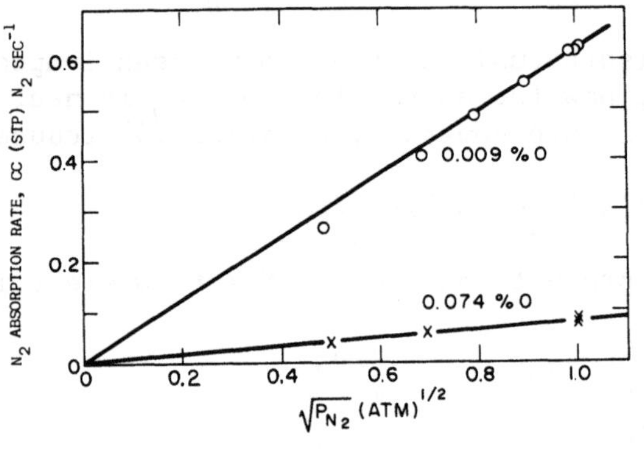

Fig. 28b. Effect of oxygen on the absorption of N_2 by liquid iron at 1600°C. (Pehlke and Elliott [25]).

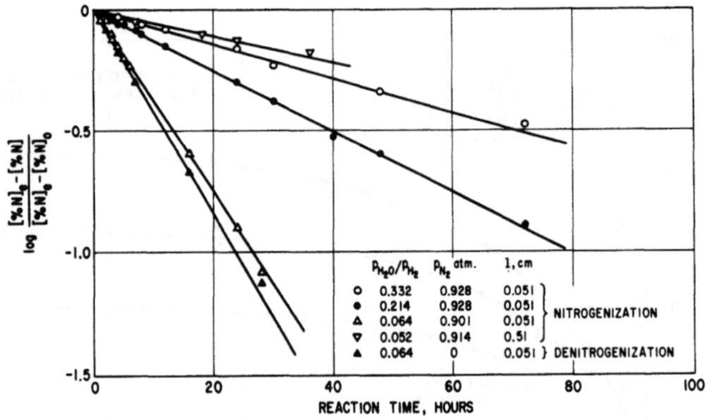

Fig. 29. Experimental results of Fig. 25 plotted in accordance with the rate equation (8) for nitrogenization and denitrogenization of iron at 1000°C. (Turkdogan and Grieveson [27]).

The experimental results given in Figs. 29 and 30 again demonstrate that the reaction is of first order with respect to nitrogen. The slopes of the lines, designated $q = \Phi\ (p_{H_2}/p_{H_2O})$, are shown in Fig. 31 to be directly proportional to the ratio p_{H_2O} as indicated by equation (8), thus supporting the general assumption used in its development and specifically the assumption that θ_0 is near one under the oxygen potentials used.

If the same reaction model applies to the rate of nitrogenization, or denitrogenization, of liquid containing more than 0.02% O, the rate should be inversely proportional to the oxygen content of the metal. The experimental results of Schenck et al [24] and Pehlke and Elliott [25], reproduced in Fig. 32 from a previous publication [30], are in general accord with this interpretation.*

This example of the rate of nitrogen transfer across an iron surface at modest oxygen potentials thus tends to confirm the utility of the rather naive model adopted for a surface at nearly full site-coverage. It should be noted that the model needed to produce the agreement with experiment illustrated in the preceding figures, especially Fig. 31, is far from complete. It leaves us with no

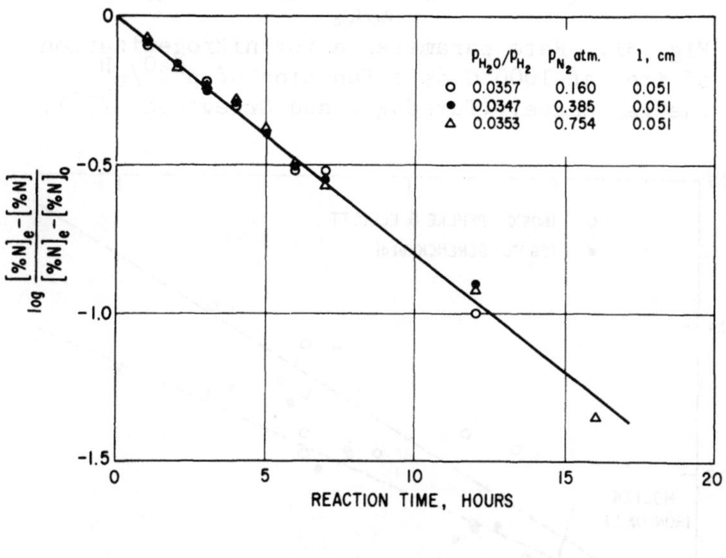

Fig. 30. Experimental results of Fig. 27 plotted in accordance with the rate equation (8) for nitrogenization of iron at 1000°C. (Turkdogan and Grieveson [27]).

* See addendum

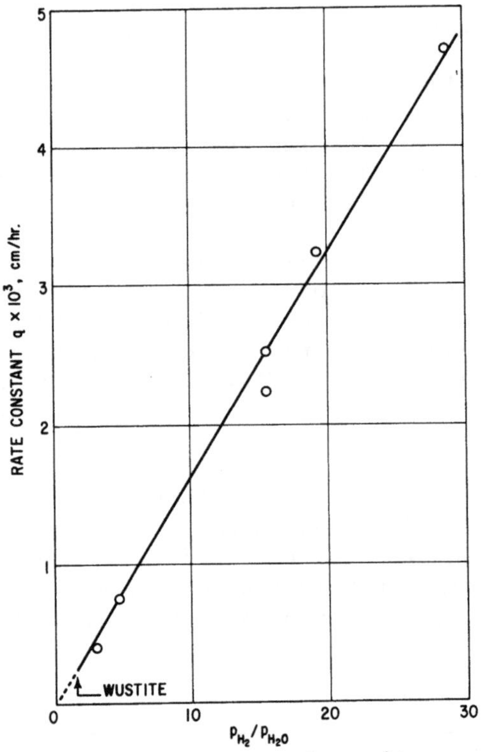

Fig. 31. Rate parameter q for nitrogenization of iron at 1000°C as a function of p^{H_2O}/p^{H_2} the gas phase. (Turkdogan and Grieveson [27]).

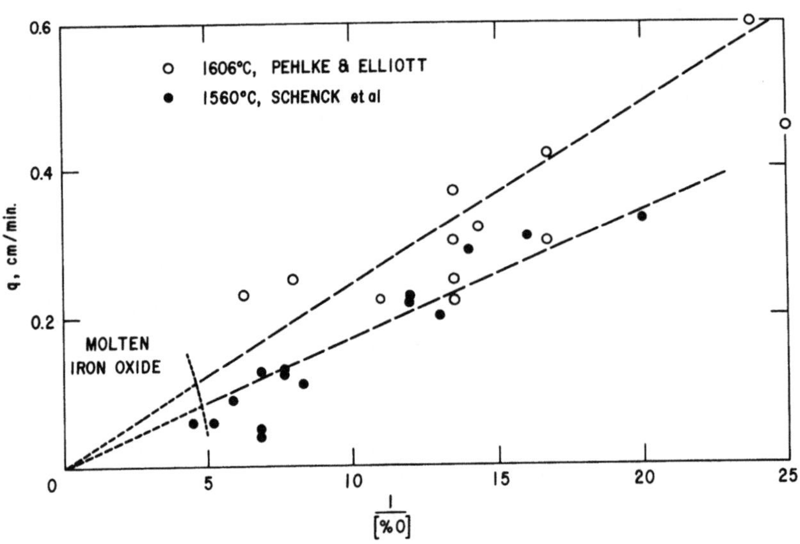

Fig. 32. Rate parameter q for nitrogenization of liquid iron as a function of oxygen content for which $\theta_O \to 1$. (Swisher and Turkdogan [30]).

well-defined or even satisfactory interpretation of the atomistics of nitrogen adsorption or desorption; clearly these processes are not controlled by the breaking (or formation) of the N-N bond of gaseous N_2 or the process would not be of first-order as observed. In fact, it is possible and perhaps probable (as suggested by Turkdogan and Grieveson [27]) that an extremely small number of surface sites not occupied by oxygen provide the only entry sites for nitrogen and that the reaction is thus controlled by diffusion through these "bottlenecks". However, the dominant effect of adsorbed oxygen has been clearly demonstrated and quantitatively explained.

RATE OF DISSOCIATION OF HYDROGEN SULPHIDE ON PYRRHOTITE

Several investigators [31-35] observed that in the oxidation of metals in oxidizing atmospheres, e.g. CO_2-CO, H_2O-H_2 or H_2S-H_2 mixtures, there is slow approach to parabolic growth rate. As an example, reference may be made to a recent investigation [35] of the growth rate of pyrrhotite on iron in hydrogen sulphide-hydrogen gas mixtures. A purified iron block of 5 mm thickness was allowed to react in a gas stream of H_2S and H_2 with $p(H_2S)/p(H_2) = 1.0$, 0.1 or 0.01 at 670, 800 or 900°C for a specified time. The cooled sample was sectioned and the change in the half-thickness of the sample, $\Delta\ell$, was measured. Typical examples of the results are given in Fig. 33, where the square of the half-thickness of iron consumed, $(\Delta\ell)^2$, is plotted against the time of reaction. After the initial reaction time of a few hours or a few days (depending on gas composition and temperature), the reaction proceeds in accordance with the parabolic rate law. If the sulphur potential were essentially the same near the surface of the sulphide layer as that in the gas phase of fixed composition, then the approach to the parabolic rate law would have been much faster.

In another series of experiments [36], the rate of sulphidation was measured by the weight gain method using 0.5 mm thick purified iron strips. The slow approach to gas-sulphide equilibrium is well demonstrated by the results in Fig. 34, where the square of the weight gain per unit area is plotted against time for gas mixtures having $p(H_2S)/p(H_2) = 1.0$ with 0, 25, 50 and 67% Ar. The value of 1.5×10^{-2} $(gs/cm^2)^2$ corresponds to almost complete sulphidation of the iron strip (0.5 mm thick to iron sulphide.

The slopes of the "linear" parts of the S-shaped curves increase with increasing $p(H_2S)$, or $p(H_2)$, for a fixed ratio $p(H_2S)/p(H_2)$. In all cases the slopes are lower than the value when gas-sulphide surface equilibrium is established, as shown by the broken line reproduced from Fig. 33.

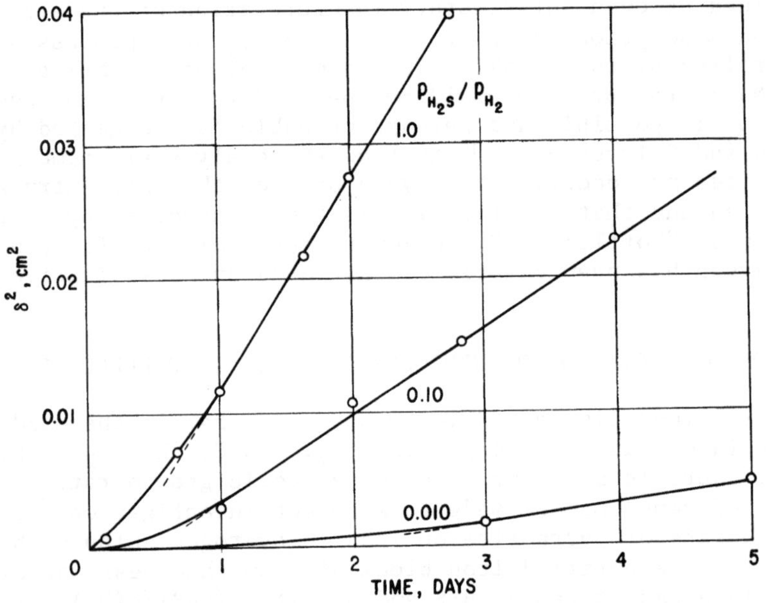

Fig. 33. Parabolic growth rate of pyrrhotite on iron in H_2S/H_2 mixtures at 800°C. (Turkdogan [35]).

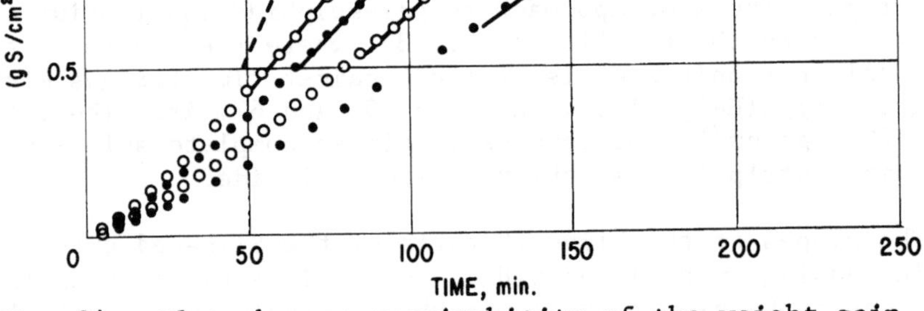

Fig. 34. Plot showing unsuitability of the weight gain method for determination of parabolic growth rate of iron sulphide on iron in H_2S-H_2-Ar Mixtures with P_{H_2S}/P_{H_2} = 1.0 at 800°C. (Worrell and Turkdogan [36]).

This argument may be illustrated in another way as shown schematically in Fig. 35a were the flux of sulphur J_S from the gas to the pyrrhotite is plotted against time. The dotted curve is for the parabolic growth rate of the pyrrhotite layer at fixed sulphur activities on each side of the sulphide layer. When the sulphide layer is thin, the diffusional flux in the pyrrhotite would be very great for such diffusion control and hence the rate would be controlled essentially by a chemical reaction at the gas-sulphide interface. As shown in Fig. 35b, the sulphur activity at the surface of the sulphide layer increases with reaction time. When the sulphur activity at the surface approaches that in equilibrium with the gas phase $(a_S)_e$, the subsequent rate of sulphidation is controlled primarily by the diffusion process.

The kinetics of the surface reaction of hydrogen sulphide with iron sulphide during the early stages of sulphidation of an iron strip has been investigated by Worrell and Turkdogan [36]. Typical examples of the results for 800°C are given in a log-log plot in Fig. 36. Within the limits of uncertainty of the results, the rate data may be represented by straight lines in these log-log plots. These rate data are for the sulphide layer thickness within the range ∿ 60 to 300 microns. The electrical double layer plays no significant role in this range of thickness; therefore, no particular significance is attached to apparently linear log-log plots. However,such a linear plot is useful in evaluating the instantaneous rate of sulphidation as a function of time.

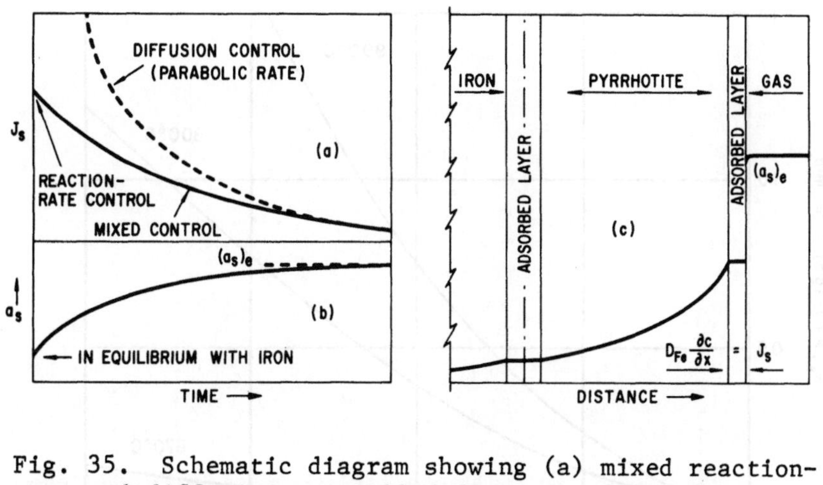

Fig. 35. Schematic diagram showing (a) mixed reaction-rate and diffusion controlled flux of sulphur from gas to pyrrohtite compared to parabolic rate of sulphidation, (b) activity of sulphur on the pyrrhotite surface as a function of time, and (c) sulphur activity profile across pyrrhotite layer at a given reaction time.

In an attempt to interpret the rate data, Worrell andTurkdogan considered a pseudosteady state between the diffusion process in the sulphide layer and the chemical reaction at the surface of the sulphide. A schematic diagram in Fig. 35c is an illustration of the assumed sulphur activity profile across the pyrrhotite layer while growing on iron at a particular reaction time. The assumption of equilibrium at the iron-pyrrhotite interface is well justified by the xray diffraction data taken from the surface of the sulphide which was adjacent to the metallic phase [35]. It is also assumed

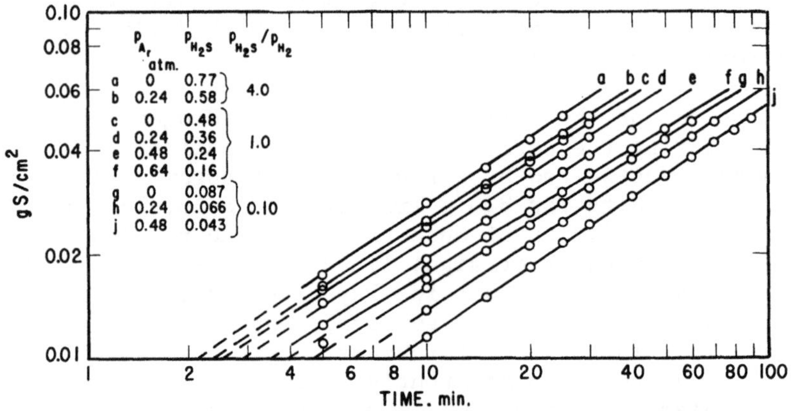

Fig. 36. Logarithmic plot of rate data for sulphidation of iron strips in H_2S-H_2-Ar mixtures at 800°C [36].

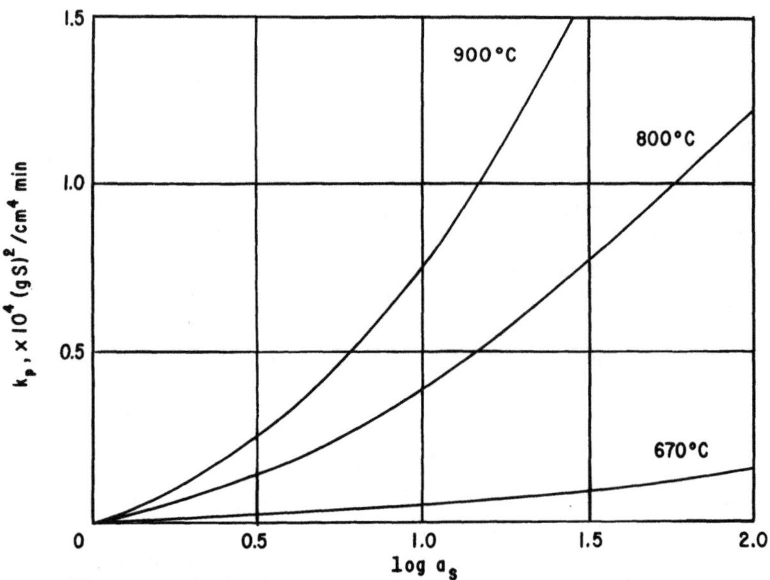

Fig. 37. Parabolic rate constant K_p as a function of sulphur activity at the gas-sulphide interface [36].

that the activity of sulphur in the adsorbed layer is essentially the same as that in the sulphide phase immediately below the adsorbed layer. The activity $(a_S)_e$ in the gas phase is that determined by the ratio $p(H_2S)/p(H_2)$. Assuming a pseudosteady state, the rate of transfer of sulphur from hydrogen sulphide to the surface of pyrrhotite, J_S, at any given instant is accompanied by an equivalent rate of diffusion of iron from the inner to the outer surface, $D_{Fe}\ \partial c/\partial x$. This may be stated in a more convenient form by the expression

$$J_S = \frac{dw}{dt} = \frac{(k_p)_t}{2w} \qquad (9)$$

where dw/dt is the instantaneous rate in units of $gS/cm^2 min$, w is gS/cm^2 gained up to time t and k_p is the parabolic rate constant for the diffusion process. The subscript t indicates that this is the instantaneous value of k_p corresponding to a particular sulphur activity at the outer surface at time t. From the experimental results dw/dt is known as a function of time (Fig. 36), and using equation (9) the corresponding values of k_p are readily computed; since k_p as a function of a_S is known [35], the activity of sulphur on the pyrrhotite surface is estimated.

Fig. 37 shows the variation of k_p with the sulphur activity* on the outer sulphide surface, determined from the parabolic growth rate [35] when there is equilibrium at iron-pyrrhotite and pyrrhotite-gas interfaces. Now, for a particular instantaneous rate dw/dt, using the linear log-log plots in Fig. 36, the value of k_p is computed from equation (9), and then the corresponding sulphur activity a_S at the sulphide surface is estimated from Fig. 37. Typical examples of such calculations are given in Fig. 38 for sulphidation of iron in H_2S-H_2 mixtures at 800°C. The sulphur activity at the pyrrhotite surface thus estimated is used in the interpretation of the reaction kinetics.

Of the several possible reaction mechanisms considered by Worrell and Turkdogan, the following was found to give the best account of the observed rates. The monolayer on the surface of pyrrhotite is assumed to consist primarily of adsorbed sulphur atoms occupying most of the surface sites, i.e. $\theta_S \rightarrow 1$. Furthermore,, as indicated in Fig. 35c, the activity of sulphur in the adsorbed layer is assumed to be the same as that in the bulk sulphide phase immediately below the surface. For this limiting case, the fraction of vacant sites on the surface is given by

$$1 - \theta_S = \frac{\phi_S}{a_S} \qquad (10)$$

*) The sulphur activity a_S is defined by the ratio p_{H_2S}/p_{H_2} in the gas phase to that corresponding to the iron-pyrrhotite equilibrium.

where ϕ_S is assumed to be a function of temperature only, in spite of the compositional changes at the surface during sulphidation, e.g. N_S/N_{Fe} from 1.00 to ~1.05.

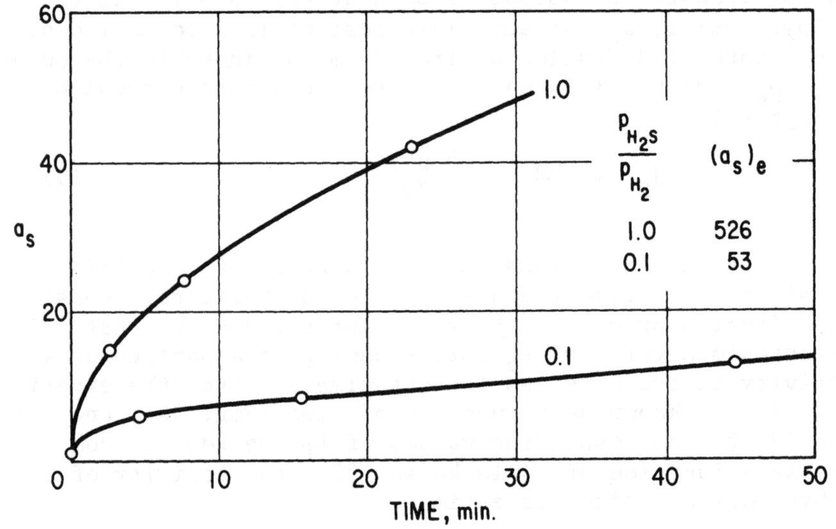

Fig. 38. Computed time dependence of sulphur activity on the surface of pyrrhotite during sulphidation of an iron strip at 800°C in H_2S-H_2 mixtures, using the results of Worrell and Turkdogan [36]; $(a_S)_e$ is the equilibrium sulphur activity.

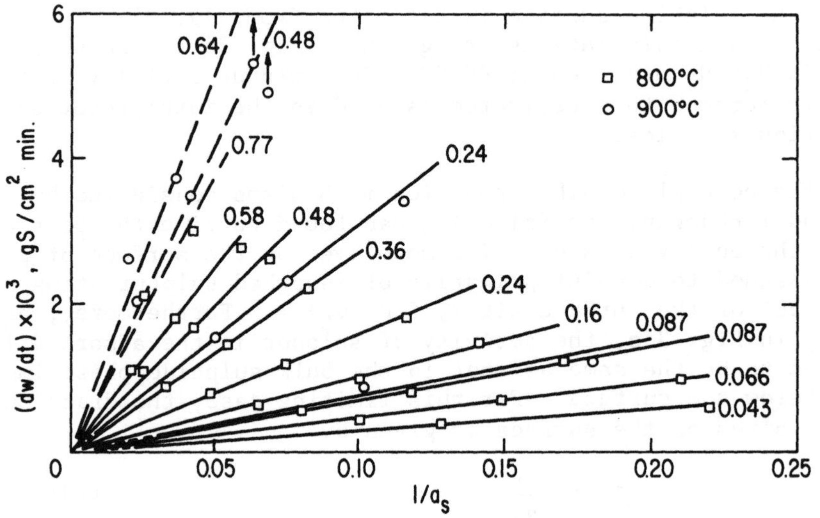

Fig. 39. Rate of sulphur transfer from H_2S-H_2 to Pyrrhotite growing on an iron strip at 800 and 900°C as a function of sulphur activity in the gas - pyrrhotite interface for indicated values of p_{H_2S} (atm).

The dissociation of adsorbed H_2S via an activated state $(H_2S)^{\dagger}$ in the chemisorbed layer is assumed to be the rate-controlling reaction, thus

$$H_2S(g) \;=\; H_2S(ads) \;=\; H_2S)^{\dagger}(ads). \qquad (11)$$

The equilibrium constant $K_{\dagger}$ for this reaction is:

$$K_{\dagger} = \frac{a_{\dagger}}{a_{H_2S}} = \frac{a_{\dagger}}{p_{H_2S}} \qquad (12)$$

where $p(H_2S)$ is the partial pressure of hydrogen sulphide in the flowing gas mixture and $a_{\dagger}$ is the activity of the adsorbed activated complex $(H_2S)^{\dagger}$. Since $\theta_{\dagger}$ (fraction of sites for the activated complex) is very small,

$$a_{\dagger} = \phi_{\dagger} \frac{\theta_{\dagger}}{1-\theta_S} \qquad (13)$$

where $\phi_{\dagger}$ is the activity coefficient of the complex for the limiting case of $\theta_S \rightarrow 1$. From equations (12) and (13), $(\theta_{\dagger})_f$ for the forward reaction is

$$(\theta_{\dagger})_f = \frac{K_{\dagger}}{\phi_{\dagger}} p_{H_2S} (1-\theta_S). \qquad (14)$$

Similarly, for the reverse reaction

$$(\theta_{\dagger})_r = \frac{K_{\dagger}}{\phi_{\dagger}} (p_{H_2S})_e (1-\theta_S) \qquad (15)$$

where the subscript e indicates the partial pressure of hydrogen sulphide in equilibrium with a_S at the surface of pyrrhotite at time t for a particular p_{H_2} in the gas.

Inserting equations (10), (14), and (15) in Eyring's absolute reaction rate, equation (2) gives

$$-\frac{dn}{dt} = \frac{kT}{h} \Gamma_o \frac{K_{\dagger}\phi_S}{\phi_{\dagger}} \frac{(p_{H_2S}) - (p_{H_2S})_e}{a_S} \qquad (16)$$

where the instantaneous rate dn/dt is in units of g-atom S/cm^2-sec. since for the range investigated $a_S \ll (a_S)_e$, the term for the reverse reaction may be omitted, and substituting more practical units of rate dw/dt, equation (16) may be simplified to

$$\frac{dw}{dt} = \Phi_S \frac{p_{H_2S}}{a_S} \qquad (17)$$

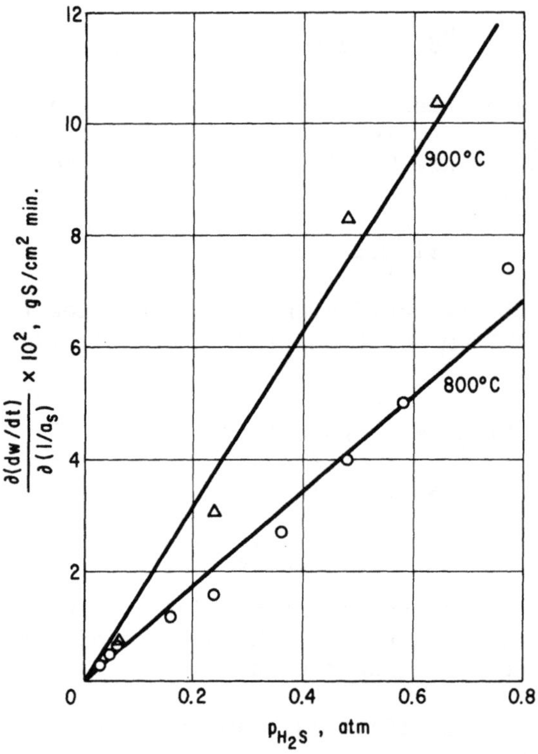

Fig. 40. Rate parameter $\partial(dw/dt)/\partial(1/a_S)$ as a function of partial pressure of hydrogen sulphide.

where for a mean temperature T_m the rate parameter

$$\Phi_S = 1920 \quad \frac{kT_m}{h} \Gamma_o \frac{K_{\dagger}\phi_S}{\phi_{\dagger}} \text{ , in gS/atm cm}^2\text{-min.}$$

Using the data of Worrell and Turkdogan, dw/dt is plotted in Fig. 39 against the reciprocal of the sulphur activity computed for the corresponding times at which w = 0.01, 0.02 and 0.04 gS/cm^2, for various gas compositions at 800 and 900°C. In most cases*, dw/dt for a given gas composition is seen to be inversely proportional to the corresponding a_S. The slopes of the lines, $\partial(dw/dt)/\partial(1/a_S)$, are shown in Fig. 40 to be approximately proportional to $p(H_2S)$. Considering the errors involved in the estimation of time-dependent a_s by an indirect method, the scatter of the data points in Figs. 39 and 40 is not too surprising. These findings are considered to be in reasonable accord with equation (17) for $\theta_S \rightarrow 1$.

The point of view presented here for the kinetics of dissociation of hydrogen sulphide on the surface pyrrhotite may be applied also to the transfer of oxygen from water vapor or carbon dioxide to the surface of the wustite phase while growing on iron. That is, the chemisorbed layer on wustite is considered to be almost saturated with adsorbed oxygen atoms and the rate of dissociation of water vapor or carbon dioxide is inversely proportional to the oxygen activity at the surface of wustite. This is approximately supported by the experimental results of Grabke [37] (electrical conductivity method) who found that the rate of oxidation of wustite in carbon dioxide-carbon monoxide mixtures is inversely proportional to the surface oxygen activity raised to a power of about 2/3.

RATE OF DECARBURIZATION OF AUSTENITE IN DRY HYDROGEN

Recently, we have investigated [21] the kinetics of decarburization of austenite in dry hydrogen over the temperature range 800 to 1140°C. Particular care was taken to ensure that the hydrogen-helium mixtures used in the experiments were adequately dried. The partial pressure of water vapor in the gas mixtures leaving the reaction tube was about 0.03 torr. A series of experiments carried out in wet hydrogen at atmospheric pressure with 1 to 4 torr water vapor indicated that in "dry" hydrogen containing 0.1 torr water vapor, the amount of carbon monoxide formed was insignificant relative to that of methane.

---

*) The results of Worrell and Turkdogan [36] for 670°C are not consistent with those for 800 and 900°C in Fig. 39. That is, at 670°C dw/dt is not proportional to 1/aS and the lines drawn through the data points do not pass through the origin of the coordinates. This apparent discrepancy is attributed to the errors involved in the estimation of the time-dependent a_S. It is in fact evident from the relationship in Fig. 37 that the evaluation of a_S from the estimate of k_p (equation 10) is subject to greater error at lower temperatures.

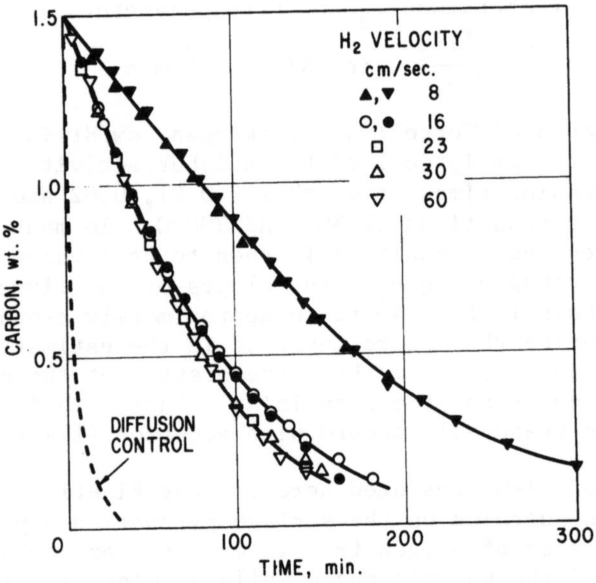

Fig. 41. Effect of gas velocity on the rate of decarburization of Austenite (0.56 mm thick strips) in dry hydrogen at 1140°C and 0.96 atm. Turkdogan and Martonic [21].

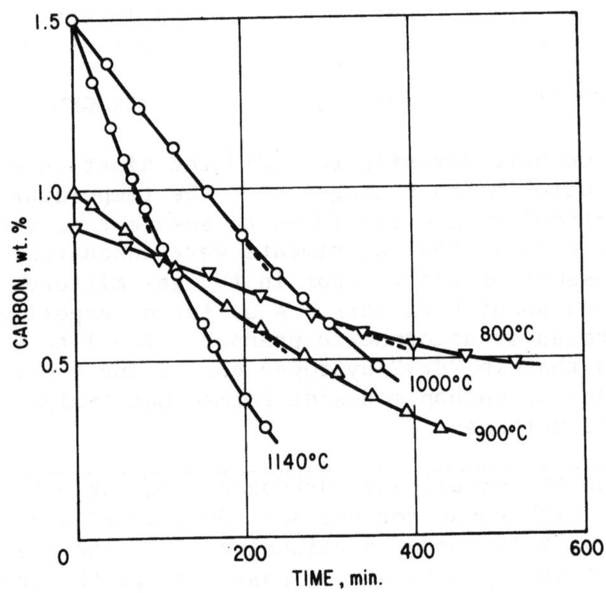

Fig. 42. Decarburization of Austenite (0.56 mm thick strips) in dry hydrogen at 0.96 atm using 50% H_2-50% He mixture flowing at 4800 ml/min. Turkdogan and Martonic [21].

A sample of an iron-carbon alloy (0.56 mm thick strip) was suspended from a silica spring in the uniform hot zone of a recrystallized alumina reaction tube flushed with a fast stream of a hydrogen-helium mixture. A typical example of the results is shown in Fig. 41 for 1140°C. At flow rates from 1800 to 4800 ml/min, for which the gas velocity in the reaction tube (∿5/4 in diam) was within 23 to 60 cm/sec at 1140°C, the rate of decarburization was independent of gas flow rate. At gas flows below 1200 ml/min, a decrease in the rate of decarburization was observed with decreasing gas flow rate. In the subsequent measurements fast gas flows were maintained to suppress the gas film resistance to transport of the reaction product from the surface of the sample.

Dotted curve in Fig. 41 is computed for diffusion-controlled decarburization using the available data [38] for diffusivity of carbon in austenite. It is evident that under the experimental conditions employed, the gas film and solid-state diffusion processes do not interfere with the measured rates of decarburization. Therefore, the rates may be interpreted in terms of a surface reaction between hydrogen and carbon giving methane as the reaction product.

Another set of experimental results is given in Fig. 42 for 800,900 and 1000°C for 1 : 1 hydrogen-helium mixtures. In all cases the rate measurements were carried out over the range of carbon concentration within the austenite field. One significant feature of the results is that for temperatures and gas compositions investigated, the rate curves approach linearity at high carbon concentrations, and departure from linearity occurs at lower carbon contents with decreasing temperature, e.g. ∿ 1.2%C at 1140°C, ∿1.0%C at 1000°C, ∿ 0.7%C at 900°C and ∿0.5%C at 800°C. This is indicative of a relatively strong chemisorption of carbon on the surface of iron.

The reaction model proposed involves the assumption that hydrogen in the gas and carbon in solution in iron are in equilibrium with adsorbed species on the surface. The activated state $(CH_x)^{\dagger}$ involved in the rate-controlling reaction may be expressed in a general form (without specifying the formation of intermediate hydrocarbons), thus

$$C\ (\text{in soln}) + \frac{x}{2}\ H_2\ (g) = (CH_x)^{\dagger} \rightarrow \text{product } CH_4 \tag{18}$$

The rate equation is derived in the usual manner, giving

$$\frac{dn}{dt} = -\frac{kT_m}{h}\ \Gamma_o\ \frac{K_{\pm}}{\phi_{\pm}}\ a_C\ \left[p(H_2)\right]^{x/2}\ (1-\theta) \tag{19}$$

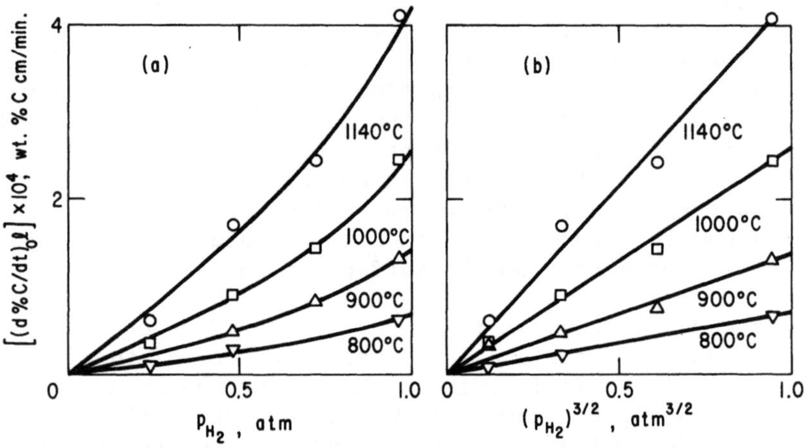

Fig. 43. Pressure dependence of rate, $(d\%C/dt)\ell = \Phi_c\,(pH_2)^{x/2}$, of decarburization of Austenite in hydrogen for the limiting case of $\theta_c \rightarrow 1$. Turkdogan and Mastonik [21].

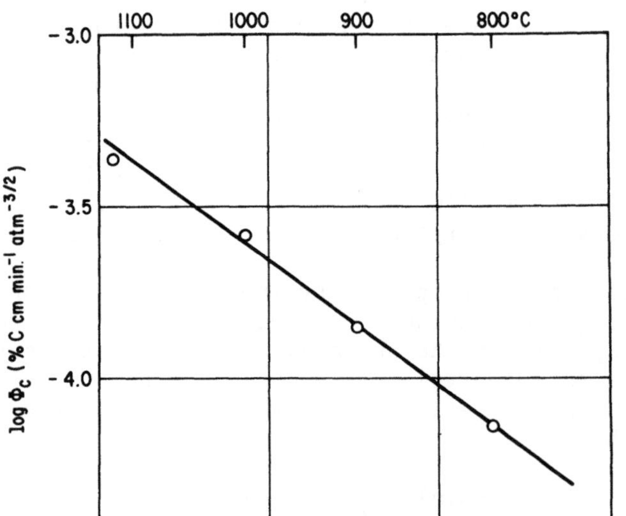

Fig. 44. Temperature dependence of the rate parameter for decarburization of Austenite for the limiting case of $\theta_c \rightarrow 1$.

where dn/dt is the instantaneous rate in g-atom C/cm^2-sec, $K_\dagger$ is the equilibrium constant for reaction (18) and $\phi_\dagger$ is the activity coefficient of the complex $(CH_x)^\dagger$.

Nothing is known about the extent of chemisorption of hydrogen and carbon on the surface of iron at elevated temperatures within the austenite range. Another assumption involved in the reaction model considered is that the fraction of sites occupied by carbon atoms is much greater than those by hydrogen atoms and hydrocarbons within the range investigated, i.e. $\theta_C >> \theta_H H + \theta_{CH} + \ldots$, and therefore $1-\theta = 1-\theta_C$. As demonstrated earlier, there appears to be a marked deviation from ideality in the adsorbed carbon monolayer on austenite; therefore we shall consider only the limiting case of almost complete coverage of the surface by carbon atoms, $\theta_C \to 1$, for which

$$1 - \theta_C = \frac{\phi_C}{a_C} \quad (20)$$

where ϕ_C is the activity coefficient of adsorbed carbon, and for this limiting case is a function of temperature only. Inserting this in equation (19) and substituting more practical units $n = (\ell\rho/60 \times 1200)$ (%C),

$$\frac{d(\%C)}{dt} = \frac{\Phi_C}{\ell}\left[p(H_2)\right]^{x/2} \quad (21)$$

where $\Phi_C = 72\text{x}10^3 \dfrac{kT_m}{h} \Gamma_o \dfrac{k_\dagger \phi_C}{\phi\dagger}$ in (%C) cm per min per atm$^{x/2}$ for a mean temperature T_m.

Equation (21) states that for the limiting case of $\theta_C \to 1$, the rate of decarburization is independent of the carbon content of iron and is directly proportional to the partial pressure of hydrogen raised to the power of x/2. The linearity of the plots in Figs. 41 and 42 at high carbon concentrations, in support of equation (21) implies that over the composition range indicated, i.e. for a_C 0.5 relative to graphite [39], the surface of iron is almost completely covered with carbon atoms.

Next to be evaluated is the exponent x/2 for the pressure dependence of the rate. As seen from the plots in Fig. 43a, the initial rate $(d\% C/dt)_o \ell$ for $\theta_C \to 1$ is not a linear function of p_{H_2}. The shape of the curves suggests that x/2 1. As shown in Fig. 43b, it appears reasonable to assume that x/2 = 3/2. That is, according to these observations and the simple reaction model assumed, the activated complex is $(CH_3)^\dagger$. If the activation energies of formation of complexes $(CH_2)^\dagger$ and $(CH_3)^\dagger$ were similar, both species would be involved in the rate-controlling reaction with $1 < x/2 < 3/2$.

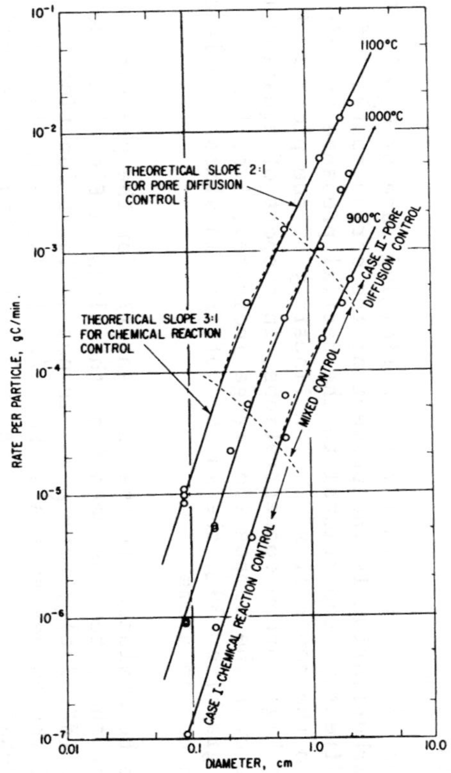

Fig. 45. Rate of oxidation of electrode graphite as a function of particle diameter for 0.96 atm CO_2. Turkdogan *et al* [40].

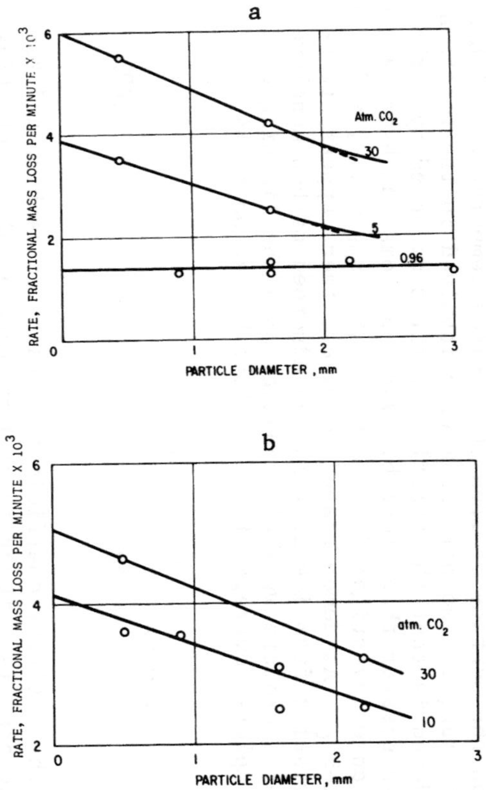

Fig. 46. Effects of particle size and CO_2 pressure in the rate of oxidation of (a) electrode graphite at 1000°C and (b) coconut charcoal at 800°C. Turkdogan and Vinters [41].

Within the limits of accuracy of the measurements, it would appear reasonable to approximate x = 3.

The slopes of the lines in Fig. 43b give the rate parameter From the temperature dependence of Φ_C in Fig. 44, an apparent heat activation for the reaction is about 17 kcal/g-atom C.

EFFECT OF CARBON MONOXIDE ON THE RATE OF OXIDATION OF CARBON BY CARBON DIOXIDE

The unusual "internal burning" phenomenon in the oxidation of graphite or other carbons (all of relatively high porosity) has stimulated much interest in this research for several decades. Although much research has been done and many papers have been published on this subject, there are many variations and inconsistencies in the interpretation of the experimental rate data. Many of the inconsistencies seem to arise from inadequate consideration of all the pertinent variables, many of which may have decisive influence on the rate-controlling processes and the ranges of their applicability.

The extent of internal burning depends on the size and type of carbon, gas pressure and temperature. With increasing temperature, gas pressure and/or size of carbon, the diffusion of gaseous reactants and products in and out of pores may become sufficiently slow, relative to gas-pore surface reaction, that a gas composition gradient may exist across the porous carbon, resulting in retardation of the gasification rate.

The effect of particle size on the rate for 0.96 atm CO_2 is shown in Fig. 45 where logarithm of rate of oxidation per particle of graphite, gC/min, is plotted against the logarithm of sphere diameter (or average size for granules). The curves drawn through the points approach linearity at small and large diameters for which the theoretical slopes are 3:1 and 2:1 respectively. For small particles there is almost complete gas diffusion within relatively large pores; therefore the rate $\propto$ square of particle diameter.

The effects of pressure and particle size on the rate of oxidation of graphite and charcoal are well demonstrated by the data in Figs. 46a and 46b.

Detailed work has been done in this laboratory on the rate of oxidation of carbons [40-42] and pore characteristics of carbons [43]. In this paper, the discussion of findings will be limited to the unusual effect of carbon monoxide on the kinetics of oxidation in carbon dioxide.

The rate of oxidation of coconut charcoal, electrode graphite and metallurgical coke granules in CO_2-CO gas mixtures at various pressures was measured within the range of 700 to 1400°C. The rate measurements were made under conditions such that almost complete internal burning prevailed, at temperatures below 1000 and 1200°C depending on the type of carbon. Conventional weight loss measurements were made in CO_2-CO mixtures flowing at 1500 cc/min STP. That is, one or two layers of sized carbon (70 to 200 mg granules ∿ 0.5mm diam) were placed on a platinum gauze suspended, within the uniform temperature zone of the reaction tube 3.7 cm diam, by a platinum wire from an automatic recording semimicrobalance. The rate measurements were continued, at a constant temperature (± 1°C), up to 10 to 20% weight loss for periods of time ranging from a few minutes to several days, depending on temperature and gas composition.

Typical examples of results are given in Fig. 47 for oxidation of electrode graphite granules in CO_2-CO mixtures at 1100°C. During the early stages of oxidation, e.g. less than 20% weight loss, departure from linearity is not marked, and sometimes not perceptible. Within the limits of uncertainty of the data, straight lines may be

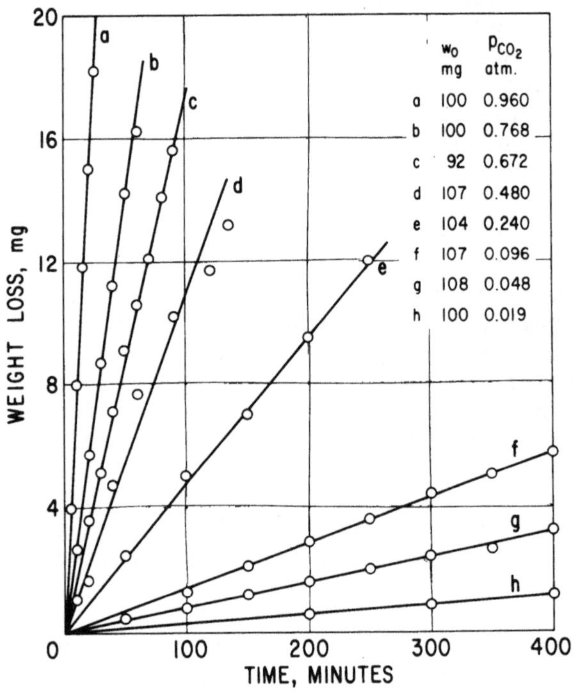

Fig. 47. Oxidation of electrode graphite granules (∿ 0.5 mm dia) at 1100°C in mixtures of CO_2-CO at 0.96 atm total pressure. Turkdogan and Vinters [41].

drawn through the points in Fig. 47, which are typical examples of all the rate measurements. The slopes of the lines, $(dw/dt)_o$, divided by the initial weight of the sample, w_o, are taken as the rate in terms of fractional mass loss per unit time, i.e. $R = 1(dw/dt)_o/w_o$ in min^{-1}.

The effect of CO_2 pressure (in the absence of CO) on the rate of oxidation of charcoal and graphite for various temperatures is shown in log-log plots in Fig. 48. An important feature common to all the results is that the data are well represented by straight lines on a log-log plot with a slope of 1/2 for pressures up to about 10 atm CO_2. That is, at pressures $p_{CO_2} < 10$ atm, the rate of oxidation is directly proportional to the square root of pressure of carbon dioxide, when little or no carbon monoxide is present in the system.

The effects of temperature and gas composition, using CO_2-CO mixtures at 1 atm, on the rate of oxidation of graphite are shown in Fig. 49 as a semilog plot; a similar behavior was observed [42] for coconut charcoal and metallurgical coke.

In many of the previous investigations carbon monoxide was noted to have a marked influence on the rate of oxidation of carbons.

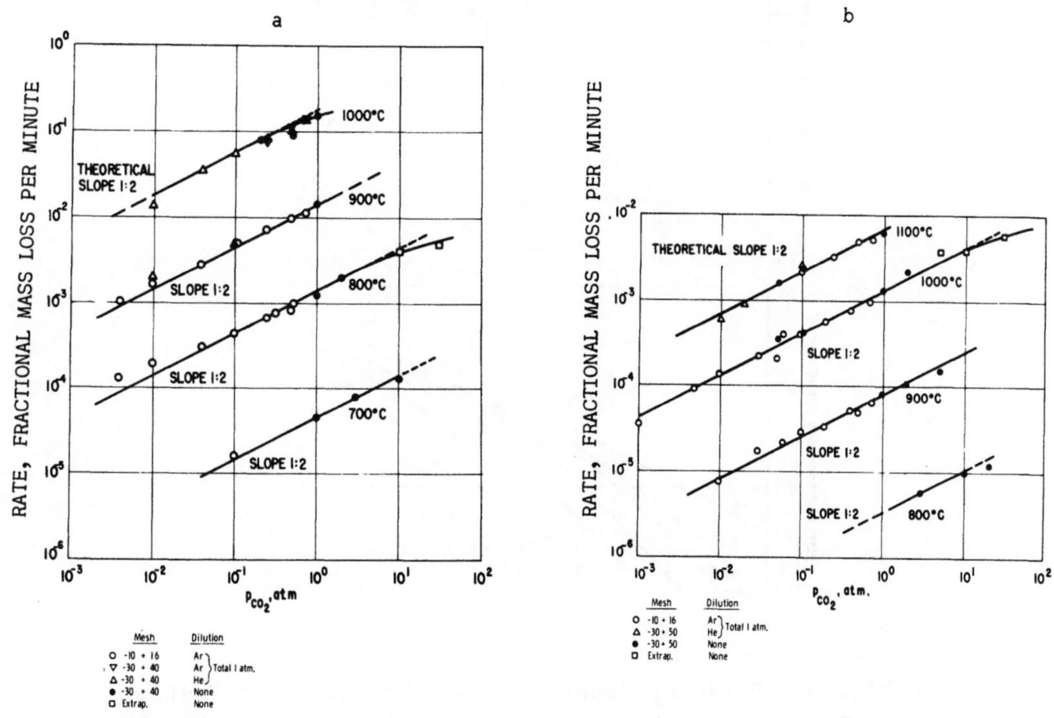

Fig. 48. Pressure and temperature dependence of rate of oxidation of (a) coconut charcoal and (b) electrode graphite [41].

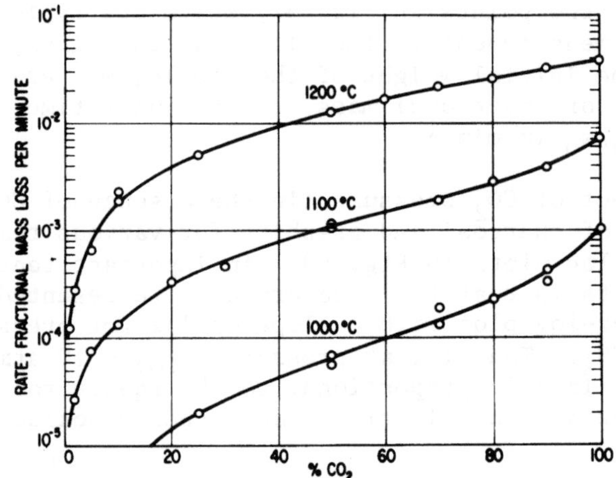

Fig.49. Effects of temperature and gas composition on the rate of oxidation of electrode graphite granules (∿0.5 mm dia) in CO_2-CO mixtures at 0.96 atm total pressure. Turkdogan and Vinters [42].

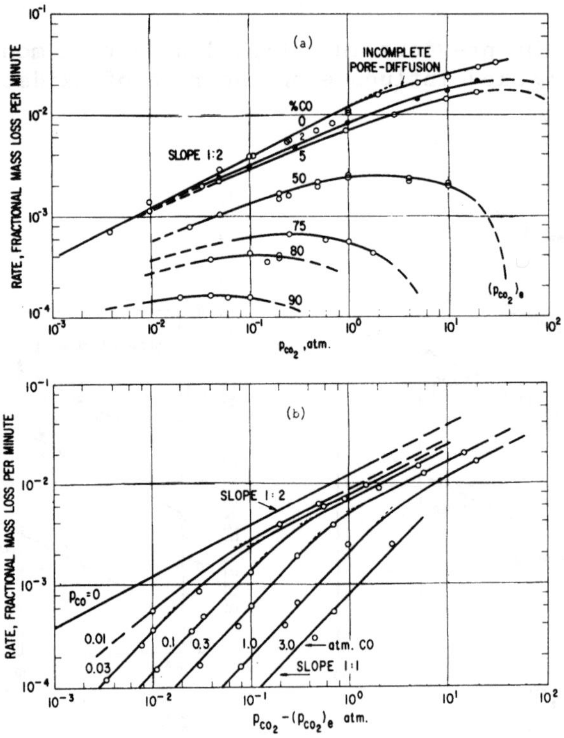

Fig. 50. Pressure dependence of the rate of oxidation of coconut charcoal granules (∿ 0.5 mm dia) in CO_2-CO mixtures at 900°C. (a) experimental results and (b) replot of (a) for various p_{CO} [42].

References to previous work on this subject are given in two recent publications [44,45]; reference should also be made to a comprehensive review paper by Walker et al [46]. None of the previous interpretations seem to account satisfactorily for the present observed effects of pressure and gas composition on the rates.

Retardation of the rate of oxidation by carbon monoxide, or by pressure at fixed p_{CO}/p_{CO_2} ratios, is brought about by: (i) dilution of carbon dioxide, (ii) the effect of the reverse reaction, $2CO = CO_2 + C$, and (iii) poisoning of the pore surface of carbon. The important effect of CO on the rate is well demonstrated by the results for coconut charcoal granules in Fig. 50a for $900^{\circ}C$ and gas mixtures containing 0, 2, 5, 50, 75, 80 and 90% CO. At higher p_{CO}/p_{CO_2} ratios, the rate first increases with increasing pressure, then decreases; the maximum rate is obtained at lower CO_2-pressures with increasing CO content of the gas mixture. Using these curves, the rates are obtained for various partial pressures of carbon monoxide, and these are plotted in Fig. 50b against $[p_{CO_2} - (p_{CO_2})_e]$ where $(p_{CO_2})_e$ is the partial pressure of carbon dioxide in equilibrium with carbon for a particular value of p_{CO} in the gas phase; this is computed using the available data [47] represented by

$$\log \frac{p_{CO_2}}{(p_{CO})^2} = \frac{8677}{T} - 8.874. \tag{22}$$

The use of (p_{CO_2}) in these plots becomes necessary only at high pressures and low temperatures ($1000^{\circ}C$).

It is to be noted that for a given CO-pressure the curves in the log-log plot in Fig. 50b approach linearity with a slope of 1:1 with decreasing $[p_{CO_2} - (p_{CO_2})_e]$. This occurs at lower CO_2 pressures with decreasing p_{CO}; for the case of $p_{CO} \simeq 0$, the square root relationship persists even at 10^{-3} atm CO_2. Similar observations were made for graphite and coke at various temperatures [42].

These experimental results indicate conclusively that the rate of oxidation of carbon in CO_2-CO mixtures is controlled (for the case of almost complete pore diffusion) by two consecutive reactions in series. On the basis of the observed CO_2-pressure dependence of the rate, it ispostulated that the activated complex for the first reaction involves two atoms of oxygen, i.e. a molecule of CO_2, and the second reaction one atom of oxygen, corresponding to 1/2 molecule of CO_2. Following are the postulated rate-controlling reactions in series.

(1) Dissociation of CO_2

$$CO_2\ (g) = CO_2(ads) = (CO_2)^{\dagger} \rightarrow C\ (gr) + 2O\ (ads) \tag{23}$$

(2) Formation of CO

$$C(gr) + O\ (ads) = (CO)^{\dagger} \rightarrow CO\ (g) \qquad (24)$$

The rates for these reactions are given by

$$R_1 = \frac{kT}{h} S\Gamma_o \frac{K_{\dagger 1}}{\phi_{\dagger 1}} (1-\theta)^x (p_{CO_2} - K_e a_0^2) \qquad (25)$$

$$R_2 = \frac{kT}{h} S\Gamma_o \frac{K_{\dagger 2}}{\phi_{\dagger 2}} (1-\theta)\ a_0 \qquad (26)$$

where S is the usable internal area per unit mass of carbon; $K_{\dagger 1}$ and $K_{\dagger 2}$ are the equilibrium constants involving activated complexes $(CO_2)^{\dagger}$ and $(CO)^{\dagger}$; $\phi_{\dagger 1}$ and $\phi_{\dagger 2}$ are the activity coefficients of the activated complexes; $(1-\theta)$ the fraction of vacant sites; a_0 is the activity of oxygen in the adsorbed layer; K_e is the equilibrium constant for the reaction 2O (ads) + C(gr) = CO_2(g). The exponent x for $(1-\theta)$ signifies that the activated complex $(CO_2)^{\dagger}$ may occupy x number of sites.

Since these two reactions are in series, $R_2 = 2R_1$, and the ratio of these rates from equations (25 and (26) after rearrangement gives

$$a_0^2 + \frac{K_{\dagger 2}\phi_{\dagger 1}}{2K_{\dagger 1}\phi_{\dagger 2} K_e} \frac{a_0}{(1-\theta)^{x-1}} - \frac{p_{CO_2}}{K_e} = 0 \qquad (27)$$

This quadratic equation is solved for a_0, thus

$$a_0 = -\frac{K_{\dagger 2}\phi_{\dagger 1}}{4K_{\dagger 1}\phi_{\dagger 2}K_e(1-\theta)^{x-1}} + \frac{1}{2}\sqrt{\left[\frac{K_{\dagger 2}\phi_{\dagger 1}}{2K_{\dagger 1}\phi_{\dagger 2}K_e(1-\theta)^{x-1}}\right]^2 + \frac{4p(CO_2)}{K_e}}. \qquad (28)$$

This equation can be simplified for two limiting cases

Case (1), $4p_{CO_2}/K_e \ll \dfrac{K_{\dagger 2}\phi_{\dagger 1}}{2K_{\dagger 1}\phi_{\dagger 2}K_e\ (1-\theta)^{x-1}}$:

For this case the term within the square root can be simplified*, giving

$$a_0 = 2\frac{K_{\dagger 1}\phi_{\dagger 2}}{K_{\dagger 2}\phi_{\dagger 1}} (1-\theta)^{x-1}\ p_{CO_2}\ . \qquad (29)$$

Now, inserting this in equation (26), and remembering that $R_2 = 2R_1$, gives for the dissociation of CO_2

$$R_1 \quad \frac{kT}{h} S\Gamma_o \frac{K_{\dagger 1}}{\phi_{\dagger 1}} (1-\theta)^x\ p_{CO_2}. \qquad (30)$$

Assuming that CO is the main adsorbed species, $1-\theta = 1-\theta_{CO}$ can be represented in terms of the partial pressure of carbon monoxide, thus

$$1-\theta_{CO} = \frac{1}{1 + (p_{CO}/\theta_{CO})}. \quad (31)$$

Inserting this in equation (30) and including the term for the reverse reaction, the net isothermal rate of oxidation of carbon controlled primarily by mechanism (1) is given by

$$R_1 = \frac{\Phi_1 \,[(p_{CO_2}) - (p_{CO_2})_e]}{(1 + p_{CO}/\phi_{CO})^x} \quad (32)$$

where $(p_{CO_2})_e$ is the partial pressure of carbon dioxide in equilibrium with graphite for a particular value of p_{CO} in the gas phase, and Φ_1 is the product of all the constants for a given mean temperature T_m:

$$\Phi_1 = \frac{kT_m}{h} \, s\Gamma_o \, \frac{K_{+1}}{\phi_{+1}} . \quad (33)$$

Values of $\Phi_1(1-\theta_{CO}) = \Phi_1/\ 1+p_{CO}/\phi_{CO}$ for each p_{CO} may be derived from this expression and the lines with a slope 1:1 in Fig. 50b. The reciprocal of the values of $\Phi_1(1-\theta_{CO})$ thus derived is plotted against p_{CO} in Fig. 51. The relationship may be assumed to be linear at pressures $p_{CO} > \sim 0.1$ atm, suggesting that the exponent x for $(1-\theta_{CO})$ is unity. From the slope and intercept of the line the values of Φ_1 and ϕ_{CO} may be obtained. It should be pointed out that equation (32) does not apply for the limiting case of $p_{CO} \rightarrow 0$; under these conditions mechanism (2) in the reaction series becomes rate controlling as discussed below.

Case (2), $4p_{CO_2}/K_e \gg \dfrac{K_{+2}\phi_{+1}}{2K_{+1}\phi_{+2}K_e(1-\theta)^{x-1}}$:

For this case equation (28) simplifies to

$$a_O = \left[\frac{p_{CO_2}}{K_e}\right]^{1/2}$$

and the rate from equation (26) is

$$R_2 = \frac{kT}{h} S\Gamma_o \frac{K^{\dagger 2}}{\phi^{\dagger 2}\sqrt{K_e}} (1-\theta)\, (p_{CO_2})^{1/2} \qquad (34)$$

Again, substituting $(1-\theta_{CO})$ from equation (31) and including the reverse reaction, the net isothermal rate of oxidation controlled by the formation of CO in this limiting case is given by

$$R_2 = \frac{\phi_2 \left[(p_{CO_2})^{1/2} - (p_{CO_2})_e^{1/2}\right]}{1 + (p_{CO}/\phi_{CO})} \qquad (35)$$

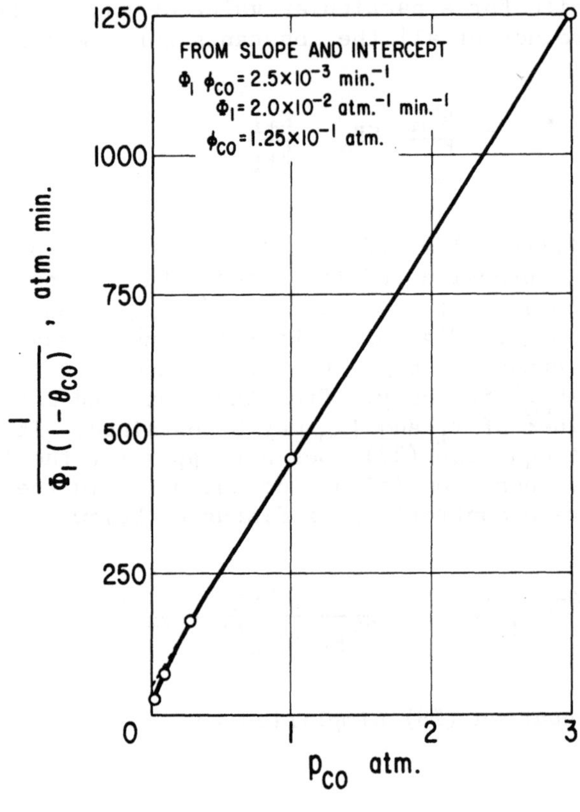

Fig. 51. Effect of p_{CO} on the parameter $\phi_1 (1 - \theta_{CO})$ derived from the lines in Fig. 50b for charcoal at 900°C. Turkdogan and Vinters [42].

and for $p_{CO} \to 0$,

$$R_2 = \Phi_2 \; (p_{CO_2})^{\frac{1}{2}} \tag{36}$$

where the rate parameter Φ_2 is the product of the constants for a mean temperature T_m, i.e.

$$\Phi_2 = \frac{kT_m}{h} S\Gamma_o \frac{K_{\dagger 2}}{\phi_{\dagger 2} \quad Ke} . \tag{37}$$

It is seen from the typical examples of the experimental results in Fig. 50a that when CO-pressure is low, the limiting case (1) is not realized even at $p_{CO_2} \simeq 10^{-3}$ atm. However, with about 10% CO there is a change in the rate-controlling reaction from mechanism (2) to (1). This would imply that the activity coefficient $\phi_{\dagger 1}$ for $(CO_2)^{\dagger}$ increases markedly with increasing coverage of the surface with adsorbed carbon monoxide. In fact, departure from linearity in Fig. 51 for $p_{CO} < \sim 0.1$ atm supports this reasoning. This effect of CO-adsorption is of course reflected on the rate parameter Φ_1 (for $p_{CO} \to 0$)> Φ_2 > Φ_1 (for ~ 0.1 atm or higher p_{CO}). This is another example of the nonideal behavior in adsorbed layers and consequences thereof.

On the basis of these reaction mechanisms, Turkdogan and Vinters [42] evaluated the thermodynamic parameter ϕ_{CO} and the rate parameters Φ_1 and Φ_2. The temperature dependence of the rate parameters Φ_1 and Φ_2 is shown in Figs. 52 and 53. Variation in the value of Φ_1, or Φ_2, with the type of carbon is attributed almost entirely to the difference in the internal surface area. In fact, the total surface area of charcoal is much greater than that of graphite and the area of coke falls in an intermediate range. Assuming that the usable internal surface areas of these carbons do not change much with the temperature of oxidation, parallel lines may be drawn through the points in Figs. 52 and 53, and represented by the following equations.

For mechanism (1)

For coconut charcoal: $$\log \Phi_1 = -\frac{13,200}{T} + 9.68 \tag{38a}$$

For metallurgical coke: $$\log \Phi_1 = -\frac{13,200}{T} + 8.60 \tag{38b}$$

For electrode graphite: $$\log \Phi_1 = -\frac{13,200}{T} + 7.68 \tag{38c}$$

The apparent heat of activation for this reaction is about 60 kcal/g-atom C. It should be emphasized once again that these values of Φ_1 do not hold for low CO contents because of the anticipated decrease in the value of the coefficient $\phi_{\dagger 1}$ as $p_{CO} \rightarrow 0$.

For mechanism (2)

For coconut charcoal: $$\log \Phi_2 = -\frac{15{,}000}{T} + 11.00 \qquad (39a)$$

For metallurgical coke: $$\log \Phi_2 = -\frac{15{,}000}{T} + 9.98 \qquad (39b)$$

For electrode graphite: $$\log \Phi_2 = -\frac{15{,}000}{T} + 8.75 \qquad (39c)$$

The apparent heat of activation for this reaction is about 69 kcal g-atom C.

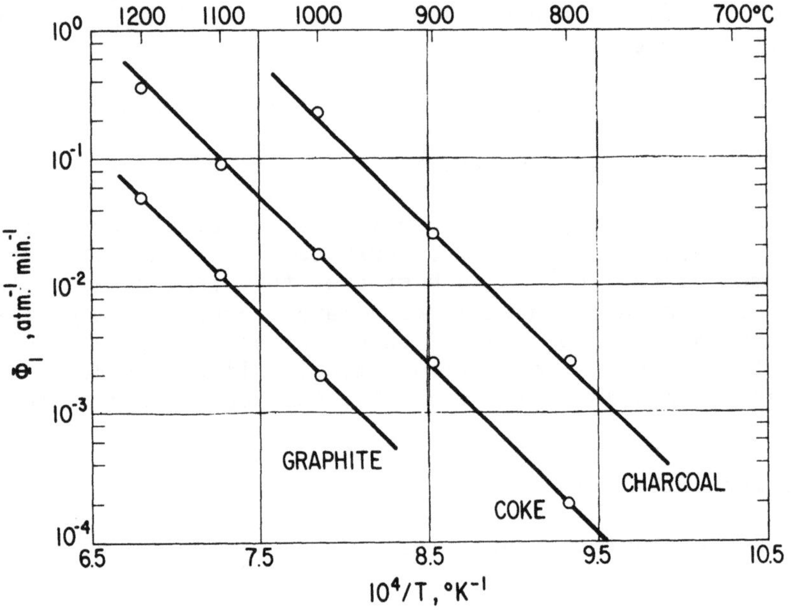

Fig. 52. Temperature dependence of the rate parameter ϕ_1 for oxidation in CO-CO_2 mixtures. Turkdogan and Vinters [42].

The apparent heat of activation for this reaction is about 69 kcal/g-atom C.

From a detailed study [43] of the pore characteristics of carbons it becomes evident that only a small fraction of the total pore surface area was accessible for the oxidation reaction. Most of the internal surface area in carbons is due primarily to the presence of micropores within the range 300 to ∿ 10 Å diam. Most of the diffusive flux of reacting gases in carbon particle takes place via large pores of diameters in the micron range. The area due to these large pores is the usable, or effective, area for reaction. When speaking of almost complete internal burning, one refers to oxidation on the surface of large interconnected pores where most of the diffusive gaseous flux can occur.

The temperature dependence of the thermodynamic parameter ϕ_{CO} for adsorption of CO estimated from the rates of oxidation [42] is shown in Fig. 54. The values of ϕ_{CO} for graphite and coke are similar but are lower than those for charcoal by a factor of about four. The x-ray diffraction patterns of these three forms of carbon indicate almost complete absence of crystallinity, i.e. graphitization, in charcoal, while coke is partially graphitized and the electrode graphite is well crystallized. Apparently stronger CO-adsorption

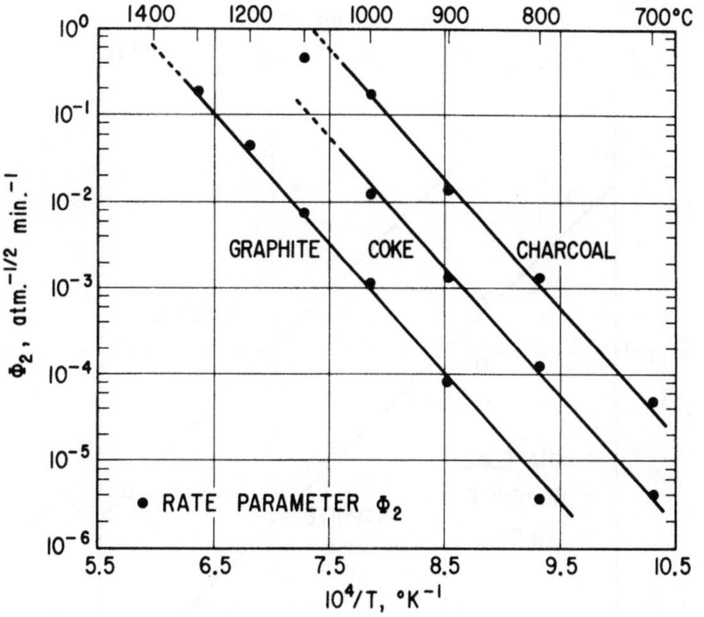

Fig. 53. Temperature dependence of the rate parameter ϕ_2 for graphite, coke, and charcoal.
Turkdogan and Vinters [42].

on graphite and coke in comparison to charcoal may be associated with the vast difference in the state of crystallinity in these carbons. The temperature dependence of ϕ_{CO} for graphite and coke may be represented by the following expression

$$\log \phi_{CO}\ (\text{atm}) = -\frac{5940}{T} + 3.46. \qquad (40)$$

The slope of the line gives an average value of about -27 kcal/mole CO for the change of enthalpy accompanying chemisorption of CO. The thermodynamic parameter ϕ_{CO} thus derived from the rate data may be used to compute the adsorption isotherms for graphite and coke for an ideal monolayer. The isotherms computed for 900 and 1100°C are shown, as examples, in Fig. 55. The strong poisoning effect of CO is evident from these isotherms. For example, half of the surface sites are occupied by adsorbed CO at 0.025 atm CO at 900°C and at 0.14 atm CO at 1100°C. It should, of course, be realized that the adsorption isotherms derived indirectly in the manner described are subject to errors.

An important outcome of this work is the finding of dual poisoning effect of carbon monoxide on the pore surface of carbon: (i) strong adsorption of CO reduces the number of vacant sites available for reaction and (ii) the activity coefficient of the activated com-

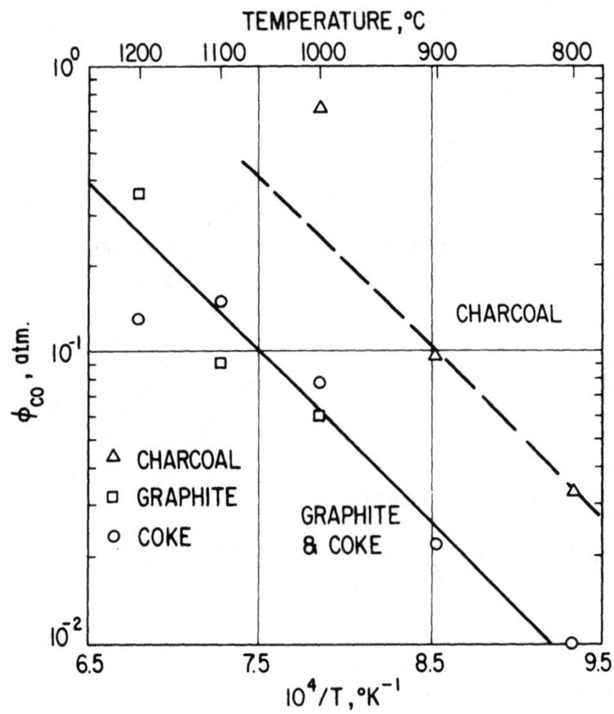

Fig. 54. Temp.dependence of ϕ_{CO} for adsorption of CO on carbon.

plex $(CO_2)^\dagger$ for the dissociation of CO_2 is increased markedly by adsorbed CO, resulting in change in the rate-controlling reaction mechanism.

CATALYTIC OXIDATION OF GRAPHITE

Much has been published in the technical literature and in patents on catalysts and catalytic reactions. Except perhaps for certain systems, the physical chemistry of catalytic reactions is not well understood. The case that we shall consider briefly here is the catalytic oxidation of carbon. In a review paper in 1923 Bancroft [48] cited interesting observations made by many investigators on the catalytic oxidation of coal, charcoal and other carbonaceous materials. A variety of additives was noted to catalyse the steam-carbon reaction, e.g. alkali and alkaline earth salts (chlorides, carbonates, sulphates, etc.), metal oxides and metals. Apparently, common salt or spelter (commercial zinc) has long been recommended for the removal of chimney soot.

As early as 1903, Schenck and co-workers [49-51] found that soot formation from carbon monoxide was catalyzed by the presence of metallic iron, and oxides having a lesser effect. In a detailed investigation of the reactivity of coke, Jones et al [52] observed that metallic iron (up to about 1% Fe) was more effective than iron oxide in accelerating the oxidation of coke in carbon dioxide. They concluded that "It must, therefore, be assumed that at the Reactivity I stage (initial fast reaction), the reducible iron is present as the metal, and that prolonged treatment with carbon dioxide at 950°C, under the conditions of the standard method for determination of reactivity, results in the conversion of all metallic iron to ferrous oxide."

Long and Sykes [53] observed that the rate of oxidation of coconut charcoal in steam or carbon dioxide decreased by about an order of magnitude upon acid extraction of impurities from the charcoal. However, relative dependence of the rate on the gas composition was not affected by this purification treatment. In a series of a *ad hoc* experiments done by Wynne-Jones [54] a marked decrease was noted in the ignition temperature of cellulose chars by the addition of alkali carbonates. For example, the ignition temperature in air decreased from about 500 to 300°C by impregnating the char with 2% Na_2CO_3. A marked increase was also noted in the rate of oxidation of Ceylon graphite in CO_2 at 800°C with iron impregnation (up to 0.25% Fe). Similar observations are reported in a more recent paper by Gallagher and Harker [55] demonstrating the catalytic effect of compounds of iron, cobalt and nickel on the oxidation of carbons

Study of the published papers on the catalytic oxidation of carbons leaves one bewildered. At present there appears to be no clear understanding of the reaction mechanisms which may be responsible for the catalytic effect. Much of the previous work was of an exploratory nature, illustrating the phenomenon, but not quantitative enough for better assessment of the causes of enhancement of the reaction rates. We are at present investigating the catalytic oxidation of carbons; the work so far indicates that the addition of trace amounts of iron or nickel to graphite enhances the rate of oxidation in steam or carbon dioxide by several orders of magnitude--an effect much greater than those reported by previous investigators.

In the course of investigation of the rate of oxidation of graphite in hydrogen-steam mixtures in this laboratory, McKewan and Rice [56] observed a remarkable catalytic effect of iron and nickel on the oxidation of graphite. When the graphite sample was in contact with metallic iron or nickel, or deliberately contaminated with a dilute chloride solution of iron or nickel, a drastic increase in the rate of oxidation was noted. This is well demonstrated in Fig. 56 by the photographs of graphite samples after oxidation in steam at 900 to 1000°C in the presence of a nickel or an iron catalyst.

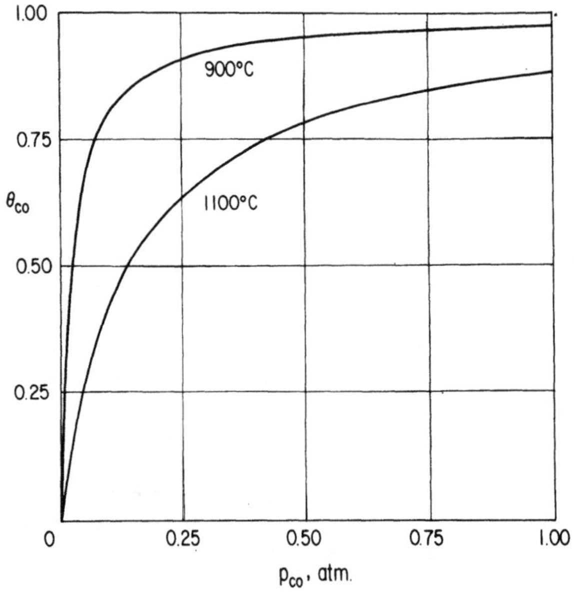

Fig. 55. CO adsorption isotherms at 900 and 1000°C for metallurgical coke and electrode graphite estimated from data for the rates of oxidation in CO_2-CO mixtures. Turkdogan and Vinters [42].

(a) During oxidation in water vapor for 30 min at 1000°C, a point contact was maintained on the surface of an electrode graphite sphere by a 0.1 mm diam nickel wire. Severe erosion of graphite over an extended area, 1 cm diam, the center of which was the point of contact with nickel wire, is indicative of enhanced rate of oxidation.

(b) One drop of 0.5% $FeCl_3$ solution was placed on electrode graphite discs and dried, then oxidized for 30 min at 1000°C. Again there was a severe erosion of the graphite.

(c) A drop of 0.5% $NiCl_2$ was placed on a section of a pyrolytic graphite, normal to the basal plane, dried and oxidized in steam for 30 min at 900°C. Enhanced oxidation is once again evident.

The catalytic oxidation of graphite and other forms of carbon is being investigated in our laboratory. In the context of the present paper a brief mention will be made of some of the preliminary results showing the marked effect of iron on the rate of oxidation of graphite in CO_2-CO gas mixtures. Electrode graphite granules (∿ 0.5 mm diam) were impregnated with iron by soaking in a 5 or 20% $FeCl_3$ solution, followed by washing, drying and heating in a stream

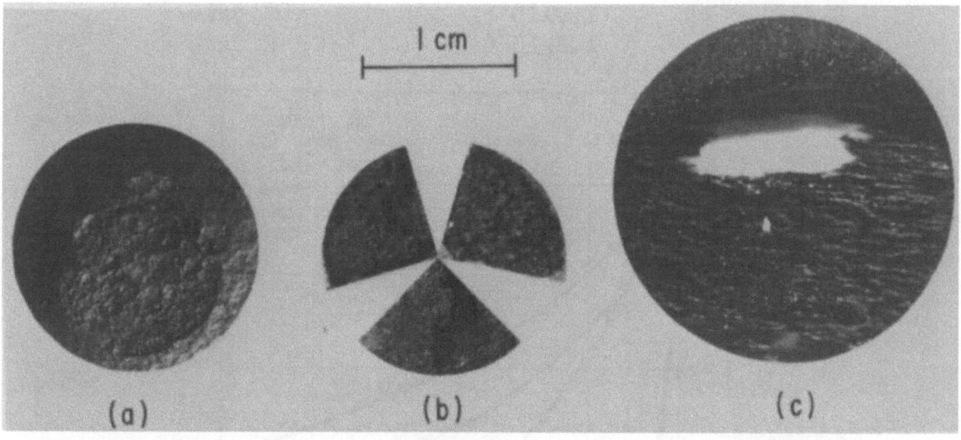

Fig. 56. The catalytic oxidation of graphite in water vapor in the presence of iron or nickel.
(a) 0.1 mm dia. nickel wire touching electrode graphite sphere during oxidation for 30 min at 1000°C.
(b) A drop of 0.5% Fe Cl_2 solution placed on electrode graphite disc; dried and oxidized for 30 min at 1000°C.
(c) A drop of 0.5% $NiCl_2$ solution placed on a section of a pyrrolytic graphite, normal to the basal plane; dried and oxidized for 30 min at 900°C.
McKewan and Rice [56].

of hydrogen for ∿ 24 hrs at 800°C. Three batches of graphite granules thus treated contained 0.01%, 0.27% and 1.0% Fe. The rate of oxidation in CO_2-CO mixtures at a total flow of 1500 ml/min was measured as described in the previous section. The typical examples of the weight loss data, as a semilog plot, are shown in Fig. 57 for 0.27% Fe-graphite granules (∿ 0.5 mm diam) oxidized in CO_2-CO mixtures at 800°C.

The linear parts of the curves for initial fast rates, denoted as stage (I), extend over the range 10 to 50% mass loss, depending on gas composition. At later stages of oxidation, the rate ultimately reaches a steady state at a lower level, designated as stage (II). Examples of the rate data in the following table illustrate the marked catalytic effect of iron on the oxidation of graphite.

Table II

Rate of oxidation of graphite granules, impregnated with iron, in 1 atm CO_2 at 800°C.

% Fe	Fast rate (I)	Slow rate (II)
	Fractional mass loss per minute	
0.00	3.5×10^{-6}	3.5×10^{-6}
0.01	4.5×10^{-4}	--
0.27	7.0×10^{-2}	2.5×10^{-3}
1.0	1.3×10^{-1}	2.0×10^{-3}

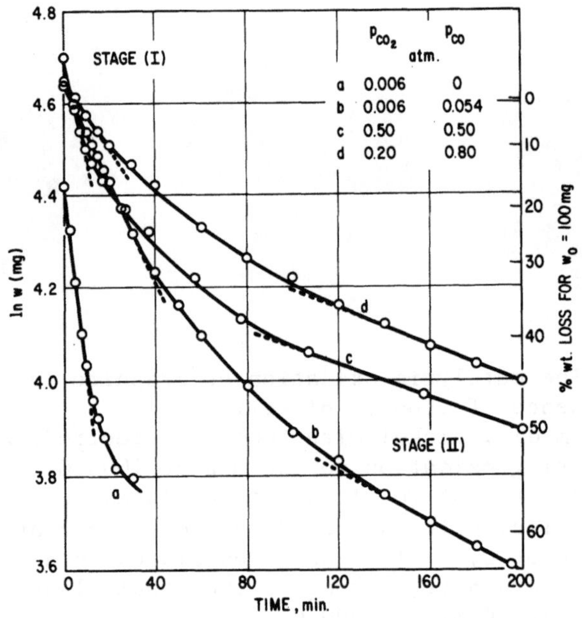

Fig. 57. Semi-log plot of weight loss data for oxidation of 0.27% Fe-graphite granules (∿ 0.5 mm dia) at 800°C in CO_2-CO mixtures [58].

The effect of gas composition at 1000°C on the rate of oxidation of graphite impregnated with 0.01% Fe is shown in Fig. 58. The broken curves reproduced from Fig. 49 for 1000, 1100 and 1200°C for comparison purposes are for graphite granules which do not contain iron. It is seen that even a trace amount of iron of the order of100 ppm is adequate to bring about a "catastrophic" oxidation of graphite. The curve for 0.01% Fe-graphite approaches a plateau at about 50% CO_2, above which the rate is essentially independent of gas composition. At these very high rates of oxidation, the reaction rate is much affected by a relatively slow gas diffusion in the pores of the carbon. It is for this reason that there is little or no change in the rate for gas compositions above 50% CO_2.

The rate of catalytic reaction of iron-impregnated graphite in hydrogen is also a few orders of magnitude greater than that for iron-free graphite. With 1% Fe-graphite granules the rate for 800°C and 1 atm H_2 is 1.4×10^{-3} fractional mass loss per minute. If most of the internal surface area ($\sim 10^4$ cm^2/g were accessible for the reaction (a gross approximation), the rate in terms of unit pore area would be 1.4×10^{-7} gC/cm^2 min. This rate is within an order of magnitude of the rate of carbon removal from graphite-saturated austenite, i.e. 53×10^{-7} gC/cm^2 min at 800°C and 1 atm H_2. If only a small fraction of the total pore area of graphite is accessible for the reaction (a more likely situation), there would be closer agreement between these two rates.

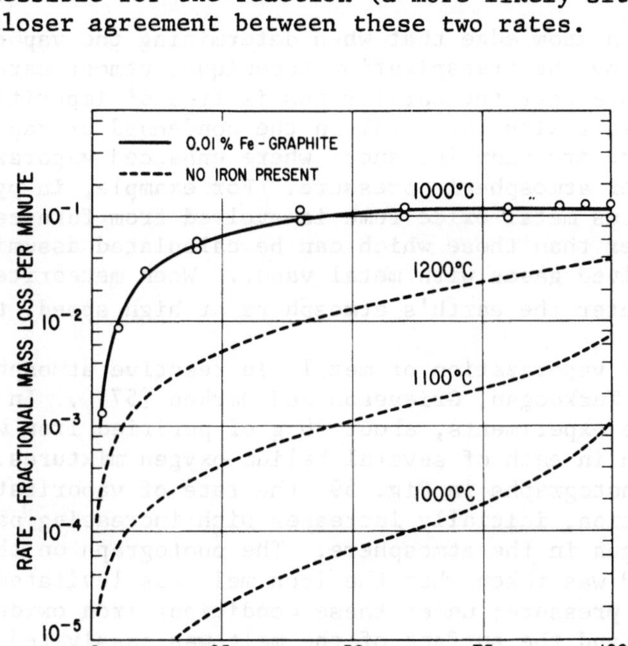

Fig. 58. Rate of catalytic oxidation of graphite granules ($\sim$0.5 mm dia), impregnated with 0.01% Fe, in CO-CO mixtures at 0.96 atm [58].

The explanation of the catalytic effect of iron, or other transition elements, on the reactivity of graphite, or other forms of carbon, must await the completion of the investigation we have undertaken. However, our tentative thoughts at present are that the metallic iron or nickel, in contact with graphite, is readily adsorbed on the pore surface to such an extent that the pore surface of graphite may resemble the surface of graphite-saturated iron where the rates of dissociation of CO_2 and evolution of CO are apparently much faster than those on an iron-free graphite surface. Subsequent transition to a lower catalytic effect of iron during oxidation may be attributed to the poisoning of the iron layer on graphite by adsorbed oxygen. The observations made by other investigators cited earlier [49-52] are in qualitative agreement with this point of view, i.e. the catalytic effect of iron oxide is less than that of the metallic iron.

VAPORIZATION OF METALS IN REACTIVE ATMOSPHERES

The kinetics of heterogeneous reactions are readily affected by environmental changes in the vicinity of an interface where a reaction is taking place, e.g. internal oxidation or reaction of vaporizing species with the surrounding gas near the surface. The effect could result in increasing the rate of diffusion-limited surface reaction or passivation (inhibition) of the surface reaction.

It is common knowledge that when determining the vapor pressures of metals by the transpiration technique, utmost care has to be taken to insure that the carrier gas is free of impurities which are likely to react with the metal in the condensed or vapor state. In practice there are many instances where enhanced vaporazation is observed under atmospheric pressure. For example, in pyrometallurgical processes metal oxide fume is evolved from furnaces at rates much higher than those which can be calculated assuming saturation of the evolved gases with metal vapor. When meteorites or similar objects enter the earth's atmosphere at high speed, they "ablate".

The rate of vaporazation of metals in reactive atmospheres was investigated by Turkdogan, Grieveson and Darken [57] . In the initial qualitative experiments, about 20 g of purified iron were melted by levitation in each of several helium-oxygen mixtures. As seen from the photographs in Fig. 59, the rate of vaporization enhanced by oxidation, initially increases with increasing partial pressure of oxygen in the atmosphere. The photograph on the extreme right in Fig. 59 was taken when the iron melt was levitated in pure oxygen at 1 atm pressure; under these conditions iron oxide fumes were not formed and the surface of the melt was passivated by a coating of molten iron oxide.

The effect of a reactive gas on the rate of vaporization of a metal can be considered as a counterflux transport process. For

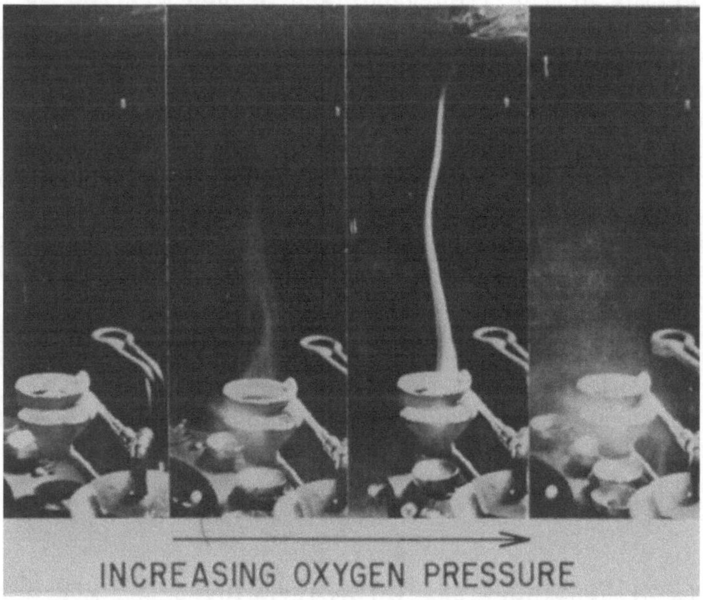

Fig. 59. Vaporization of levitated iron melts in helium and oxygen atmospheres. Turkdogan Grieveson and Darken [57].

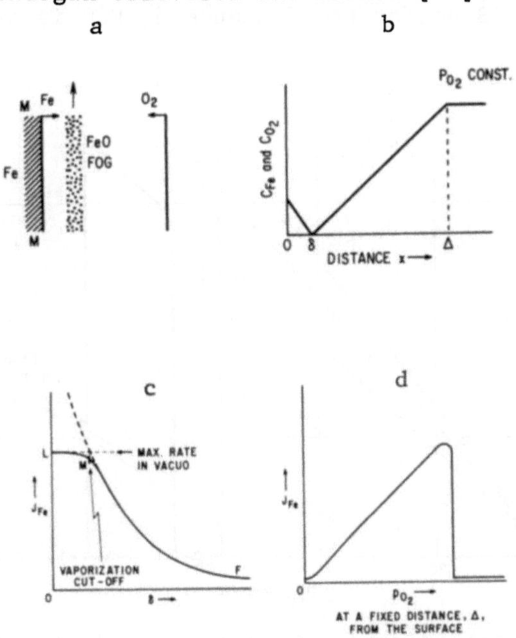

Fig. 60. Schematic diagram showing the variation of J_{Fe} with δ and $p(O_2)$. Turkdogan, Grieveson and Darken [57].

example, when iron vaporizes in a stream of argon-oxygen mixture, the counterflux of the iron vapor and oxygen will be as shown in Fig. 60a. At some short distance from the surface of the metal, iron vapor and oxygen will react forming an iron oxide mist. The formation of an iron oxide, which is in a state of fine subdivision, in the gas phase provides a sink for the iron vapor and oxygen, resulting in the counterflux of these two gaseous species. When a steady state of counterflux is established under these boundary conditions in an isothermal system, the concentration profiles for iron vapor, C_{Fe}, and oxygen, C_O, close to the metal surface are as shown schematically in Fig. 60b. At a particular distance δ, an iron oxide fog is formed and beyond a particular distance Δ, which is the thickness of the aerodynamic boundary layer, the oxygen partial pressure is considered as being maintained constant. Under these boundary conditions and for the simplified case of $\delta \ll \Delta$, by Fick's law, the counterflux of iron vapor and oxygen for a steady state is given by

$$J_{Fe} = -2\,J_{O_2} = 2m\,\frac{p_{CO_2}}{RT} \tag{41}$$

where D_O is oxygen-inert gas interdiffusivity and $m = D_O/\Delta$ is the average film mass-transfer coefficient for the transport of oxygen through the diffusion boundary layer which is calculable for known flow conditions. Since at the distance δ, FeO is assumed to be

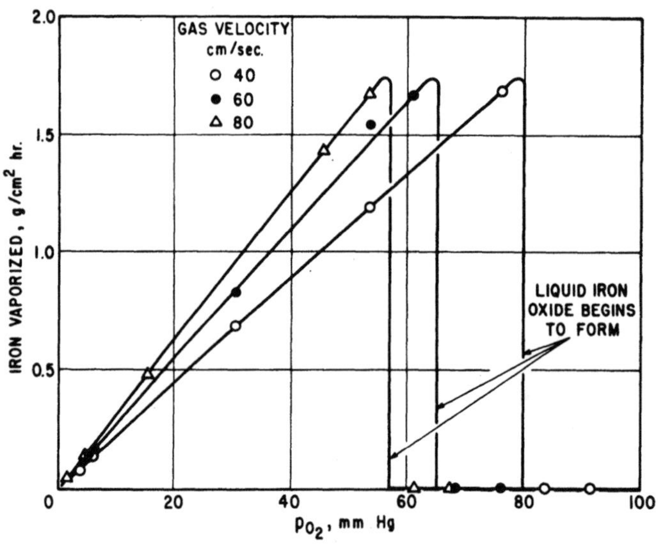

Fig. 61. Relationship showing the effects of oxygen partial pressure of Ar-O_2 mixtures and the velocity of the gas stream on the rate of vaporization of liquid iron at 1600°C. Turkdogan, Grieveson and Darken [57].

formed, the counterflux of iron vapor, J_{Fe}, is twice that of oxygen, $J_{Fe} = - 2J_{O_2}$. Since the partial pressures p'_{Fe} and p'_{O_2} at $x = \delta$ are much lower than the respective values p_{Fe} and p_{O_2} at $x = 0$ and $x = \Delta$, equation (41) is given in a simplified form. Another simplifying assumption made is that the formation of a layer of an iron oxide fog at δ does not interfere with the suggested mechanism, in practice there is sufficient forced or free convection to remove the layer of fog (by vertical motion as shown in Fig. 60a).

According to equation (41), for a given temperature and film mass-transfer coefficient, the rate of vaporization of iron, or any other metal reacting with oxygen, is expected to increase linearly with increasing partial pressure of oxygen. In other words, increasing oxygen partial pressure decreases the distance δ through which the iron vapor is transported. However, there are two serious restrictions to this mechanism of increasing the rate of vaporization by increasing oxygen partial pressure: (I) Maximum free vaporization cannot be exceeded and (II) for given metal and temperature, ther is a minimum oxygen partial pressure below which the metal oxide fog does not form.

(I) Maximum rate of vaporization: Under no conditions can the rate of evaporation exceed that in vacuo. Using the Langmuir equation, the rate of evaporation in vacuo is given by

$$J_{max} = \frac{p_i}{\sqrt{2\ \pi FT\ M_i}} \tag{42}$$

where p_i and M_i are the vapor pressure of metal (i) and its molecular weight respectively. When this limiting value of the flux is approached, the distance δ approaches the mean-free path, and the plot of J_{Fe} against δ will be as shown by curve FML in Fig. 60c. On further increase in oxygen partial pressure at Δ, i.e. decrease in δ, the flux of oxygen toward the surface of the metal is greater than the equivalent counterflux of metal vapor, resulting in the oxidation of the metal surface. Since the vapor pressure of the metallic species of the oxide is much lower than that of the metallic phase, the vaporization should become vanishingly small when the surface is coated with an oxide layer. This phenomenon of passivity or inhibition was subsequently extended by Wagner [59] to other more complex situations illustrating the generality of the concept.

(II) Normal rate of vaporization: Depending on the oxygen potential of the oxide (fôg) formed at δ, there is a minimum critical oxygen partial pressure below which an oxide fog cannot be formed. Below this critical oxygen partial pressure, the isothermal rate of vaporization of the metal is given by the equation

$$J_{min} = \frac{D_i}{\Delta} \frac{p_i}{RT} \tag{43}$$

where D_i is the diffusivity of the metal vapor.

Taking the two foregoing restrictions into account, the relation between J_{Fe} and p_{O_2} expected from equation (41) is shown in Fig. 60d.

To test the validity of this concept, Turkdogan *et al* [57] measured the rate of vaporization of several metals, e.g. liquid copper, nickel, cobalt, iron, manganese and solid chromium, in streams of argon-oxygen mixtures under conditions such that the mass transfer coefficient m was known for the flow conditions employed. As seen from an example of the results in Fig. 61, the rate of vaporization of iron increases with increasing partial pressure of oxygen until a mazimum rate is obtained, beyond which vaporization almost ceases. The maximum rate of vaporization, obtained immediately before the occurrence of the vaporization cut-off, observed experimentally, is shown in a log-log plot in Fig. 62 to be in excellent agreement with the theoretical free rate of vaporization.

This example illustrates well the extent to which the rate of a heterogeneous reaction can be affected by other reactions occurring in the vicinity of the reacting surface. Another example of a related phenomenon is the enhancement of diffusion-limited vaporization rates by a convection-condensation process brought about

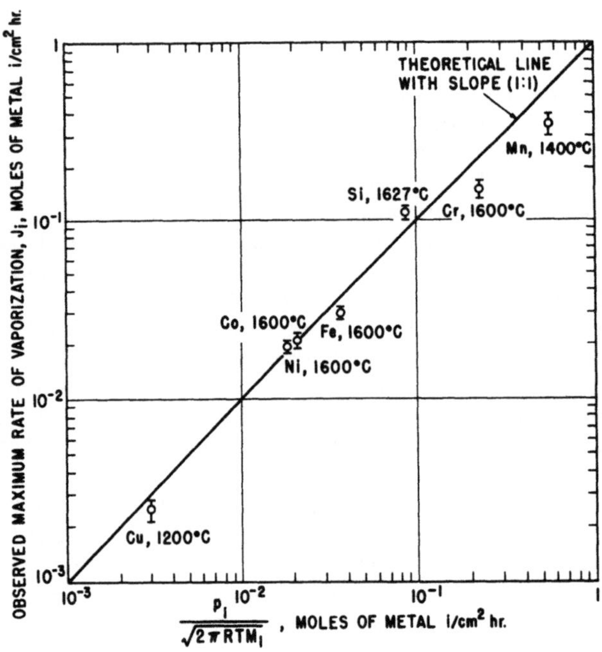

Fig. 62. Observed maximum rate of vaporization in $Ar+O_2$ mixtures related to the Langmuir equation for vaporization in vacuo. Turkdogan, Grieveson and Darken [58].

by a steep temperature gradient in the gas phase near the metal surface. For further details on this work, reference should be made to previous publications [60,61].

CONCLUDING REMARKS

Our knowledge of the kinetics of heterogeneous reactions at elevated temperatures is still far inferior to our knowledge of the corresponding equilibria, in spite of the rather extensive amount of research on reaction rates. Many of the reasons are quite obvious. Equilibrium, in contrast to kinetics, is unique for a given system and in most cases independent of trace amounts of extraneous substances. Heterogeneous reactions take place in the two-dimensional solutions of the interface; compositions and chemical potentials here present extreme difficulties to experimental determination. Then there is the ever-present experimental difficulty of eliminating or properly accounting for the transport (diffusion) phenomena almost always associated with the surface reaction. Many more phases may coexist during reaction than at equilibrium, as, for example, in the case of hematite reduction. Porosity, as in the case of carbon oxidation, introduces vast complications in reaction rate which have no counterpart at equilibrium.

Obviously, an improved knowledge of the thermodynamics of the essentially two-dimensional solutions, known as chemisorbed layers, would be a big step forward in aiding our interpretation of the kinetics of heterogeneous reactions. However, there is little indication of rapid progress in this direction, though we do hope for some illumination from the field-ion microscope atom-probe and from Auger spectroscopy.

However, in spite of these difficulties and deficiencies we feel that the phenomenological approach adopted in the examples here cited does lead to some insight on surface and interface kinetics, especially where strong adsorption is involved. It has been shown (1) that with sufficient care in the design of experiment the effects of transport (diffusion) and of chemical reaction rate may be separated and evaluated, and (2) that adsorbed monolayers frequently approach complete coverage and exert a very pronounced effect on the reaction rate. A summary of the chemical reaction rates herein discussed in some detail is given in Table III. A feature brought out clearly by this tabulation is the importance of strong chemisorption which is reflected in the nonunity value of the $(1-\theta)$ term (equations 4 and 5) and leads to an inverse proportionality between reaction rate and activity of the strongly adsorbed species (via equation 1).

Table III

Summary of Reaction Rates Discussed

Overall Reaction	Activated Complex Includes (besides substrate)	Strongly Adsorbed Species	Forward Rate at $\theta \approx 1$ Proportional to
$\frac{1}{2} N_2(g) \rightarrow \underline{N}(Fe)$	N	O,S,CO	a_N/a_O
$C(Fe) + 2H_2(g) \quad CH_4(g)$	CH_3	C	$p_{H_2}^{3/2}$
$H_2S(g) + Fe \rightarrow H_2(g) +$ "FeS"	H_2S	S	p_{H_2S}/a_S
$H_2O(g) + Fe \rightarrow H_2(g) +$ "FeO"	H_2O	O?	p_{H_2O}/a_O
$CO_2(g) + Fe \rightarrow CO(g) +$ "FeO"	CO_2	O?	p_{CO_2}/a_O
$C(g) \quad + CO_2(g) \rightarrow 2CO(g)$	CO_2	CO	p_{CO_2}/p_{CO}

ACKNOWLEDGMENTS

The authors wish to thank J. L. Bomback and G. J. W. Kor of this Laboratory for their work involving scanning electron microscopy and thermal etching respectively.

ADDENDUM

Kinetics of Nitrogen Transfer through an Iron Surface

Recently H. J. Grabke [Ber. Bunsenges. physik. Chem., 72, 533, 541 (1968)] investigated the reaction of nitrogen with a 5 mμ thick iron foil using the electrical resistivity technique. In the absence of oxygen, or other strongly adsorbed impurities, it was found that at low nitrogen activities, e.g. < 1 atm N_2, the rate of nitrogenization of alpha and gamma iron is proportional to a_N^2 (second order type), and at high nitrogen activities (as in NH_3-H_2 mixtures) the rate is proportional to a_N (first order type). The general form of the rate equation for the reaction involving two atoms of nitrogen, with the complex occupying a single site, is in terms of the present formulation: forward rate $\propto a_N^2 (1-\theta)/\phi_\dagger$. Substituting $1-\theta$ as deduced from Eq. 1 for nitrogen adsorption, it is found that: rate $\propto \phi_N a_N^2/\phi_\dagger(\phi_N+a_N)$. Thus at low nitrogen activity, i.e. $(1-\theta)\rightarrow 1$, rate $\propto a_N^2$, and at high nitrogen activity, i.e.

$(1-\theta) \rightarrow 0$, rate $\propto a_N$. The form of the rate equation simplified for these two limiting cases is in accord with Grabke's observations. When there is a strong surface poisoning for example by oxygen adsorption, our data indicate that rate $\propto a_N$. These findings imply that there are two consecutive reactions in series in the transfer of nitrogen through the surface of iron into the gas phase, or vice versa: (i) transfer of nitrogen from the bulk lattice to the adsorbed layer involving an atom of nitrogen in the activated state and (ii) desorption of nitrogen from the adsorbed layer involving two nitrogen atoms in the activated complex. In both cases the complex occupies a single site. On the basis of the experimental observations, it is deduced that when there is a preferential strong oxygen adsorption such that $1-\Theta \rightarrow 0$, the first step in the denitrogenization controls the rate. On the other hand, when the surface is essentially free of adsorption species, the second reaction in the series exerts a greater resistance to the reaction. The rate for mechanism (ii) at $1-\Theta \rightarrow 1$ is many times that for mechanism (i) at $1-\Theta \rightarrow 0$.

REFERENCES

1. W. M. McKewan: in *Steelmaking, The Chapman Conference*, pp. 141-155, M.I.T. Press, Cambridge, Mass., (1965).
2. R. H. Spitzer, F. S. Manning and W. O. Philbrook: Trans. Met. Soc. AIME, 236, 726 (1966).
3. R. H. Olsson and W. M. McKewan: Trans. Met. Soc. AIME, 236, 1518, (1966).
4. W. M. McKewan and R. G. Olsson: to be published.
5. G. V. Smith, J. A. MacMillan and E. J. Dulis: Trans. ASM, 43, 692, (1951).
6. H. H. Podgurski: *Annals of the N. Y. Acad. of Sci.*, 58, 959 (1954).
7. B. B. Rath and H. Hu: Trans. Met. Soc. AIME, 245, 1577, (1969).
8. J. W. Cahn: Acta Met. 10, 789, (1962).
9. K. Lücke and H. P. Stüwe: *Recovery and Recrystallization of Metals*, Interscience (AIME), 1963).
10. P. Kozakevitch and G. Urbain: Rev. Met., 58, 517, (1961).
11. F. A. Holden and W. D. Kingery: J. Phys. Chem., 59, 557, (1955).
12. H. A. Wriedt: Trans. Met. Soc. AIME, 245, 43, (1969).
13. K. Schwerdtfeger, P. Grieveson and E. T. Turkdogan: Trans. Met. Soc.AIME, in press.
14. B. E. Sundquist: Acta Met., 12, 505,(1964).
15. S. S. Brenner: Surface Sci., 2, 496, (1964).
16. J. Moreau and J. Benard: Acta Met., 10, 247, (1962).
17. E. D. Hondros: Proc. Roy. Soc. (London), A286, 479, (1965).
18. F. Cabané-Brouty: J. Chem. Phys., 62, 1056, (1965).

19. S. J. Gregg: *The Surface Chemistry of Solids*, 2nd ed., Rheinhold, N. Y., 62(1961).
20. J. A. Morrison, J. M. Los, and L. E. Drain: Trans. Faraday Soc., 47, 1023, (1951); 48, 840 (1952); J. Chem. Phys., 19, 1963, (1951).
21. E. T. Turkdogan and L. J. Martonik: to be published.
22. G. Naeser: Stanl u. Eisen, 68, 375, (1948).
23. W. A. Fischer and A. Hoffmann: Arch. Eisenhuttenw., 31, 215, (1960).
24. H. Schenck, M. G. Frohberg and H. Heinemann: Arch. Eisenhuttenw., 33, 593, (1962).
25. R. D. Pehlke and J. F. Elliott: Trans. Met. Soc. AIME, 227, 844 (1963).
26. P. Kozakevitch and G. Urbain: Rev. Met., 60, 143 (1963).
27. E. T. Turkdogan and P. Grieveson: J. Electrochem. Soc., 114, 59, (1967).
28. P. Grieveson and E. T. Turkdogan, Trans. Met. Soc. AIME, 230, 407, (1964).
29. H. H. Podgurski, F. N. Davis, and R. P. Smith: J. Vac. Sci. Tech., 4, 186, (1967).
30. J. H. Swisher and E. T. Turkdogan: Trans. Met. Soc. AIME, 239, 602 (1967).
31. K. Hauffe and H. Z. Pfeiffer: A. Metallk., 44, 27, (1953).
32. F. S. Pettit, R. Yinger, and J. B. Wagner: Acta Met., 8, 617, (1960).
33. W. W. Smeltzer: Trans. Met. Soc. AIME, 218, 674, (1960).
34. E. T. Turkdogan, W. M. McKewan and L. Zwell: J. Phys. Chem., 69, 327, (1965).
35. E. T. Turkdogan: Trans. Met. Soc. AIME, 242, 1665, (1968).
36. W. L. Worrell and E. T. Turkdogan: Trans. Met. Soc. AIME, 242, 1673, (1968).
37. H. -J. Grabke: Ber. Bensenges. physik. Chem., 69, 48, (1965).
38. R. P. Smith: Acta Met. 1, 578,(1953); Trans. Met. Soc. AIME, 230, 476, (1964).
39. R. P. Smith, J. Am.Chem. Soc., 68, 1163 (1946).
40. E. T. Turkdogan, V. Koump, J. V. Vinters, and T. F. Perzak: Carbon 6, 467,(1968).
41. E. T. Turkdogan and J. V. Vinters: Carbon, 7, 101, (1969).
42. E. T. Turkdogan and J. V. Vinters: Carbon, in press.
43. E. T. Turkdogan and R. G. Olsson, and J. V. Vinters, to be published.
44. K. Hedden and A. Löwe: Carbon 5, 339, (1967).
45. H.-J. Grabke: Ber.Bunsenges. physik. Chem., 70, 664 (1966).
46. P. L. Walker, F. Rusinko, and L. G. Austin: Advances in Catalysis 11, 133, (1959).
47. "JANAF Thermochemical Tables", PB168, 370 (1965) and PB168, 370-1 (1966), U.S. Dept. Commerce, Washington, D. C.
48. W. D. Bancroft: J. Phys. Chem. 27, 801, (1923).
49. R. Schenck and F. Zimmermann, Ber. 36, 1231 (1903).
50. R. Schenck: Ber., 36, 3663 (1903.

51. R. Schenck and W. Heller: Ber. 38, 2139 (1905).
52. J. H. Jones, J. G. King, and F. S. Sinnatt: Fuel Research Tech. Paper No.25 (1930).
53. F. J. Long and K. W. Sykes: J. Chem. Phys. 47, 361 (1950).
54. W. F. K. Wynne-Jones: 6th Conf. British Coke Res. Assn. 8 (1953).
55. J. T. Gallagher and H.Harker: Carbon 2, 163 (1964).
56. W. M. McKewan and B. B. Rice, to be published.
57. E. T. Turkdogan, P. Grieveson and L. S. Darken: J. Phys. Chem. 67, 1647 (1963).
58. E. T. Turkdogan and J. V. Vinters, to be published.
59. C. Wagner: Corrosion Science 5, 751 (1965).
60. E. T. Turkdogan: Trans. Met. Soc. AIME 230, 740 (1964).
61. E. T. Turkdogan and K. C. Mills: Trans. Met. Soc. AIME 230,

DISCUSSION OF PAPER

C. E. Birchenall: I have difficulty with the concept of one of the constituents of the product of a reaction serving as a poison for the reactions. In the reaction path for the formation of FeS, adsorbed S seems likely to be farther along the path to incorporation in the product than the activated complex involving H_2S. Sulfidation of Fe in S_2 is diffusion controlled (possibly after a short time), so the incorporation of S at maximum rate permitted by its availability is indicated. How does the availability of adsorbed S slow the sulfidation of Fe in H_2S? How is the S retained in the adsorbed state short of the final state of $S^{=}$, to which it goes quickly in S_2?

L. D. Darken: Perhaps the concept of "poison" is a bit subjective as illustrated in our example of the decarburization of austenite (Fig.24). The "poisoning" effect referred to here is simply the leveling off of the rate at high carbon contents which we attribute to fillage of the surface sites by carbon. At high carbon levels the rate no longer continues to increase with carbon content as it did with low carbon levels. This effect is anticipated from the equations given in the text.

I see no especial difficulty in reconciling the sulfidation of iron by S_2 vapor with that by H_2S-H_2 mixtures discussed in our paper. These are two entirely different reactions. The rate constant for the forward reaction in the former case might perhaps be written

$$R_1 = k_1 p_{S_2}^{1/2}/a'_S$$

and in the latter as

$$R_2 = k_2 p_{H_2S}/a''_S$$

Evidently k_1 is very much greater than k_2 so that in the former case the surface and subsurface are essentially in equilibrium with the gas, whereas such equilibrium is not obtained in the latter case. Admittedly, a'_S must be larger than a''_S (in using this simplifying equation, we are assuming the surface to be essentially fully covered with sulfur in both cases); i.e., the fact that $a'_S > a''_S$ is over-balanced by the fact that $k_1 >> k_2$.

The arguments presented in the paper avoid to considerable extent the detailed atomistics of the adsorbed layer and are thus independent of the difference between S and $S^=$ providing the electrical conductivity is adequate, as it certainly is in this case. The treatment here is very similar to that for the oxidation of iron to produce a wüstite layer.

F. D. Richardson: The catalytic effects of iron and nickel on the attack of carbon by CO_2 are very interesting from the standpoint of certain metallurgical reactions. In the case of the Ni wire it appeared that the carbon was preferentially attacked at considerable distances from the nickel. How can this be explained? Do these effects also occur in the attack of carbon by H_2O?

How was the iron distributed in the carbon and what was its state of subdivision? Does the iron affect the poisoning action of adsorbed CO as well as the attack by CO_2?

E. T. Turkdogan: The Catalytic effect of iron and nickel on the oxidation of graphite was observed under all conditions of oxidation in CO_2 and H_2O. For the case illustrated in Fig. 56a, the eroded area due to rapid oxidation was about 1 cm diam, while the nickel wire in contact with graphite at the center of the reacted region was 0.1 mm diam. We believe that through rapid surface diffusion of nickel or iron, an adsorbed layer of metal could be formed on the surface of carbon and the enhanced rate of oxidation may be attributed to the presence of such an adsorbed layer. In the rate measurements using 0.5 mm diam granules, the graphite granules were impregnated with iron by soaking in a 5 or 20% $FeCl_3$ solution, followed by washing, drying and heating in a stream of hydrogen for ~ 24 hours at 800°C. Such a treatment is expected to give a uniform distribution of finely subdivided iron particles in graphite granules. In fact, the chemical analysis and x-ray examination of different samples confirmed the uniformity of iron impregnation. Carbon monoxide retards the rate of catalytic oxidation of graphite in a manner similar to oxidation in the absence of iron.

B. C. H. Steele: There is evidence that at lower temperatures (400 - 600°C) the enhanced oxidation of graphite in the presence of transition metals can be attributed to the formation of liquid metal carbonates. Could this mechanism also be important at the higher temperatures involved in your studies.

It is relevant to note that the catalytic deposition of carbon from carbon monoxide in the presence of iron or nickel is a well known process in refractories technology, and can be responsible for the catastrophic disintegration of alumino-silicate materials.

E. T. Turkdogan: Numerous additives are known to catalyze the oxidation of carbons to varying degrees. Some of the additives such as alkali salts could be molten at the temperature of oxidation. However, the presence of a liquid phase in the pores of the carbon is not a necessary condition for the catalytic effect. With iron or nickel impregnation of graphite, no liquid carbonate is is expected to form at the oxidation temperatures employed.

A. W. D. Hills: The evidence presented by Dr. Turkdogan may not necessarily prove that gas phase mass transfer did not affect the rate at which hydrogen removed carbon from austenite. In principle, the mass transfer coefficient should increase with increasing gas flow rate, but the increase that can be achieved under normal experimental conditions, especially if gravimetric techniques are being used, is not very great. Thus the establishment of a region in which the rate of a gas/solid reaction is relatively unaffected by changes in gas flow can merely mean that sufficient gas is being provided to sustain the reaction.

A reaction can only be assumed to be unaffected by gas phase mass transfer if the chemical rate constant calculated from the measured reaction rate is shown to be very much smaller than mass transfer coefficient established at the surface of the sample under the experimental conditions. Mass transfer coefficients under the flow conditions used by Dr. Turkdogen cannot be calculated from the literature, so they must be measured directly. Has Dr. Turkdogan made any such measurements?

E. T. Turkdogan: As demonstrated in Fig. 41, at gas flow velocities from 23 to 60 cm/sec, the rate of decarburization is no longer affected by the gas flow rate. This is indeed a good indication that the gas film transport does not impose a limitation on the rate of the reaction. This argument can be substantiated by other quantitative evidence. If the gas film transport of methane from the surface of the sample were rate-controlling, the rate of decarburization should have been proportional to the product $a_C(p_{H_2})^2$. Such a relationship is not observed experimentally. Furthermore, the temperature coefficient of the rate at high carbon contents, corresponding to 27 kcal/g-atom C, is much too high for a gas film transport control. The analytical solution of the average mass transfer coefficient for flow over a flat plate[1] can be approximated with reasonable assurance to the present experimental condition for high gas velocities, thus

$$\frac{\bar{m}L}{D} = 0.664 \quad (Sc)^{1/3} \quad (Re)^{1/2} \tag{1}$$

where

- $\bar{m}$ = average mass transfer coefficient
- L = length of the strip in the direction of gas flow
- D = interdiffusivity
- Sc = Schmidt number
- Re = Reynolds number.

Substituting the corresponding expressions for Sc and Re, the average mass transfer coefficient for the present experimental condition is given by

$$\bar{m} = 0.664 \quad \left(\frac{D^4}{\nu}\right)^{1/6} \quad \left(\frac{u}{L}\right)^{1/2} \tag{2}$$

where

- ν = kinematic viscosity
- u = gas velocity away from the surface.

For the case of gas film transport control, the rate of decarburization in a stream of dry hydrogen would be given by

$$\text{Rate} = \bar{m}AK \frac{a_C (p_{H_2})^2}{RT} \tag{3}$$

where

- A = surface area of the strip, 7.5 cm^2
- aC = activity of carbon in the alloy at any given time
- K = equilibrium constant = $p_{CH_4}/a_C(p_{H_2})^2$

At 1140°C, D (CH_4-H_2)= 32.9 cm^2/sec, $\nu(H_2)$ = 14.1 cm^2/sec, $K = 4 \times 10^{-3}$ atm^{-1}. Using these values and L = 2.5 cm for the length of the strip (0.056 cm thick), the rate equation for decarburization at 1140°C in 1 atm H_2 flowing at u = 60 cm/sec is given by

$$\frac{d(\%C)}{dt} = 0.25 \, a_C, \quad (\%C)/\text{min} \tag{4}$$

From the known thermodynamics[2] of the iron-carbon alloys, the activity of carbon relative to graphite can be represented by the following exponential function in terms of wt.% C.

$$a_C = 0.231 \quad (\%C) \quad e^{0.385\,(\%C)} \tag{5}$$

Inserting this in equation (4) gives the instantaneous rate for an assumed gas film transport process as a function of carbon content of the sample, thus

$$\frac{d(\%C)}{dt} = 5.77 \times 10^{-2} (\%C)\, e^{0.385(\%C)}, \quad (\%C)/\text{min} \qquad (6)$$

The dotted middle curve in Fig. A is computed using this equation for the gas film transport control. The top curve drawn by broken lines is calculated for rate controlled by diffusion of carbon in austenite. Comparison of these computed rate curves with the experimental conditions employed in this work, the diffusional processes do not have much retarding effect on the rate of reaction over a wide composition range. Another quantitative evidence in support of this conclusion is obtained from the heat effect on the rate. The main temperature effect on the transport-controlled rate in equation (3) is that associated with the equilibrium constant K, e.g. $\sim$ -18 kcal/mole CH_4 for the heat of formation of methane from hydrogen and graphite. For given carbon and hydrogen activities, the temperature dependence of the gas transport-controlled reaction would have corresponded to about - 15 or - 20 kcal; that is, the rate should have increased with decreasing temperature. This is quite contrary to the experimentally observed temperature effect. All the evidence cited above support the view that, under the experimental conditions employed, the rate of decarburization of austenite in dry hydrogen must be controlled almost entirely by a surface reaction as discussed in the paper.

1. E. R. G. Eckert and R. M. Drake: *Heat and Mass Transfer*, McGraw-Hill Book Co., New York (1959).
2. R. P. Smith: J. Am. Chem. Soc. 68, 1163-75 (1946).

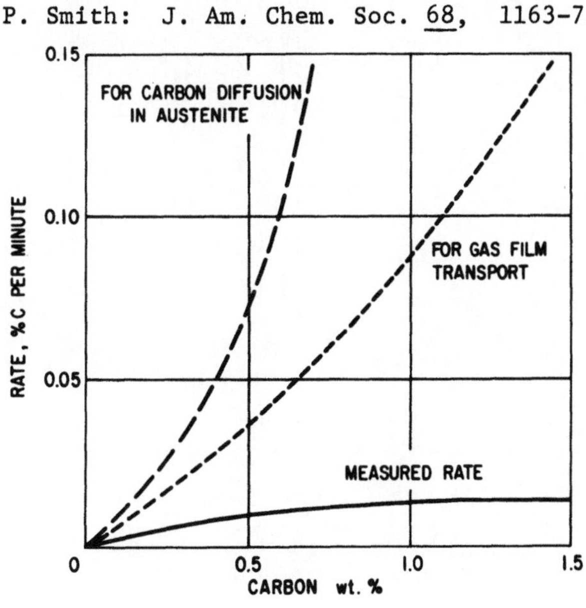

Fig. A.

THE CONCEPT OF THE THERMODYNAMIC ACTIVITY OF ATOMIC SPECIES IN HETEROGENEOUS KINETICS

Carl Wagner

Max-Planck-Institut für physikalische Chemie

Göttingen, Germany

It is shown how the thermodynamic activity of atomic species such as hydrogen or oxygen on the surface of a catalyst or a reacting solid can be determined experimentally and how these values can be evaluated in conjunction with other data in order to test tentatively assumed mechanisms of heterogeneous reactions and to establish the rate determining step of consecutive reactions.

INTRODUCTION

In most experimental investigations of reactions involving gaseous reactants and products at the surface of a catalyst the rate is determined as a function of partial pressures or concentrations of reactants and products in the gas phase. Since the individual steps involve adsorbed reactants and intermediates, it is appropriate to consider the rates of individual steps as functions of the surface concentrations of adsorbed species rather than volume concentrations of gaseous species. Tentatively, one may assume equilibrium for adsorption and desorption of reactants and products. In general, several species are present in the adsorption layer. Thus one has to deal with the problem of multicomponent adsorption. In most cases, complete information cannot be attained, since even in a non-reacting multicomponent system rather intricate measurements are required in order to obtain all pertinent data. In a reacting system, there are still greater difficulties.

In addition, the following problems are encountered. Conventional methods yield the total amount of oxygen adsorbed on a metal but do not differentiate between molecular and atomic adsorbed oxygen.

A differentiation between these species is especially important when equilibrium between these species is not established. Further, when one tries to determine the amount of oxygen adsorbed on the surface of a solid metal oxide, one must take in account that an increase in oxygen partial pressure in the gas phase results not only in adsorption but also in changes in the diffuse electrical double layer next to the surface and further in changes in the bulk composition of the metal oxide.

The aforementioned difficulties may at least in part be overcome by introducing the concept of the thermodynamic activity of adsorbed atomic species such as hydrogen, oxygen, or sulfur. In the thermodynamics of liquid or solid solutions, it is usual to define the activity a_i of component i in a co-existing ideal gas phase, multiplied by an appropriately chosen constant whose value depends on the choice of the standard state at which $a_i = 1$. Similarly, the thermodynamic activity of an adsorbed species may be defined by considering the equilibrium between atoms in the adsorption layer and atoms in a hypothetical ideal gas phase denoted as "virtual atmosphere" by Takaishi [1] and setting the thermodynamic activity a_i of an atomic species i proportional to the partial pressure p_i of atoms of type i. In the virtual gas phase, diatomic molecules of element i may be present in equilibrium with diatomic molecules in the adsorption layer. The recombination of atoms to molecules in the virtual gas phase is supposed to be suppressed, if one has a state of non-equilibrium in the adsorption layer. Since the partial pressure of hydrogen, oxygen, or sulfur atoms in the virtual gas phase cannot be measured directly, indirect measurements are needed. Appropriate methods are described in what follows. Subsequently, it is shown how experimental values of the thermodynamic activity of adsorbed atomic species can be evaluated in conjunction with other data in order to test tentatively assumed mechanisms to establish the rate determining step in a sequence of consecutive reactions.

METHODS FOR THE EXPERIMENTAL DETERMINATION OF THE THERMODYNAMIC ACTIVITY OF ATOMIC SPECIES IN THE ADSORPTION LAYER IN A REACTING SYSTEM

If a catalyst is exposed to a reacting mixture for a sufficiently long time and the external conditions are kept constant, e.g., in a flow reactor, thermodynamic equilibrium between atoms in the surface and atoms in the bulk of the catalyst is established and, therefore, the thermodynamic activity of an atomic species in the bulk becomes equal to its activity in the adsorption layer. Consequently, the activity of an atomic species in the adsorption layer of a catalyst may be obtained by determining a bulk property which is related to its activity. In particular, one may determine the chemical composition of the bulk catalyst, e.g., the content of dissolved oxygen, nitrogen, or hydrogen in a metallic cayalyst, or the deviation from the ideal metal-to-oxygen ratic in an oxide phase expressed in terms of an excess or a deficit of metal. Since molecules are practically not dissolved by metals or metal compounds usually serving as catalysts, the presence of molecules on the surface not in equilibrium with atoms on the surface does not affect the composition of the bulk catalyst.

Alternatively, one may determine a physical property which is a function of composition, e.g., lattice constants, the electrical resistivity or conductivity, the magnetic susceptibility, optical absorption, etc. First one may measure a property at predetermined values of the activity a_i of an atomic species i under equilibrium conditions. Second one may measure the same property under steady-state conditions of a catalytic process. Upon comparing the two sets of measurements, one obtains readily values of $a_i(st)$ under steady-state conditions of a catalytic process.

The following special procedures are cited:

(1) Under equilibrium conditions the hydrogen activity a_H in palladium is given by the square root of the hydrogen partial pressure $p(H_2)$ in a coexisting gas phase:

$$a_H = [p_{H_2}(eq)]^{1/2} \qquad (1)$$

where molecular hydrogen of atmospheric pressure is the standard state at which $a_H = 1$. At 180°C the electrical resistivity of palladium increases about proportionately to the hydrogen content, which under equilibrium conditions is approximately proportional to $p(H_2)^{1/2}$. Thus, comparing the electrical resistance of a Pd foil in a reacting mixture, e.g., $C_2H_4 + H_2$, with the electrical resistance in pure hydrogen of given pressure, one obtains the hydrogen activity in and on palladium while the hydrogenation of ethylene proceeds [2]. Likewise one may determine the hydrogen activity in palladium during the decomposition of formic acid, $HCOOH = H_2+CO_2$ [3].

(2) Under equilibrium conditions the conductivity of many oxide phases, e.g., Cu_2O, FeO, CoO, NiO, or ZnO, at elevated temperatures is a function of the oxygen partial pressure in the ambient gas phase inasmuch as the oxygen partial pressure determines the metal excess or deficit in the metal oxide and thereby the concentration of electrical carriers, especially excess electrons and electron holes [4]. Thus, after appropriate calibration, the steady-state oxygen activity $a_O(st)$ may be deduced from measurements of the conductance of a metal oxide foil used as catalyst while an oxygen transfer reaction, e.g., $2N_2O = 2N_2 + O_2$ or $CO_2 + H_2 = CO + H_2O$, proceeds at the surface of the metal oxide [5,6].

(3) Likewise, conductivity measurements permit the determination of the steady-state sulfur activity $a_S(st)$ in Ag_2S used as catalyst for the reaction $H_2 + 1/2S_2 = H_2S$ or $COS + H_2 = CO + H_2S$ [7].

(4) The steady-state oxygen activity $a_O(st)$ in a metal oxide may also be obtained with the help of measurements of the emf E of the cell:

Ni, NiO (reference electrode)	ZrO_2(+CaO)	Metal Oxide	Reacting Mixture
Pt lead		Pt lead	

where $ZrO_2(+O_2)$ is a solid electrolyte which exhibits predominantly ionic conductance due to the motion of oxygen ions via oxygen ion vacancies. Then one has according to Nernst:

$$a_O(st) = a_O(\text{reference electrode}) \cdot \exp \frac{2\,EF}{RT} \tag{2}$$

This procedure has not been used so far but may be helpful if the conductance of a metal oxide, such as Fe_3O_4, Co_3O_4, or CuO, is virtually independent of oxygen activity.

(5) Likewise, the steady-state sulfur activity a_S in Ag_2S used as catalyst for a sulfur transfer reaction, e.g., $COS + H_2 = CO + H_2S$, may be deduced from the measurement of the emf E of the cell [7]:

Ag	AgI	Ag_2S	Reacting Mixture
Pt lead		Pt lead	

where AgI is a solid electrolyte exhibiting virtually only cationic conduction. According to Nernst one may calculate the steady-state activity of silver $a_{Ag}(st)$ in Ag_2S as:

$$a_{Ag}(st) = \exp\,(-EF/RT) \tag{3}$$

Since Ag_2S shows only small deviations from the ideal stoichiometric composition, the product $a^2_{Ag} \cdot a_S$ is virtually constant. thus:

$$a_S(st) = a_S(Ag, Ag_2S) \cdot \exp\,(2\,EF/RT) \tag{4}$$

where $a_S(Ag,Ag_2S)$ is the sulfur activity in Ag_2S coexisting with metallic silver.

THE ACTIVITY OF ATOMIC OXYGEN DURING OXYGEN TRANSFER REACTIONS

1. The Decomposition of N_2O on NiO as Catalyst

During the catalytic decomposition of N_2O in the presence of a large excess of O_2 on NiO as catalyst at 715°C, the electrical conductivity of NiO has been found to be slightly higher than in oxygen alone [5]. In essence, the steady-state oxygen activity $a_O(st)$ on the surface of NiO is, therefore, determined by the equilibrium:

$$O_2(g) \rightleftarrows O_2(ad) \rightleftarrows 2O\ (ad) \qquad (5)$$

Since the electrical conductivity of NiO increases with increasing oxygen activity and the electrical conductivity during the decomposition of N_2O is higher than in oxygen alone, it follows that:

$$a_O(st) \gtrsim a_O\ [eq,\ reaction\ (5)] \qquad (6)$$

At $P(O_2) = 0.3$ atm, $p(N_2O) = 0.1$ atm, and 715°C, e.g., $a_O(st) \cong 1.017\ a_O(eq)$ has been observed. Although the deviation between $a_O(st)$ and $a_O(eq)$ is rather small, this deviation yields a decisive clue regarding the prevailing mechanism.

The rate of N_2O decomposition on metal oxides is in essence proportional to the partial pressure of N_2O. Thus one may assume that the decomposition of adsorbed N_2O molecules into N_2 and adsorbed oxygen atoms is the primary step,

$$N_2O(ad) \rightarrow N_2\ (ad) + O(ad) \qquad (7)$$

Since the following recombination of oxygen atoms to oxygen molecules has a finite rate, the steady-state concentration of adsorbed oxygen atoms is greater than the equilibrium concentration.

On the other hand, the primary step may be the reaction of an adsorbed N_2O molecule with an adsorbed oxygen atom:

$$N_2O(ad) + O(ad) \rightarrow N_2(ad) + O_2(ad) \qquad (8)$$

with a steady-state concentration of adsorbed oxygen atoms below the equilibrium concentration.

The experimental result stated in Eq. (6) indicates that predominantly reaction (7) rather than reaction (8) is the primary and rate-determining step.

2. The Oxidation of CO on Silver as Catalyst

During the oxidation of CO to CO_2 in the presence of a large excess of molecular oxygen on silver as a catalyst at 400°C, the oxygen activity $a_0(st)$ has been found to be slightly lower than in oxygen alone [8]:

$$a_0(st) \lesssim a_0 \text{ [eq, reaction (5)] } = p(O_2)^{1/2} \tag{9}$$

where values of $a_0(st)$ have been obtained from measurements of the rate of internal oxidation of silver containing 1 atomic percent Cd under conditions where diffusion of oxygen in the alloy is the rate determining step of the oxidation of cadmium dissolved in silver. Under these conditions, the rate of internal oxidation is inversely proportional to the diffusion path and proportional to the concentration of oxygen atoms in silver next to the surface, which in turn is proportional to the steady-state activity $a_0(st)$ in the surface.

In view of the experimental result stated in Eq. (9) it may be assumed that the reaction:

$$2\ CO + O_2 \rightarrow 2\ CO_2 \tag{10}$$

occurs by virtue of the steps:

$$O_2(g) \rightleftarrows O_2(ad) \rightleftarrows 2O(ad) \tag{11}$$

$$2\ \{CO(g) \rightleftarrows CO(ad)\} \tag{12}$$

$$2\ \{CO(ad) + O(ad) \rightarrow CO_2(ad) \rightleftarrows CO_2(g)\} \tag{13}$$

A quantitative analysis of rate data and values of $a_0(st)$ at various compositions of the reacting CO-O_2 mixture has revealed further details of the above scheme. It has also been shown that in addition to reaction (13) the reaction:

$$CO(ad) + O_2(ad) \rightarrow CO_2(ad) + O(ad) \tag{14}$$

takes place though only to a minor extent.

3. The Oxidation of SO_2 on Vanadium Oxide Catalysts

The conversion of SO_2 into SO_3:

$$2SO_2 + O_2 = 2\ SO_3 \tag{15}$$

may be catalyzed either by platinum, or by vanadium oxide catalysts, especially a melt of potassium sulfate with dissolved vanadium oxides on a support of alumina. In this melt the ratio of tetravalent to pentavalent vanadium as a measure of the oxygen activity in the catalyst has been determined after quenching by means of titration, first in SO_2-SO_3-O_2 mixtures under equilibrium conditions, and second in SO_2-SO_3-O_2 mixtures during the reaction 2 SO_2+O_2 = 2 SO_3 [9, 10]. The ratio of tetravalent to pentavalent vanadium in the melt during the reaction 2 SO_2+O_2 = 2 SO_3 has been found to be determined in essence by the SO_2/SO_3 ratio in the reacting mixture but to be virtually independent of the oxygen partial pressure. Thus the rate-determining step is the oxidation of tetravalent to pentavalent vanadium by molecular oxygen,whereas virtually equilibrium for the reactions effecting transfer to oxygen from the melt to SO_2 under formation of SO_3 is established. For further elucidation of the mechanism of the overall reaction (15), one has to explore the mechanism of the oxidation of tetravalent to pentavalent vanadium by molecular oxygen.

4. The Gasification of Graphite in CO_2-CO and H_2O-H_2 Mixtures

Investigations on the gasification of graphite in CO_2-CO mixtures [11-14] have led to the conclusion that the overall reaction:

$$C(s) + CO_2(g) + 2\,CO(g) \qquad (16)$$

takes place by virtue of the consecutive reactions:

$$CO_2(g) \rightleftarrows CO_2(ad) = CO(ad) + O(ad) \qquad (17)$$

$$O(ad) + C(s) = CO(ad) \qquad (18)$$

$$2\,\{CO(ad) = CO(g)\} \qquad (19)$$

The rate of reaction (17) has been determined independently [14] by measuring the rate of formation of ^{14}CO from $^{14}CO_2$ due to the reaction:

$$^{14}CO_2(ad) = {}^{14}CO(ad) + O(ad) \qquad (20)$$

followed by the reaction:

$$^{12}CO(ad) + O(ad) = {}^{12}CO_2(ad) \qquad (21)$$

At 1000°C and CO_2/CO ratios between 0.2 and 2, the rate of the overall reaction (16) has been found to be a function of the CO_2/CO ratio but to be independent of the total pressure P between 0.1 and 1 atm. In view of the results obtained for the rate of formation of ^{14}CO from $^{14}CO_2$, reaction (18) is the rate-determining step whereas virtually equilibrium for reaction (17) is established. Thus, for molecular oxygen of atmospheric pressure as the standard state, one has:

$$a_0(st) \cong K_{23}(p_{CO_2}/p_{CO}) \tag{22}$$

for $a_0 = 1$ in molecular oxygen of atmospheric pressure with K_{23} as the equilibrium constant of the reaction:

$$CO_2(g) = CO(g) + 1/2\ O_2(g) \tag{23}$$

For the gasification of graphite in H_2O mixtures:

$$C(s) + H_2O(g) = CO(g) + H_2(g) \tag{24}$$

one may assume an analogous mechanism involving the following steps [14]:

$$H_2O(g) \rightleftarrows H_2O(ad) = H_2(ad) + O(ad) \tag{25}$$

$$O(ad) + C(s) = CO(ad) \tag{18}$$

$$CO(ad) = CO(g) \tag{19}$$

$$H_2(ad) = H_2(g) \tag{26}$$

In reaction (25), hydrogen atoms and hydroxyl radicals occur eventually as intermediates. This modification, however, is immaterial if virtually equilibrium for reaction (25) is established and reaction (18) is the rate-determining step. Under these conditions, one has:

$$a_0 \cong K_{28}(p_{H_2O}/p_{H_2}) \tag{27}$$

where K_{28} is the equilibrium constant of the reaction:

$$H_2O(g) = H_2(g) + 1/2\ O_2(g) \tag{28}$$

Under conditions where the gasification of graphite in CO_2-CO mixtures and further the gasification of graphite in H_2O-H_2 mixtures is determined essentially by the rate of step (18), one has, therefore, to expect the same rate of reactions (16) and (23) if the oxygen activities according to Eqs. (22) and (27) are equal. This conclusion has been confirmed experimentally by Grabke [14].

THE ACTIVITY OF ATOMIC HYDROGEN DURING THE DECOMPOSITION OF FORMIC ACID

On palladium as catalyst at 180°C, formic acid decomposes mainly into carbon dioxide and hydrogen:

$$HCOOH \rightarrow CO_2 + H_2 \tag{29}$$

In principle, the decomposition may be a one-step process. Alternatively, adsorbed hydrogen atoms may be formed in consecutive steps followed by recombination to molecular hydrogen [3]:

$$HCOOH(g) \rightleftarrows HCOOH(ad) \tag{30}$$

$$HCOOH(ad) \rightleftarrows HCOO(ad) + H(ad) \tag{31}$$

$$HCOO(ad) \rightarrow CO_2(ad) + H(ad) \tag{32}$$

$$2\ H(ad) \rightleftarrows H_2(ad) \tag{33}$$

$$H_2(ad) \rightleftarrows H_2(g) \tag{34}$$

$$CO_2(ad) \rightleftarrows CO_2(g) \tag{35}$$

With the help of measurements of the electrical resistance of a Pd foil used as catalyst in $HCOOH$-H_2 mixtures with a large excess of hydrogen, it has been shown that equilibrium for reactions (33) and (34) is nearly established. Actually, the steady-state hydrogen concentration in the catalyst is slightly higher than the equilibrium concentration [3], i.e.:

$$a_H(st) \gtrsim a_H\ [eq,\ reactions\ (33)\ and\ (34)] \tag{36}$$

Consequently, adsorbed hydrogen atoms are formed at least in part when HCOOH decomposes. Then one may ask to what extent H_2 molecules are formed in a single step according to Eq. (29) and to what extent the consecutive reactions (30) to (35) are instrumental. This problem may be solved by an independent measurement of the exchange rate of reaction (33) defined as the rate of the forward or the backward rate of reaction (33) under equilibrium conditions where the net rate of Eq. (33) vanishes. For this purpose, one may conduct relaxation measurements, i.e., one may measure the electrical resistance of the palladium foil as a function of time upon a sudden quasi-differential increase of the hydrogen partial pressure in the $HCOOH$-H_2 mixture. The corresponding increase of the concentration of hydrogen atoms in the bulk catalyst and accordingly the increase in the electrical resistance is delayed because of the finite rate of the phase boundary reaction (33) and diffusion of hydrogen atoms in palladium. In the case of a sufficiently thin

Pd foil, the delay is due exclusively to the phase boundary reaction (33).

A quantitative evaluation of the experimental value $[a_H(st)-a_H(eq)]$ and the exchange rate of reaction (33) leads to the conclusion that at HCOOH partial pressure lower than 0.02 atm two hydrogen atoms are formed for each HCOOH molecule decomposed [3]. Thus, in the terminology suggested by Horiuti [15] the stoichiometric number of reactions (31) and (32) is practically equal to one. Consequently, under the conditions stated above, the decomposition of formic acid occurs predominantly by virtue of reactions (30) to (35).

THE OCCURRENCE OF DIFFERENT PHASES OF METAL-OXYGEN SYSTEMS DURING THE CATALYSIS OF THE REACTION $2\ H_2 + O_2 = 2\ H_2O$

When a H_2-O_2 mixture is passed over copper at 250°C, the catalytic reaction

$$2\ H_2 + O_2 = 2\ H_2O \tag{37}$$

takes place. The surface of the solid may be either metallic copper, Cu_2O, or CuO. Occurrence of different phases is readily indicated by visual observations. In hydrogen involving 0.2% O_2 the catalyst has a bright appearance characteristic of metallic copper, whereas in the presence of 2% O_2 the catalyst is discolored because some oxide phase, probably Cu_2O, has been formed. The bright color is restored by subsequent treatment in hydrogen involving 0.2% O_2. According to available thermodynamic data for 250°C, Cu is stable if $a_O \leq 3.5 \cdot 10^{-14}$ atm$^{1/2}$, Cu_2O is stable if $3.5 \cdot 10^{-14} < a_O < 1.4 \cdot 10^{-9}$ atm$^{1/2}$. In the presence of 0.2% O_2 with copper as the stable phase, $a_O(st)$ is, therefore, much lower than the oxygen activity $a_O = 4.5 \cdot 10^{-2}$ atm$^{1/2}$ for equilibrium with oxygen molecules. On the other hand, in the presence of 2% O_2 with Cu_2O as the stable phase, $a_O(st)$ is much greater than the oxygen activity of the H_2O-H_2 couple, which is $a_O = 1.6 \cdot 10^{-25}$ atm$^{1/2}$ at $p(H_2O/p(H_2) = 10^{-3}$.

Differences in the catalytic activity of the various phases of a metal-oxygen system such as NiO or CuO have important consequences on the performance of a flow reactor, especially the concentration profile under steady-state conditions. The following typical situations are envisaged [16].

(1) The phase which is stable in the section next to the entrance of a flow reactor may have a high catalytic activity, whereas the phase which is stable further down-stream at a different composition of the gas stream may have a low catalytic activity. Thus a long terminal section of the flow reactor is needed for practical completion of the reaction.

(2) Conversely, the phase which is stable in the section next to the entrance may have a low catalytic activity, whereas the phase which is stable further down-stream may have a high catalytic activity. In this case the gradient of the gas composition in the first section of a flow reactor is rather low. Once the gas stream has reached the critical composition for transition to the catalytically more active phase, practical completion of the reaction may be accomplished over a relatively short length of the terminal section of a flow reactor.

REFERENCES

1. T. Takaishi, Z. Naturforschg 11a, 297 (1956).
2. C. Wagner and K Hauffe, Z. Elektrochem. 45, 309 (1938).
3. H. J. Schönnagel and C. Wagner, Ber. Bunsenges. physik Chem. 69, 699 (1965).
4. F. A. Kroeger, *The Chemistry of Imperfect Crystals*, (North-Holland Publishing Company, Amsterdam, 1964).
5. C. Wagner and K. Hauffe, Z. Elektrochem. 44, 172 (1938).
6. S. Stotz, Ber. Bunsenges. physik. Chem. 70, 37 (1966).
7. E. Bechtold, Ber. Bunsenges. physik. Chem. 69, 328 (1965).
8. L. Imre, Ber. Bunsenges. physik. Chem. 72, 863 (1968).
9. P. Mars. and J. G. H. Maessen, *Proceedings of International Congress on Catalysis*, Amsterdam, 1, 266 (1964).
10. P. Mars and J. G. H. Maessen, J. Catalysis 10, 1 (1968).
11. P. L. Walker, F. Rusinko, and L. G. Austin, *Advances in Catalysis*, 11, 134 (1959).
12. S. Ergun, Bull. Bureau of Mines 589 (U.S. Government Printing Office, Washington, D.C., 1952).
13. S. Ergun, J. Phys. Chem. 60, 480 (1956).
14. H. J.Grabke, Ber. Bunsenges. physik. Chem. 70, 664 (1966).
15. J. Horiuti, J. Research Institute of Catalysis, Hokaido Univ. 1, 8 (1948.
    J. Horiuti and T. Nakamura, Z. physik. Chem., N.F. 11, 358 (1957).
    J.Horiuti, J. Catalysis 1, 199 (1962).
16. C. Wagner, to be published.

DISCUSSION OF PAPER

T. B. King. I am interested in Dr. Wagner's experiment on the catalytic effect of copper on the reaction:

$$H_2(g) + O_2(g) = H_2O(g) \qquad (1)$$

Dr.Johannes Picaird and I did some work two years ago on the low temperature oxidation of iron in H_2-O_2 gas mixtures. It appeared that iron would oxidize at low temperatures (up to 500°C) in H_2-O_2 mixtures for which the calculated pH_2/pH_2O (assuming completion of reaction 1) was higher than the equilibrium value for the formation of FeO. We have thought the effect to be due to competition for oxygen between reaction 1 above (as catalyzed by Fe) and the reaction:

$$Fe(s) \quad + \quad 1/2O_2 \quad = \quad FeO(s)$$

Pre-reacting of the H_2, O_2 mixtures on a platinum catalyst prevented oxidation of the iron.

C. Wagner. The behavior of iron in H_2-O_2 mixtures is also being investigated at present. At 250°C, iron in H_2 with 0.2% O_2 remains bright, whereas iron in H_2 with 2% O_2 is oxidized presumably to Fe_3O_4. Once a layer of Fe_3O_4 has been formed, adsorbed oxygen atoms react in part with hydrogen to form water vapor and in part with iron ions and electrons, diffusing across the Fe_3O_4 layer to form more Fe_3O_4.

DISSOCIATION OF GASEOUS MOLECULES ON SOLID SURFACES AT ELEVATED TEMPERATURES

W. L.Worrell and H. I. Kaplan

School of Metallurgy and Materials Science

University of Pennsylvania, Philadelphia, Pennsylvania

Resistivity measurements using thin foils is one of the most promising experimental techniques for measuring the rates of dissociation of gaseous molecules on solid surfaces at elevated temperatures. Recent studies using this technique are summarized. Measurements of the rate of dissociation of H_2S *on an iron-sulfide (FeS) surface at* $670^{\circ}C$ *are used to illustrate the details of the experimental technique. Two different approaches are used to derive similar rate equations for the dissociation of* H_2S*. The assumptions involved in each approach are discussed in detail. Initial experimental results, which are interpreted in terms of the derived rate-equations, indicate that the rate of dissociation of* H_2S *on a FeS surface may depend on the activity of sulfur at the surface. However, more experimental data over a wider range of sulfur activities are necessary to establish the exact form of the rate-equation.*

INTRODUCTION

One of the most useful concepts in high-temperature kinetic studies is that of the rate-controlling step. A chemical reaction can be divided into individual steps or paths. The slowest step which determines the overall rate is called the rate-controlling step.

In many high-temperature reactions, diffusion of the reactant or product in a condensed phase controls the rate of the overall reaction. However, there are also examples in which the overall reaction rate is influenced or controlled by the gas-solid surface.

In these cases one must determine the rate and the mechanism of the reaction at the surface. During the past five years, C. Wagner and his associates have developed several experimental techniques for investigating gas-solid surface reactions [1]. One of the most promising techniques is the relaxation method using thin foils [2].

In the relaxation method, one measures the resistance change of a thin foil with time. The rate at which the resistance changes is a measure of the rate of the gas-solid reaction occurring at the surface of the foil. For a valid measurement, there must be no contribution from either diffusion in the foil or mass-transport in the gas phase. For 5 to 50 microns (μ) thick foils, the contribution of solid-state diffusion to the overall reaction rate is usually negligible. Mass transport in the gas phase is minimized by using high velocity flow rates of gas mixtures across the foil.

The relaxation method has been used to investigate a variety of high-temperature gas-solid surface reactions [1, 2, 3, 4, 5, 6, 7]. It is the purpose of this paper to review the published literature and to report our use of this technique to measure the rate of dissociation of $H_2S(g)$ on FeS.

LITERATURE REVIEW

Using resistance changes to measure rates of chemical reactions was proposed in the 1930's by Doehlemann [8] and Wagner [9]. However, reproducible, quantitative results using this method have only recently appeared in the literature. These recent studies use high-purity metal foils and the highly sensitive four- probe direct-current technique for the resistance measurements.

There are six published investigations using the relaxation technique [2,3,4,5,6, 7]. Two of these are summarized below. One is a study of the wüstite-water-gas reaction; the other is a measurement of the rate of nitridation and denitridation of iron in NH_3/H_2 gas mixtures.

Stotz [1] has measured the transfer rate of oxygen in CO_2/CO-H_2/H_2O gas mixtures on the surface of FeO at 900°C. He first showed that the resistance of the wüstite (FeO) thin foils (15 μ thick) is a reproducible function of the oxygen activity of the FeO. The following reactions on the FeO surface were studied:

$$CO_2\ (g) \rightarrow CO\ (g) + O\ (ad) \qquad (1)$$

$$H_2\ (g) + O\ (ad) \rightarrow H_2O\ (g) \qquad (2)$$

where (ad) indicates that oxygen is adsorbed on the surface of FeO.

Combining reactions (1) and (2) gives the water gas reaction:

$$CO_2 \text{ (g)} + H_2 \text{ (g)} \rightleftarrows CO \text{ (g)} + H_2O \text{ (g)} \qquad (3)$$

The rates of reactions (1), (2) and (3) are symbolized by v_1, v_2 and v_3 respectively. The rate of reaction (1) was obtained by measuring the resistance change of the wüstite foil after the CO_2/CO gas ratio was changed. The rate of reaction (2) was determined using a psychometric method.

At the steady-state oxygen activity in and on the wüstite foil, the amount of oxygen supplied by reaction (1) must equal that removed by reaction (2). Thus at the steady-state oxygen content, the rates v_1 and v_2 are equal. Agreement between the measured oxygen content of the wüstite foil and the values determined by equating the measured rates of reactions (1) and (2) was very good.

Stotz [1] also showed that the rate of oxygen supplied to FeO by reaction (1) decreases with increasing oxygen activity. Fig. 1 shows the variation of the rate constant for reaction (1) with activity of FeO. This rate constant is directly proportional to a_O^m; m, which is the slope of the line in Fig. 1, varies from -0.5 to -0.7. These values agree with the results of Grabke [10] who measured the dissociation rate of CO_2 on FeO using radioactive carbon in the CO_2.

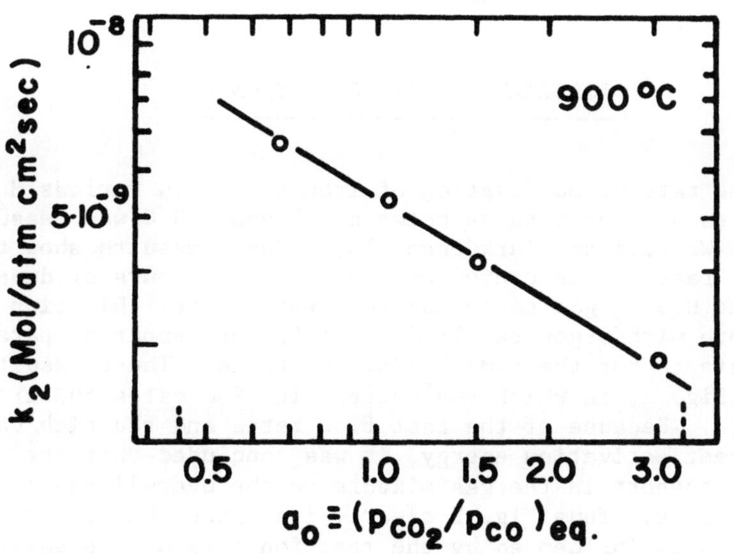

Fig. 1. The rate constant for equation (1), k_1, plotted versus oxygen activity of FeO (after Stotz) [1].

Grabke [2,3] has used the relaxation method to measure the rates of nitridation and denitridation of iron in various mixtures of NH_3/H_2. Using thin foils (5μ thick) of iron, the following reaction was studied at temperatures between 300 and 550°C:

$$NH_3\ (g) \rightarrow 3/2\ H_2\ (g) \ + \ \underline{N}\ \ (\text{in iron}) \qquad (4)$$

The measured resistance was directly proportional to the concentration of nitrogen in the iron. The measured rates of reaction (4) obeyed the following equation:

$$v_4 = \left[k_r' \ p_{NH_3}/p_{H_2}\right]^{\left(\frac{3-r}{2}\right)} - k_r[\underline{N}] \ \ (p_{NH_2})^{r/2} \qquad (5)$$

where v_4 is the rate of reaction (4) in moles nitrogen/cm^2 sec, k_r' is the forward rate constant, k_r is the rate constant for the reverse reaction, and r is either 2 or 3 depending on whether the decomposition of NH_2 (ad) or NH_3 (ad) is rate controlling.

Grabke [5] also has measured the rate of carburization and of decarburization of iron foils in CH_4/H_2 atmospheres at temperatures between 800 and 1040°C. Additional studies include measurements of the dissociation rate of CO_2 on various oxides [6, 7] and the dissociation rate of H_2S on metallic surfaces [11].

DISSOCIATION RATE OF H_2S ON FeS

The rate of sulfidation of iron to FeS in various H_2S/H_2 atmospheres at temperatures between 670 and 900°C was measured recently by Worrell and Turkdogan [12]. Their results show that the parabolic rate law is obeyed only after a few hours or days depending on the H_2S/H_2 gas ratio and the temperature. Dilution of the gas mixture with argon resulted in different "apparent parabolic" rate constants for the same H_2S/H_2 gas ratio. These results are shown in Fig. 2, in which the dotted line indicates the true parabolic rate. Because of the fast flow rates and the high value of the apparent activation energy, it was concluded that the influence of mass transport in the gas mixture on the overall reaction rate was negligible. Thus Fig. 2 clearly indicates that the rate of sulfidation is influenced by the reaction rate at the gas-sulfide surface.

In the analysis of these results, an instantaneous steady-state between the diffusion of iron in the sulfide layer and the rate of sulfur transfer from the gas to the sulfide was assumed. Using this assumption and the previously measured data for the diffusion of iron in FeS [13], values for the time dependent sulfur activity at the gas-sulfide surface were calculated. Reasonable agreement was obtained between these calculated values of a_S at the surface and the derived rate equation for the dissociation of H_2S on the sulfide surface.

However, it is clearly worthwhile to measure directly the dissociation rate of H_2S on the FeS surface in the absence of solid-state diffusion in the sulfide. Thus we are measuring the dissociation rate of H_2S on a thin foil of FeS using the relaxation method. Our initial results will be used to illustrate not only the details of the experimental method but also its use in determining the correct mechanism for the surface reaction.

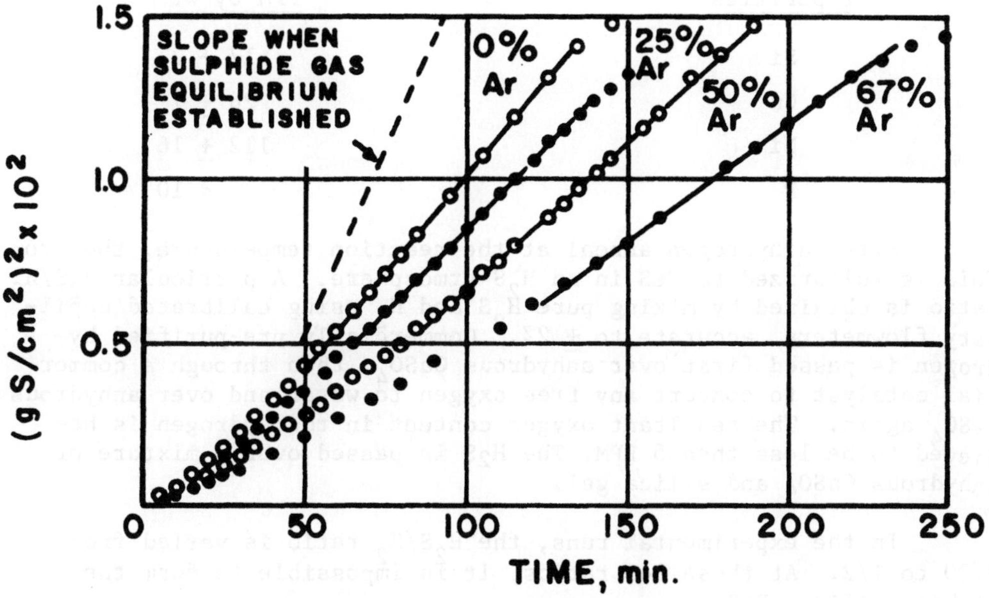

Fig. 2. Plot showing unsuitability of the weight gain method for determination of parabolic growth rate of iron sulfide on iron in hydrogen sulfide-hydrogen-argon gas mixtures, with pH_2S/pH_2 = 1.0 at 800°C. (After Worrell and Turkdogan) [12].

EXPERIMENTAL METHOD

The experimental equipment has been described previously [14]. A silicon-carbide tubular element furnace with a two-inch constant temperature zone is used. The temperature control is $\pm 1^{\circ}C$. The temperature is measured using a Pt, Pt-10% Rh thermocouple, which is placed within 1/4 inch of the foil.

The resistance of each foil is measured using a direct-current potential-probe technique with four gold leads spot-welded to each specimen. The FeS foils are prepared from zone-refined polycrystalline iron foils. The impurity content of the iron is considerably above that specified by the supplier. The analyzed impurities are given in Table I; the impurity content of all analyzed samples is within the uncertainty limits shown in Table I.

Table I

Analysis of Impurities in Iron Foil

Impurities	PPM by Wt
Si	17 ± 2
Cu	26 ± 7
Ni	112 ± 16
P	< 10

After a hydrogen anneal at the reaction temperature, the iron foil is sulfurized to FeS in an H_2S atmosphere. A particular H_2S/H_2 ratio is obtained by mixing pure H_2S and H_2 using calibrated capillary flowmeters accurate to $\pm$ 2%. Commercially pre-purified hydrogen is passed first over anhydrous $CaSO_4$, then through a commercial catalyst to convert any free oxygen to water and over anhydrous $CaSO_4$ again. The resultant oxygen content in the hydrogen is believed to be less than 5 PPM. The H_2S is passed over a mixture of anhydrous $CaSO_4$ and silica gel.

In the experimental runs, the H_2S/H_2 ratio is varied from 1/20 to 1/2. At these low ratios, it is impossible to form the higher sulfide, FeS_2.

The experimental gas flow rates at $670^{\circ}C$ are 10-13 cm/sec across a FeS foil of approximately 2 sq. cm. total surface area. The effect of any mass transport in the gas mixture should be negligible because an extremely small amount of sulfur is transferred from the H_2S to the sulfide. Under our experimental conditions, only 0.006 to 0.01 at.% of the H_2S in the flowing gas reacts with the sulfide before the new equilibrium composition is reached.

The thickness of the FeS foil is 2.7 times that of the initial iron foil. This number is calculated in the usual manner; an average composition of FeS is assumed. Our typical FeS foil is approximately 50μ thick. Using the data for diffusion of iron in FeS [13], diffusion equilibrium is calculated to occur in less than five seconds in a 50μ thick FeS foil.

In an experimental run, a calibration line relating the resistance of the FeS foil to the equilibrium sulfur activity (a_S) is established. The measured calibration line at 670°C is shown in Fig. 3. In Fig. 3, the standard state for sulfur ($a_S \equiv 1$) is chosen

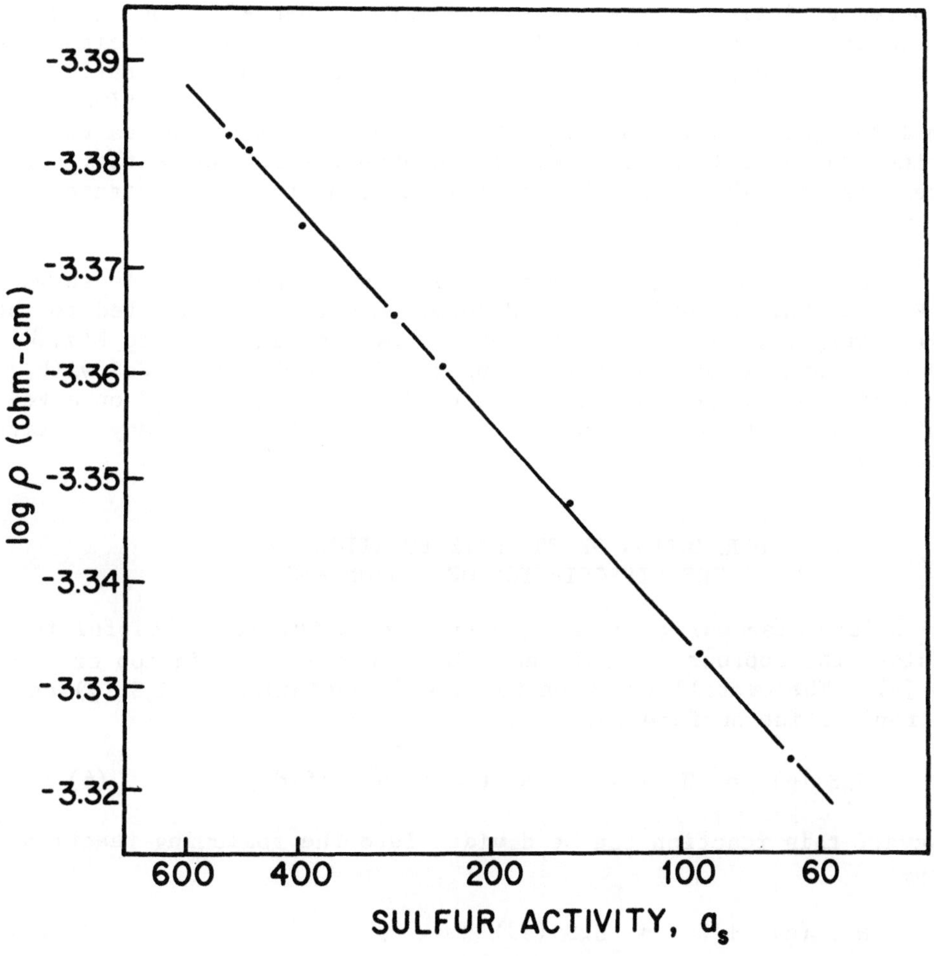

Fig. 3. Experimental results for the relationship between log of resistance and equilibrium sulfur activity at 670°C for an iron sulfide foil approximately 50μ thick.

to be the iron-iron sulfide equilibrium. Thus FeS is the stable phase when $a_S > 1$.

The equilibrium constant for the FeS $\rightleftarrows$ 1/2 S_2 (g) + Fe reaction at 670°C is 2.85×10^{-6} when the standard state for sulfur is diatomic sulfur gas at 1 atm. This constant can be used to convert the sulfur activities in Fig. 3 back to the usual ones.

After the calibration line is determined, the sulfide foil is equilibrated at a particular H_2S/H_2 ratio. A new ratio is suddenly introduced by changing the liquid level in the hydrogen sulfide capillary flowmeter. The resulting change in resistance is measured until the sulfide foil equilibrates with the new H_2S/H_2 ratio. The expected variation of the resistance with a change in the H_2S/H_2 gas ratio is shown schematically in Fig. 4. It should be emphasized that it takes approximately thirty seconds for a new gas ratio to be extablished at the gas-sulfide surface. At 670°C the time for complete equilibration between the FeS foil and the new gas ratio varies from 5 to 10 minutes depending upon the new gas ratio. No rate data are taken from the initial portion of the resistance versus time curves.

One of the measured resistance versus time curves is shown in Fig. 5. This experimentally determined curve is converted to the sulfur-weight gain versus time curve shown in Fig. 6 using Fig. 3 and the thermodynamic data relating the iron deficit in FeS with the sulfur activity [13]. The rate of dissociation of H_2S on a FeS surface at 670°C is the slope of the curve in Fig. 6 at any particular time.

DERIVATION OF THE RATE EQUATION FOR THE DISSOCIATION OF H_2S ON FeS

Before discussing our experimental results, it is helpful to consider the appropriate rate-equation for the dissociation of H_2S (g). The overall reaction for the dissociation of H_2S (g) on an iron sulfide surface is:

$$H_2S \text{ (g)} = H_2 \text{ (g)} + \underline{S} \text{ (in iron sulfide)} \qquad (6)$$

However, this reaction can be divided into the following reaction steps:

$$H_2S \text{ (g)} + \square = H_2S \text{ (ad)} \qquad (7.1)$$

$$H_2S \text{ (ad)} + \square = HS \text{ (ad)} + H \text{ (ad)} \qquad (7.2)$$

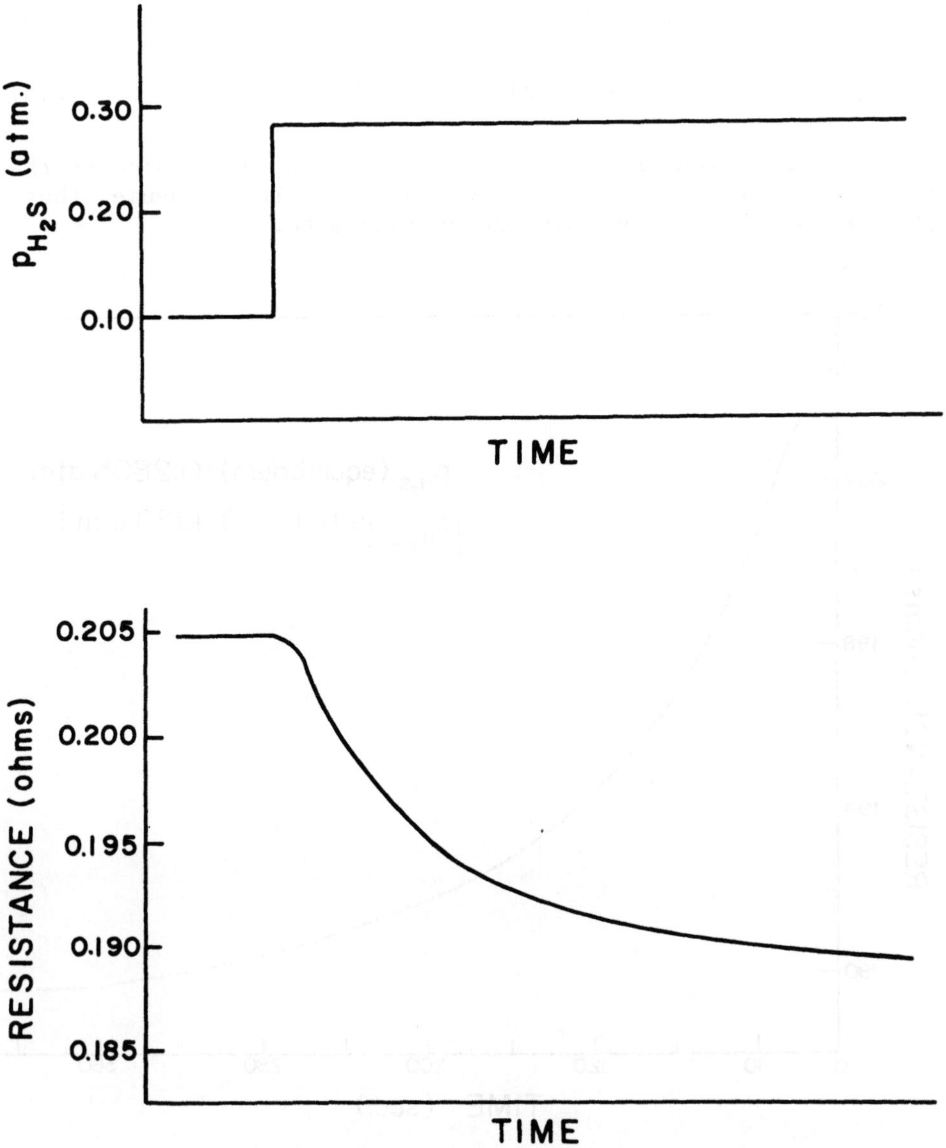

Fig. 4. Schematic representation of the change in resistance of an iron sulfide sample after a change in gas composition.
Top: gas composition versus time.
Bottom: resistance versus time.

$$HS\ (ad) + \square = H\ (ad) + S\ (ad) \tag{7.3}$$

$$S\ (ad) = \underline{S} + \square \tag{7.4}$$

$$2H\ (ad) = H_2\ (g) + 2\ \square \tag{7.5}$$

where $\square$ is a single vacant surface site and (ad) indicates that the species is adsorbed on the FeS surface. It is assumed that all adsorbed species occupy only one surface site.

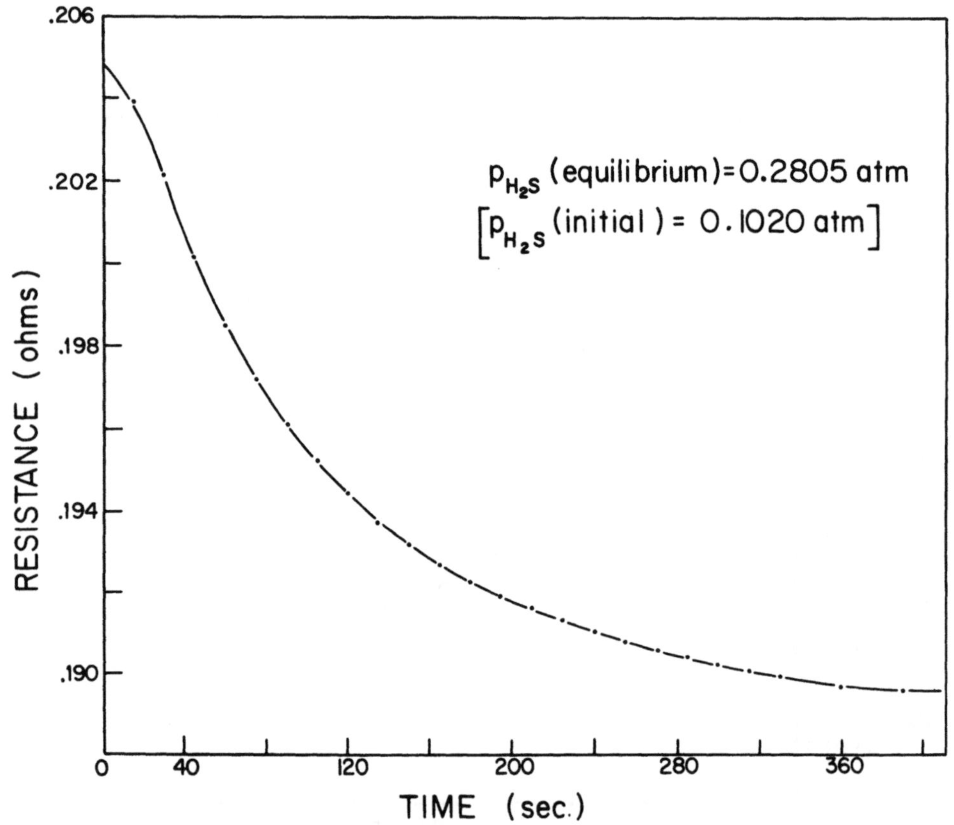

Fig. 5. Experimental results for resistance versus time at 670°C after the pressure of H_2S in the gas is changed from 0.102 atm. to 0.2805 atm. (Total pressure is one atmosphere).

In general, one of the reaction steps is the slowest step and is the rate-controlling step in the overall reaction. The adsorption and desorption of gaseous molecules at elevated temperatures is usually quite rapid; thus (7.1) and (7.5) should be in equilibrium. In thin foils of FeS one also would expect reaction step (7.4) to equilibrate quickly. Thus the overall rate of reaction (6) is determined by the rate of dissociation of either H_2S (ad) or HS (ad). A rate-equation will be derived first assuming that step (7.2) is rate-controlling. Then the rate-equation considering step (7.3) as the slowest step will be derived.

Worrell and Turkdogan [12] have showed that the activated-complex approach leads to the following rate-equation for the forward reaction of step (7.2):

$$dw/dt = q\ P_{H_2S}\ \theta_{\square} \qquad (8)$$

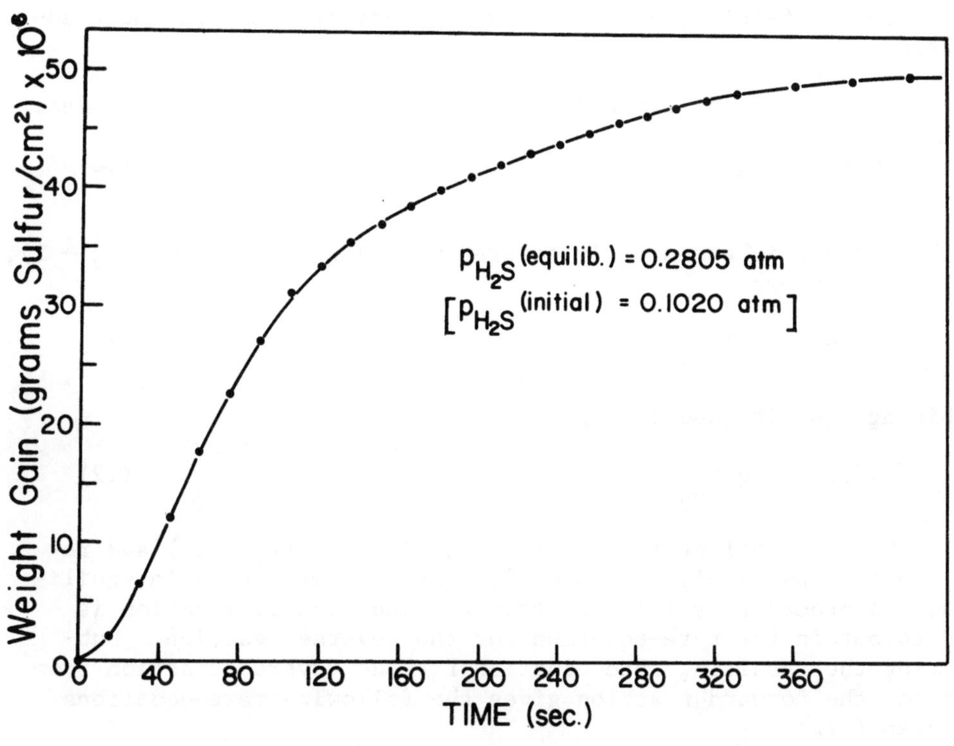

Fig. 6. Calculated weight gain of an iron sulfide sample plotted versus time from the experimental results presented in Fig. 5. (The calculations were made using the calibration curve in Fig.3 and the available thermodynamic data relating the concentration of iron vacancies in FeS to an equilibrium sulfur activity [13].

where dw/dt is the rate in units of gm S/min, q is a constant at constant temperature, $P(H_2S)$ is the partial pressure of H_2S in the gas mixture, and $\theta_\square$ is the fraction of vacant single sites on the FeS surface.

Turkdogan and Grieveson [16] have shown that with certain assumptions:

$$a_i = k_i \theta_i/\theta_\square \tag{9}$$

where a_i is the activity of the adsorbed species i, k_i is a temperature dependent proportionality constant, and θ_i is the fraction of single surface sites occupied by species i. In the derivation of Eq. (9), a constant heat of adsorption is assumed, and there are no preferred surface sites for adsorption. Eq. (9) is considered valid for three limiting cases: (1) When there is an essentially bare surface, ($\theta_\square \rightarrow 1$), (2) when one adsorbed species essentially covers the surface ($\theta_i \rightarrow 1$), (3) when there is ideal Langmuir adsorption.

For an essentially bare surface ($\theta_\square \rightarrow 1$), Eq. (8) becomes:

$$dw/dt = q P_{H_2S} \tag{10}$$

If the surface is essentially covered with adsorbed sulfur ($\theta_S \rightarrow 1$), Eq. (9) becomes:

$$a_S = k_s 1/\theta_\square \tag{11}$$

Combining Eqs. (8) and (11) gives:

$$dw/dt = q P_{H_2S}/a_S \tag{12}$$

We now consider the reverse reaction of step (7.2) and recall that steps (7.3), (7.4) and (7.5) are assumed to be in equilibrium. A procedure similar to that for the forward reaction is used to obtain the rate-equation for the reverse reaction. Subtracting the resulting rate-equation for the reverse reaction from that for the forward reaction gives the following rate-equations for step (7.2):

$$\text{If } \theta_\square \rightarrow 1 \quad dw/dt = q\left[1 - \frac{a_S}{(a_S)_e}\right] P_{H_2S} \tag{13}$$

$$\text{If } \theta_S \rightarrow 1 \quad dw/dt = q\left[\frac{1}{a_S} - \frac{1}{(a_S)_e}\right] P_{H_2S} \tag{14}$$

where $(a_S)_e$ is the sulfur activity in equilibrium with the experimental H_2S/H_2 gas ratio. The activity of sulfur at the surface is the same as that in the gas mixture $[a_S = (a_S)_e]$ when dw/dt is zero.

If the experimental rate data follow Eq. (13), one would conclude that there are essentially no adsorbed molecules on the surface of FeS ($\theta_{\square} \rightarrow 1$). One also would conclude that H_2S is adsorbed on the FeS surface as a neutral molecule (see step 7.1) and that its rate of dissociation is unaffected by the number of atomic and electronic defects in FeS.

If Eq. (14) adequately describes the experimental rate data, one could conclude that the FeS surface is essentially covered with sulfur ($\theta_S \rightarrow 1$) and that the few adsorbed H_2S molecules are neutrally charged. However, Eq.(14) can be obtained under different assumptions concerning the adsorption of H_2S.

Non-stoichiometric FeS can be pictured as containing both Fe^{++} and Fe^{+++} cations. Thus it is possible for electrons to be supplied to any adsorbed species by:

$$Fe^{++} \rightleftarrows Fe^{+++} + e^- \qquad (15)$$

For FeS one could rewrite adsorption step 7.1 as:

$$H_2S\ (g) + \square + m\ e^- \rightleftarrows H_2S^{-m}\ (ad) \qquad (16.1)$$

where m is the number of electrons transferred to an H_2S adsorbed molecule. At elevated temperatures, step (16.1) should be in equilibrium; thus:

$$a_{H_2S^{-m}} \ \alpha\ a_{\square}\ (a_{e^-})^m\ P_{H_2S} \qquad (17)$$

where $a(H_2S)^{-m}$ is the activity of the adsorbed H_2S^{-m} molecule, $a_{\square}$ is the activity of the vacant single surface sites, and a_{e^-} is the activity of the electrons in FeS.

Now we may consider that the dissociation of H_2S^{-m} is the rate-controlling step.

$$H_2S^{-m}\ (ad) + \square = HS^{-m}\ (ad) + H\ (ad) \qquad (16.2a)$$

Alternatively, we could postulate that the adsorbed H_2S molecule is neutrally charged but that it requires m electrons to dissociate. Thus we would rewrite step (16.2a) as:

$$H_2S\ (ad) + \square + m\,e^- = HS^{-m}\ (ad) + H\ (ad) \qquad (16.2b)$$

The derived rate-equation is the same whether we consider step (16.2a) or step (16.2b) as the rate-controlling step. In step (16.2a) we picture the adsorbed H_2S molecule as negatively charged; in step (16.2b) the neutrally adsorbed H_2S molecule must combine with electrons before it dissociates.

It should be noted that the concept of a negatively adsorbed molecule has been suggested in the literature. For example, oxygen has been reported to adsorb on solid silver as O_2^- [17]. Grabke [6] also has suggested that the rate of dissociation of CO_2 on an FeO surface could be controlled by a reaction similar to that of step (16.2).

The forward reaction rate for step (16.2a) may be written as:

$$dw/dt = k'\ a_{H_2S^{-m}}\ a_{\square} \qquad (18)$$

At this point it is helpful to compare the forward rate equation for the dissociation of neutral H_2S (m = o) obtained from Eq. (18) with Eq. (8) which was obtained using the activated - complex approach. With m = o and step 7.1 in equilibrium, the forward rate is proportional to $P(H_2S)\ a_{\square}^2$ using Eq. (18). The $a_{\square}^2$ term arises because we have assumed that each species (H_2S, HS, H) occupies one surface site. However in Eq. (8), the forward rate is proportional to $\theta_{\square}$ and not $\theta_{\square}^2$. In obtaining Eq. (8) it was assumed implicitly that the dissociation of the activated complex ($H_2S^{\ddagger}$) requires only one surface site. This could be interpreted to indicate that after the dissociation of H_2S, only HS is chemisorbed on the surface and that H is physically adsorbed on the chemisorbed layer. If it had been assumed that the dissociation of H_2S requires two surface sites, then Eq. (8) would have had a $\theta_{\square}^2$ term and would have been identical to that obtained from Eq. (18).

Combining Eqs. (18) and (17), we obtain:

$$dw/dt = k\ a_{\square}^2\ (a_{e^-})^m\ P_{H_2S} \qquad (19)$$

where k' and k are proportionality constants. Assuming that there is very little adsorption on the surface of iron sulfide ($\theta_{\square} \to 1$), then $a_{\square} \approx 1$.

The equilibrium between the adsorbed sulfur on the surface

and the internal defect structure of FeS may be written using Kröger's notation [18] as:

$$S\,(ad) + 2\,e^- = V_{Fe}'' + S_s \qquad (20)$$

where V_{Fe}'' is an iron cationic vacancy, and S_s is a sulfur anion in the lattice. It can be assumed that the activity of the sulfur anion is constant.

The concentration of iron vacancies in FeS $[V_{Fe}'']$ is a function of the equilibrium sulfur activity [13, 19]:

$$[V_{Fe}''] \;\alpha\; a_S^n \qquad (21)$$

Assuming that the activity of the iron vacancies equals $[V_{Fe}'']$, one obtains from reaction (20) that:

$$(a_{e^-})^2 \;\alpha\; \frac{[V_{Fe}'']}{a_S} \;\alpha\; a_S^{\,n-1} \qquad (22)$$

where a_S is the activity of sulfur adsorbed at the FeS surface. For thin foils, step 7.4 is in equilibrium, and the activity of sulfur at the surface and in the interior of FeS is the same.

Recalling that $a_\square \approx 1$ and combining Eqs. (22) and (19), we obtain for the forward reaction of step (16.2):

$$dw/dt = k\; P_{H_2S}\; a_S^{\frac{(n-1)\,m}{2}} \qquad (23)$$

Considering the reverse reaction for step (16.2), we obtain the following rate-equation for the overall reaction:

$$dw/dt = k \left[1 - \frac{a_S}{(a_S)_e} \right] a_S^{\frac{(n-1)\,m}{2}} \; P_{H_2S} \qquad (24)$$

The assumptions used in deriving Eq. (24) are that the FeS surface is essentially bare $(\theta_\square \to 1)$, that the activity of iron vacancies in FeS is equal to their concentration, and that m electrons are required for the dissociation of the adsorbed H_2S molecule.

In our experimental range of sulfur activities at 670°C, we calculate that $n \approx 1/5$ using the equilibrium data of Turkdogan [13]. If we assume that two electrons are required for the dissociation of the adsorbed H_2S molecule, we obtain from Eq. (24) :

$$dw/dt = k \left[1 - \frac{a_S}{(a_S)_e} \right] a_S^{-4/5} \quad P_{H_2S} \tag{25}$$

Eq. (14) can be written as:

$$dw/dt = q \left[1 - \frac{a_S}{(a_S)_e} \right] a_S^{-1} \quad P_{H_2S} \tag{14}$$

Eq. (14) and Eq. (25) are practically identical. Indeed it is doubtful whether one could ever experimentally differentiate between an $a_S^{-4/5}$ dependence and an a_S^{-1} dependence in the measured rate data at constant H_2S (g) pressure. Thus if our measured rate data follow Eq. (14), we could conclude that the FeS surface is essentially covered with sulfur ($\theta_S \rightarrow 1$). In accepting this interpretation, one must accept the assumptions used in obtaining Eq. (9). Alternatively, we could conclude that the FeS surface is essentially bare ($\theta_\square \rightarrow 1$) and that two electrons are required for the dissociation of H_2S.

Up to this point we have considered the dissociation of H_2S (ad) as the rate-controlling step. Now if we assume that the dissociation of HS (ad) is rate-controlling, we obtain the following rate-equations for step (7.3):

$$\text{If } \theta_\square \rightarrow 1 \qquad dw/dt = q \left[1 - \frac{a_S}{(a_S)_e} \right] \frac{P_{H_2S}}{(P_{H_2})^{1/2}} \tag{26}$$

$$\text{If } \theta_s \rightarrow 1 \qquad dw/dt = q \left[\frac{1}{a_S} - \frac{1}{(a_S)_e} \right] \frac{P_{H_2S}}{(P_{H_2})^{1/2}} \tag{27}$$

If $\theta_\square \rightarrow 1$ and m electrons are necessary for the dissociation of HS (ad),we obtain:

$$dw/dt = k \left[1 - \frac{a_S}{(a_S)_e} \right] a_S^{\frac{(n-1)\ m}{2}} \frac{P_{H_2S}}{(P_{H_2})^{1/2}} \tag{28}$$

The assumptions involved in obtaining Eqs. (13), (14) and (24) also apply to Eqs. (26), (27) and (28) respectively; indeed the only difference in the derived rate-equation when reaction step 7.3 is the rate controlling step, is the dependence on $1/(P_{H_2})^{1/2}$. Otherwise

Eqs. (26),(27) and (28) are identical to Eqs. (13), (14) and (24) respectively.

EXPERIMENTAL RESULTS

It is possible to vary experimentally the pressure of hydrogen while the pressure of H_2S is constant. Using this technique, preliminary results indicate that the measured reaction rates are independent of the pressure of hydrogen and that the dissociation of H_2S (ad) rather than HS (ad) is the rate-controlling step.

The experimentally determined rates for the dissociation of H_2S (ad) on an iron-sulfide surface are the measured slopes of the curve in Fig. 6. To compare these measured rates with the rate-equations derived in the previous section, the results are plotted as a function of the sulfur activity at the FeS surface. The sulfur activity at the surface is obtained using the previously measured calibration line (Fig. 3).

The measured rates of reaction (6) are plotted versus $1/a_S$ in Fig. 7;the same results are plotted versus a_S in Fig. 8. It is evident that either representation fits the data equally well; thus one cannot differentiate between Eqs. (13) and (14). Similar results have been obtained from all experimental runs.

The inability of the data to distinguish between a $1/a_S$ and an a_S dependence is believed due to the small experimental range of sulfur activity. In Figs. 7 and 8, a_S varies from 320 to 490. Measurements over a wider range of sulfur activity are necessary to establish the exact form of the rate equation. Experimental work is continuing with this objective. Measurements also have been initiated under pure hydrogen conditions where only the rate of the reverse reaction is measured.

CONCLUSIONS

Preliminary results from this study using thin foils of FeS support earlier work [12] which indicated that the dissociation of H_2S (ad) is the rate-controlling step.

In the derivation of an equation for the rate of dissociation of H_2S (ad) on a FeS surface, two different assumptions yield almost identical rate equations.

The rate of dissociation of H_2S (ad) may depend on the activity of sulfur at the FeS surface. However, more experimental data over a wider range of sulfur activities are necessary to establish

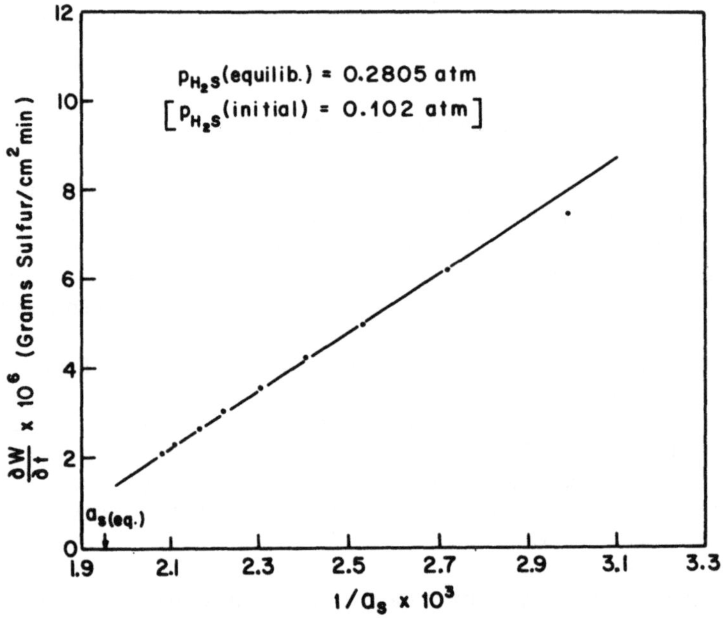

Fig. 7. Slope of the weight gain versus time curve in Fig. 6 plotted versus $1/a_S$ for iron sulfide at 670°C.

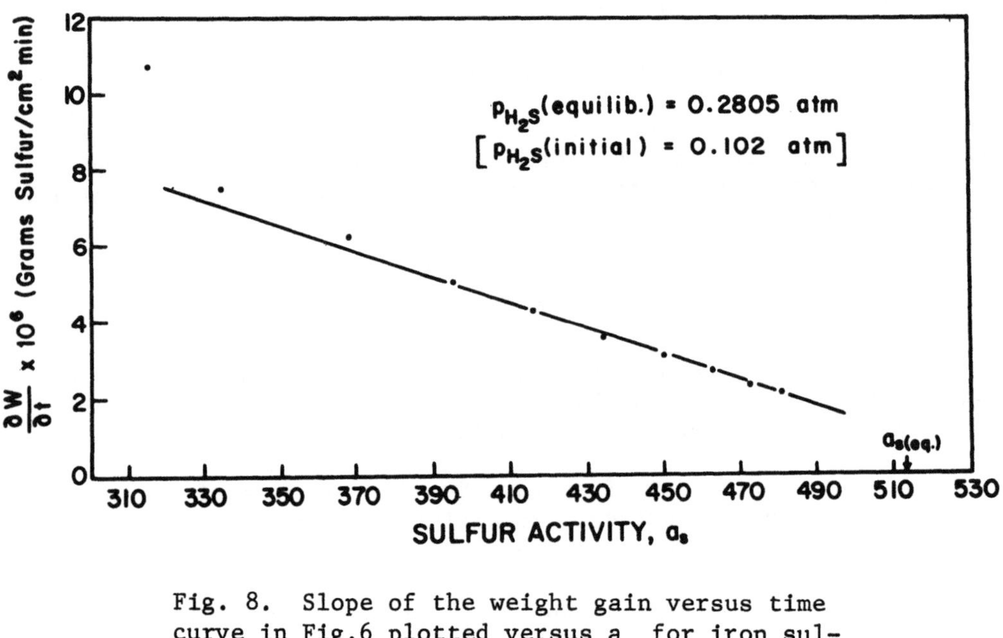

Fig. 8. Slope of the weight gain versus time curve in Fig.6 plotted versus a_S for iron sulfide at 670°C.

the exact form of the rate equation.

It is also concluded that resistivity measurements using thin foils is a very promising experimental technique for measuring and for interpreting the dissociation of gaseous molecules on solid surfaces elevated temperatures.

ACKNOWLEDGEMENTS

This study, a contribution from the Laboratory for Research on the Structure of Matter, University of Pennsylvania, was supported by the Advanced Research Projects Agency, Office of the Secretary of Defense. The authors wish to thank Dr. G. R. Belton and Dr. H. J. Grabke for their helpful comments.

REFERENCES

1. C. Wagner: *Molecular Processes on Solid Surfaces*, E.Drauglas, R. D. Gretz, R. I. Jaffe, Eds., McGraw Hill Book Co., New York, 1969, p. 17-30.
2. S. Stotz: Ber. Bunsenges Physik Chem. 70, 37-44 (1966).
3. H. J. Grabke: Ber. Bunsenges Physik Chem. 72, 533-541 (1968).
4. H. J. Grabke: Ber. Bunsenges Physik Chem. 72, 541-548 (1968).
5. H. J. Grabke: Ber. Bunsenges Physik Chem. 69, 409-414 (1965).
6. H. J. Grabke: Ber. Bunsenges Physik Chem. 69, 48-57(1965).
7. S. Stotz: Ber. Bunsenges Physik Chem. 70, 769-775 (1966).
8. E. Doehlemann, Z. Electrochem 42, 561 (1936).
9. C. Wagner and K. Hauffe: Z. Electrochem. 44, 172 (1938).
10. H. J. Grabke: Ber. Bunsenges Physik Chem. 71, 1067 (1967).
11. H. J. Grabke: To be Published.
12. W. L. Worrell and E. T. Turkdogan: Trans. AIME 242, 1673 (1968).
13. E. T. Turkdogan: Trans. AIME 242, 1665 (1968).
14. H. I. Kaplan and W. L. Worrell: To be published in the *Proceedings of the Advanced Study Institute on the Chemistry of Extended Defects in Non-Metallic Solids*, April, 1969, L. Eyring and M. O'Keefe, Eds., North American Holland, Publisher.
15. Hengstenberg, et al: *Messen und Regeln in der Chem. Technik*, Springer-Verlag, Berlin, Göttingen, Heidelberg (1957).
16. E. T. Turkdogan and P. Grieveson: J. Electrochem. Soc. 114, 61 (1967).

17. S.V. Gerei, K. M. Kholyavenko and M. Ya. Rubanik: Ukr. Khim. Zh. 31, 166-177 (1965).
18. F. A. Kröger: *The Chemistry of Imperfect Crystals*, North Holland Publishing Co., Amsterdam, 1964.
19. H. J. Grabke and W. Burgmann, Jr.: To be published.

DISCUSSION OF PAPER

S. E. Khalafalla: In relaxation kinetics, one usually obtains the rate coefficients of the forward and backward reaction (their summation) or a function thereof, by observing the manner at which the new equilibrium state is reached. This is clearly seen in your second slide. I don't see why you took a step backward and transformed this data to sulfur weight gain versus time and then obtained the rate instead of the rate constant.

W. L. Worrell: Expressing our experimental results in this manner enables us to compare directly our measured data with our derived rate equations. Also, one of the basic objectives of our experimental study is to determine if the measured rate of dissociation of H_2S on FeS depends only on the pressure of H_2S or if it is a function of the activity of sulfur at the surface. If we had assumed initially that the reaction rate was independent of the activity of sulfur, the measured rate constant would not have been constant at a given temperature and pressure of H_2S.

F. D. Richardson: From his experience of these solid state studies of sulfidation by H_2S, would Dr. Worrell care to comment on the possible influence of adsorbed sulfur in a liquid metal on the kinetics of absorption of sulfur from mixtures of $H_2 + H_2S$. Would one expect different rate constants in the presence of very small amounts of sulfur from those obtained when the surface is saturated.

W. L. Worrell: The rate of dissociation of H_2S on solid FeS is affected by the activity of sulfur at the surface. An identical rate-equation is obtained using two different approaches. In the first approach the surface of FeS is assumed to be essentially covered with sulfur ($\theta_S \to 1$); in the second approach the adsorbed H_2S molecule is assumed to be $H_2S^{=}$ (ad). If the second approach is correct, the dependence of the reaction rate upon the activity of sulfur results from the defect equilibrium in solid FeS, and one would not expect similar behavior for the dissociation of H_2S on liquid metals.

However, if surface coverage with sulfur proves to be the correct interpretation of the experimental results for FeS, one would predict similar effects for the dissociation of H_2S on liquid metals. If there is essentially no adsorption on the liquid metal surface ($\theta_{\square} \rightarrow 1$), the forward reaction rate should be directly proportional to the pressure of H_2S and independent of the activity of sulfur at the surface. If the liquid metal surface is essentially covered with adsorbed H_2S ($\theta_{H_2S} \rightarrow 1$), the forward reaction rate should be independent of both the pressure of H_2S and of the activity of sulfur. If the liquid metal surface is essentially covered with sulfur ($\theta_S \rightarrow 1$), the forward reaction rate should be directly proportional to the pressure of H_2S and inversely proportional to the activity of sulfur at the surface.

C. B. Alcock: You represented your change in sulfur potential as a step function of time. How could you demonstrate that this change occurred within a small time when compared with the whole time-scale of the experiment?

W. L. Worrell: The representation of the change in sulfur potential as a step function in Figure 4 is purely schematic. A finite time is necessary to establish the new H_2S/H_2 ratio in the gas mixture flowing past the FeS foil. At 670°C our results indicate that this time is approximately 30 seconds while five to ten minutes are necessary for completion of the surface reaction. This is indicated by the initial portion of the measured resistance versus time curve shown in Figure 5. Thus no rate data are obtained using the initial portion of the resistance versus time curves.

A. W. D. Hills: Have you measured the actual temperature of the foil during its reaction? What was the magnitude of the heat of the reaction?

W. L. Worrell: Our temperature-measuring thermocouple is less than 1 cm.from the foil. In our experiments only about .01 to .006% of the total H_2S which flows past the foil dissociates on the sulfide. The change in composition of the FeS foil during the reaction is extremely small. Thus the magnitude of the heat of the reaction should be negligible, and one can safely ignore its effect on the temperature of the foil.

ELECTROCHEMICAL STUDIES OF HETEROGENEOUS KINETICS AT ELEVATED TEMPERATURES

B. C. H. Steele

Metallurgy Department, Imperial College

London, S. W. 7

All *electrochemical* reactions are *heterogeneous* reactions as they occur at the interface between two adjoining phases, one of which is a predominately ionic conductor (the electrolyte), and the other a predominately electronic conductor (the electrode).

At high temperatures the electrolyte phase will invariably consist of a fused salt, a molten silicate, or a solid ionic conductor. The properties and experimental arrangements of electrochemical cells incorporating these materials have been the subject of recent reviews [1-7] and the present contribution is therefore confined to a brief survey of those kinetic studies amenable to investigation by high temperature electrochemical techniques. The classification scheme adopted to summarise the available techniques should also provide a useful guide for future applications of electrochemical methods.

KINETIC PARAMETERS

Electrochemical reactions involve a transfer of electrical charge (i.e. a flow of current) across the interface of the type:

$$O + ne = R \qquad (1)$$

where O and R represent the oxidised and reduced form of the electroactive species. At equilibrium the rates of the forward and backward reactions across the interface are equal, and the electrode exhibits a characteristic equilibrium potential (E_e). The magnitude of the equilibrium reaction rates is often designated [8, 9] as an apparent exchange current density I^o (amps. cm^{-2}), which is related to the rate constant K^o (cm. sec^{-1}) by

$$I^o = n\,FK^o\ C_o^{1-\alpha}\,C_r \qquad (2)$$

where C_o and C_r represent the concentration of the oxidised and reduced species at the electrode, and α is the transfer coefficient which may be regarded as a measure of the symmetry of the energy barrier for the cathodic and anodic reactions. However, when a net reaction (i.e. current flow) occurs at the interface then in general the electrode potential is displaced from the equilibrium value, and the associated reaction kinetics can be followed by observing how the variables, *current* and *potential*, change with time. The current-voltage-time relationship will be a function of the following processes:

(i) Kinetics of the over-all charge-transfer process (may produce charge-transfer polarisation);

(ii) Kinetics of the heterogeneous or homogeneous chemical reactions associated with the charge transfer process (may produce reaction polarisation);

(iii) Kinetics of mass transfer of reactants or products of the electrode reaction (may produce concentration polarisation).

The sum of the different types of polarisation is the total over-voltage (η) and is equal to the difference between the experimental electrode potential and its equilibrium value.

For a net reaction occurring at the interface then in the absence of mass transfer effects, the current density is given by,

$$I = I^o \left(\exp\left[-\frac{\alpha nF}{RT}\right] - \exp\left[-(1-\alpha)\frac{nF}{RT}\right]\right) \qquad (3)$$

For small over-voltage values, $\eta << \frac{RT}{nF}$, then

$$I_{\eta\to 0} = -\frac{nF}{RT}\ \eta\ I^o \qquad (4)$$

and it follows that a reversible electrode system which is required to exhibit negligible over-voltage must obviously possess a very large exchange current.

For the molten salt and solid electrolyte systems involved in high temperature electrochemistry the charge-transfer and coupled chemical reactions are usually very fast* producing large exchange

* Exceptions to this general observation are provided by certain reactions involving gas-gas ion electrode processes (e.g. O_2/O^{2-}). The overall kinetics of these processes may be controlled by the rates of change transfer, resulting in activation over potential. Further details are provided by Inman et. al. in refs. 3 and 5.

currents and negligible over-voltages. Selected values for kinetic parameters associated with molten and solid electrolytes are given in Table 1 together with typical data for aqueous systems.

Table 1

KINETIC PARAMETERS

Electrolyte	Electrode Process	Electrode	Exchange Current I(amp.cm^{-2})	Rate Constant K(cm.sec^{-1}	Ref.
Molten $LiNO_3$-KNO_3 - $NaNO_3$ (140°C)	$Ni^{2+} \rightarrow Ni^o$	Hg	-	6×10^{-3}	43
Molten $NaNO_3$-KNO_3 (250°C)	$Ag^+ \rightarrow Ag^o$	Ag	1.6		44
Molten LiCl-KCl (410°C)	$Ag^+ \rightarrow Ag^o$	Ag	0.1	4×10^{-2}	45
Molten LiCl-KCl (450°C)	$Zn^{2+} \rightarrow Zn^o$	Pt	15	3×10^{-1}	46
$Zr_{0.85}Ca_{0.15}O_{1.85}$ (1000°C)	$O_2 \rightarrow O^{2-}$	Pt	0.1-1		47
1M aq. $ZnSO_4$ (25°C)	$Zn^{2+} \rightarrow Zn^o$	Zn	2×10^{-5}		48
1N aq. HCl (25°C)	$H^+ \rightarrow \frac{1}{2} H_2$	Pt	10^{-4}		48
0.2N aq. HNO_3 (25°C)	$\frac{1}{2} O_2 \rightarrow O^{2-}$	Pt	2×10^{-10}		48

The rate constant K^o, which may be regarded as the velocity of penetration of the energy barrier at the electrode/electrolyte interface, can be compared with the average velocity, $\overline{v}$, $\overline{v} = 2\,(D/t)^{1/2}$, of a diffusing particle. Taking as typical values, $K^o = 10^{-2}$ cm sec^{-1}, and $D = 3 \times 10^{-6}$ cm^2 sec^{-1} for a liquid, then for times greater than 4 secs. $K^o/\overline{v} > 10$, and the electrode process is clearly diffusion controlled. The mass transfer process only becomes insignificant at time intervals less than 10^{-5} sec, and thus to obtain information about these electrode kinetics it is necessary to employ relaxation methods in which the electrode is subjected to a perturbation and the response followed for only a very short time. These specialised techniques have been reviewed elsewhere and will not be included in the present discussion.

The preceding comments indicate that high temperature electrochemical reactions are usually mass transport controlled and the expression [3] for the current must be modified to:

$$I = I^{o} \left(\left(\frac{C_o}{C_o^o} \right) \exp \left[- \frac{\alpha nF}{RT} \right] - \left(\frac{C_r}{C_r^o} \right) \exp \left[\frac{-(1-\alpha)nF}{RT} \right] \right) \tag{5}$$

where C_o^o and C_r^o are the initial bulk concentrations of the oxidised and reduced form of the electroactive species, and η is now the mass transfer over-voltage. The current-potential-time relationships for mass transport control can in principle be evaluated from this expression providing it is known how the quantities (C_o/C_o^o) or (C_r/C_r^o) change with time. The manner in which these quantities vary thus provides a convenient classification for the various techniques available for investigating many high temperature transport processes.

It is also necessary to consider whether the charging process at the electrolyte/electrode interface is likely to influence significantly the characteristics of the appropriate current-potential-time relationship. Examination of the problem [10] suggests that the double-layer charging process usually becomes unimportant for times greater than 10^{-2} sec, although difficulties may persist when the electrode potential continually changes with time, or when the current is very small.

CLASSIFICATION OF EXPERIMENTAL TECHNIQUES

In formulating a classification scheme for the techniques used in high temperature electrochemistry it was assumed that the response of one electrode to an appropriate excitation could be recorded and that a suitable unpolarised (reference) electrode could be devised to complete the cell configuration. The techniques may then be conveniently classified into four principal types of measurement.

potentiometric
steady state
controlled potential (potentiostatic)
controlled current (galvanostatic)

The electrochemical methods are generally applicable to electrodes and electrolytes in both the solid and liquid state, and thus by appropriate combination of electrochemical technique and experimental arrangement it is possible to obtain a variety of mass transport data for electrolyte, electrode, and adjoining phases. Investigations of phase boundary reactions are also feasible at the interface between the electrode and its environment. It is also useful operationally to divide the various reactions that have been studied into the following two types:

Type 1: Processes involving only the electrolyte and electrode phases.

Type 2: Processes involving interaction of the electrode with adjoining phases other than the electrolyte phase; e.g. interaction of the electrode with its environment.

The techniques and types of reactions are summarised in the adjoining schematic diagram (Fig. 1), and examples of the various processes that can be investigated are provided in subsequent sections.

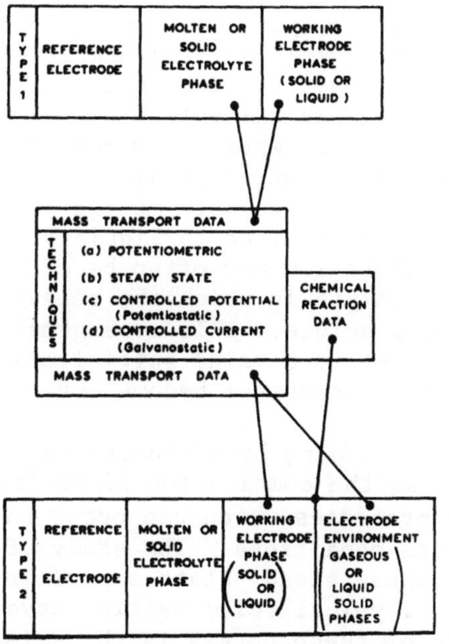

Fig. 1. Classification of electrochemical techniques.

PROCESSES INVOLVING ONLY ELECTRODE AND ELECTROLYTE PHASES

Potentiometric Measurements

Potentiometric or zero current measurements involve no Faradaic reaction, and the electrochemical cell is being used essentially as a potential probe to monitor reactions proceeding within the electrolyte or electrode phase. Potentiometric titrations and other chemical reactions involving molten salt solvents have, for example, been investigated [1, 2, 3]. The technique can also provide information about the rate at which an electrolyte or electrode system initially not in equilibrium returns to its equilibrium state, and an appropriate example [11] is provided by studies of glass-ceramic crystallisation by e.m.f.measurements. In general, however, this type of measurement involving only the electrode and electrolyte phases is not very relevant to high temperature kinetic studies of heterogeneous systems.

Steady-state Measurements

In steady-state experiments the current-voltage relationships are independent of time, and it is immaterial whether the potential is controlled and the resulting current measured or vice-versa. It follows that for any given applied voltage the concentration of the electroactive species at the electrode surface must also be independent of time, and thus the flux of electroactive species must be constant and equal to the rate at which these species are consumed at the electrode. Mass transport can proceed by three methods, migration, diffusion, and convection. In migration, the driving force that operates to move the species to (or away from) the electrode surface arises when an electrical potential gradient is established across the electrolyte phase, whereas transport by a diffusion process requires a chemical potential gradient. These two types of mass transport are applicable to both molten and solid phases whereas convection transport is limited to liquid systems. When appropriate, transport by migration may often be eliminated by using an inert supporting electrolyte in which the relevant electro-active species is present as a solute. This arrangement decreases the electrical potential gradient to a value sufficiently small so that diffusion and/or convection processes become current-limiting.

Mass transport proceeding by diffusion alone produces a time-dependent flux, whereas the combination of diffusion and convection (natural or forced) processes which can occur in liquid systems may establish steady-state conditions. The study of current-potential relationships under these steady-state experimental conditions is termed voltammetry. Typical voltammetric curves are depicted in Fig. 2 and a feature of these curves is the existence of a limiting current (i_ℓ) which arises when the concentration of the electroactive species at the electrode becomes zero. Although rigorous quantitative treatment of mass transport involving a combination of diffusion and convection is difficult, progress has been made for certain systems of which the rotating disc electrode [12, 13] probably provides the best example. Approximate treatments involving the Nernst diffusion layer concept [9, 13] are usually applied, resulting in the following expressions for reversible electrode systems,

$$E = E_{\frac{1}{2}} + \frac{RT}{nF} \ell n \left(\frac{i_\ell}{i}\right) \tag{6}$$

and

$$E = E_{\frac{1}{2}} + \frac{RT}{nF} \ell n \, (i_\ell - i)$$

when the electrodeposit is at unit activity (7)

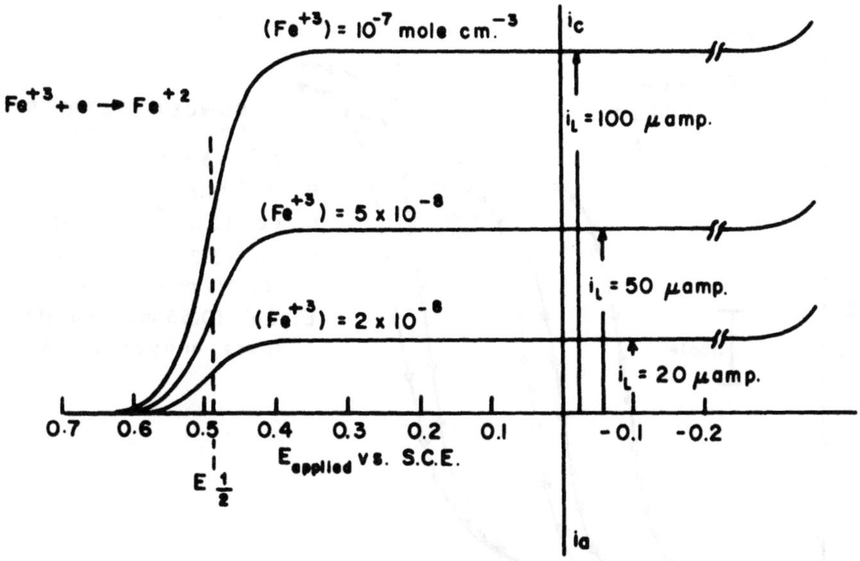

Fig. 2. Voltametric current potential curves for Fe^{3+} in 1*M* $HClO_4$. After Murray and Reilley [9].

Although polarography must be formally regarded as a controlled potential technique (next section) measurement of the average polarographic current as a function of potential also produces curves (polarograms) in accordance with equations (6) and (7). Examples of polarographic waves in a molten nitrate electrolyte at 145°C [14] are shown in Fig. 3. These data were analysed, and diffusion coefficients for the various cation solutes were calculated in accordance with the Ilkovic equation [9].

Steady state measurements involving blocking electrodes can also be devised to measured electronic conductivity in solid electrolyte materials [15, 16].

Whilst voltammetric measurements can only provide data about relevant rate constants in certain favourable circumstances they can be usefully employed to obtain general preliminary information about electrochemical processes including values for the number of electrons (n) apparently involved in the electrode reaction. Polarography has also been used to check that mass transport is the rate controlling step in kinetic studies involving liquid metal/aqueous electrolyte interfaces [17].

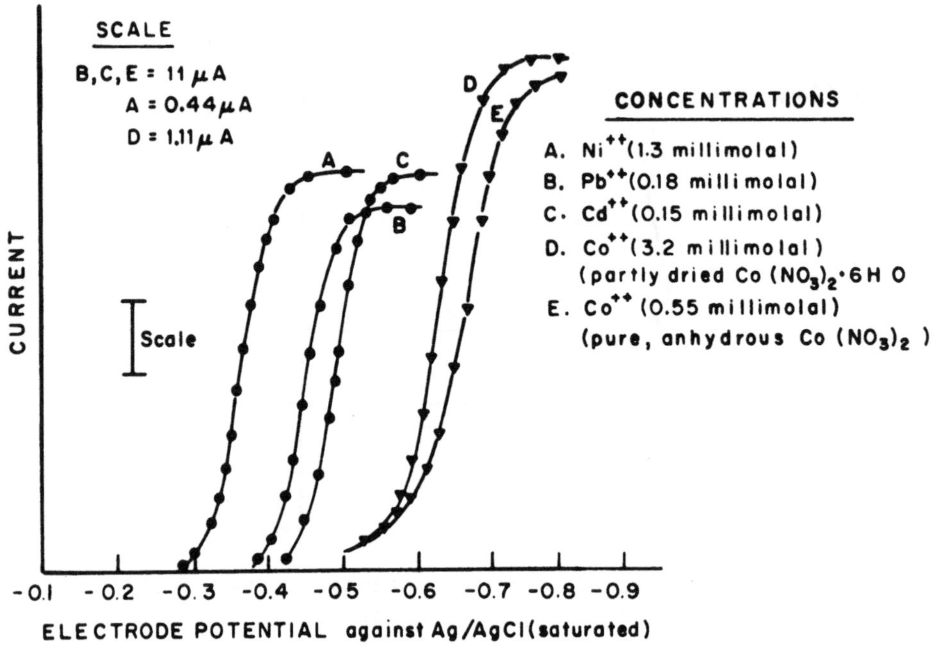

Fig.3. Polarographic reduction waves of Ni^{++}, Pb^{++}, Cd^{++}, and Co^{++} ions at a dropping mercury electrode in the $LiNO_3 + NaNO_3 + KNO_3$ melt at 145°C. After Inman et. al. [14].

Controlled Potential (Potentiostatic) Measurements

This technique in its simplest form is termed *Chronoamperometry* [9] and involves the sudden application of a potential of sufficient magnitude to produce an electrode reaction followed by measurement of the resulting current as a function of time. By arranging the experimental conditions so that an electroactive species is transported solely by a diffusion process without any contribution from convection or migration, then the relevant kinetic parameters can be evaluated from the observed current-time relationships. For a situation involving a plane electrode, and assuming linear diffusion into a semi-infinite phase, solution of the diffusion equations for the appropriate boundary conditions produces the well-known *Cottrell* equation:

$$i_{(t)} = \frac{nFA\ (C - C^{o})\ D^{\frac{1}{2}}}{\pi t^{\frac{1}{2}}} \tag{8}$$

where C is the concentration of the electroactive species established at the electrode surface by the imposed potential, C^o is the bulk concentration and D is the chemical diffusion coefficient.*

Chronoamperometric studies have provided transport data for selected cations dissolved in molten electrolytes [1, 2], and more recently the technique has been applied to systems incorporating solid electrolytes. Oxygen diffusion coefficients [5, 6, 18, 19], for example, have been determined in liquid and solid silver, and in liquid and solid copper using zirconia solid solutions as the electrolyte material, and the diffusion of silver in solid silver-gold alloys has been studied with the aid of solid AgBr [20].

A schematic diagram of the experimental arrangement used by Rickert for the potentiostatic measurement of oxygen diffusion in solid silver is shown in Fig. 4 and Fig. 5 indicates that the associated data can be analysed in terms of the Cottrell equation i.e. $i \propto (1/t^{\frac{1}{2}})$. Evaluation of the diffusion coefficient from this expression requires information about the concentration term. In situations where a finite concentration is established at the electrode surface by the imposed potential then a knowledge of the activity-composition relationship must in general be available, although the situation is simplified for systems obeying Henry's Law. Using a configuration involving a solid oxide electrolyte tube Rickert [5] was able to calculate the diffusion coefficient for oxygen in solid silver without prior knowledge of the oxygen solubility. If necessary the required activity-composition relations may also be obtained electrochemically using coulometric titrations and emf measurements which have been reviewed elsewhere [7, 21].

Potentiostatic measurements can be employed to measure diffusion coefficients across a wide range of alloy compositions by applying sequential potential steps although it is necessary to examine how long to wait between successive steps. It is also feasible to use the electrochemical method to measure diffusion values corresponding to both inward and outward diffusion into a phase. Both these techniques were employed by Raleigh and Crowe [20] in their study of the Ag-Au system and examples of their results are reproduced in Figs. 6 and 7. This investigation also demonstrated another useful feature of the electrochemical technique as diffusion coefficients as low as 10^{-14} cm^2 sec^{-1} were determined from diffusion profile penetrations of the order of 100 Å.

Experiments with controlled potentials are currently being undertaken at Imperial College [22] to determine oxygen chemical

* When the experimental arrangements involve mass transport along a concentration gradient the derived diffusion coefficient is often termed a chemical diffusion coefficient in contrast to the self-diffusion coefficient measured in the absence of a concentration gradient.

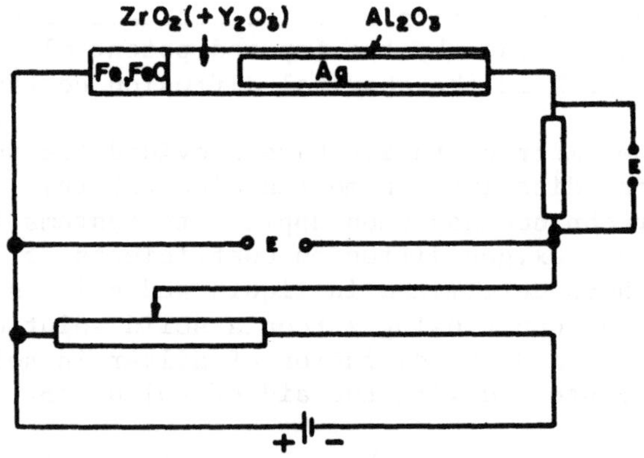

Fig. 4. Arrangement for potentiostatic measurement of oxygen diffusion in silver. After Rickert [5].

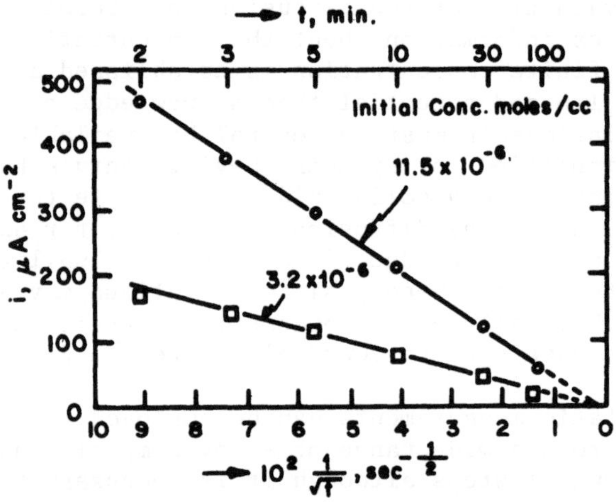

Fig. 5. Current density i as a function of $t^{-1/2}$ for the diffusion of oxygen in silver in the linear geometry at T = 800°C. After Rickert [5].

diffusion coefficients in non-stoichiometric oxides (UO_{2+x}, CeO_{2-x}, Nb_2O_{5-x}). These oxides are incorporated into the cell as an electrode phase and are in contact with a solid oxide electrolyte ($Zr_{0.8}Y_{0.2}O_{1.9}$). The composition of the oxide is adjusted to a predetermined value either by equilibration with an appropriate gas mixture or by coulometric titrations, and then the electrode is perturbed by the application of a controlled potential. Typical results for UO_{2+x} are depicted in Fig. 8, and the derived oxygen diffusion coefficients are of the magnitude expected from published data. However, a similar investigation on non-stoichiometric Nb_2O_{5-x} produced the results shown in Fig. 9. Although linear plots were

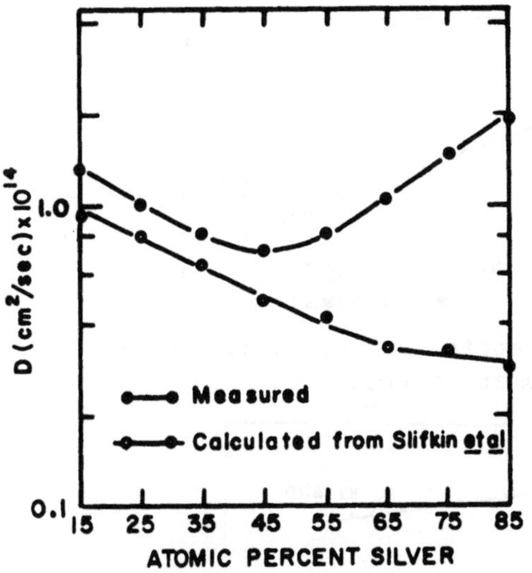

Fig. 6. Calculated and measured chem. diffusion coefficients of Ag into Ag-Au alloys at 394°C. After Raleigh and Crowe [20].

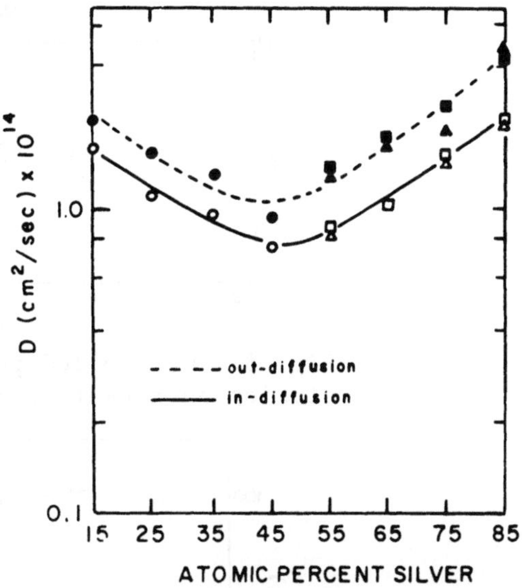

Fig. 7. Comparative in-diffusion and out-diffusion D values. After Raleigh and Crowe [20].

obtained, the observed currents do not tend towards zero at infinite time as required by the Cottrell equation. Leakage of oxygen from the surrounding gaseous environment along the electrolyte electrode interface is probably responsible for the significant current still observed after long periods of time.

Russian workers [23] have attempted to investigate the transport properties of silver in the solid electrolyte sodium chloride; again the conditions of the simple Cottrell relationship do not appear to have been satisfied.

Controlled Current (Galvanostatic) Measurements

In controlled current experiments the response of the electrode potential to a controlled excitation of the current is observed as a function of time. The simplest application of this technique is termed *D.C. Chronopotentiometry*, [9], and involves the application of a constant current to an electrode and subsequent measurement of the potential-time relationship. Solution of the diffusion equations with the appropriate boundary conditions for one-dimensional diffusion into a semi-infinite phase produces the expression,

$$C_{(t)} = C^{o} - \frac{2it^{\frac{1}{2}}}{nFAD^{\frac{1}{2}}\pi^{\frac{1}{2}}}$$

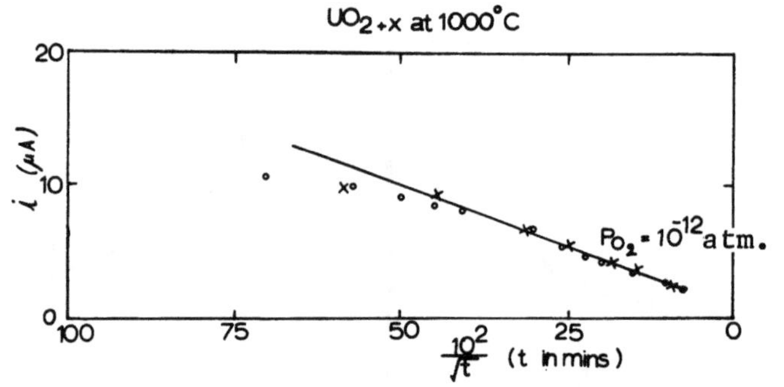

Fig. 8. Controlled potential oxygen diffusion studies on non-stoichiometric UO_2.

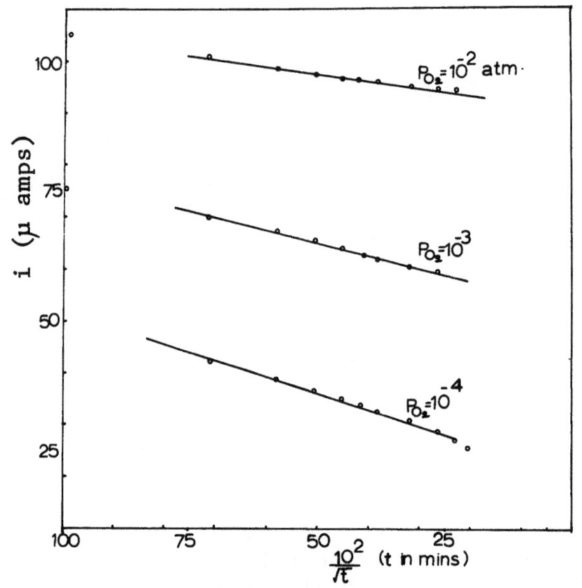

Fig. 9. Controlled potential oxygen diffusion studies on non-stoichiometric Nb_2O_5 pre-equilibrated at different oxygen pressures.

where $C_{(t)}$ is the concentration of the species at the electrode surface at time t, and C^o is the original concentration. During the electrolysis the concentration of the electroactive species at the electrode surface eventually falls to zero at some time, τ. At this time, called the *Transition Time*, the potential suddenly changes to a value corresponding to the reaction of some other species because the diffusion of the initial species no longer can support the impressed current. The transition time (τ) is related to the other variables by the equation:

$$\tau^{\frac{1}{2}} = \frac{nFAD^{\frac{1}{2}}\pi^{\frac{1}{2}}C^o}{2i} \qquad (9)$$

There are many reports in the literature of chronopotentiometric experiments involving fused salt electrolytes [1, 2, 5]. Most of these are concerned with studying the transport and stability of selected species dissolved in the electrolyte melt. It is also possible, however, to use molten electrolytes to investigate the transport properties of solutes dissolved in the electrode phase. Using anodic chronopotentiometry Van Norman [24], for example, studied the diffusion coefficients of zinc and lithium in liquid bismuth with the aid of fused lithium chloride-potassium chloride eutectic. Typical chronopotentiograms were obtained, as shown in Fig. 10 and at 450°C the derived diffusion coefficients for zinc and lithium were 6.5 and 2.2 x 10^{-5} $cm^2 sec^{-1}$ respectively.

More recently, investigators have employed solid electrolytes in chronopotentiometric studies, Rapp [25], for example, using a solid state cell consisting of pellets of $FeO-Fe_3O_4$ (reference electrode), $Zr_{0.85}$ $Ca_{0.15}$ $O_{1.85}$ (solid electrolyte) and Cu (working electrode) measured the diffusion of oxygen in solid copper in the temperature range 900 - 1030°C. The diffusion of oxygen in solid copper in the temperature range 900 - 1030°C. The diffusion coefficients were derived from the slope of a current-time function instead of from equation (9).

For appropriate systems, chronopotentiometry can also be employed to investigate mass transport processes in solid electrolyte phases. Inman [26],for example, has obtained data for the diffusion

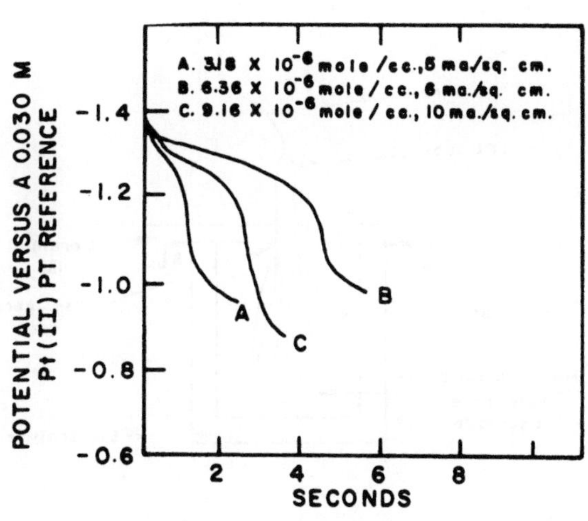

Fig. 10. Experimental curves from a constant current study of the diffusion of zinc in liquid bismuth using a fused chloride electrolyte. From Van Norman [24].

of silver in solid lithium sulphate using this technique, and the writer is currently using the same method to study the transport properties of transition metal cations in the solid electrolyte $Zr_{0.8}Y_{0.2}O_{1.9}$.

For solid phases chronopotentiometric methods do not offer any advantages over potentiostatic techniques in separating the concentration and diffusion terms, and are more subject to interference from double-layer charging and electronic cell currents and give a mean value of D when compositional variation is involved.

PROCESSES INVOLVING ELECTRODE AND ADJOINING PHASES (ENVIRONMENT)

Potentiometric (Zero Current) Measurements

Electrochemical cells used as potential probes, can be incorporated into a variety of ingeneous experimental arrangements to enable many different types of high temperature processes to be investigated. In these studies *solid* electrolytes possess many advantages as they can often be fabricated in the most appropriate configuration. For example, the compact completely solid state oxygen meter [27] depicted in Fig. 11 has been developed at Imperial College and can provide a continuous record of the oxygen partial pressure developed in many high temperature reactions.

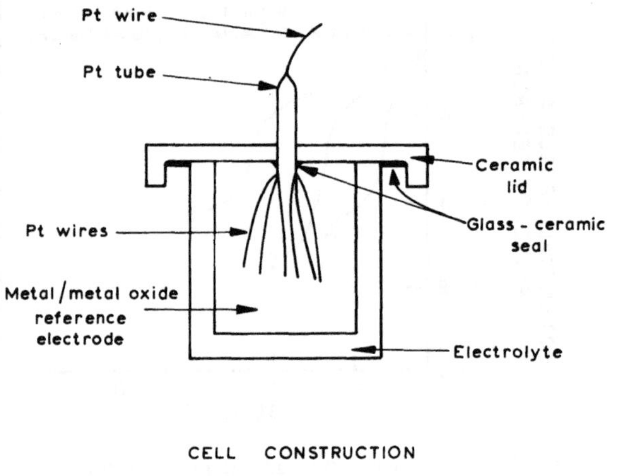

Fig. 11. Solid state oxygen meter for the measurement of oxygen pressures during high temperature reactions.

The electrode (usually platinum) in the above type of oxygen meter essentially functions as an inert electronic conductor. However, the electrode can be part of the reacting system. The mass transfer of oxygen from jets into metals (electrode phase) has been followed by Davenport et al [28] using stablilzed zirconia electrolyte tubes, and a number of investigators [29, 30] have also studied de-oxidation reactions in liquid iron using emf measurements and solid oxide electrolytes. As an example the results of Olette are reproduced in Fig. 12 which shows the oxygen content as a function of time after the de-oxidant has been added.

Reactions proceeding in solid electrodes may also be followed as in the experiments designed by Rickert [31, 32] to investigate the reaction of Ag and Cu with liquid sulphur. The experimental arrangement is shown in Fig. 13 and incorporates small electrochemical probes of Ag/AgI in contact with the reaction product Ag_2S or Cu/CuBr in contact with Cu_2S. These probes were used to examine the degree of local equilibrium at the metal/scale and scale/sulphur interfaces during the scaling reaction.

Oxygen diffusion coefficients in liquid and solid metals can also be determined by using solid oxide electrolytes of appropriate

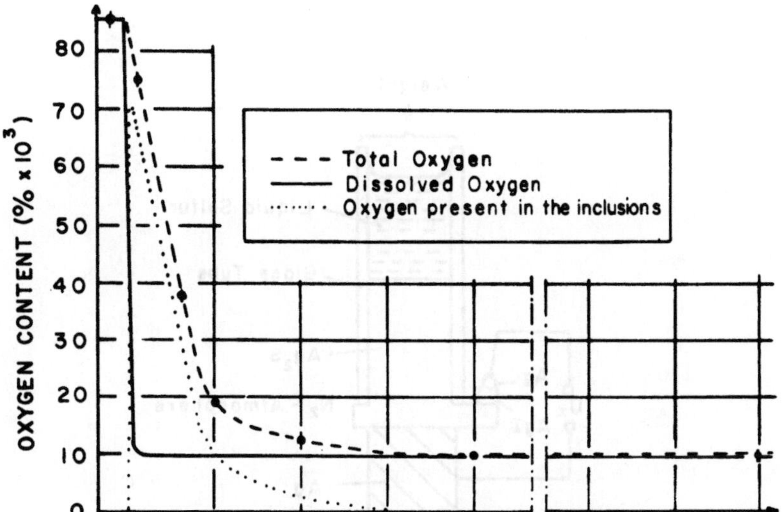

Fig. 12. Use of a solid electrolyte to measure rates of deoxidation of liquid iron. Results of Gatellier, Torsell and Olette [29].

configurations to monitor the rate of absorption or desorption of oxygen from a surrounding gas phase. It is important in this type of experiment to check that mass transport in the electrode phase is rate controlling and is not influenced by the gas-metal interfacial reaction or mass transport in the gas phase. Masson and Whiteway [33], for example, used the apparatus depicted in Fig. 14 to measure oxygen diffusion in liquid silver contained in an impervious zirconia-line electrolyte tube. The dissolution of oxygen from the gas phase could be analysed satisfactorily in terms of the relevant diffusion equations. However, when the liquid silver, saturated with oxygen, is exposed to a stream of argon, oxygen is evolved more rapidly than predicted by diffusion in the metal phase. Typical results are shown in Fig.15. The relatively rapid evolution of oxygen was attributed to convection due to density gradients established in the liquid when oxygen was removed from the surface layer. In the initial stages of oxygen evolution the rate was apparently determined by mass transport in the gas phase.

Investigations of phase boundary reactions may also be accomplished using potentiometric measurements. Schmalzried and Tare [34] used the cell represented in Fig. 16 to measure the reaction rate constant of the phase boundary reaction associated with the oxidation of iron to wüstite by CO-CO_2 gas mixtures in the temperature range 850-975°C. Additional details of this investigation are given in the following paper by Fitterer and Cassler.

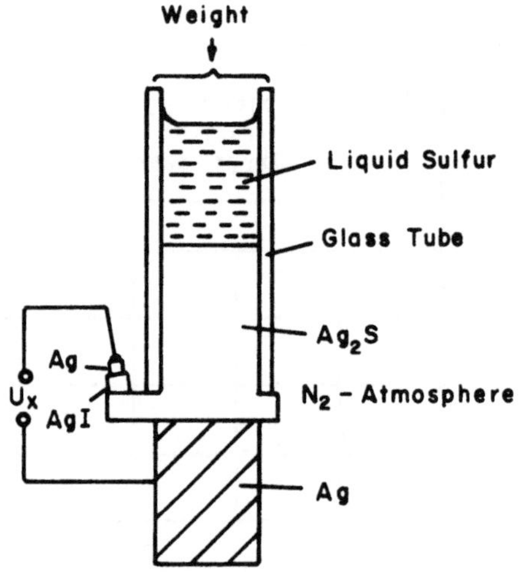

Fig. 13. Cell used by Rickert to study the reaction of silver with liquid sulphur [31].

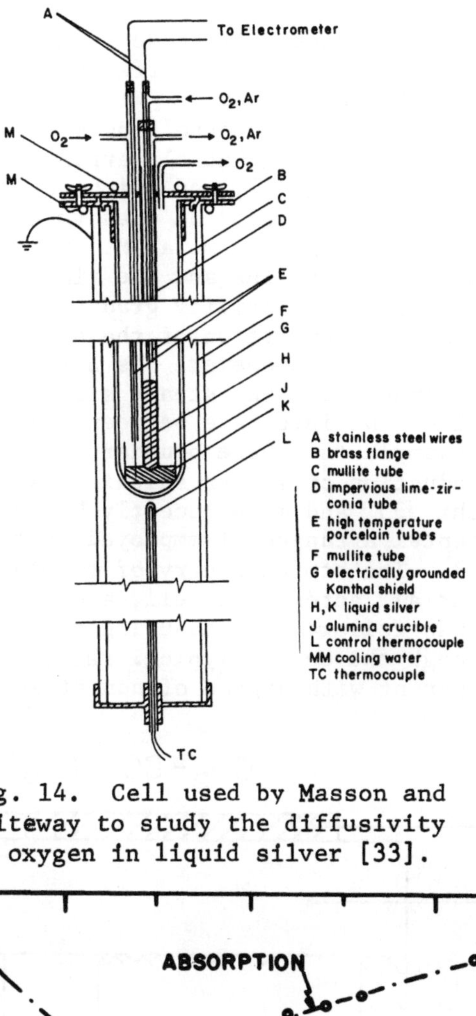

Fig. 14. Cell used by Masson and Whiteway to study the diffusivity of oxygen in liquid silver [33].

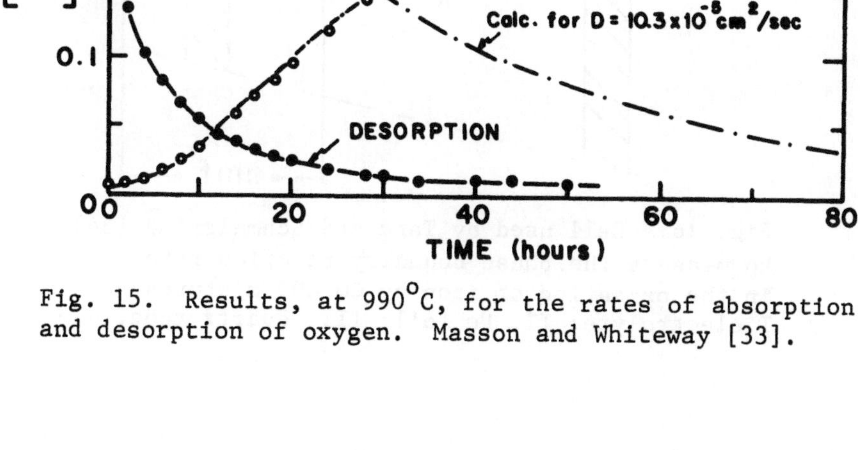

Fig. 15. Results, at 990°C, for the rates of absorption and desorption of oxygen. Masson and Whiteway [33].

Steady-State Measurements

A variety of electrolytes, electrodes, and electrode environments may be combined to investigate different types of reactions by means of steady state measurements. With reference to the electrode phase exhibiting the rate-controlling step, Schmalzried [34] and Raleigh [20] have mentioned that all arrangements of the type depicted schematically in Fig. 17, and which consist of separate cells with a common electrode could be used to establish steady state diffusion fluxes through the metal M. Measurements of the magnitude of the steady-state flux and the time required to attain it after the imposition of an activity gradient enable both the diffusion coefficient and concentration of the diffusant to be calculated. This type of experiment, however, has not so far been reported in the literature. Indeed all the steady state measurements so far reported have involved the interface between the electrode and its invironment. For example the phase boundary reaction involved in the oxidation or reduction of silver sulphide has been studied by Wagner and Kobayashi [35] and more recently by Roy and Schmalzried [36]. The experimental arrangement employed by the latter workers is shown in Fig. 18. The stoichiometry of the Ag_2S was controlled by imposing a potential across the cell, and the rate of the phase boundary reaction was followed by measuring the steady-state current of silver ions through the cell. Typical results at 300°C showing the variation of current with degree of non-stoichiometry of Ag_2S

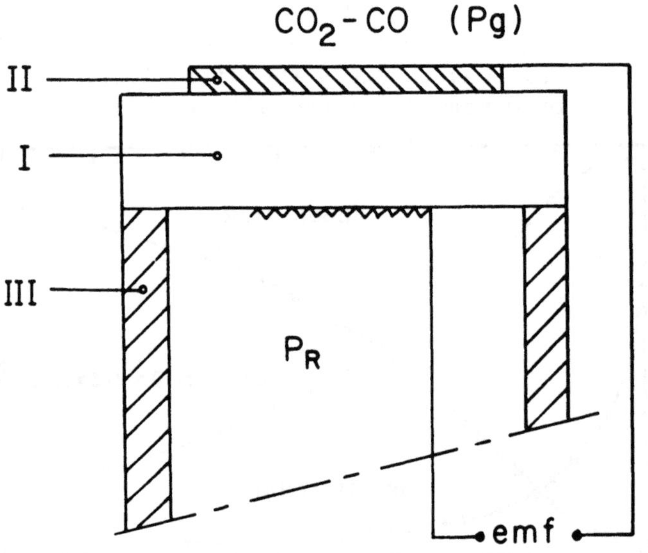

Fig. 16. Cell used by Taré and Schmalzried [34] to measure the phase boundary reaction rate in the oxidation of iron by CO/CO_2 mixtures. I, electrolyte; II, Fe foil; III, quartz tube.

and the flow rate of H_2S are reproduced in Fig. 19. It was concluded that the phase boundary reaction should be formulated as:

$$H_2(g) \;+\; S^{2-}(ad) \;=\; H_2S(g) \;+\; 2e(Ag_2S), \text{ and}$$

at high sulphur activities in Ag_2S the H_2S was oxidised to H_2S_2. Analysis of the data requires that electrical or chemical potential gradients should not exist across the silver sulphide electrode phase. This condition, however, is easily fulfilled as the chemical diffusion coefficient of Ag in Ag_2S at the operating temperature is very large, approximately $10^{-2}cm^2sec^{-1}$.

The attractive kinetic properties of the cell, Ag/AgI/Ag_2S, have been exploited by Rickert [5] and others to examine a variety of evaporation and condensation processes. The evaporation of sulphur from silver sulphide has been followed using the experimental arrangement shown in Fig. 20 and in which two processes were occurring simultaneously. First silver was removed from the Ag_2S by the passage of current. This initially produced a decrease in the silver chemical potential in Ag_2S and a corresponding increase of the sulphur potential. Secondly, sulphur evaporated at an increasing rate from the Ag_2S surface as the sulphur potential was raised. Eventually a steady state was achieved in which the amount of sulphur being evaporated was equivalent to the amount of silver withdrawn from the sulphide by the passage of current, and under these conditions the silver and sulphur chemical potentials within the Ag_2S also attained a constant value. These potentials were measured by means of the small Ag/AgI potentiometric probe located on top of the Ag_2S pellet. The observed rates of evaporation were much lower than those predicted by the Hertz-Langmuir expression and were also a function of the activities of silver and sulphur established in the Ag_2S. It was concluded that formation of S_2 molecules on the surface was the rate-determining step. Similar studies have been carried out for the evaporation of selenium from silver selenide and iodine from cuprous iodide. These investigations of course only examine the evaporation of one component as the other component (silver) is removed through the cell and thus differs from normal evaporation processes which involve the loss of both metallic and non-metallic components.

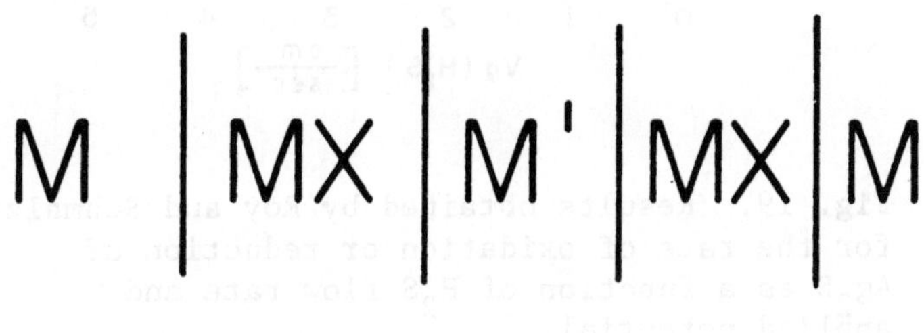

Fig. 17. Cell arrangement for steady-state measurement of diffusivities.

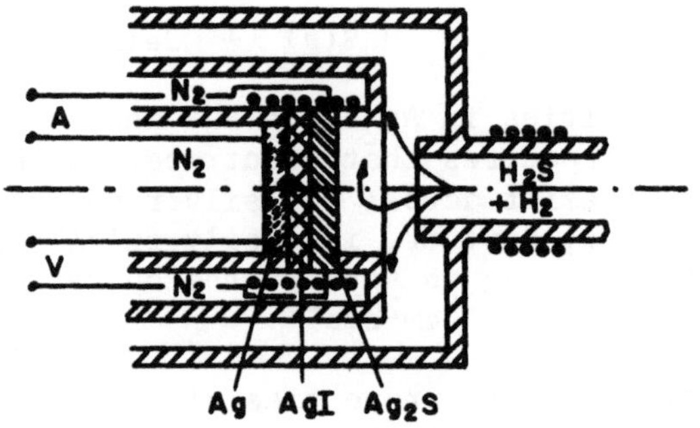

Fig. 18. Cell used by Roy and Schmalzried to measure the rate of the phase boundary reaction in the oxidation and reduction of Ag_2S [34].

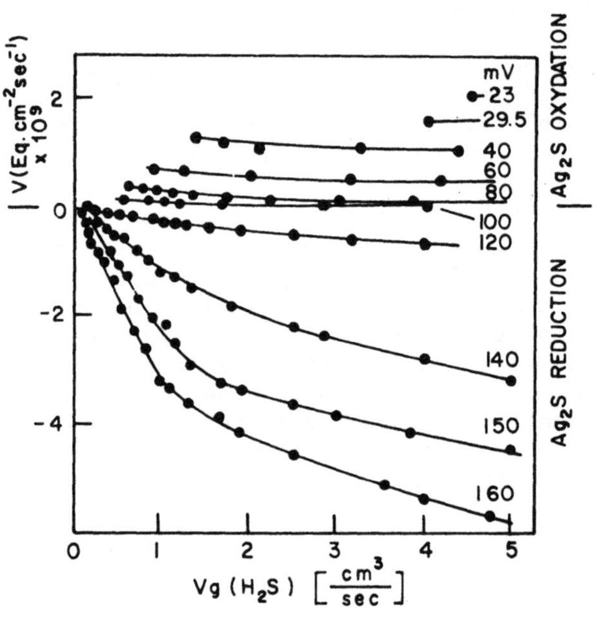

Fig. 19. Results obtained by Roy and Schmalzried for the rate of oxidation or reduction of Ag_2S as a function of H_2S flow rate and applied potential.

In addition to the 'free' evaporation studies Ratchford and Rickert [37] have constructed the electrochemical Knudsen cell depicted in Fig. 21 to investigate pressures of the various gaseous sulphur species in equilibrium with a range of compositions of Ag_2S. The total rate of diffusion of sulphur from the Knudsen cell was determined by the current passing through the electrochemical cell and the voltage measured the silver and sulphur activities in the Ag_2S. Under conditions in which the loss of one molecular species is predominant, the cell current will indicate the evaporation rate of this species. When there are many molecular species, which is the situation for sulphur and selenium vapour then it is advantageous to combine the electrochemical Knudsen cell with the mass spectrometer [38]. This combination makes it possible to obtain accurate thermodynamic data for all the vapour species. Moreover using this

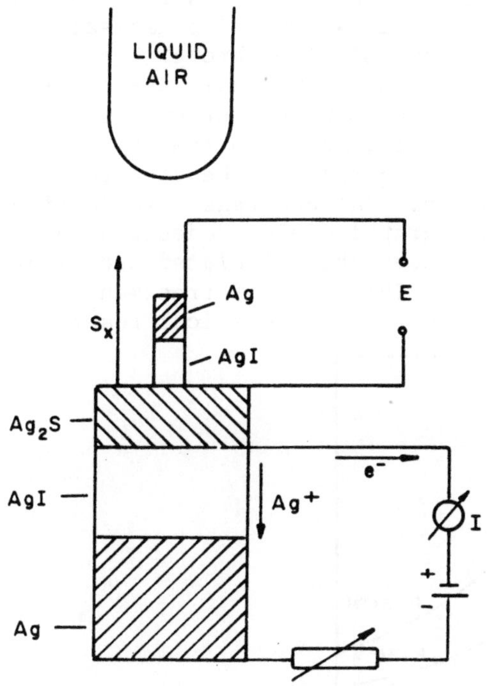

Fig. 20. Experimental arrangement for electrochemical study of free vaporization of sulphur from Ag_2S. Rickert [5].

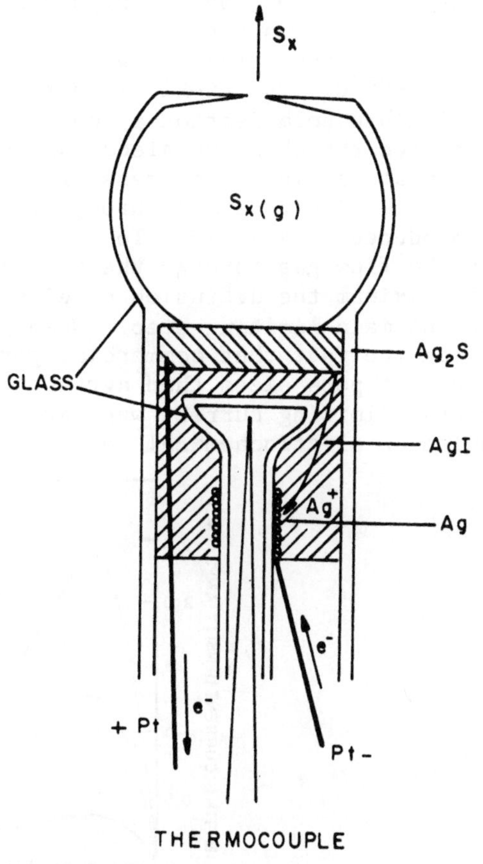

Fig. 21. Electrochemical Knudsen cell used by Ratchford and Rickert [37].

technique it is possible to distinguish between molecular fragments and primary particles,as well as to measure ionisation cross-sections in the spectrometer ion source.

It is also appropriate in this section on steady-state measurements to comment upon kinetic studies associated with the development of high temperature fuel cells, batteries and electrolytic cells. Characteristics of these types of cells are often investigated in terms of steady or pseudo-steady state voltage-current curves and limiting currents are determined for various experimental conditions. Theoretical limiting currents calculated from available information for the various mass transport processes, are then compared with the observed limiting currents and the particular rate-determining step identified. A suitable example is provided by measurements [39] on the high temperature lithium-chlorine battery. This system consists of a molten Li-electrode, molten LiCl electrolyte, a porous carbon or graphite chlorine electrode and operates at about 650°C. The Li-$LiCl_2$ electrode is very reversible and exhibits negligible polarisation at current densities of 10 amps cm^{-2}, and as expected the chlorine electrode is the main source of polarisation. Calculations involving mass transport of chlorine through the porous graphite electrode and through the electrolyte produces the curves reproduced in Fig. 22. To the left of the maximum current is limited by the flow gas through the porous material while to the right of the maximum the diffusion of dissolved Cl_2 through the electrolyte is the main limiting step. When the calculations take into consideration the fact that commercial porous graphite electrodes possess a range of pore sizes then agreement between the calculated and experimental limiting current was very good. This type of treatment in which an electrochemical cell is considered as a chemical reactor

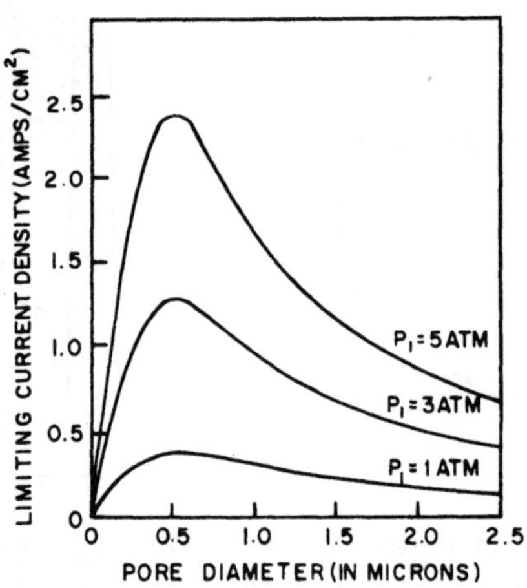

Fig. 22. Calculated effect of graphite electrode porosity on limiting current density in $Li-Cl_2$ fuel cell. Swinkels [39].

is an aspect of electrochemical engineering which has been reviewed elsewhere [40].

Potentiostatic and Galvanostatic Measurements

Potentiostatic or galvanostatic measurements involving the electrode and adjoining phases, other than the electrolyte phase, have so far received little attention although many different types of experiments are possible. For example the growth of a reaction product layer on a metal electrode can be followed by a cell of the type illustrated in Fig. 23. When the oxidation product, MeO, is sufficiently thick so that diffusion in MeO controls the oxidation rate, then the current in a potentiostatic experiment should correspond to the rate of growth of the oxide when subjected to a constant oxygen partial pressure established by an imposed voltage. The potentiostatic experiment is generally preferred to a galvanostatic experiment as the former is likely to produce less polarisation of the reference electrode and is more directly analogous to a typical chemical oxidation experiment.

Mrowec and Rickert [5, 41] have employed the cells depicted in Fig. 24 to study the formation of nickel sulphide.

Initially a galvanostatic technique was used and a constant current was passed through the cell thus maintaining a definite constant rate of NiS formation, and the potential of sulphur at the Ag_2S/NiS interface measured as a function of time. Subsequently potentiostatic experiments measured the growth of NiS when exposed to a constant sulphur activity, and typical results are given in Fig. 25. The results are in accordance with the assumption of a parabolic rate law and the relevant rate constant ($k = 5 \times 10^{-9}$ $cm^2 sec^{-1}$ at 400°C) can be derived from the slope of the line. Information about the defect structure of NiS was also obtained from the measurements. Preliminary experiments had established that the flow of electrons in NiS was not rate-controlling and that electron equilibrium was maintained at the phase boundary and in the reaction

$$\text{OXYGEN REFERENCE ELECTRODE} \,|\, Zr_{0.85}Ca_{0.15}O_{1.85} \,|\, MeO \,|\, Me$$

Fig. 23. Cell for the study of the rate of growth of an oxide layer on a solid metal.

layer. It was also demonstrated that interdiffusion or ternary compounds were not present at the Ag_2S-NiS interface.

To supplement their galvanostatic experiments already mentioned, Rapp and Pastorek [25] have also employed the more complicated cell arrangement depicted in Fig. 26 to measure oxygen diffusion coefficients in solid copper. Both potentiostatic and galvanostatic measurements were made.

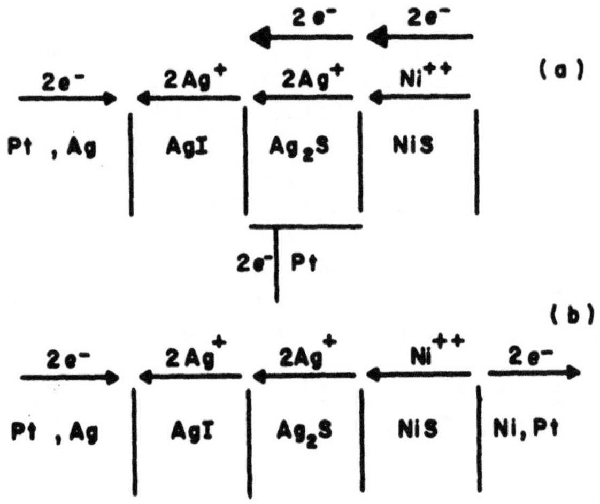

Fig. 24. Cell used by Mrowec and Rickert [41] for the study of the formation of solid NiS on metallic nickel for transport of (a) Ni ions and electrons in NiS and (b) Ni ions only.

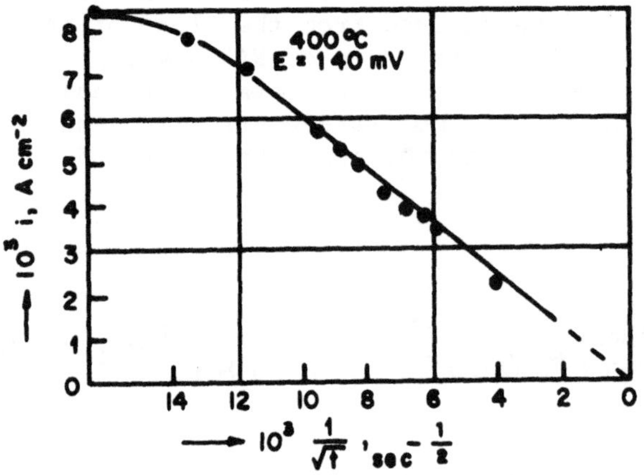

Fig. 25. Potentiostatic results for the growth of NiS at constant sulphur activity Mrowec and Rickert [41].

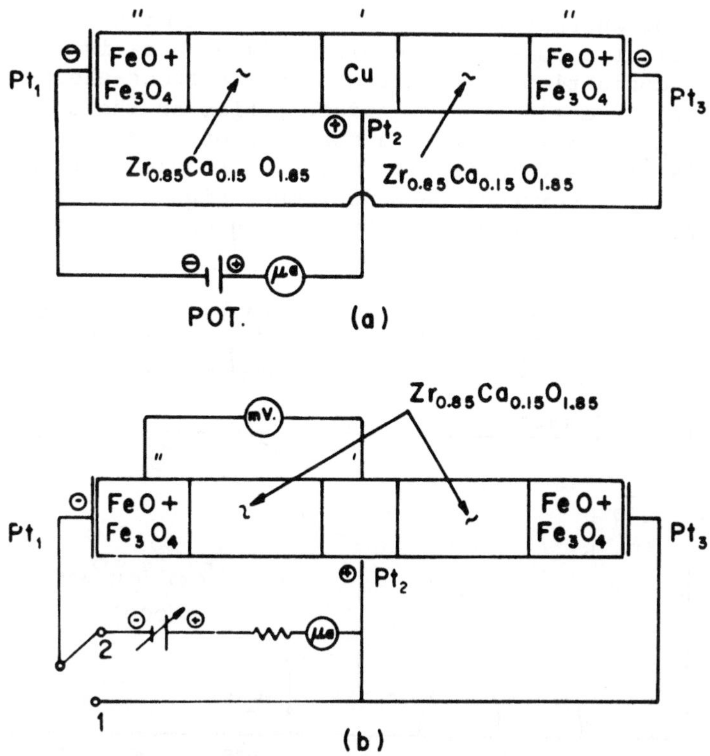

Fig. 26. Experimental arrangements used by Rapp and Pastorek [25] for (a) potentiostatic and (b) galvanostatic studies of the diffusivity of oxygen in solid copper.

CONCLUSIONS

The wide variety of high temperature heterogeneous reactions amenable to investigation by electrochemical cells has been surveyed in the previous sections. It is apparent that cells incorporating solid electrolytes are particularly useful for these investigations as the electrolyte material can often be fabricated in the most appropriate configuration. It is relevant, however, to emphasise some of the experimental problems associated with the use of solid electrolytes. There is, for example, very little quantitative information regarding the current densities that can be tolerated by the various reference electrodes (e.g. Fe/FeO) which have been employed. Other problems may arise due to the higher resistance often encountered with solid electrolytes, and intimate solid electrode-electrolyte contacts may be difficult to realise experimentally [42]. The high mobility of electrons in ionic crystals may also produce a significant electronic conductivity, and it is always important, therefore, to examine whether the imposition of a chemical or electrical potential gradient across a solid electrolyte phase is likely to result in mixed conduction at either of the electrode-electrolyte interfaces. Ionic and electronic conductivities of the solid oxide electrolyte $Zr_{0.85}Ca_{0.15}O_{1.85}$ as a function of chemical and electrical potentials are illustrated in Fig. 27. Providing these problems are considered and overcome, then many more high temperature reactions will undoubtedly be studied by techniques involving electrochemical cells.

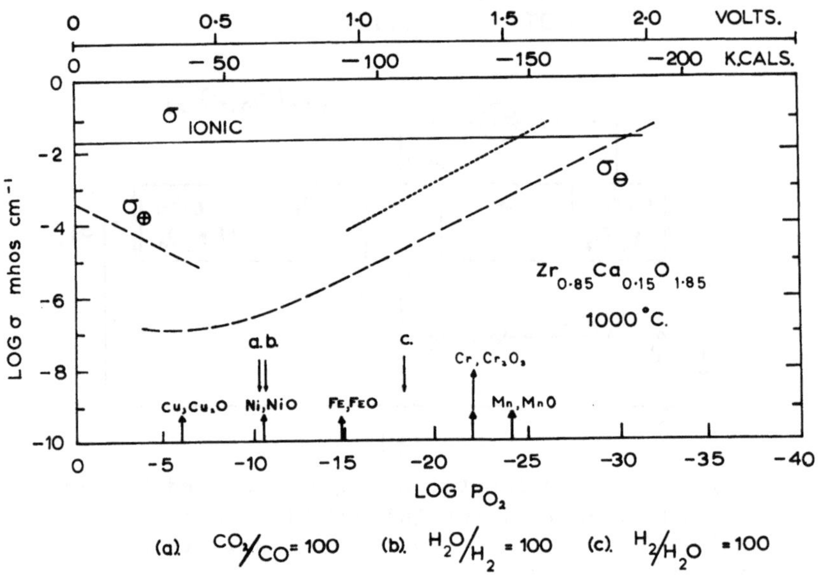

Fig. 27. Ionic and electronic conductivities of the calcia-zirconia electrolyte as a function of $P(O_2)$.

REFERENCES

1. *Molten Salt Chemistry*, Edit. by M. Blander, Interscience, New York, 1964.
2. B. R. Sundheim: *Fused Salts*, McGraw-Hill, New York, 1964.
3. A. D. Graves, G.J. Hills and D. Inman in *Advances in Electrochemistry and Electrochemical Engineering*, Vol. 4 edit. by P. Delahay, Interscience, New York, 1966.
4. J.D. Corbett and F. R. Duke, in *Technique of Inorganic Chemistry*, Vol. 1, edit. by H. B. Jonassen and A. Weissberger, Interscience, New York, 1963.
5. *Electromotive Force Measurements in High Temperature Systems*, edit by C.B. Alcock, Institute of Mining and Metallurgy, London, 1968.
6. D.O. Raleigh, in *Progress in Solid State Chemistry*, Vol. 3, Macmillan, New York, 1966.

7. R. A. Rapp and D. A. Shores in *Techniques in Metals Research*, Vol. IV, edit. by R. A. Rapp, Interscience, New York, 1970.
8. P. Delahay: *Double Layer and Electrode Kinetics*, Wiley, London, 1965.
9. R. W. Murray and C. N. Reilley: *Electroanalytical Principles*, Interscience Reprint, New York, 1963.
10. P. Delahay, in *Advances in Electrochemistry and Electrochemical Engineering*, Vol. 1, edit. by P. Delahay, Interscience, New York, 1961.
11. N. K. Dertev, A. F. Borisov, V. I. Zadumin and A. I. Trushkov: Isvestya Akad. Nauk S.S.S.R. Neorganicheskie Materialy 1, 957 (1965).
12. V. G. Levich: *Physicochemical Hydrodynamics*, Prentice Hall, Inc., New Jersey, 1962.
13. A. C.Riddiford, in *Advances in Electrochemistry and Electrochemical Engineering*, Vol. 4, Edit. by P. Delahay, Interscience, New York, 1966.
14. D. Inman, D. G. Lovering and R. Narayan: Trans. Farad Soc. 63, 3017 (1967).
15. C. Wagner: Proc. Int. Comm. Electrochem. Thermo. and Kinetics (CITCE) 7th meeting Lindau (1955), Butterworth Scientific Publ., London,1957.
16. J.W.Patterson, E. C. Bogren and R. A. Rapp: J. Electrochem. Soc. 114, 752 (1967).
17. W. F. Porter, F. D. Richardson and K. N. Subramanian, in *Heat and Mass Transfer in Process Metallurgy*, edit. by A. W. D. Hills, Inst. of Mining and Metallurgy, London, 1967.
18. H. Rickert and R. Steiner: Z. phys. Chem. 49, 127 (1966).
19. H. Rickert and A. A. El. Miligy: Z. Metallkunde 59, 635 (1968).
20. D.O. Raleigh and H. R. Crowe, J. Electrochem Soc. 116, 40 (1969).
21. B.C. H. Steele in *Mass Transport in Oxides*, Edit. by J. B. Wachtman and A. D. Franklin, N.B.S. Special Publication 296, 1968.
22. J. Chen. Metallurgy Dept. Imperial College, London. Private communication.
23. S. V. Karpachev and V. P. Obrosov: Elektrokhimiya 5, 486 (1969).
24. J.D.Van Norman: Anal. Chem. 33, (7), 946 (1961).
25. R. L.Pastorek and R. A. Rapp: Trans. AIME 245, 1711 (1969).
26. D. Inman: Metallurgy Dept. Imperial College, London, Private communication.
27. Prov. Brit. Patent 42576/69.
28. W. G. Davenport, D. H. Wakelin and A. V. Bradshaw, in *Heat and Mass Transfer in Process Metallurgy*, edit by A. W. D. Hills, Inst. of Mining & Metallurgy, London 1967.
29. C.Gatellier, K. Torsell and M. Olette: Comptes Rendus (Series C) 266, 1753 (1968).
30. W. A. Fischer and W. Ackermann : Archiv Eisenhuttenw. 37, 43 (1966).
31. H.Rickert: Werkstoffe and Korrosion 19, 869 (1968).

32. D. Donner and H. Rickert: Z. physik. Chem. N. F. 60, 11 (1968).
33. C. R. Masson and S. G. Whiteway: Canad. Met. Quart. 6, 199 (1967).
34. H. Schmalzried: Archiv fur Eisenhuttenwesen 39, 531 (1968). and Trans. AIME 236, 444 (1966).
35. H.Kobayashi and C. Wagner: J. Chem. Phys. 26, 1609 (1957).
36. P. Roy and H. Schmalzried: Ber. Bunsenges. phys. Chem. 71, 200 (1967).
37. R. J. Ratchford and H. Rickert: Z. Elektrochem 66, 497 (1962).
38. J. Droward,P. Goldfinger, D. Detry, H. Rickert and H. Kellet in *Advances in Mass Spectrometry IV*, edit. by E. Kendrick, London, 1968.
39. D.A. J. Swinkels: J. Electrochem. Soc. 113, 6 (1966). Ibid.115, 994 (1968).
40. *Advances in Electrochemistry and Electrochemical Engineering*, Edit. by C. W. Tobias, Interscience, New York, 1962.
41. S. Mrowec and H. Rickert: Z. phys. Chem. 28, 422 (1968). and Ber Bunsenges. phys. Chem. 66, 14 (1962).
42. D. O.Raleigh: J. Phys. Chem. Solids 29, 261 (1968).
43. J. E. B. Randles and W. White: Z. Elektrochem. 59, 666 (1955).
44. D. Inman, J. O'M. Bockris and E. Blomgrem: J. Electroanal. Chem. 2, 506 (1961).
45. D. C. Hill, G. J. Hills, L. Young and J. O'M. Bockris: J. Electroanal.Chem. 1, 79 (1959).
46. H.A.Laitinen, R. P. Tischer and D. K. Roe: J. Electrochem. Soc. 107, 546 (1960).
47. H. A. Spacil 103, in ref. 5.
48. J. O'M. Bockris, in *Modern Aspects of Electrochemistry* Vol. 1 edit. by J. O'M. Bockris, Butterworth, London, 1962.

DISCUSSION OF PAPER

F. D. Richardson. Towards the end of your lecture you referred to the difficulties of making good solid + solid contacts. At the beginning you gave us a Table of Exchange current densities with one for the Pt + O_2 + ZrO_2.CaO electrode. Was the value based on a real area of contact between the Pt and the ZrO_2 or on a normal area?

B. C. H. Steele. The exchange current value quoted in Table 1 for Pt $(O_2)/Zr_{0.85}\ Ca_{0.15}\ O_{1.85}$ electrode system was based on the macro geometrical area. No attempt was made to correct for the actual contact area which would be much smaller, and thus the quoted figure for the exchange current represents a minimum value.

T. B. King. Esin has reported that, in experiments on molten oxides, in which transition times were measured for solutes present at relatively low concentrations,the results were quite different depending upon whether the working electrode was solid or liquid. Now, the species being studied is present in the electrolyte and is simply discharged at the working electrode. According to Esin, convection in the electrolyte occurs much more readily when the electrode is liquid and it is therefore more difficult to measure a transition time.

Since we have encountered a similar effect in our own work, I wonder if you could comment on your own experience.

B. C. H. Steele. I have no direct experience of measuring kinetic parameters for liquid electrode-fused salt systems, but it is generally considered that the effect of convective mass transport for unstirred liquid metal/fused salt interfaces only remain insignificant provided that transition times do not exceed about 1 second. This conclusion is based on experimental results and contrasts with the longer transition times (many secs.) generally permitted in comparitive studies involving solid electrodes. It would appear therefore that convection within the liquid electrolyte is modified by the presence of a liquid electrode.

S. E. Khalafalla. Stirring at liquid metal electrodes is a common phenomena in electrochemistry. Von Stackelberg studied this phenomena at the dropping mercury electrode and attributed it to local variation in surface tension at the electrode/electrolyte interface. The change in surface tension is in turn, attributed to changes in current density at peripheral zones from those at the main central part of the electrode, and is assumed to follow the Lippmann electrocapillary curve. In polarographic practice one usually uses gelatin or other surface active substance to equalize the surface tension and hence remove the anamolous phenomenon of current voltage maxima. Perhaps similar surface tension depressents could be used to prevent convective transfer and hence harmonize the results obtained with liquid and solid electrodes.

A. W. D.Hills. The effect described by Professor King may be due to some form of convective flow. Such a flow would distort the surface of a molten electrode,increasing its surface area, but would have no comparable effect on a solid electrode.

A STUDY OF THE KINETICS OF THE SURFACE OXIDATION OF METALS BY SOLID ELECTROLYTE TECHNIQUES

G. R. Fitterer and C. D. Cassler

Metallurgical and Materials Engineering Department

University of Pittsburgh

The solid electrolyte technique which was suggested by Tare and Schmalzried has been used successfully with certain modifications for the determination of the oxidation rate of iron under various conditions. Gas mixtures with ratios of from 0.5 to 5.0 of CO_2/CO were used at temperatures from 900 to 1100°C. This technique uses thin metal foil samples, and the specimens in this work varied in thickness from 0.4 to 9.5 mils.

At all temperatures and at all gas ratios, the oxidation was found to follow a linear trend up to this thickness. A tendency for the rate to follow a parabolic law is slightly evident in thick foils, low temperatures and CO_2/CO ratios.

The specific reaction rate constant for the oxidation of iron in CO_2-CO mixtures was determined to vary with temperature from 2×10^{-8} to 2×10^{-7} per square centimeter of surface per second, per mole over the range of 900 to 1100°C. This is an excellent check of the rate constant determined ty Tare and Schmalzried by this method and is slightly larger than the constant determined by Grabke, using a radioactive technique.

A new and simplified procedure was developed for the determination of the rate constant from solid electrolyte data.

The much referred to development of solid electrolyte techniques by Kiukkola and Wagner [1] has nucleated a vast amount of metallurgical research of various types.

Investigations of this nature have extablished the free energies of various oxides, [1, 2, 3, 4] and alloy phase boundaries [5, 6, 7,

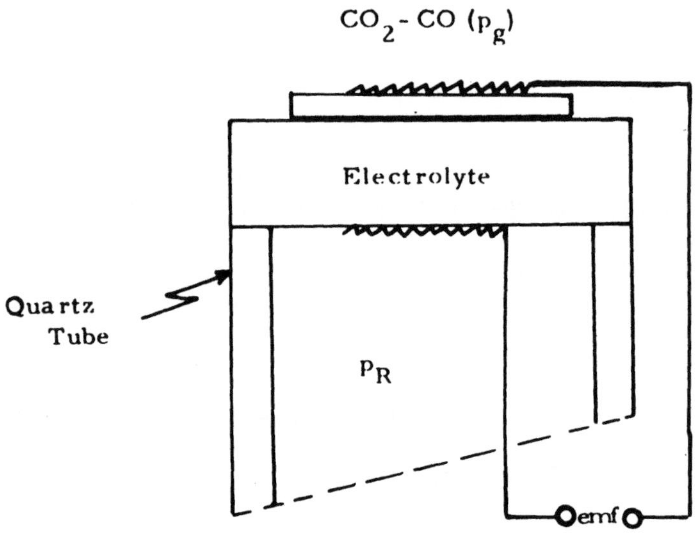

Fig. 1. Schematic design of the Taré and Schmalzried [10] solid electrolyte cell.

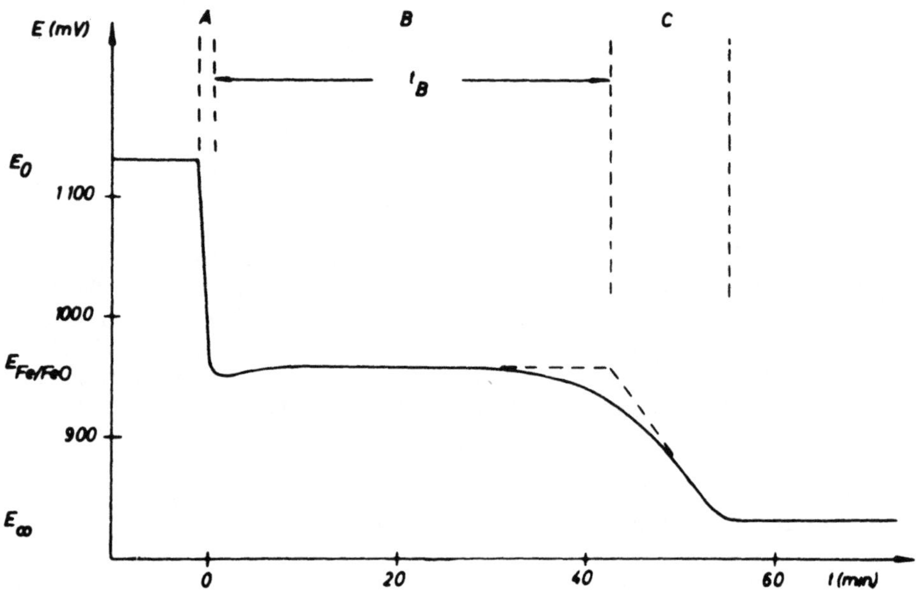

Fig. 2. Trends in cell voltage during oxidation of iron to Wustite, Taré and Schmalzried [10].

8]. Also, the diffusion rates of oxygen in solid [9] and liquid [10] metals have all been recorded in the 10-12 years following the initial publication. Recently, solid electrolyte techniques for the study of the kinetics of high temperature reactions have been developed. Rickert's [9] research in this field is well known.

The work about to be described is an extension of the interesting technique suggested by Tare and Schmalzried [11] for the study of the kinetics of oxidation of solid metals and alloys. As shown schematically in Fig. 1, the cell used in their work consisted of an electrolyte disc, on one side of which is pressed a metal foil of known thickness. A gaseous mixture (i.e., CO_2 and CO) of known equilibrium oxygen content is directed against the metal foil and oxygen or air on the opposite side of the electrolyte serves as the reference state. Platinum leads and thermocouples permit one to measure the changes in the cell voltage during the oxidation of the metal at the selected temperature.

Figure 2 according to Tare and Schmalzried, schematically illustrates the trend of the cell potential with time. Prior to heating to the experimental temperature, a voltage (E_o) is imposed against the cell so as to just prevent oxidation. Also, the CO_2/CO ratio is held slightly on the reducing side during heating. When thermal equilibrium has been attained, the gas ratio is suddenly changed to the desired level. Thus, during the initial period indicated as A in the figure, oxygen is adsorbed on the surface; and as in the case of the oxidation of iron, diffuses into the foil until the formation of wüstite is nucleated at the low point of the curve. Wüstite continues to form until the entire foil is transformed to oxide (i.e., in period-B) after which the emf drops to a lower value (period-C) depending upon the oxygen differential. By means of this technique, a linear reaction rate constant for the oxidation of iron to wüstite was determined by these authors.

The current work has involved an attempt to verify and extend the Tare and Schmalzried conclusions regarding the oxidation of iron and also to modify the technique where necessary so as to study the oxidation characteristics of more refractory metals and alloys. These latter studies are currently in progress.

EXPERIMENTAL TECHNIQUE

The cell used in the current work is shown in Fig. 3. It is designed with two distinct electrical circuits. Circuit A-B is similar to the Tare and Schmalzried concept, and it is through this circuit that the changes in the cell voltage were measured during the oxidation of the metal foil.

The second circuit (i.e., A-C) is designed to give the basic cell voltage, without the metal foil. The reaction involved is

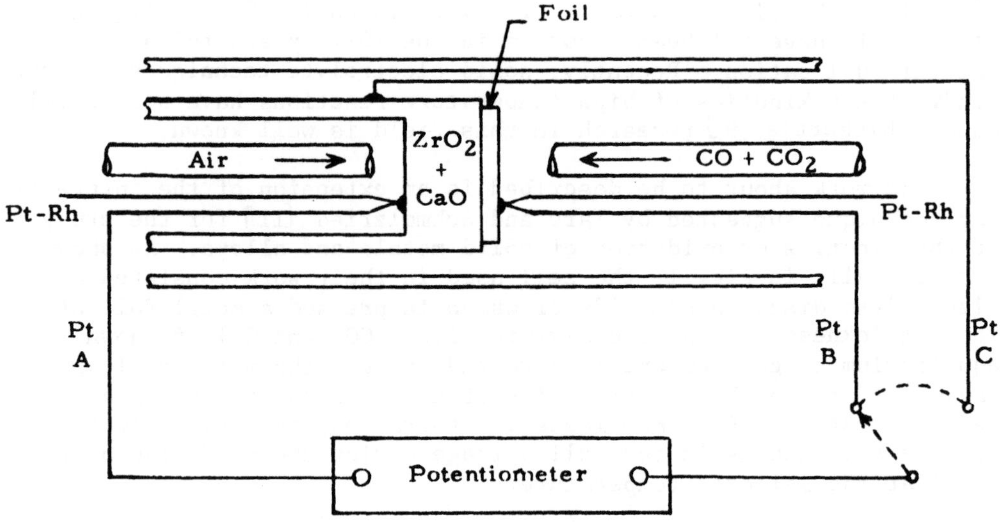

Fig. 3. Experimental apparatus (schematic)

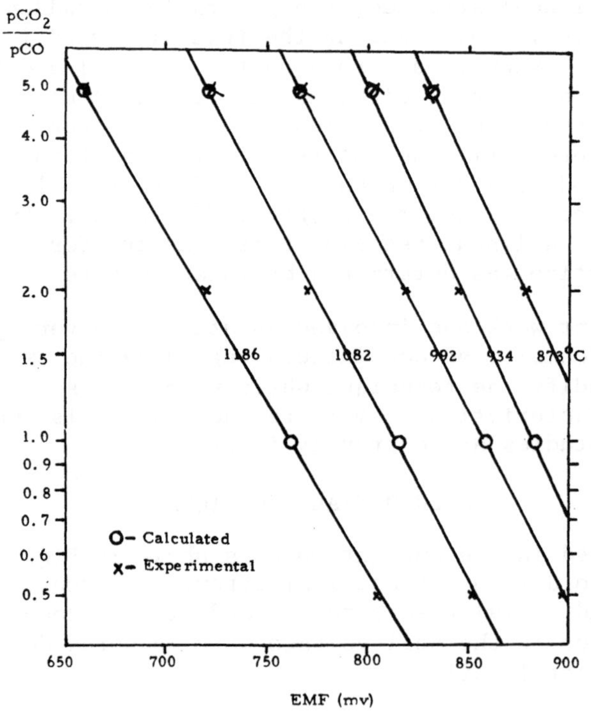

Fig. 4. Calibration of the solid electrolyte cell for the reaction: $CO + 1/2\ O_2$ (air) $= CO_2$.

$$CO + 1/2O_{2(air)} = CO_2 \quad (1)$$

Inasmuch as the standard free energy of this reaction is well known, a family of curves, as shown in Fig. 4, may be calculated in accordance with the van't Hoff isotherm;

$$\Delta G_R = -neF = \Delta G^o + RT\ln J \quad (2)$$

wherein
- e = the cell voltage
- n = 2
- F = 23,066 cal/volt or the Faraday equivalent
- ΔG^o = -67,500 + 20.75T according to Kubaschewski, Evans and Alcock [12].
- R = 1.986 cal/°K
- T = °K
- J = $pCO_2/pCO \sqrt{pO_2}$ where pO_2 = 0.21 at.(air).

Thus, J expresses the experimental concentrations.

Preliminary tests were made using circuit A-C at various temperatures from 873 to 1186°C and with CO_2/CO ratios as indicated in the figure. As shown, the experimental data correlate quite satisfactorily with the thermodynamically calculated lines. The maximum deviation of these experimental values from the calculated lines is within ± 0.1%. Later, circuit A-C was used as a reference to check the cell behavior with metal foil during the experiments. Sudden alterations of the CO_2/CO ratio resulted in corresponding responses in the cell voltage within less than 15 seconds at all potential levels.

A billet of (99.99 +%) iron powder was prepared from iron carbonyl by Magnetics, Inc. of Butler, Pennsylvania. This billet was then rolled to 0.100 inch strip at 2200°F and then cold reduced to foils (0.004" and 0.0043" thick) on a Sendzimer planetary mill. Some of the carbonyl iron billet material was also sliced and polished down to various thicknesses so as to extend the experimental data to heavier "foil".

All samples were recrystallization annealed in hydrogen, then washed and cleaned with acetone prior to the oxidation experiments. The rolled material presented different types of surfaces and grain orientations than the samples which had been polished down from bulk specimens.

The initial CO_2/CO ratio was adjusted at the start to a point which was very slightly on the reducing side according to prior thermodynamic calculations.

As in the Tare and Schmalzried procedure, the CO_2/CO ratio was then quickly adjusted to the desired experimental oxidizing value.

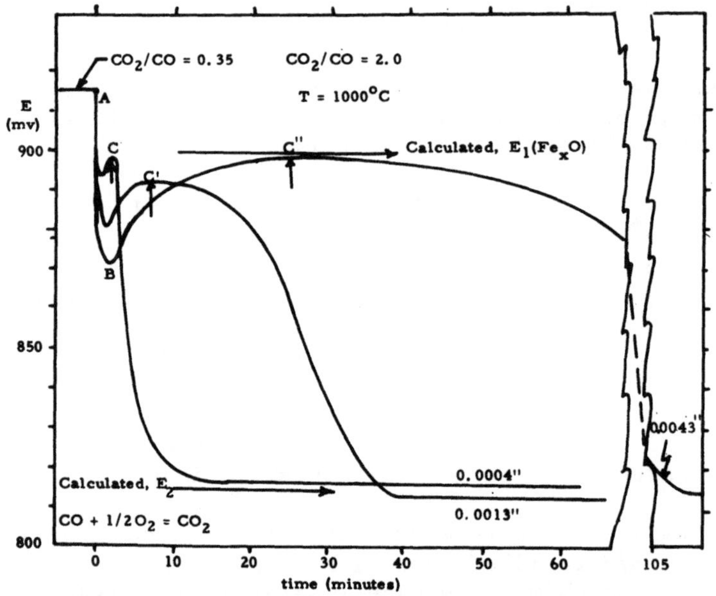

Fig. 5. Typical curves of voltage vs. time in oxidation of iron to wustite.

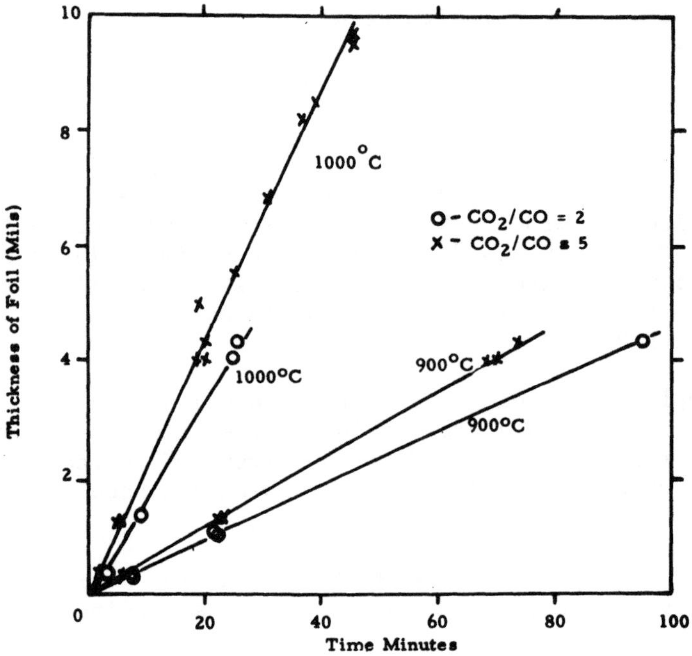

Fig. 6. Thickness of iron foil and time to reach peak voltage at different temperatures and CO_2/CO ratios.

The temperature was also held at a constant level throughout each experiment. As will be shown, the cell voltage varied according to the oxidation trends during a given experiment in a manner generally similar to the Tare and Schmalzried curve in Fig. 2 but sufficiently different to require an explanation.

INSTRUMENTATION

The cell voltage was recorded potentiometrically on a Honeywell, Electronic - 19 instrument backed up with a constant voltage source so that only the top 100 millivolts were recorded during any given experiment.

Specially calibrated (Manostat) flow meters were used to adjust the CO_2/CO ratio throughout all experiments.

EXPERIMENTAL RESULTS

The experiments were designed to provide several "families" of curves which would represent the trends in the oxidation of iron under various conditions.

The variables were:

1. The temperature (i.e., 900, 1000 and 1100°C).
2. The CO_2/CO ratios (i.e., 0.5, 1.0, 2.0 and 5.0).
3. The foil thickness (ten steps from 0.0004" to 0.0096").

The three curves shown in Fig. 5 are condensed reproductions of experimental curves and are typical of all of those obtained in the oxidation of iron. At an operating temperature of 1000°C and a CO_2/CO ratio of 2/1, the trends are similar and vary only because of the thickness of the foil. In each case, the initial CO_2/CO ratio was held at 0.35 as the samples were heated to the experimental temperature.

When the gas ratio was changed to 2/1, the emf dropped (as at Point-B) until the formation of wüstite was nucleated, then the voltage rose to the Points-C, as shown in the figure. All three of the C-values or peak voltages are within less than 1% of the calculated voltage for the formation of wüstite under these conditions.

These curves all differ from the Tare and Schmalzried schematic curve of Fig. 2, however, in that in this case, as soon as the foil had been oxidized to wüstite, the voltage started to drop. This continued in every case until the basic cell voltage for the formation of CO_2 was reached, as in Fig. 4. This point will be discussed in detail shortly. Experimental data for all tests are given in Table I.

Table I -- Experimental Data

Test No.	Temp. (%C)	Foil Thickness (mils)	CO_2/CO	Time to Reach Peak (Min.)	E^o (Volts)	E'(Peak) (Volts)	$k_1 x 10^{-8}$ $(sec^{-1}cm^{-2})$
1	900	0.4	2.0	6.0	0.938	0.918	0.862
2	900	0.4	5.0	6.5	0.945	0.895	2.00
3	900	1.0	5.0	22.0	0.948	0.916	0.935
4	900	1.3	2.0	23.0	0.950	0.923	0.970
5	900	1.3	5.0	22.5	0.943	0.870	2.79
6	900	1.3	5.0	24.5	0.935	0.862	2.48
7	900	4.0	2.0	70.0	0.948	0.921	1.00
8	900	4.0	5.0	68.0	0.940	0.901	1.52
9	900	4.3	2.0	95.0	0.943	0.923	0.582
10	900	4.3	5.0	75.0	0.943	0.900	1.59
11	1000	0.4	0.5	6.5	0.907	0.890	0.700
12	1000	0.4	0.5	3.5	0.906	0.889	1.28
13	1000	0.4	1.0	2.0	0.902	0.883	2.55
14	1000	0.4	1.0	4.0	0.906	0.890	1.07
15	1000	0.4	1.0	2.0	0.908	0.900	1.06
16	1000	0.4	1.0	3.0	0.918	0.900	1.58
17	1000	0.4	2.0	2.0	0.916	0.898	2.37
18	1000	0.4	2.0	2.0	0.904	0.876	3.76
19	1000	0.4	5.0	2.0	0.906	0.885	2.82
20	1000	0.4	5.0	2.0	0.902	0.868	4.60
21	1000	1.3	2.0	8.5	0.914	0.893	2.12
22	1000	1.3	5.0	5.0	0.905	0.873	5.58
23	1000	1.3	5.0	5.0	0.916	0.891	4.05
24	1000	4.0	2.0	24.5	0.913	0.900	1.41
25	1000	4.0	5.0	20.0	0.904	0.880	3.22
26	1000	4.0	5.0	18.5	0.920	0.900	2.84
27	1000	4.3	2.0	25.0	0.905	0.893	1.37
28	1000	4.3	5.0	20.0	0.905	0.883	3.16
29	1000	5.0	5.0	19.0	0.906	0.888	3.14
30	1000	5.5	5.0	20.5	0.908	0.895	2.31
31	1000	6.9	5.0	30.0	0.906	0.893	1.98
32	1000	8.2	5.0	37.0	0.906	0.896	1.47
33	1000	8.5	5.0	39.5	0.914	0.898	2.24
34	1000	9.5	5.0	45.0	0.909	0.888	2.94
35	1000	9.6	5.0	51.5	0.910	0.889	2.61
36	1100	0.4	1.0	1.5	0.877	0.862	2.73
37	1100	0.4	2.0	0.5	0.860	8.837	12.9
38	1100	0.4	5.0	0.5	0.864	0.817	26.5

In these studies, the time from the start of the oxidation to the peaks of the voltage curves (i.e., from points A to C of Fig. 5) was selected as the total time for the oxidation of the foil. These time periods have been plotted against the thickness of the foil in Fig. 6. Here, the effects of temperature and CO_2/CO ratio are clearly shown and it is apparent that the oxidation of iron under these conditions is linear for foils up to 0.0096" thickness. The variation in temperature is more effective than changes in CO_2/CO ratios within these ranges.

No attempt was made in this work to determine any variations in stoichiometry of the wüstite (i.e., the value of x in Fe_xO). This had been studied in considerable detail by coulometric titration using solid electrolyte cells by Rizzo [13]. Instead, it was assumed that the composition of the oxide as indicated by the peak emf at the C-points did not deviate much from the line A-B, as shown in Fig. 7 by Darken and Gurry [14].

Kinetics and the Determination of the Rate Constant

In a cell of this type, the cell potential is a function of the reaction free energy

$$\Delta G_R = -neF \tag{3}$$

The reaction free energy in turn is a function of the equilibrium dissociation oxygen pressure of the oxide in the contact with the metal on the one hand and the oxygen pressure of the gas phase as determined by the CO_2/CO ratio at the experimental temperature.

In their study of the kinetics of oxidation iron, Tare and Smalzried [11] used a modified relationship suggested by Kubayaski and Wagner [15], as shown in Eq. 4. The time period selected in their work was the length of the plateau of B-section of their curves (Fig. 2).

$$k = \frac{\theta \cdot d_{Fe}}{t \cdot M(1-x)} \left[pCO_2 \left(1 - \frac{pCO}{pCO_2} \quad \frac{pCO_2}{pCO}_{Fe/FeO} \right) \right]^{-1} \tag{4}$$

where; θ = foil thickness (cm)
d = density of iron
t = time (seconds)
M = Atomic Weight of iron
1-x = Composition $Fe_{(1-x)}O$ in equilibrium with iron

This relationship was also used by Pettit and Wagner [16] in a somewhat different form to describe a weight-gain study of the oxidation of iron.

In the current work, the rate constant was considered as the time required for the cell voltage to change from the initial voltage at the start (i.e., Point A of Fig. 5) to that at point-C. This peak voltage was assumed to indicate the complete oxidation of the foil to wüstite at a point along line A-B of Fig. 7, depending on the temperature and CO_2/CO ratio. As shown in Fig. 5, no true voltage plateau region was indicated in any recordings of this work. When very low CO_2/CO ratios were used at low temperatures, however, a plateau type recording was approximated. Even in these cases, however, the "plateau" was not entirely flat and a peak voltage was evident. The Tare and Schmalzried plateau may be explained by the fact that their experimental temperatures and CO_2/CO ratios were lower than in most of the current work. Thus, it may be that an apparent plateau was indicated over much longer oxidizing time periods.

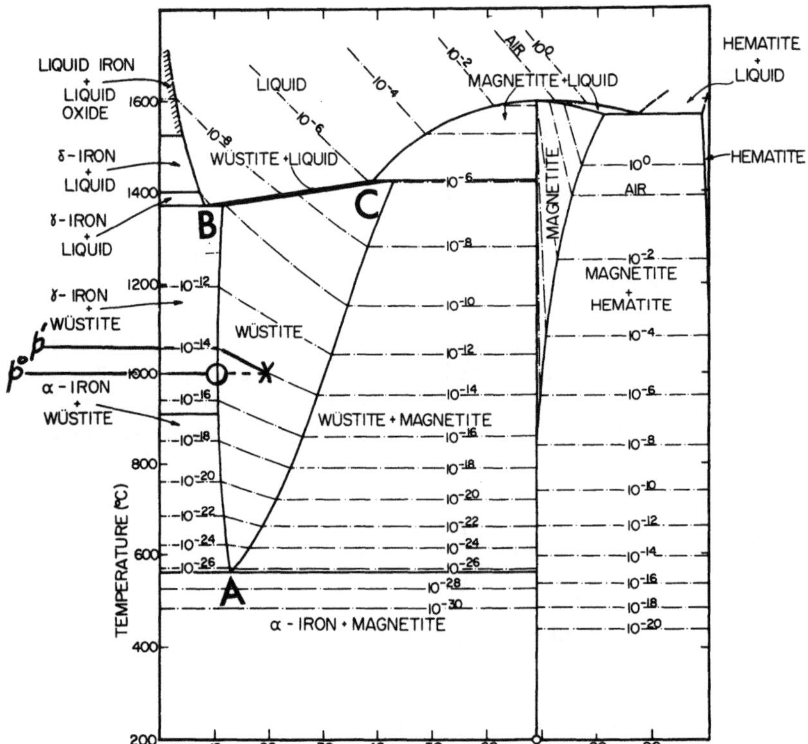

Fig. 7. The $FeO-Fe_2O_3$ system according to Darken and Gurry [14].

Eight of their ten experiments indicated an oxidizing period of more than 72 minutes. In the current work, only two of the thirty-eight experiments extended beyond this length of time--in fact, more than one half of the samples in this work were completely oxidized in less than 8 minutes.

In this work, the rate constant was determined from the following equation;

$$k = \frac{\theta d}{tM} \ln \frac{Eo}{E_o - \Delta E} \qquad (5)$$

where;
- θ = foil thickness (cm)
- d = density of iron
- t = time (seconds)
- M = Atomic Weight of iron
- E^o = Equilibrium electrode potential
- ΔE = the difference between the experimental (E') and the equilibrium potential (E^o)

$$k = (8.1 \times 10^{-4}) \frac{\theta}{t} \log \frac{Eo}{E'} \qquad (6)$$

where; $E' = E_o - \Delta E$ or the peak voltage at point-C.

This approach to the problem is similar to that proposed by Taylor [17] for reactions in which one of the reactants was being consumed. In this case, the iron is progressively transformed to oxide.

In the temperature range of 850-975°C, Tare and Schmalzried reported an average rate constant of 3.5×10^{-8}. In this work at 900 and 1000°C, an average of thirty five experiments indicates a rate constant of 2.38×10^{-8}. Grabke [18] using electrical resistance measurements and radioactive tracer techniques indicated specific rate constants of 1.59×10^{-9} and 1.81×10^{-9}, respectively, at 800°C.

The reason for the lack of a "plateau" in the current work may best be explained by reference to Fig. 7. For example, if in a given experiment at 1000°C the pCO_2/pCO ratio is adjusted so that the resulting oxygen pressure is slightly less than 10^{-15}, no oxidation of iron will occur. However, with a $p^o(O_2)$ of 10^{-15} atmospheres, oxidation will proceed to form wüstite as indicated by point-0 on Line A-B. Such a ratio should exhibit a perfect plateau at this temperature.

If however, the CO_2/CO ratio were adjusted so that $p'(O_2)=10^{-14}$, the oxidation would proceed beyond line A-B to point x at 1000°C.

Thus, the e.m.f. of the cell would increase to a peak (point-C) when the oxidation had reached the A-B line. It would then decrease with the diffusion of additional oxygen into the wustite. It is believed that this at least partially explains the lack of plateaux in the current work. A CO_2/CO ratio of about 12/1 would be required to increase the oxygen content of wüstite to line A-C at this temperature and the formation of magnetite. This ratio was just beyond the limit of the flow meters and thus was never attained.

Activation Energy

The experimental values for the specific rate constant were averaged for each temperature and the activation energy was calculated from the averages to be 38,720 calories over the temperature range of from 900 to 1100°C.

Pettit and Wagner [16] found an unexplained change in the activation energy for this same reaction at the α to γ transformation in iron. Above this temperature, their work indicated an activation energy of 51,000 calories and below it, a value of 22,000 calories was reported.

The value reported in the current work was averaged over this temperature range and the value of 38,720 is approximately the average of the above two values (i.e. 36,500 calories).

Forward Reaction Rates

Edmiston and Grace [22] reported an activation energy of27,400 calories for the reduction of wüstite with hydrogen in this temperature range.

Their values were determined by calculations of the forward and backward reaction rates close to the equilibrium level by a method suggested by Prigogine, Outer and Herbo [19] and developed further by others [20, 21]. This procedure was investigated in the current work with identical results for the activation energy which is shown as the slope of the line in Fig. 8.

Weight Gain

To date, most of the kinetic data regarding the oxidation of metals have been obtained by thermo-gravimetric methods. These, of course, are bulk oxidation procedures, since oxidation of the specimen occurs on all exposed surfaces as contrasted with the linear oxidation on only one exposed surface as in the current research.

However, the experimental data from the method used in the current work can be translated into weight gain values. The weight of each specimen is known and by assuming that the foils were transformed completely to wüstite at point-C in the emf-time curve, the gain in weight of each specimen was calculated. These results are shown in Table 2 and in Fig. 9 where they may be compared with the results of Pettit and Wagner [16]. The experimental values from this work with a 5/1 ratio of CO_2/CO at 1000°C compare favorably with Pettit and Wagner's results with a 1/1 ratio at 1100°C.

It has been pointed out by Pettit and Wagner [16] and others that the oxidation of iron progresses from a linear to a parabolic rate after the oxide film has progressed beyond a certain thickness. This is somewhat evident from the trends shown in Fig. 9. Kofstad [23] has suggested the term "general parabolic" to describe this type of oxidation behavior. Such reactions are apparently "interface

Table II - Gain in Weight

$\frac{CO_2}{CO}$	t °C	Foil Thickness (in.)	time	$\sqrt{t}$	$\frac{W}{A}$ $\frac{gms}{cm^2}$ x10^3
2/1	900	0.0004	360	19.0	0.90
"	"	0.0013	1380	37.2	2.90
"	"	0.0040	4200	65.0	8.96
"	"	0.0043	5700	75.5	9.65
5/1	"	0.0004	390	19.7	0.90
"	"	0.0010	1320	36.3	2.20
"	"	0.0013	1350	36.8	2.90
"	"	0.0013	1470	38.4	2.90
"	"	0.0040	4080	63.7	8.96
2/1	1000	0.0004	120	11.0	0.90
"	"	0.0004	120	11.0	0.90
"	"	0.0013	510	22.6	2.90
"	"	0.0040	1470	38.3	8.96
"	"	0.0043	1500	38.6	9.65
5/1	"	0.0004	120	10.9	0.90
"	"	0.0013	300	17.3	2.90
"	"	0.0040	1200	34.7	8.96
"	"	0.0043	1500	38.8	9.65
"	"	0.0050	1140	33.6	11.20
"	"	0.0055	1230	35.0	12.20
"	"	0.0069	1800	42.3	15.50
"	"	0.0082	2220	47.0	18.60
"	"	0.0085	2380	48.8	18.80
"	"	0.0095	2700	51.8	21.30
"	"	0.0096	3090	55.5	21.40

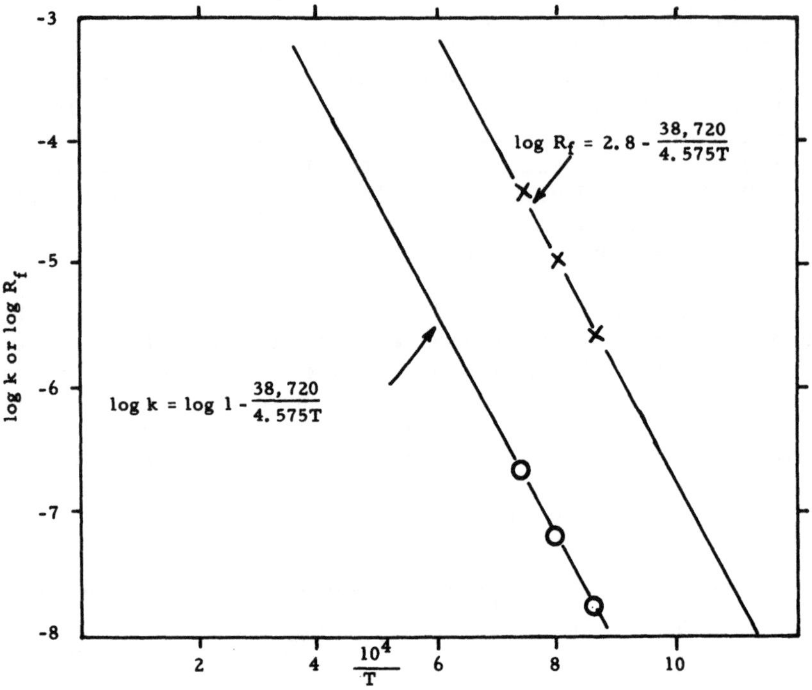

Fig. 8. Comparison of the forward reaction rate with the specific rate constant.

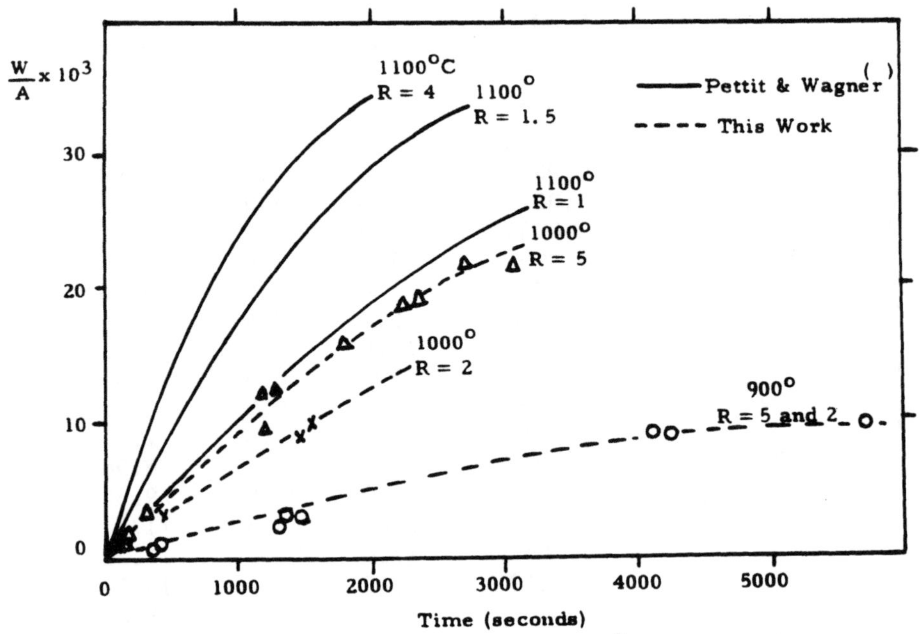

Fig. 9. Weight gain of iron per cm^2 at different temperatures and CO_2-CO ratios.

controlled and linear during the initial stages and diffusion-limited or parabolic after extended oxidation."

SUMMARY AND CONCLUSIONS

The solid electrolyte technique which was proposed by Tare and Schmalzried has been verified and promises to develop into an effective and useful tool for the determination of oxidation rates of metals and alloys.

In the current work, the linear rate constant proposed by these authors has been checked. In addition, the experimental data correlate reasonably well with the weight-gain method used by others for a study of the oxidation of iron to wüstite.

This work has verified the fact that the oxidation of iron in CO_2-CO atmospheres is linear at low oxide thicknesses, then tends to become parabolic in thicker oxide films when the oxidation rate becomes diffusion controlled,particularly below the alpha-gamma transformation in iron.

The solid electrolyte method promises to be quite useful because of its rapid response and the lack of need for high vacuum to prevent oxidation during the heating cycle.

The oxidation in this procedure progresses from only one surface and is unique in this respect. Also, the cell voltage gives one a clue as to which oxide is being formed at various stages.

Current and future work is directed toward developing the method into a rapid procedure for evaluating the oxidizing characteristics of metals and alloys under various oxidizing conditions.

The oxidizing characteristics of pure nickel and TD nickel as well as nickel and TD nickel alloys with chromium are being studied.

ACKNOWLEDGEMENTS

The assistance of the National Aeronautics and Space Administration under NASA Grant NGR 39-011-067 in the support of this work is gratefully appreciated.

REFERENCES

1. K. Kiukkola and C. Wagner: J. Electrochem. Soc., 104, 379 (1957).
2. Y. Matsushita and K. Goto: *Symposium on Thermodynamics, Proceedings Series,* Vol. 1, IAEE, Vienna, 1966, P. 111.

3. W. L. Worrell: ibid., p. 131.
4. G. B. Barbi: J. Phys. Chem., 68, (5), 1025 (1964).
5. R. A. Rapp and F. Maak: Acta Met. 10, 63 (1962).
6. K. Schwerdtfeger and A. Muan: Acta Met. 13, 509 (1965).
7. R. J. King and G. R. Fitterer: Pt-Co System. Presented at the AIME Institute of Metals Meeting, Pittsburgh, May, 1969.
8. L. A. Pugliese and G. R. Fitterer: The Cr-Ni System, ibid.
9. H. Rickert: *Electromotive Force Measurements in High Temperature Systems*, C. B. Alcock, ed., American Elsivier Publishing Co., 1968.
10. F. G. Arcella and G. R.Fitterer: Diffusion of Oxygen in Liquid Lead, Presented at the AIME Annual Meeting, Washington, D. C., Feb., 1969.
11. V. B. Tare and H.Schmalzried, Trans AIME 236, 444 (1966).
12. O. Kubaschewski, E. Ll. Evans and C. B. Alcock: *Metallurgical Thermochemistry*, 4th ed., Pergamon Press, 1967.
13. H. Rizzo, Thermodynamics of the Fe-O System, Ph.D. Thesis, University of Utah, 1968.
14. L. S. Darken and R. W. Gurry: J. Am. Chem. Soc., 67, 1398 (1945).
15. H. Kubayashi and C.Wagner: J. Chem. Phys. 26, (6), 1609, (1957).
16. F. S. Pettit and J. B. Wagner: Acta. Met. 12, 35, (1964).
17. H. A. Taylor: J. Phys. Chem. 34, 2761, (1930).
18. H. J. Grabke: Ber. Bunsinger. phys. Chem. 69, (1), 46, (1965).
19. I. Prigogine, P. Outer and Cl. Herbo: J. Phys. and Colloid Chem. 52, 321, (1948).
20. M. Manes, L. J. E. Hofer and S. Weller: J. Chem. Phys. 18, 1355, (1950).
21. S. W. Benson: *The Foundations of Chemical Kinetics*, McGraw Hill, 1960.
22. W. A. Edmiston and R.E. Grace: Trans. AIME 236, 1547 (1966).
23. Per Kofstad, *High Temperature Oxidation of Metals*, John Wiley and Sons, 1966.

VAPORIZATION CHEMISTRY IN THE OXIDATION OF CARBON, SILICON, CHROMIUM, MOLYBDENUM AND NIOBIUM

Earl A. Gulbransen and Sven A. Jansson

Westinghouse Research Laboratories

Pittsburgh, Pennsylvania 15235

Six types of oxidation processes occur in the oxidation of materials at high temperature. Reactions involving volatile species are important in evaluating and understanding the use of materials at high temperature. Thermochemical diagrams are used to analyze the complex equilibria found in the oxidation of carbon, silicon, chromium, molybdenum, and niobium. The results are compared with experimental oxidation data.

INTRODUCTION

The kinetics of oxidation of carbon, silicon, chromium, molybdenum, and niobium have been studied extensively over a wide temperature range and a wide pressure range. At least six types of oxidation phenomena occur: (1) At low temperatures, mechanisms based on the diffusion of oxygen and metal species through a compact oxide film, (2) At moderate and high temperatures, a combination of oxide film formation and oxide volatility, (3) At moderate and high temperatures, the formation of volatile metal and oxide species at the metal-oxide interface and transport through the oxide lattice or through mechanically formed cracks in the oxide layer, (4) At moderate and high temperature, the direct formation of volatile oxide gases, (5) At high temperature, the gaseous diffusion of oxygen through a barrier layer of volatilized oxides, and (6) At high temperature, spalling of metal and oxide particles. In addition, if the oxide melts, one may have transport of oxygen to the metal interface by convection currents in the liquid.

This paper presents thermochemical analyses of the role of

volatilization processes in the oxidation of these five metals. Predictions are made of four important phenomena. One of these is where formation of gaseous species over the condensed oxide phases begins to affect the kinetics of oxidation. This is taken as the temperature where the vapor pressure of only one species reaches 10^{-9} atm or more. The choice is based on comparisons between normally occurring oxidation rates and the probable rates of volatilization as influenced by vapor pressure, diffusion in a boundary layer and gas flow rates. Experiments confirm the correctness of this choice.

Secondly, the temperature at which the pressure of volatile, oxide species generated at the metal-oxide interface equals the external gas pressure. These are conditions where the oxide may fracture. It also is a situation where species may diffuse extensively in fissures in the oxide film from the gas-metal interface towards the outside of the oxide. The species may condense or react with the oxidant as they reach the outer regions of the oxide film. This is a vapor deposition process, producing oxide away from the metal. Alternatively, the species may diffuse away from the specimen and react in another portion of the reaction system. All of these processes clearly can affect the kinetics of oxidation.

The third phenomenon is where complete volatilization of the reaction products begins to occur. This characteristically can be expressed as combinations of temperature and gas pressure. The absence of any condensed reaction product will eliminate solid state diffusion as an important reaction step and limit the considerations to the processes of adsorption, desorption, and surface reaction.

The fourth important phenomenon is where gas diffusion of reactant oxygen through the layer of gaseous or condensed (smoke) products becomes rate controlling. Experiments have shown also that for many systems volatilization tends to become very important as the vapor pressure of volatile species in equilibrium with the condensed phase exposed to the gas reaches one percent of the total pressure.

In addition to these four phenomena which can be referred to as "thermochemical effects", the normal processes of diffusion of reactant species through an oxide layer and of oxide spalling, due to the build-up of mechanical stresses in the oxide or at the metal-oxide interface, must be taken into account when the mechanisms of metal oxidation are discussed. The formation of molten oxides also affects the reaction. The discussions are limited here to the role of the "thermochemical effects". Oxygen pressures in the practical range of 10^{-6} to 1 atm are generally considered.

THERMOCHEMICAL ANALYSES

Thermochemical analyses can be applied to evaluate the stable condensed oxide phases, the nature of the volatile species and the equilibrium pressures of the several oxide species as a function of temperature and oxygen pressure. Until recently, thermochemical analyses had limited application to oxidation processes where volatile species were formed due to (1) lack of available data at high temperatures on both the condensed and volatile oxide species, and (2) lack of a simple method for treating the complex equilibria. Since 1960, theoretical calculations from optical spectra of volatile species, mass spectrometer studies of volatile species, and the use of computer techniques for compilation of data have greatly aided thermochemical analyses. Free energy, enthalpy and equilibrium constant data have been collected in the form of tables for many compounds and molecular species over a wide range of temperature, e.g., the JANAF Tables [1] and Schicks compilation [2].

Vaporization chemistry and thermochemical analyses have been used previously to interpret the mechanisms of oxidation of silicon [3,4], chromium [5] and molybdenum [6]. Since these studies were made, graphic methods have been developed for analyzing the complex equilibria found in the high temperature oxidation of metals. This development, together with new compilations of thermochemical data, have made possible new insights in our understanding high temperature oxidation mechanisms. Complete thermochemical phase diagrams can now be developed for many elemental-oxygen systems over a wide range of temperature.

For considerations of metal-oxygen systems that form condensed phases as well as metal or oxide gaseous species, the most useful type of diagram is where for a chosen temperature the logarithms of the pressures of these species $\{\log p(Me_xO_y)\}$ are plotted against the logarithm of the oxygen pressure $\{\log p(O_2)\}$. Phase boundaries between condensed phases are shown as lines at the appropriate oxygen pressures. The diagrams were first used by Kellogg [7] and by Jansson and Gulbransen [8]. Gulbransen [9] in a recent paper used $\log p(Me_xO_y)$ vs. $1/T$ diagrams to evaluate the oxidation of refractory metals.

Table 1 shows the melting and boiling points for metals and oxides and the chemical formulae for the major volatile species in each system.

Tables 2-6 shows the thermochemical data for the several element-oxide systems in units of logarithms of the equilibrium constant, $\log K_p$, for the several equilibria involving condensed and gaseous oxide species. Here (s), (ℓ) and (g) refer to the solid, liquid, and gaseous states.

TABLE 1

Physical and Chemical Properties of C, Cr, Si, Mo and Nb and Their Oxides

Element			Condensed Oxide			Major
Formula	M.P. °C	B.P. °C	Formula	M.P. °C	B.P. °C	Volatile Species
C	--	4200[a]	CO	-207	-190	CO, CO_2
			CO_2	--	-78.5[a]	
Cr	1900	2642	Cr_2O_3[b]	2280	--	Cr, CrO_3
Si	1412	3219	SiO_2	1610	--	SiO, SiO_2
Mo	2617	4692	MoO_3[b]	797	1155	$(MoO_3)_x$
Nb	2468	4768	Nb_2O_5[b]	1512	--	NbO_2

a. sublimes
b. stable phase for temperatures of 1273-2273°K and for log $P(O_2)$ = $0 \rightarrow -4$
c. high quartz → liquid

TABLE 2

Thermochemical Data for the Carbon-Oxygen System [1]

Equation	Equilibrium Constant	log K_p at T°K					
		1000	1200	1400	1600	1800	2000
$C(s) \rightleftarrows C(g)$	$K_p = p_C$	-29.200	-22.953	-18.493	-15.150	-12.552	-10.476
$2C(s) \rightleftarrows C_2(g)$	$K_p = p_{C_2}$	-33.381	-26.118	-20.941	-17.068	-14.063	-11.665
$3C(s) \rightleftarrows C_3(g)$	$K_p = p_{C_3}$	-31.158	-24.268	-19.360	-15.691	-12.847	-10.579
$3C(s) + O_2(g) \rightleftarrows C_3O_2(g)$	$K_p = p_{C_3O_2}\, p_{O_2}^{-1/2}$	7.884	7.086	6.512	6.079	5.740	5.466
$2C(s) + 1/2\ O_2(g) \rightleftarrows C_2O(g)$	$K_p = p_{C_2O}\, p_{O_2}^{-1/2}$	-8.609	-6.107	-4.325	-2.992	-1.959	-1.135
$C(s) + 1/2\ O_2(g) \rightleftarrows CO(g)$	$K_p = p_{CO}\, p_{O_2}^{-1/2}$	10.459	9.479	8.771	8.234	7.811	7.469
$C(s) + O_2(g) \rightleftarrows CO_2(g)$	$K_p = p_{CO_2}\, p_{O_2}^{-1}$	20.680	17.243	14.785	12.940	11.504	10.353

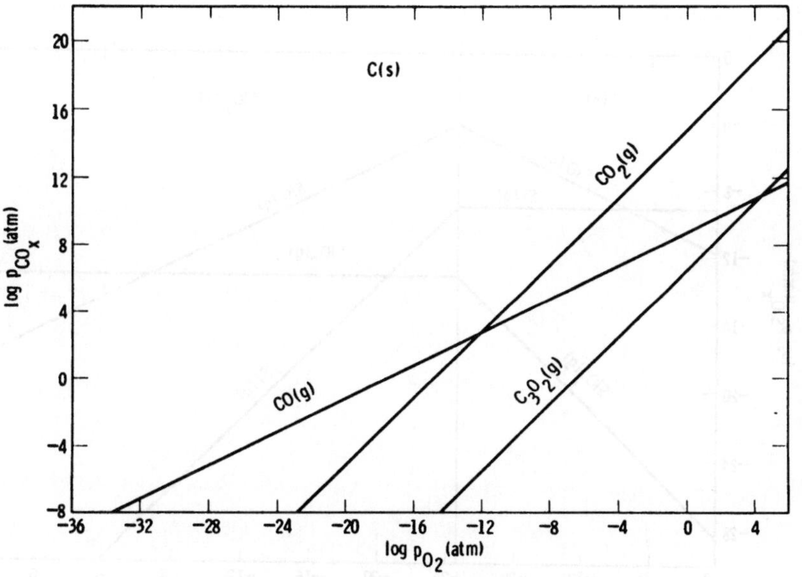

Fig. 1—Thermochemical diagram for the carbon-oxygen system at 1400°K

TABLE 3

Thermochemical Data for Silicon Oxygen Reactions [1]

Equation	Equilibrium Constant	log K_p at T°K					
		800	1000	1200	1400	1600	1800
$Si(s,\ell) \rightleftarrows Si(g)^{a)}$	$K_p = p_{Si}$	-21.650	-15.786	-11.885	-9.107	-7.029	-5.518
$2Si(s,\ell) \rightleftarrows Si_2(g)$	$K_p = p_{Si_2}$	-28.634	-20.994	-15.916	-12.302	-9.603	-7.713
$3Si(S,\ell) \rightleftarrows Si_3(g)$	$K_p = p_{Si_3}$	-30.655	-22.443	-16.998	-13.133	-10.256	-8.333
$Si(s,\ell) + 1/2\ O_2(g) \rightleftarrows SiO(g)$	$K_p = p_{SiO}\ p_{O_2}^{-1/2}$	11.151	9.787	8.864	8.194	7.683	7.178
$Si(s,\ell) + O_2(g) \rightleftarrows SiO_2(s,\ell)^{a,b)}$	$K_p = p_{O_2}^{-1}$	49.985	38.150	30.277	24.663	20.460	17.096
$Si(s,\ell) + O_2(g) \rightleftarrows SiO_2(g)$	$K_p = p_{SiO_2}\ p_{O_2}^{-1}$	20.128	16.114	13.433	11.513	10.070	8.845

a) Melting points
Si : 1685°K
SiO_2: 1883°K

b) Transition point
SiO_2: 848°K

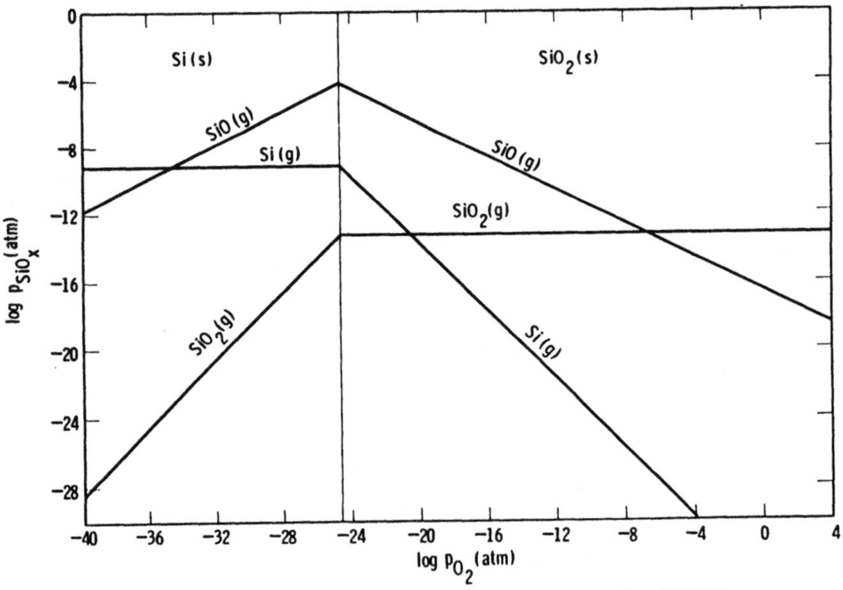

Fig. 2—Thermochemical diagram for the silicon-oxygen system at 1400°K

TABLE 4

Thermochemical Data for the Chromium-Oxygen System [2, 10]

Equation	Equilibrium Constant	log K_p at T°K					
		1000	1200	1400	1600	1800	2000
$Cr(s,\ell) \rightleftarrows Cr(g)$[a]	$K_p = p_{Cr}$	-13.006	-9.604	-7.191	-5.398	-4.026	-2.865
$Cr(s,\ell) + 1/2\ O_2(g) \rightleftarrows CrO(g)$	$K_p = p_{CrO}p_{O_2}^{-1/2}$	-5.296	-3.532	-2.292	-1.380	-0.695	-0.165
$Cr(s,\ell) + O_2(g) \rightleftarrows CrO_2(s)$	$K_p = p_{O_2}^{-1}$	Decomposes above 800°K					
$Cr(s,\ell) + O_2(g) \rightleftarrows CrO_2(g)$	$K_p = p_{CrO_2}p_{O_2}^{-1}$	5.512	4.762	4.208	3.776	3.416	3.111
$Cr(s,\ell) + 3/2\ O_2(g) \rightleftarrows CrO_3(s)$	$K_p = p_{O_2}^{-3/2}$	Gaseous above 800°K					
$Cr(s,\ell) + 3/2\ O_2(g) \rightleftarrows CrO_3(g)$	$K_p = p_{CrO_3}p_{O_2}^{-3/2}$	11.604	9.085	7.275	5.906	4.821	3.939
$2Cr(s,\ell) + 3/2\ O_2(g) \rightleftarrows Cr_2O_3(s)$	$K_p = p_{O_2}^{-3/2}$	45.78	35.92	28.89	23.74	19.55	16.40

a) Melting point for Cr(s) is 2173°K

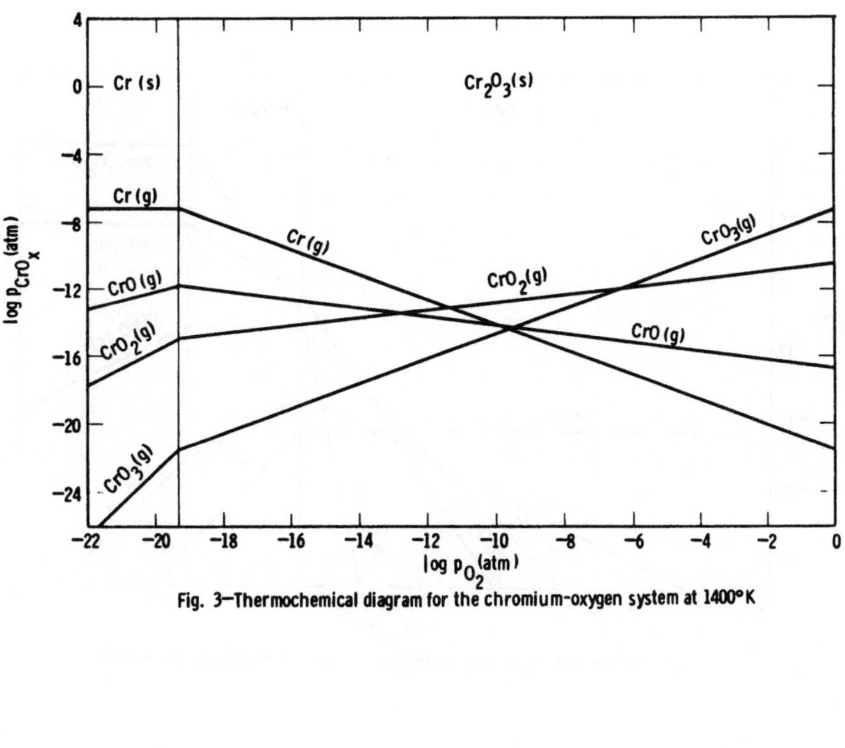

Fig. 3—Thermochemical diagram for the chromium-oxygen system at 1400°K

TABLE 5

Thermochemical Data for the Molybdenum-Oxygen System [2, 11]

Reaction	Equilibrium Constant	log K_p at T°K						
		600	800	1000	1200	1400	1600	1800
$Mo(s) \rightleftarrows Mo(g)$	$K_p = p_{Mo}$	-49.657	-35.293	-26.690	-20.908	-16.892	-13.845	-11.483
$Mo(s) + O_2(g) \rightleftarrows MoO_2(s)$	$K_p = p_{O_2}^{-1}$	41.569	28.836	21.240	16.203	12.640	9.984	7.943
$Mo(s) + 1/2\ O_2(g) \rightleftarrows MoO(g)$	$K_p = p_{MoO}p_{O_2}^{-1/2}$	-26.823	-18.925	-14.207	-11.078	-8.855	-7.200	-5.924
$Mo(s) + O_2(g) \rightleftarrows MoO_2(g)$	$K_p = p_{MoO_2}p_{O_2}^{-1}$	1.758	1.688	1.631	1.581	1.535	1.490	1.446
$Mo(s) + 3/2\ O_2(g) \rightleftarrows MoO_3(s,\ell)$[a]	$K_p = p_{O_2}^{-3/2}$	51.491	35.401	25.806	19.719	15.544	12.4[b]	10.0[b]
$Mo(s) + 3/2\ O_2(g) \rightleftarrows MoO_3(g)$	$K_p = p_{MoO_3}p_{O_2}^{-3/2}$	26.473	19.084	14.651	11.695	9.580	7.950	6.749
$MoO_3(s,\ell) \rightleftarrows 1/3(MoO_3)_3(g)$[c]	$K_p = p_{(MoO_3)_3}^{1/3}$	-4.616	-2.327	-0.955	-0.345	-0.0393	0.1646	0.3676
$Mo(s) + 3/2O_2(g) \rightleftarrows 1/3(MoO_3)_3(g)$	$K_p = p_{(MoO_3)_3}^{1/3}p_{O_2}^{-3/2}$	46.875	33.07	24.851	19.374	15.505	12.565	10.37

a) Melting point for MoO_3 is 1070°K

b) Estimated

c) Reference (11)

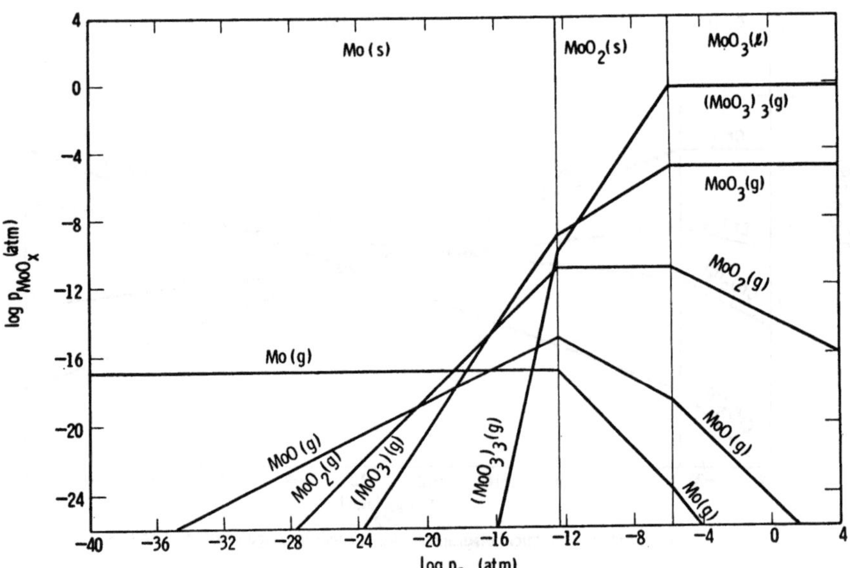

Fig. 4—Thermochemical diagram for the molybdenum-oxygen system at 1400°K

TABLE 6

Thermochemical Data for the Niobium-Oxygen System [2]

Equation	Equilibrium Constant	log K_p at T°K						
		1000	1400	1800	2000	2200	2600	3000
$Nb(s,\ell) \rightleftarrows Nb(g)^a$	$K_p = p_{Nb}$	-29.708	-18.969	-13.017	-10.940	-9.243	-6.640	-4.783
$Nb(s,\ell) + 1/2\ O_2(g) \rightleftarrows NbO(s,\ell)^a$	$K_p = p_{O_2}^{-1/2}$	16.755	10.723	7.400	6.245	5.306	4.061	3.121
$Nb(s,\ell) + 1/2\ O_2(g) \rightleftarrows NbO(g)$	$K_p = p_{NbO}\ p_{O_2}^{-1/2}$	-5.114	-2.350	-0.848	-0.334	0.081	0.704	1.100
$Nb(s,\ell) + O_2(g) \rightleftarrows NbO_2(s,\ell)^a$	$K_p = p_{O_2}^{-1}$	32.006	20.390	14.009	11.789	9.980	7.392	5.496
$Nb(s,\ell) + O_2(g) \rightleftarrows NbO_2(g)$	$K_p = p_{NbO_2}\ p_{O_2}^{-1}$	11.750	8.564	6.794	6.171	5.660	4.866	4.231
$2Nb(s,\ell) + 5/2\ O_2(g) \rightleftarrows Nb_2O_5(s,\ell)^a$	$K_p = p_{O_2}^{-5/2}$	76.398	48.372	32.943	27.877	23.776	17.551	12.981

a) Melting Points

Nb: 2741°K

NbO: 2218°K

NbO_2: 2270°K

Nb_2O_5: 1785°K

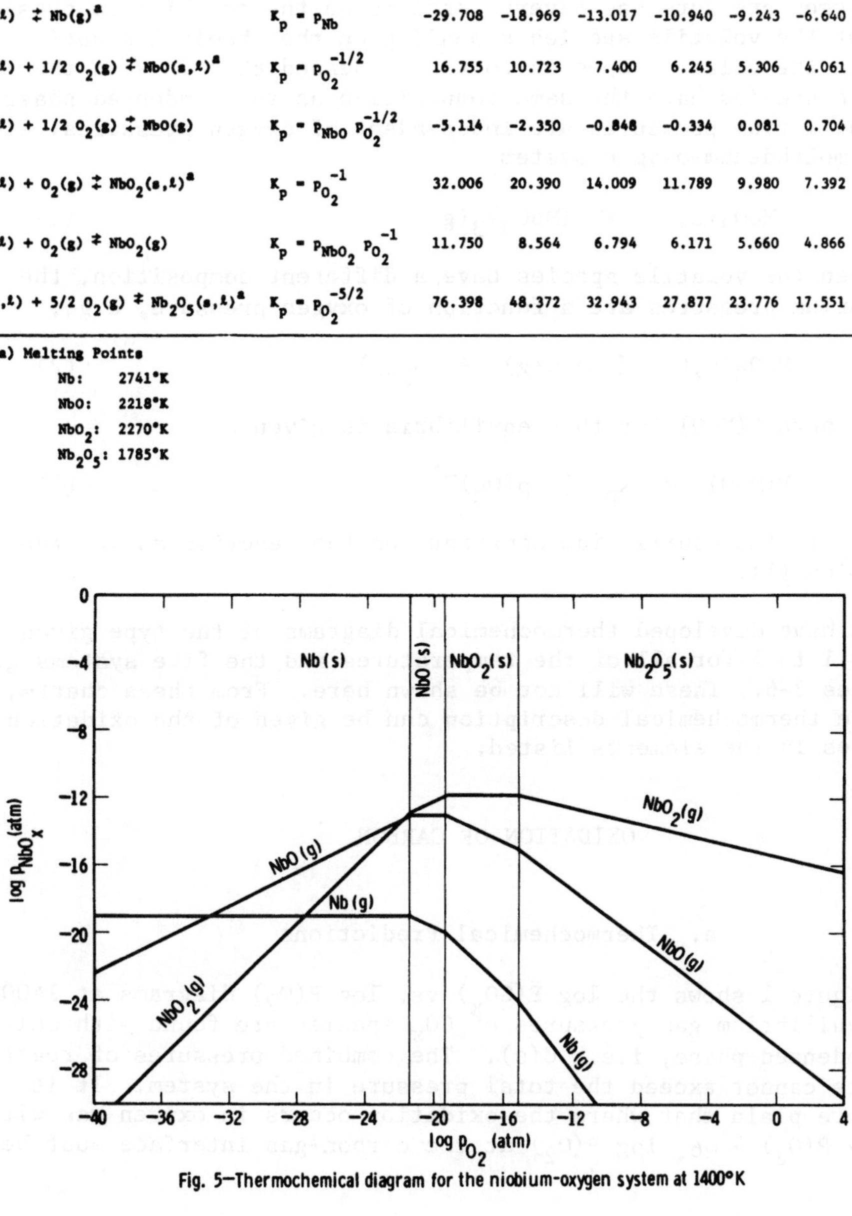

Fig. 5—Thermochemical diagram for the niobium-oxygen system at 1400°K

Figures 1-5 shows typical log P(MO_x) vs. log P(O_2) diagrams at 1400°K for log P(O_2) = -40 to + 4 for the five systems. systems. Wide variations can be seen in the diagrams. The stable condensed phases for high oxygen pressures of log P(O_2) = 0 to -6 of interest in normal oxidation processes are C(s), Cr_2O_3(s), SiO_2(s,ℓ), MoO_3(s,ℓ), and Nb_2O_5(s,ℓ) for temperatures between 600° and 2000°K. Equilibrium oxygen pressures between the various condensed phases are shown by the vertical lines.

Oxygen pressure has several effects on the equilibrium pressures of the volatile species depending on the chemical equation relating the volatile species to the condensed phase. When the volatile species have the same composition as the condensed phase, the equilibrium pressures are independent of oxygen pressures, e.g., in the molybdenum-oxygen system,

$$3MoO_3(s,\ell) \rightleftarrows (MoO_3)_3(g) \tag{1}$$

When the volatile species have a different composition, the equilibrium pressures are a function of oxygen pressure, e.g.,

$$MoO_3(s,\ell) \rightleftarrows MoO(g) + O_2(g) \tag{2}$$

The pressure P(MoO) for this equilibria is given by

$$P(MoO) = K_p \cdot p(O_2)^{-1} \tag{3}$$

where K_p is the equilibrium constant for the reaction as written in expression (1).

We have developed thermochemical diagrams of the type given in Figures 1 to 5 for all of the temperatures and the five systems given in Tables 2-6. These will not be shown here. From these charts, a complete thermochemical description can be given of the oxidation processes in the elements listed.

OXIDATION OF CARBON

a. Thermochemical Predictions

Figure 1 shows the log P(CO_x) vs. log P(O_2) diagrams at 1400°K. High equilibrium gas pressures of CO_x species are found with only one condensed phase, i.e., C(s). The combined pressures of reaction products cannot exceed the total pressure in the system. It is therefore plain that where the oxidation occurs in oxygen gas with $0 > \log P(O_2) > -6$, log P(O_2) at the carbon-gas interface must be approximately -18 to -29 in order for equilibrium to occur. CO(g) will be the dominating species throughout this range. This condition corresponds to the end of the reaction in a closed vessel, i.e., when no oxygen is left to diffuse to and react with the carbon. In cases where there is a sufficiently large supply of $O_2(g)$ in the system, the steady state rate at gas diffusion control conditions will correspond to the interdiffusion of $O_2(g)$ and CO(g). This means a mass balance where one molecule of $O_2(g)$ diffuses to the surface of the carbon as two molecules of CO(g) diffuse in the opposite direction. This does not represent equilibrium in the gas at the carbon surface. Where gas diffusion is not important, the processes of chemical adsorption of oxygen, desorption of CO(g) (and perhaps of some $CO_2(g)$) or else the surface reaction must be rate determining.

The adsorption and desorption processes have been described theoretically by Glasstone, Laidler and Eyring [12]. Chemical processes are characterized by high temperature coefficients of reaction. Gas diffusion processes have low temperature coefficients.

b. Experimental Results

Gulbransen and coworkers [13,14] have studied the kinetics of oxidation of high purity graphite from 300° to 1500°C. Table 7 and Fig. 6 show the kinetic data for an oxygen pressure of 38 torr. Below 725°C the data fall on a straight line and show an activation energy of 39 kcal/mole. A mobile adsorption of oxygen was proposed as the mechanism of oxidation. Above 725°C a sharp break occurs in the plot to a nearly temperature independent reaction rate. The transition temperature varies with oxygen pressure, sample size and flow rate. These observations and the low temperature coefficient of reaction support the proposed gas diffusion of oxygen through a barrier layer as the mechanism of reaction.

The rate of oxidation at the transition temperature is: log dn'/dt = 16.7 for a 6.3 cm^2 area specimen. For 1.19 cm^2 area specimens, log dn'/dt = 17.5 for an oxygen pressure of 38 torr and a transition temperature of 800°C. Using small specimens, reaction rates at transition of log dn'/dt = 19 have been noted at 38 torr oxygen pressure.

Experiments at 76 torr showed gas diffusion to become rate determining as the reaction rate became approximately 4×10^{17} atoms of carbon per second at 1000°C.

Gas diffusion control phenomena were avoided by use of a flow system. This gave higher reaction rates [15].

TABLE 7

Rates of Oxidation of Graphite 600°C - 1500°C
38 Torr of Oxygen, 6.3 cm^2 Area Specimens

Temp. °C	Rate dn'/dt (initial) atoms of C/cm^2-sec	Temp. °C	Rate dn'/dt (initial) atoms of C/cm^2-sec
600	1.65×10^{15}	1000	5.42×10^{16}
700	2.96×10^{16}	1200	7.88×10^{16}
800	5.17×10^{16}	1400	9.53×10^{16}
900	5.47×10^{16}	1500	9.53×10^{16}

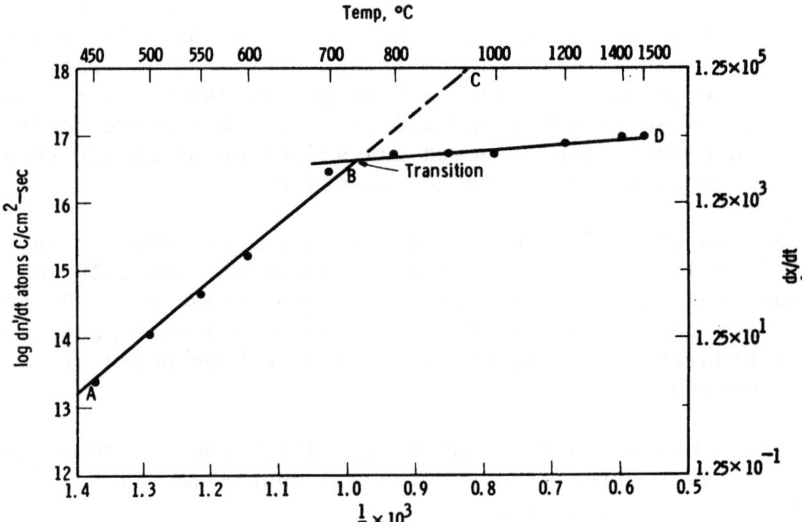

Fig. 6—Oxidation of graphite, 38 torr O_2, 6.2 cm^2 area
ABC chemical control, BD gas diffusion control

c. Comparison of Theoretical and Experimental Data

There is excellent agreement between predicted and observed phenomena. Gas diffusion was indeed observed to become rate-determining under certain conditions. The conditions where transition occurs depends naturally on the combined effects of high reaction rate at the C(s) surface and the transfer rate for CO(g) to diffuse from the specimen and O_2(g) to diffuse through the reactant layer towards the specimen surface. The maximum rate of reaction of the carbon surface cannot be foreseen on the basis of thermochemical analyses.

The characteristic feature of this system is the formation of gaseous reaction products that do not condense. It is clear from the thermochemical considerations and it has been demonstrated that flow conditions will aid in eliminating the gas diffusion step.

OXIDATION OF SILICON

a. Thermochemical Predictions

Figure 2 shows a log $P(SiO_x)$ vs. log $P(O_2)$ diagram at 1400°K. Strong peaks in the SiO(g) pressures occur at oxygen pressures found at the Si(s)-SiO_2(s) interface, i.e., with log P_{SiO} = -4.2 for log $P(O_2)$ = -24.6. The silicon-oxygen system is unusual due to the formation of high SiO(g) pressures at the metal-oxide interface.

Figure 7 shows a log $P(SiO_x)$ vs. 1/T plot of the volatile species SiO_x at the oxide-gas interface for an oxygen pressure of 1 atm. SiO_2(g) is the only important species at temperatures up to the melting point of silicon at 1412°C, log $P(SiO_2)$ < -9. No important volatilization of SiO_2(g) should occur before this temperature is reached. Figure 8 shows a plot of log P_{SiO} and log P_{Si} vs 1/T at the Si(s)-SiO_2(s) interface. Comparison is made with log P_{CrO} and log P_{Cr} from the chromium-oxygen system at the Cr(s)-Cr_2O_3(s) interface.

Although low SiO(g) and SiO_2(g) pressures are found at the oxide-gas interface at normal oxygen pressures, volatile species will be important for three reasons. First, high SiO(g) pressures which develop at the Si(s)-SiO_2(s) interface can lead to a rapid transfer of SiO(g) through a porous oxide film or to rupturing of the oxide film when P_{SiO} > $P(O_2)$. Second, very low SiO_2(g) and SiO(g) pressures exist over SiO_2(s) crystals formed when SiO(g) reacts with excess O_2(g) to form SiO_2(s). Third, holes in the SiO_2(s) can be formed upon reaction with Si(s) to form SiO(g). Silicon is unique among the metals considered here in all of these respects. The effective species in a gas barrier layer which limits oxygen diffusion can be SiO_x(g) or else SiO_2(s) smoke, which would offer less resistance to oxygen diffusion than gaseous oxide barriers. If smoke is formed, higher transition rates of reaction are predicted than those for the oxidation of carbon.

b. Active and Passive Oxidation of Silicon

Wagner [3] has developed a theory to interpret the transition between active (rapid) oxidation of silicon found at low oxygen pressures and passive (slow) oxidation found at higher oxygen pressures. Wagner gives an equation for the maximum oxygen pressures, $P(O_2)$, for which a bare silicon surface can exist as

$$P(O_2) = 1/2 \quad D(SiO)/D(O_2)^{1/2} P(SiO)$$

Here P(SiO) is the equilibrium pressure of SiO(g) at the Si(s)-SiO_2(s) interface and D(SiO/D(O_2) is the ratio of diffusion constants for SiO(g) and O_2(g) in the boundary layer and equal to 0.64. Passive oxidation would occur when $P(O_2) > 0.64\ P(SiO)$ and active oxidation when $P(O_2) < 0.64\ P(SiO)$.

C. Experimental Study

Gulbransen, Andrew and Brassart [4] confirmed Wagner's [3] prediction of active and passive oxidation processes for the

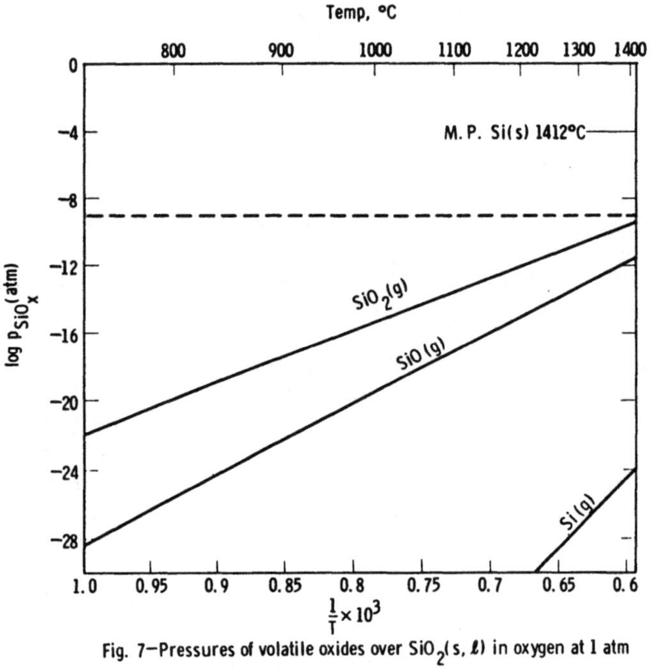

Fig. 7—Pressures of volatile oxides over SiO_2(s, ℓ) in oxygen at 1 atm

oxidation of silicon at temperatures of 1100°-1300°C. A dynamic flow system was used at low oxygen pressures. The oxidation data in units of atoms of silicon reacting per cm^2-sec are shown in Table 8. Weight losses indicating active oxidation were found for all of the oxidation experiments at 9×10^{-3} and 4×10^{-2} torr and for the 1200° and 1300°C experiments at 10^{-1} torr oxygen pressure. A slow weight gain of 1.13×10^{16} atoms of silicon reacting per cm cm^2-sec was found at 1100°C and 10^{-1} torr oxygen pressure. At 1200°C and 10^{-1} torr a weight loss of 3.5×10^{18} atoms of silicon per cm^2-sec was found. The ratio of slow to rapid oxidation was 1 to 300. Table 8 also shows that the rates of oxidation in the active region were independent of temperature and nearly proportional to pressure or to the flow rate of oxygen.

d. Comparison of Theoretical and Experimental Data

Figure 9 shows a log P_{SiO} vs. 1/T plot for SiO(g) pressures at the Si(s)-SiO_2(s) interface together with the experimental data for active (A) and passive (P) types of oxidation. The solid line given in Fig. 9 shows equilibrium values of log P_{SiO} at the metal-oxide interface. Points to the left of this line are for the condition

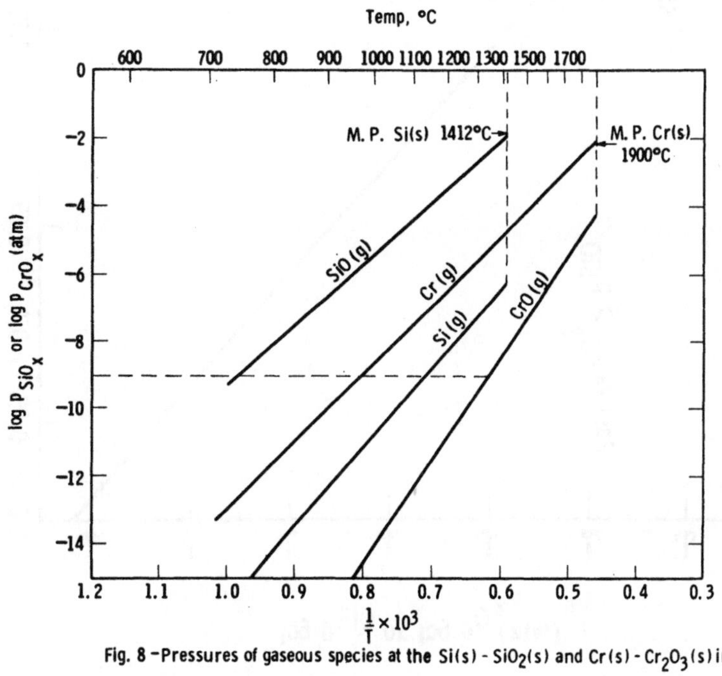

Fig. 8 -Pressures of gaseous species at the Si(s) - SiO_2(s) and Cr(s) - Cr_2O_3(s) interfaces

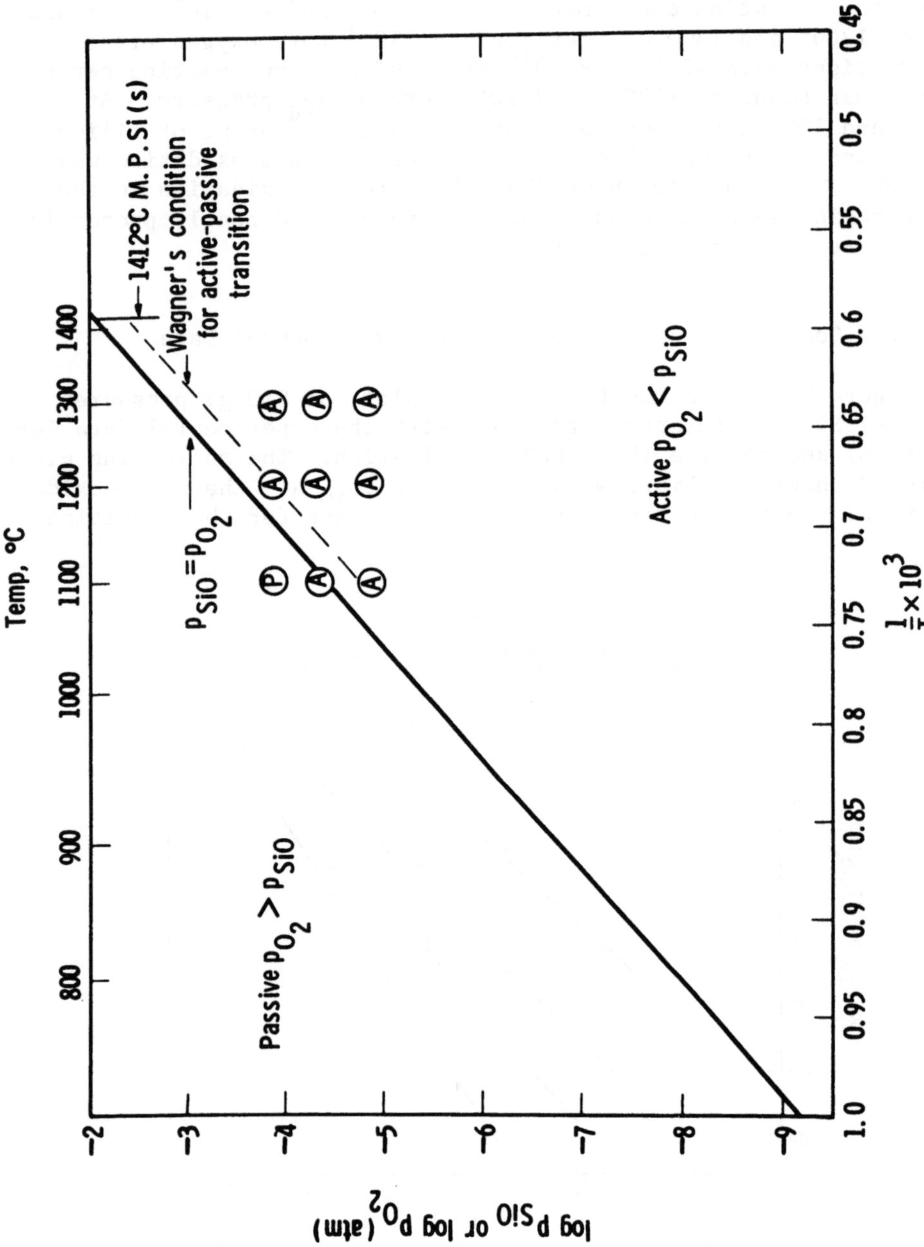

Fig. 9—Active and passive regions for the oxidation of silicon

TABLE 8

Rates of Oxidation of High Purity Silicon
1100° - 1300°C

p(torr)	Temp °C	Flow, atoms O/sec	Weight Loss dn'/dt atoms Si/cm^2-sec
9×10^{-3}	1100	8.6×10^{17}	2.96×10^{17}
9×10^{-3}	1200	4.3×10^{17}	1.56×10^{17}
9×10^{-3}	1200	1.0×10^{18}	3.53×10^{17}
9×10^{-3}	1300	7.3×10^{17}	2.45×10^{17}
4×10^{-2}	1100	3.9×10^{18}	1.19×10^{18}
4×10^{-2}	1200	4.1×10^{18}	1.09×10^{18}
4×10^{-2}	1300	3.9×10^{18}	1.17×10^{18}
1×10^{-1}	1100	1.4×10^{19}	1.13×10^{16}*
1×10^{-1}	1200	1.4×10^{19}	3.51×10^{18}
1×10^{-1}	1300	1.4×10^{19}	3.24×10^{18}

* Weight Gain

$P(O_2) > P(SiO)$ and points to the right of the line $P(O_2) < P(SiO)$. The line also divides the active and passive types of oxidation. A dashed line representing Wagner's equation for log $P(O_2)$ for maintaining a bare silicon surface is also shown in the figure.

The low temperature dependence of the kinetic data of Table 8 and the high rates of oxidation indicate gas diffusion of oxygen to be the rate controlling reaction mechanism. The simple criteria of $P(O_2) = P(SiO)$ fits the transition conditions between active and passive oxidation for the limited number of experimental data points. For active oxidation $P(O_2) < P(SiO)$ and for passive oxidation $P(O_2) > P(SiO)$.

Our interpretation of the active and passive regions of oxidation is as follows: When $P(SiO) > P(O_2)$, SiO(g) forms at the

$Si(s)-SiO_2(s)$ interface consuming the oxide film. The $SiO(g)$ diffuses rapidly through the porous oxide or ruptures the oxide film to react with $O_2(g)$ to form $SiO_2(s)$ smoke. This smoke formation may occur within the porous oxide film, in the gas phase or at other surfaces in the reaction system. When $P(O_2) > P(SiO)$ a compact oxide forms which limits further reaction and represses the solid phase reaction of $Si(s)$ with $SiO_2(s)$. When $P(SiO) > P(O_2)$ rapid reaction occurs before the barrier layer of $SiO(g)$ or $SiO_2(s)$ smoke limits the diffusion of oxygen to the surface.

Although Wagner's analysis is essentially correct, the thermochemical data of Figs. 2 and 7 show that $SiO(g)$ may not exist as any equilibrium vapor in contact with $O_2(g)$. The barrier layer can also be $SiO_2(s)$ smoke. This latter fact and the formation of high $SiO(g)$ pressures at the metal-oxide interface are the most significant differences from Wagner's model.

Available data do not allow any definite conclusions to be drawn about the comparative protectiveness of the barrier layers of reaction products over carbon and silicon. The indications from flow

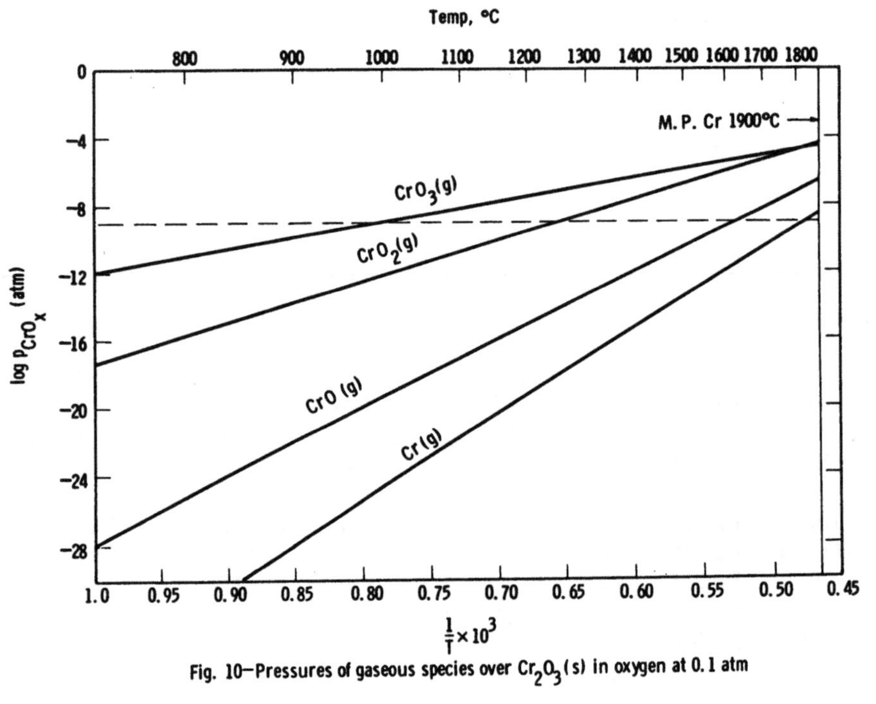

Fig. 10—Pressures of gaseous species over $Cr_2O_3(s)$ in oxygen at 0.1 atm

condition experiments are, however, that silicon oxidizes more rapidly than carbon. We believe that this is a result of $SiO_2(s)$ smoke formation.

OXIDATION OF CHROMIUM

a. Thermochemical Predictions

Jansson and Gulbransen [8] have made a thorough thermochemical study of the chromium-oxygen-carbon system. The results as applied to the oxidation of chromium in oxygen will be described here.

Figure 3 shows a log $P(CrO_x)$ vs. log $P(O_2)$ diagram at 1400°K for the chromium-oxygen system. High pressures of volatile species occur at both the $Cr(s)-Cr_2O_3(s)$ and $Cr_2O_3(s)-O_2(g)$ interfaces.

Figure 3 shows a plot of log $P(CrO_x)$ vs. 1/T at the $Cr(s)-Cr_2O_3(s)$ interface. At 1000°C, log P_{Cr} = -8.7. Figure 8 also shows data for log P(CrO), log P(SiO) and log P(Si) at the $Si(s)-SiO_2(s)$ interface. Although important pressures of Cr(g) form at the $Cr(s)-Cr_2O_3(s)$ interface, these pressures are lower at the same temperature than those found for SiO(g) at the $Si(s)-SiO_2(s)$ interface. We predict the major effect of the relatively high Cr(g) pressures at temperatures of 1000° to 1200°C is the increased transfer of chromium atoms through pores or cracks in the oxide film. The active-passive transition in the oxidation of chromium occurs at lower oxygen pressures or at higher temperatures than for the silicon case.

Figure 10 shows a plot of log $P(CrO_x)$vs. 1/T for log $P(O_2)$ = -1 at the $Cr_2O_3(s)-O_2(g)$ interface. At 1000°C log $P(CrO_3)$ = -9 and at 1800°C log $P(CrO_3)$ = -4.65. Direct oxide volatility is predicted for 1000°C and higher temperatures. Log $P(CrO_3)$ increases slowly with temperature. Oxidation with complete oxide volatility does not occur for log $P(O_2)$ = -1 at temperatures below the melting point of chromium, since log $P(CrO_3) < -1$ throughout that temperature range.

We predict on the basis of these calculations, that the kinetics of Cr(s) oxidation should be affected by both the effects of Cr(g) formation, with diffusion in fissures and pores in oxide, and of external $CrO_3(s)$ formation, reacting to form $Cr_2O_3(s)$ at other portions of the reaction system. For log P(Cr) and log $P(CrO_3)$ = -9, the points of transition are expected at 960°C and 1020°C, respectively, when log $P(O_2)$ = -1.

b. Experimental Studies

No complete study has been made on the oxidation of chromium which can be used to evaluate the thermochemical predictions.

Gulbransen and Andrew [5] made a study on the kinetics of oxidation of chromium up to 1100°C using a sensitive weight change method alone. Table 9 shows a summary of the parabolic rate law constants and Fig. 11 shows a plot of the logarithms of the parabolic rate law constants, A, vs 1/T. The transition found near 960°C was related to the beginning influence of chromium volatilization. They proposed that metal vapor could penetrate through portions of the oxide film and short-circuit the diffusion process.

TABLE 9

Parabolic Rate Law Constants for Oxidation in Chromium

Temp. °C	A $(g/cm^2)^2\ sec^{-1}$	Temp. °C	A $(g/cm^2)^2\ sec^{-1}$
700	8.22×10^{-15}	1000	4.51×10^{-12}
750	3.14×10^{-14}	1000	4.93×10^{-12}
800	6.95×10^{-14}	1025	1.15×10^{-11}
850	1.18×10^{-13}	1050	1.81×10^{-11}
900	3.02×10^{-13}	1075	3.08×10^{-11}
950	5.50×10^{-13}	1100	5.28×10^{-11}
975	1.20×10^{-12}		

c. Comparison of Theoretical and Experimental Data

The thermochemical analyses by Jansson and Gulbransen [8] allow a better understanding of the experimental data. It is thus clear that Cr(g) will volatilize through a porous oxide film if the external pressure is low. The metal vapor can react with oxygen within, at the surface of, or outside the oxide film to form Cr_2O_3(s) or CrO_3(g). In a sufficiently oxidizing environment, CrO_3(g) is the only likely gaseous species outside the oxide surface. The temperatures at which volatilization should become important, i.e., for log P(CrO_3) equal to -9, compare very closely with those for the transitions observed by Gulbransen and Andrew [5], see Fig. 11. The break C occurs at 960°C where log P(Cr) at the metal-oxide interface reaches the value of -9, and the break B occurs at approximately 1020°C where log P(CrO_3 becomes -9 for the experimental oxygen pressure, log P(O_2) = -1.

OXIDATION OF MOLYBDENUM

a. Thermochemical Predictions

Figure 4 shows the log P(MoO_x) vs. log P(O_2) diagram for 1400°K. This figure indicates that the molybdenum-oxygen system is characterized by high pressures of $(MoO_3)_3$(g) over the MoO_3(s,ℓ) condensed phase, whereas the pressures of volatile species at the Mo(s)-MoO_2(s) interface are very low. The melting point of MoO_3(s) is low or 1070°K (797°C).

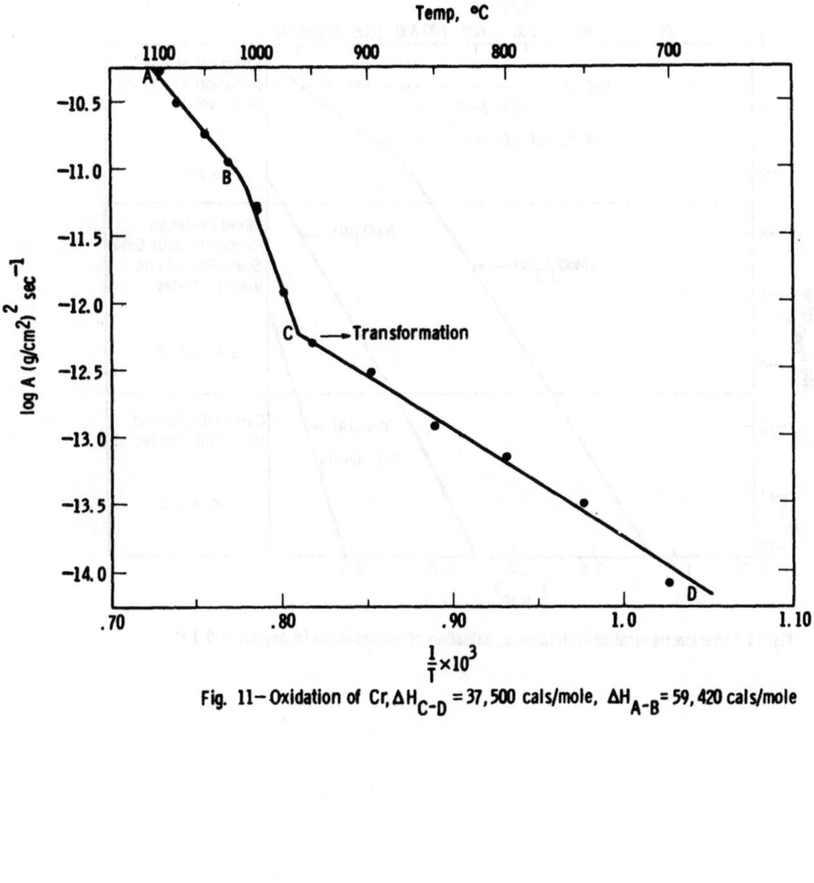

Fig. 11—Oxidation of Cr, ΔH_{C-D} = 37,500 cals/mole, ΔH_{A-B} = 59,420 cals/mole

Figure 12 shows a log $P(MoO_x)$ vs. 1/T plot for the volatile oxide species of molybdenum for log $P(O_2)$ = -1. The horizontal line for log P(MoO) = -9 indicates where oxide volatilization can first be expected to occur. For $(MoO_3)_3(g)$, this corresponds to a temperature of ∿ 475°C. The break at ∿ 800°C indicates the melting point for $MoO_3(s)$. Steady state reaction at any point requires that 9/2 $O_2(g)$ are transported inwards for each $(MoO_3)_3(g)$ being formed and diffusing outward in a boundary layer. $(MoO_3)_3(g)$ can be in equilibrium with the gas mixture at the oxide surface under mass balance conditions. This is different than for the case of carbon. It gives a critical value of log $P(MoO_3)$ = -1.74, corresponding to a temperature of ∿ 830°C.

The horizontal line for log $P(MoO_x)$ = -3 in Fig. 12 refers to experimental observations of important volatilization at 1% of the total pressure.

The high and log $P(O_2)$ independent equilibrium vapor pressures of $(MoO_3)_3(g)$ and other polymer species of the trioxide allow the gaseous products to diffuse away from the specimen without further reaction in the gas phase. Condensation, i.e., smoke formation, will occur at lower temperature regions of the reaction system but

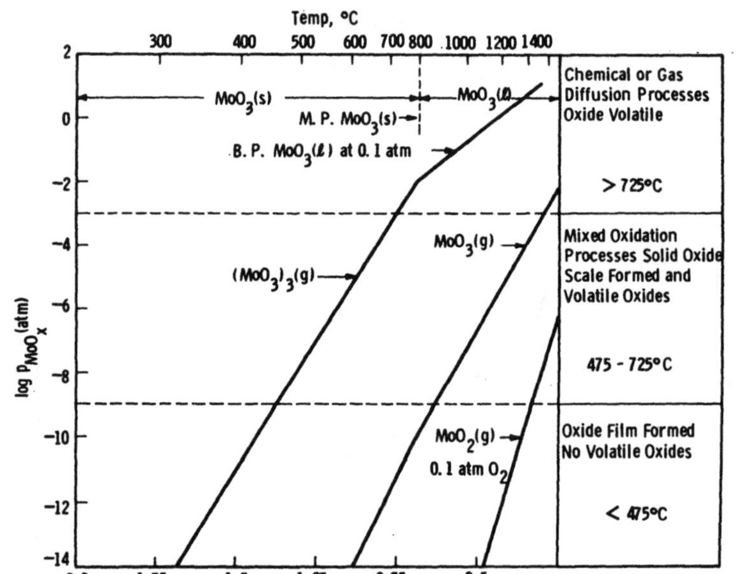

Fig. 12 –Thermochemical considerations, oxidation of molybdenum in oxygen at 0.1 atm

not in the immediate vicinity of the specimen. It is clear that the oxidation of molybdenum should occur in a similar manner to that of carbon. In particular, we would expect gas diffusion to become rate determining at sufficiently high temperature and specimen area, and for small reaction vessel volumes. For the same reaction conditions, we would expect molybdenum to react approximately three times more rapidly than carbon before gas diffusion becomes rate controlling, since $(MoO_3)_3(g)$ species are formed from molybdenum but not for carbon. $(MoO_3)_4(g)$ species would give a factor of 4.

b. Experimental Studies

Two studies have been made in our laboratories to evaluate the oxidation behavior of molybdenum. Gulbransen and Wysong [16] studied the low temperature oxidation kinetics. Adherent and compact oxide films were formed at temperatures up to 450°C. Measurable oxide volatility was found at 475°C where $\log P(MoO_3)_3 = -9$.

Gulbransen, Andrew and Brassart [6] studied the oxidation of molybdenum from 550°C to 1700°C. At 600°C both oxide film formation and oxide volatility was the major process where $\log P(MoO_3)_3 > -3$.

Gulbransen, Andrew and Brassart also studied the oxidation of molybdenum where oxidation occurs by volatilization alone and the transition to gas diffusion controlled oxidation. Table 10 shows the kinetic data for an oxygen pressure of 76 torr and for cylindrical specimens of 1.215 cm^2 area. A plot of these kinetic data indicate a transition at 975°C to a temperature independent mechanism of oxidation. Smaller specimens have higher transition temperatures and reaction rates. The activation energy for the temperature dependent reaction was 19 kcal/mole throughout the temperature range of 700 to 1400°C. The data of Gulbransen, Andrew and Brassart [6] show that gas diffusion control occurred for values of approximately 1.3×10^{18} atoms of molybdenum reacting per second at 1000°C.

c. Comparison of Theoretical and Experimental Data

The onset of volatility was observed at 475°C as predicted from the thermochemical analyses. No effect on the reaction rate was observed as the melting point of $MoO_3(s)$ was reached. Since the equilibrium vapor pressure of $(MoO_3)_3$ is approximately 0.015 atm at the melting point, this indicates that $MoO_3(s,\ell)$ does not exist as an outer layer on the metal under these conditions. Instead it is likely that $MoO_2(s)$ is formed with $(MoO_3(g)$ (and other polymer species) evolving from its surface upon reaction with $O_2(g)$, see Fig. 4.

TABLE 10

Summary Kinetic Data for Oxidation of Molybdenum

76 Torr of Oxygen, 1.215 cm^2 Area Specimens

Furnace Temp. °C	Calculated Temp. °C	cm'/dt atoms/cm°-sec	log dn'/dt
700	-	1.493×10^{17}	17.17
800	829	4.30×10^{17}	17.63
900	957	1.124×10^{18}	18.05
1000	1047	1.080×10^{18}	18.03
1100	1135	1.099×10^{18}	18.04
1200	1227	1.049×10^{18}	18.02
1400	1418	1.280×10^{18}	18.11
1600	1614	1.112×10^{18}	18.05

Comparison of reaction rates that give gas diffusion are possible for carbon and molybdenum oxidation at 76 torr $O_2(g)$ and 1000°C. For carbon the value was 4×10^{17} and for molybdenum it was 1.3×10^{18} atoms of metals per second. This is a factor of 3.2, indicating that the products evolved from the molybdenum specimen were a mixture of $(MoO_3)_3(g)$ and $(MoO_3)_4(g)$, and that these species do indeed provide a lower gas diffusion resistance than CO(g), to the extent that represents their degree of polymerization.

OXIDATION OF NIOBIUM

a. Thermochemical Predictions

Figure 5 shows the log $P(NbO_x)$ vs. log $P(O_2)$ diagram at 1400°K. This diagram shows low volatility for niobium oxides over Nb_2O_5(s) for an oxygen pressure of log $P(O_2)$ = 0 and higher but still unimportant pressures of NbO_2(g) over NbO_2(s) and of NbO(g) over NbO(s).

Figure 13 shows the volatility of the several oxides for log $P(O_2)$ = 0. NbO_2(g) is the major volatile species at all temperatures.

Two dashed lines are drawn on the figure at log $P(NbO_x)$ = -9 and -2 to predict oxide volatility. Some volatility of $NbO_2(g)$ is expected over $Nb_2O_5(\ell)$ at temperatures above 1620°C. Mixed oxidation processes with oxide formation and oxide volatilization would occur. Complete oxide volatility is not expected up to the melting point of niobium of 2415°C.

At temperatures above the melting point of 1512°C for $Nb_2O_5(s)$, transfer of niobium as volatile oxides can occur through cracks or pores in the inner oxide phases and into the $Nb_2O_5(\ell)$ phase, possibly with smoke formation of the outer oxide surface or in the gas phase.

Oxide volatility is expected to have no deleterious effects on the oxidation resistance of niobium up to the melting point of $Nb_2O_5(s)$.

b. Experimental Studies

No applicable experimental data are known to exist.

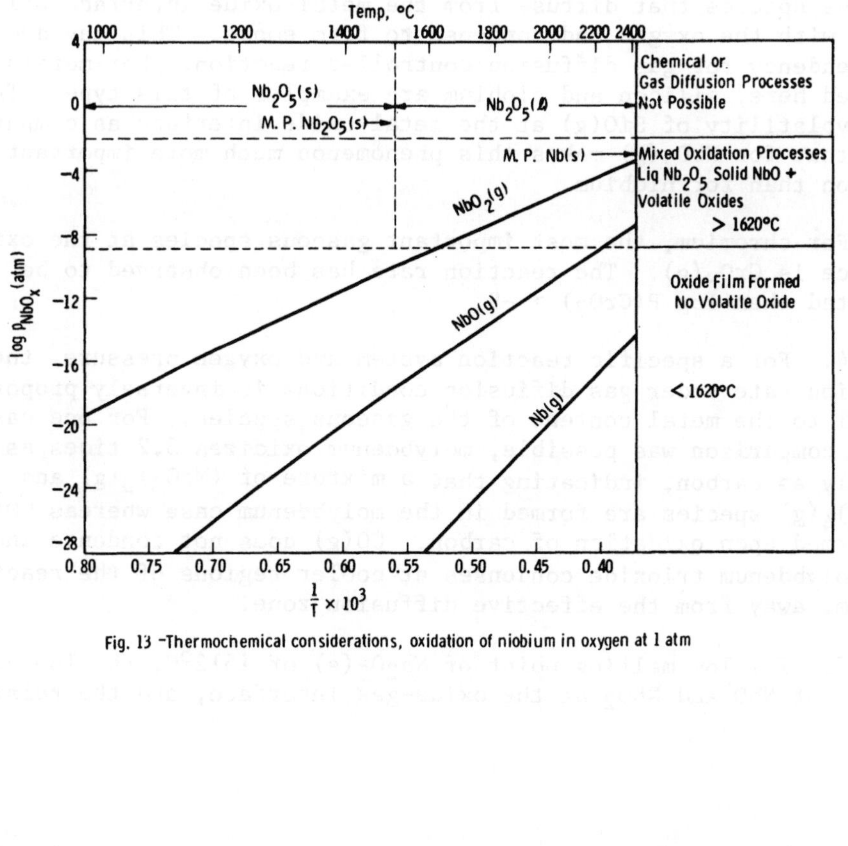

Fig. 13 -Thermochemical considerations, oxidation of niobium in oxygen at 1 atm

CONCLUSIONS

1. Thermochemical analyses can be used to predict the nature and equilibrium pressure of volatile species, and the composition and stability of condensed phases in metal-oxygen systems. This information can be used to predict the characteristics of the oxidation process as far as volatilization is important.

2. The pressure of volatile species at the metal-oxide interface can affect the oxidation reaction. If the volatile species are formed by reaction between metal and oxide, the condensed oxide phase can be used up. If the pressure of gaseous species at the interface exceeds the external pressure in the reaction system, it may cause the oxide to crack. $SiO(g)$ is an example of these phenomena. Silicon is the only metal among those considered here that exhibits this behavior.

Volatile species at the metal-oxide interface may also affect the oxidation rate by short-circuiting the transport of metal species through an oxide film. This occurs in the chromium-oxygen system at temperatures where $\log P(C_r) > -9$.

3. The equilibrium pressure of gaseous species at the oxide-gas interface determines the conditions where oxide volatility becomes important. If the equilibrium pressure is very low, any gaseous species that diffuse from the metal-oxide interface will react with the oxygen and condense to form smoke. This may decrease the tendency for gas diffusion controlled reaction. For metals considered here, silicon and niobium are examples of this type. The high volatility of $SiO(g)$ at the metal-oxide interface as compared with that for $NbO_2(g)$ makes this phenomenon much more important for silicon than for niobium.

For chromium, the most important gaseous species at the oxide surface is $CrO_3(g)$. The reaction rate has been observed to be affected when $\log P(CrO_3) > -9$.

4. For a specific reaction system and oxygen pressure, the reaction rate under gas diffusion conditions is inversely proportional to the metal content of the gaseous species. For one case where comparison was possible, molybdenum oxidizes 3.2 times as rapidly as carbon, indicating that a mixture of $(MoO_3)_3(g)$ and $(MoO_3)_4(g)$ species are formed in the molybdenum case whereas $CO(g)$ is formed upon oxidation of carbon. $CO(g)$ does not condense and the molybdenum trioxide condenses at cooler regions of the reaction system, away from the effective diffusion zone.

5. The low melting point of $Nb_2O_5(s)$ of 1512°C, the low volatility of NbO and NbO_2 at the oxide-gas interface, and the relatively

high pressures of NbO_2(g) and NbO(g) at the internal interfaces characterize the oxidation of niobium.

REFERENCES

1. JANAF Thermochemical Data, Dow Chemical Company, Midland, Michigan, including Supplement No. 30 dated December 31, 1968.
2. H. Schick, *Thermodynamics of Certain Refractory Compounds*, 2 Academic Press, 1966, New York and London.
3. C. Wagner, J. Applied Phys. 29, 1295 (1958).
4. E. A. Gulbransen, K. F. Andrew and F. A. Brassart, J. Electrochem. Soc. 113, 834 (1966).
5. E. A. Gulbransen and K. F. Andrew, J. Electrochem. Soc. 104, 334 (1957).
6. E. A. Gulbransen, K. F. Andrew and F. A. Brassart, J. Electrochem. Soc. 110, 952 (1963).
7. H. H. Kellogg, Trans. Met. Soc. AIME 236, 602 (1966).
8. S. A. Jansson and E. A. Gulbransen, paper presented at the Fourth International Congress on Metallic Corrosion, September, 1969, Extended Abstracts, p. 77.
9. E. A. Gulbransen, *Corrosion* (in press).
10. C. E. Wicks and F. E. Block, Bureau of Mines Bulletin #605, Washington, U. S. Government Printing Office, 1963.
11. E. A. Gulbransen, K. F. Andrew and F. A. Brassart, J. Electrochem. Soc. 110, 242 (1963).
12. S. Glasstone, K. J. Laidler, and H. Eyring, *Theory of Rate Processes*, McGraw-Hill Book Company, New York (1941).
13. E. A. Gulbransen and K. F. Andrew, Ind. Eng. Chem. 44, 1034 (1952).
14. E. A. Gulbransen, K. F. Andrew and F. A. Brassart, J. Electrochem. Soc. 110, 476 (1963).
15. E. A. Gulbransen, K. F. Andrew, F. A. Brasart and A. L. Feild, Jr., J. Nucl. Mat. 13, 40 (1964).
16. E. A. Gulbransen and W. S. Wysong, Trans. AIME 175, 628 (1948).

DISCUSSION OF PAPER

S. E. Khalafalla. The decrease in slope of the Arrhenius plot at higher temperatures is understandable on the basis of a change in the controlling mechanism from surface to gas diffusion. However, in the oxidation of chromium, you showed that the Arrhenius plot goes to higher slopes at higher temperatures and then reverts to the low temperature slope at very high temperatures. You explained this situation by a change in chemical reaction mechanism. I believe that at the transition point in this case, the Arrhenius curve should undergo a continuous variation, exhibiting a maximum followed by a minimum before it shoots up again with higher activation energies.

E. A. Gulbranson. The decrease in slope of the Arrhenius plot is due to the increased volatility of CrO_3 from the Cr_2O_3 film formed on chromium. This volatility acts to decrease the parabolic rate law constants from what would be expected. No minimum is expected.

J. Drowart. 1) With respect to the oxidation of Cr, it may be noted that CrO_3 polymers have been identified.

2) It seems of interest to point out that the behavior of compounds such as $MoSi_2$,ZrB_2 can also be predicted or rationalized by thermodynamic considerations analogous to those you presented for the elements. At low oxygen pressure, volatile species are formed and the overall rate of vaporization in oxygen may be high. At high oxygen pressure, if one of the constituents of the compound forms a refractory oxide, protection is achieved if this oxide adheres to the surface.

3) The thermodynamic predictions imply that equilibrium between the impinging gas and the substrate is established. It would seem to me that this may generally not be the case at low total pressures, but will be valid when the pressure is of the order of 1 torr and higher (either in a pure gas or in mixtures of an inert and a reactive gas). Under static conditions blanketing by reaction products probably then leads to some protection against oxidation.

E. A. Gulbranson. We would be glad to receive new information on the formation of CrO_3 polymers. We agree with the other comments of Dr. Drowart on the application of thermochemistry to the behavior of such compounds as $MoSi_2$ and ZrB_2 and to the use of thermochemical predictions for reactions occurring at low oxygen pressures.

STUDIES OF VAPORIZATION MECHANISMS AT ELEVATED TEMPERATURES

Gerd M. Rosenblatt

Department of Chemistry, Pennsylvania State University

University Park, Pennsylvania

The retarded vaporization of substances with small vaporization coefficients is a kinetic process of practical and theoretical interest at elevated temperatures. Studies of the retarded vaporization of As_4 and Sb_4 from (111) cleavage faces of arsenic and antimony single crystals have shown that initially slow, increasing vaporization rates coincide with the time required for thermal etch pits on the surface to intersect. Following the kinetic induction period, vaporization proceeds at a steady-state rate. Kinetic and microscopic results are in accord with the vaporization mechanism for arsenic and antimony in which the rate-limiting step occurs on a ledge of molecular dimensions (presumably at a kink site) and is associated with atomic rearrangement to form a tetrahedral vapor molecule. Observations of the vaporization of alkali halide single crystals are in marked contrast to those for arsenic and antimony. Accumulated results for KCl and NaCl (100) faces suggest that impurities, dislocations, and surface diffusion may all play a role in the vaporization of alkali halides. Retarded vaporization rates can be modified under some conditions; the steady-state vaporization rate of arsenic, but not of antimony, increases after exposure to 10^{-4} - 10^{-5} torr O_2 or bombardment with fast neutrons. The vaporization coefficient of arsenic, 7×10^{-5} at 550°K, is much smaller than that of antimony, 0.2 at 650°K. This can be rationalized as reflecting differences in equilibrium enthalpies of vaporization and in atomization energies of the gaseous tetramers.

VAPORIZATION KINETICS

Vaporization is a kinetic process of both practical and theoretical interest at elevated temperatures. All materials vaporize at sufficiently high temperatures. Understanding vaporization processes in solids would appear to be a prerequisite for elucidating the mechanisms of many more complex gas-solid reactions. A number of recent reviews summarize the current status of vaporization research and some of the developments which have contributed to understanding of the vaporization process [1-11].

Vaporization is a first-order kinetic process proportional to the surface area of the sample:

$$dN/dt = J_v A \text{ molecules/sec} \tag{1}$$

The rate constant J_v represents the flux of evaporating molecules. The molecular flux incident on a surface in equilibrium with its vapor is:

$$J_{i,E} = P_E \,(2\pi mkT)^{-\frac{1}{2}} \tag{2}$$

where P_E is the equilibrium vapor pressure, k is Boltzmann's constant, and T is the absolute temperature. At equilibrium the net vaporization flux J_v is zero, and for every molecule which condenses one vaporizes. If every molecule which hits the surface condenses then Equation (2) represents the gross condensation rate and the gross vaporization rate under equilibrium conditions. If the gross rate of vaporization remains constant as one goes from equilibrium to vacuum conditions then Equation (2) also gives the vaporization flux into vacuum J_v^o. However, the vacuum vaporization flux

$$J_v^o = \alpha_v^o \; P_E \,(2\pi mkT)^{-\frac{1}{2}} \tag{3}$$

may be less than that computed from Equation (2) by a factor α_v^o, known as the (vacuum) vaporization coefficient; for example, it is possible that under equilibrium conditions not every molecule which hits the surface condenses. When net vaporization occurs under conditions of partial saturation, $0<P<P_E$, the vaporization coefficient is defined by:

$$J_v = \alpha_v \,(P_E - P) \;\; (2\pi mkT)^{-\frac{1}{2}} \tag{4}$$

Vaporization coefficients, in principle at least, may vary with ambient pressure, so that $\alpha_v(P) \neq \alpha_v^o$, without violating the principle of microscopic reversibility; for example, the number of sites at which vaporization occurs may change with the extent of vapor saturation.

In treating experimental vacuum-vaporization data it is often convenient to define a Langmuir vapor pressure P_L which is computed from the measured flux (wN_{Av}/MAt) by:

$$P_L = (w/At) \quad (2\pi RT/M)^{1/2} \tag{5}$$

where w is the weight loss from area A in time t, N_{Av} is Avogadro's number, R is the gas constant, and M is the molecular weight of the vapor species. Comparison of Equations (3) and (5) shows that the vaporization coefficient can be obtained by comparing Langmuir and equilibrium vapor pressures:

$$\alpha_v{}^o = P_L/P_E \tag{6}$$

The temperature dependence of Langmuir vapor pressures and of vaporization coefficients may be expressed similarly to that of equilibrium vapor pressures. The standard thermodynamic relationship $\Delta G^o = -RT\ln P_E(\text{atm}) = \Delta H^o - T\Delta S^o$ may be written:

$$R \ln (P_E/P_o) = -\Delta H^o/T + \Delta S^o \tag{7}$$

where $P_o = 1.013 \times 10^6$ dyn cm^{-2} represents the standard atmosphere. By analogy, activation enthalpies and entropies may be defined by:

$$R \ln (P_L/P_o) \equiv -\Delta H^*/T + \Delta S^* \tag{8}$$

so that

$$R \ln \alpha_v{}^o \equiv -(\Delta H^* - \Delta H^o)/T + (\Delta S^* - \Delta S^o) \tag{9}$$

This definition of activation enthalpy and entropy differs from those based upon the Arrhenius relationship:

$$J_v{}^o = A \exp (-E_a/RT) \tag{10}$$

or upon absolute rate theory:

$$J_v{}^o = \kappa(kT/h) \; e^{\Delta S\dagger/R} \; e^{-\Delta H\dagger/RT} \tag{11}$$

If activation parameters are obtained from experimental data by least-squares fit to Equations (8), (10), or (11) assuming ΔH^*, ΔS^*, A, E_a, $\Delta S\dagger$, and $\Delta H\dagger$ are constants independent of temperature over the experimental temperature range [12], the parameters are related by the following equations:

$$\Delta H^* = E_a + RT/2 \tag{12}$$

$$\Delta H^* = \Delta H\dagger + 3RT/2 \tag{13}$$

$$\Delta S^*/R = \ln A - \ln P_o + 1/2 + \ln (2\pi mkT)^{1/2} \quad (14)$$

$$\Delta S^*/R = \Delta S^\dagger/R + \ln (kT/P_o) + 3/2 + \ln (2\pi mkT/h^2)^{1/2} + \ln \kappa \quad (15)$$

The second and third terms on the right hand side of Equations (14) and (15) are associated with a change in units (standard state) from molecules/cm^3 to atmospheres while the fourth term is associated with the gain of one translational degree of freedom upon going from two-dimensional surface units (molecules/cm^2) to three dimensional gaseous units (molecules/cm^3 or atmospheres).

The vaporization coefficients of most metals and other simple substances are close to unity [1, 7, 11]. However, some molecular and ionic systems vaporize at rates appreciably lower than are predicted from their equilibrium vapor pressures and Equation (2). Such systems may have vaporization coefficients very much less than one. The vaporization of substances with small vaporization coefficients may be called "retarded vaporization" and has particular import for the behavior of materials at elevated temperatures.

As Equation (9) indicates, the study of retarded vaporization yields activation enthalpies and/or entropies which differ from equilibrium quantities and which can be used to gain insight into the vaporization mechanism. Retarded vaporization can arise from the presence of a slow step in the overall vaporization process. When a slow step is present, it may be possible to vary parameters affecting that step in order to gain further mechanistic insight. On the other hand, it is difficult to obtain mechanistic information from kinetic studies on systems having essentially unit vaporization coefficients which yield only thermodynamic parameters. Kinetic studies of retarded vaporization can also provide data needed to correlate and predict vaporization coefficients, and therefore vaporization rates, and to modify vaporization rates by catalysis or controlled "poisoning". Equilibrium vapor pressures at elevated temperatures are often measured by Knudsen or Langmuir techniques [13]. Low vaporization coefficients can lead to errors in these measurements.

The following sections of this paper discuss mechanistic conclusions derived from experimental studies of retarded vaporization in our laboratory, where arsenic and antimony have been chosen as prototype molecular systems. These results are compared with recent work on ionic systems, specifically the alkali halides. Some data on the enhancement of vaporization rates are also discussed, as is the correlation of vaporization coefficients in similar molecular systems.

ARSENIC AND ANTIMONY

Table I summarizes kinetic results on the steady-state vaporization of As_4 and Sb_4 from arsenic and antimony single crystal wafers, cleaved to expose primarily (111) faces [14, 15]. However,

Table I. Steady-state Vaporization Kinetics of As_4 and Sb_4 from As (111) and Sb (111) Wafers.[a]

	Arsenic[b]	Antimony[c]
Temperature range (°K)	505-580	606-697
Langmuir vapor pressure, P_L (atm)	6.6×10^{-10}	1.2×10^{-9}
Vaporization coefficient, α_V°	7.1×10^{-5}	0.17
Activation enthalpy, ΔH^* (kcal/mol)	43.9	49.7
Excess activation enthalpy, $\Delta H^* - \Delta H^{\circ}$ (kcal/mol)	9.1	2.3
Activation entropy, ΔS^* (cal/deg mol)	37.9	35.7
Excess activation entropy, $\Delta S^* - \Delta S^{\circ}$ (cal/deg mol)	-2.3	0

a. Data at average temperature of measurements, 550°K for arsenic and 650°K for antimony.
b. Data from reference [14] adjusted to revised As_4 equilibrium vapor pressures.
c. From reference [15].

before reaching steady-state, cleaved arsenic and antimony single crystals show an "induction period" [16]. Freshly cleaved samples vaporize initially with slow, but increasing, rates which rise to the steady-state value. Once the induction period is past vaporization rates become constant and reproducible. The induction period is much more pronounced with antimony crystals than with typical arsenic crystals cleaved at room temperature as it depends upon dislocation density [17]. Cleaving arsenic at liquid-nitrogen temperature to decrease the number of dislocations introduced, greatly magnifies the induction period.

During the induction period the surface topography undergoes major changes. Microscopic observation of vaporizing crystals shows that upon vaporization a plane, cleaved surface develops small, shallow, symmetric triangular pits which grow until the surface looks like a jagged mass of intersecting planes. Typical changes on antimony are illustrated in Fig. 1. These surface changes occur only upon mass transport. The pyramidal pits have sides of constant slope (although there are discontinuities in slope on some antimony pits) which intersect the (111) cleavage surface at angles of 8-13° with arsenic and 2-9° with antimony. There is a more-or-less constant density of pits on a given surface, that is, new pits do not continually appear. However, observation of different size pits on many surfaces shows that the pits do not all start growing at exactly the same time. Time-lapse cinemicrophotography shows that large and small pits grow at the same constant rate expected if vaporization from a pit is proportional to its area. A number of experiments indicate that the pits are associated with dislocations, apparently primarily spiral dislocations. These experiments include vaporization of surfaces previously etched with CP-4 chemical etchant known to produce pits at dislocations on these semimetals [18-20], vaporization of originally mated cleavage surfaces, vaporization of surfaces stressed locally with a diamond point, and reducing pit density by cleaving at low temperature.

During the induction period vaporization occurs solely, or primarily, from growing thermal etch pits. After intersection the rate becomes constant. The weight loss required to attain steady-state which is computed from the measured pit slopes and pit densities, assuming all vaporization occurs from pits, agrees with that observed. Reducing the dislocation density lengthens the induction period; starting pits by chemically etching with CP-4 shortens it. Steady-state vaporization rates are independent of dislocation density.

The kinetic results in Table I indicate that the retarded vaporization of arsenic and antimony may be described as primarily an energy or enthalpy effect. Vaporization coefficients increase exponentially with temperatures; preexponential factors are between 0.3 and 1. The kinetic results and the observed surface changes, as illustrated below, are consistent with a vaporization mechanism in which the slow step is associated with rearrangement of four nearest-neighbor atoms in the metal lattice to form a tetrahedral vapor molecule.

In accord with the terrace-ledge-kink model [1-5, 21-28] (Fig.2), the pits are presumed to be made up of (111) terraces separated by ledges of molecular dimensions. (Interference contrast microscopy does not reveal ledges under conditions which would reveal ledges higher than 20-50Å [17].) The ledges spiral around the sides of pits at spiral dislocations or form concentric loops on the sides

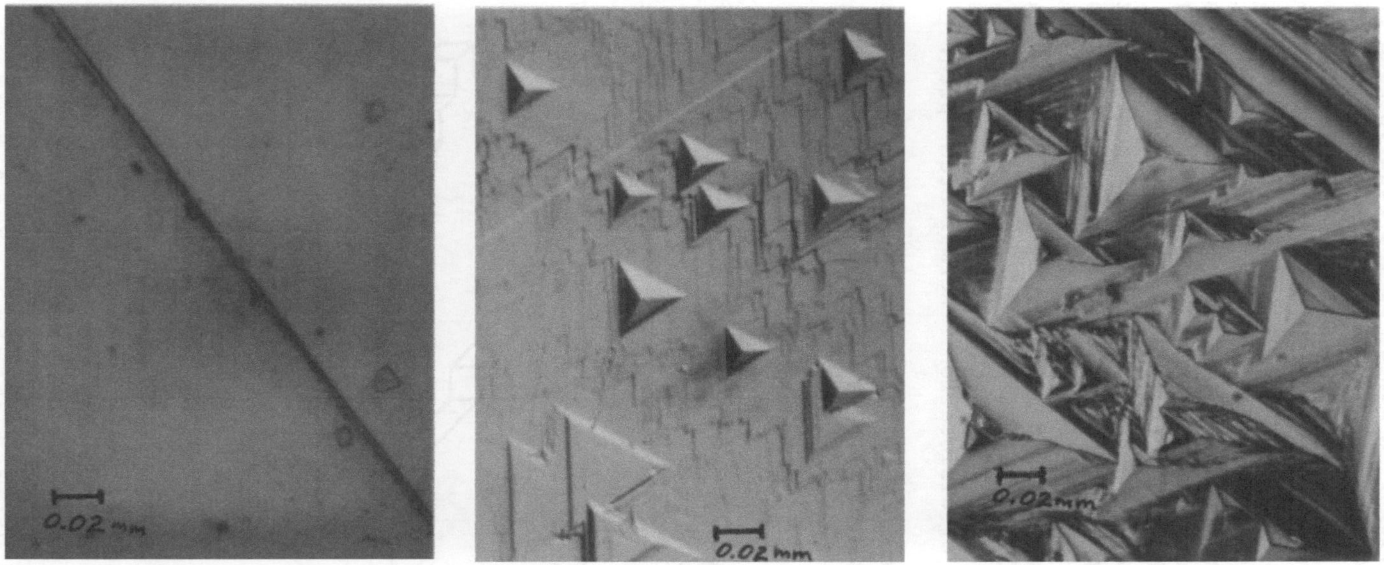

1a 1b 1c

Fig. 1. Surface changes occuring during vaporization of Sb (111) cleavage face. (a) Freshlv cleaved antimony crystal. (b) isolated pits during kinetic induction period. (c) intersected surface after extensive vaporization. The photographs are of different crystals.

of pits at edge dislocations. A number of the observations suggest that the rate of vaporization is directly proportional to available ledge length: (a) Pits have sides of constant slope, which corresponds to a terrace distance between ledges λ which is independent of the distance between the ledge and the pit center. (b) Steady-state vaporization rates are independent of dislocation density, as expected if after pit intersection the surface contains a constant length of ledge independent of the number of pit centers. (c) Time-lapse motion pictures show that the sides of pits grow at the constant rate expected if the vaporization rate is proportional to ledge length (which is proportional to pit area). If it is assumed that ledges contain a constant density of kinks then these observations imply that the vaporization rate is directly proportional to kink concentration. All the data can be interpreted by assuming that the slow molecular rearrangement process occurs at kink sites on ledges, on the sides of dislocation "thermal etch" pits.

The vaporization studies do not yield any information about what happens to the tetrahedral molecule after formation at a ledge.

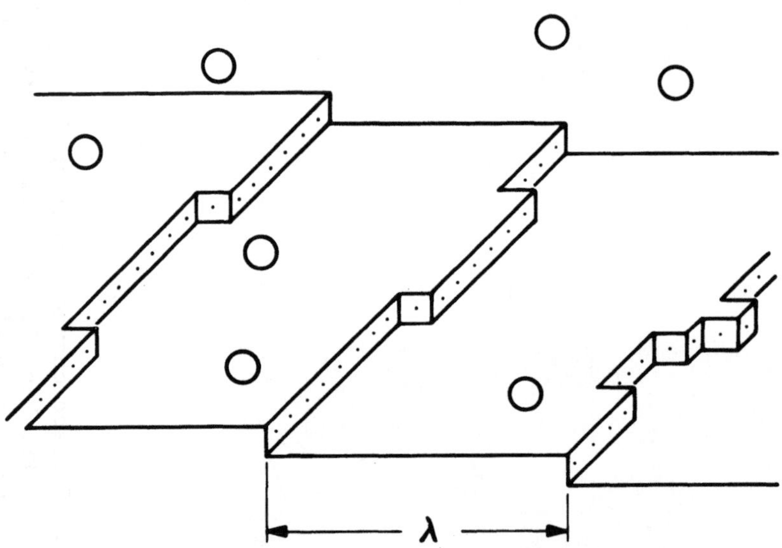

Fig. 2. Schematic representation of crystal surface at steady state showing terraces, ledges, and kinks. For arsenic and antimony the circles represent tetrahedral vapor molecules and it is presumed likely that the ledges are two atom layers high, the spacing between equivalent (111) atom planes. The average spacing between ledges is symbolized by λ.

However, observations of the condensation of As_4 on GaAs [29] and of the epitaxial growth of antimony [30] and very approximate estimates of As_4 surface lifetimes assuming attractive energies of 5-15 kcal/mol, suggest that As_4 and Sb_4 molecules may diffuse on the surface long enough to equilibrate before desorbing into the vapor [31]. If, however, molecules "pop" into the vapor directly from kink sites the subliming molecules might contain excess translational and/or vibration-rotational energy [32].

It might be mentioned briefly that the mechanism postulated for vaporization of arsenic and antimony suggests that measured pit slopes reflect ledge spacings λ near the core of dislocation etch pits. Cabrera and Levine [33] have derived the equation $1/\lambda = -(kT/19\gamma\Omega) \ln (\bar{n}_s/n_{se})$ for the ledge spacing near the core of a spiral dislocation [1, 2, 27]. In this equation γ is the surface free energy, Ω is the molecular volume, $\bar{n}_s$ is the average number of molecules on the terraced surface, and n_{se} is the number of molecules on the surface under equilibrium conditions. When the speed at which molecules leave ledges or kinks is rate-determining, the number of molecules on the surface compared to that at equilibrium, $\bar{n}_s/n_{se}$, equals the vaporization coefficient, α_v. Although the Cabrera-Levine model may not be applicable if molecules leave ledges at a nonequilibrium rate, this equation predicts λ on As and Sb of ~24 and ~320Å respectively, assuming surface free energies γ of 110 and 216 erg/cm^2. For comparison, measured pit slopes yield ~18 and ~43Å for As and Sb, assuming ledge heights equal to lattice interplanar spacings. Assuming applicability of the Cabrera-Levine equation when the rate at which molecules leave ledges is rate determining, leads to vaporization coefficients which decrease with increasing vapor saturation according to $\alpha_v(P) = \alpha_v^o \ln (P/P_E)/\ln \alpha_v^o$, if α_v^o is very small and if spiral dislocations are the primary ledge sources. This result appears consistent with available data [17].

ALKALI HALIDES

The behavior of arsenic and antimony is in marked contrast to that recently reported for alkali halides, particularly sodium and potassium chloride. Lester and Somorjai [34] report that the overall vaporization coefficient of NaCl varies from ∿ 0.5 to 1.0 as the dislocation density is increased from ∿ 10^6 to 10^7 cm^{-2}. The activation energy is the same as the equilibrium enthalpy of vaporization, indicating that the below unity vaporization coefficient of NaCl crystals of lower dislocation density is temperature independent. A similar temperature-independent vaporization coefficient of 0.72 was reported earlier for KCl crystals [35]. On the other hand, Rothberg, Eisenstadt, and Kusch [36] measured temperature-dependent vaporization coefficients for NaCl, Na_2Cl_2, and lithium

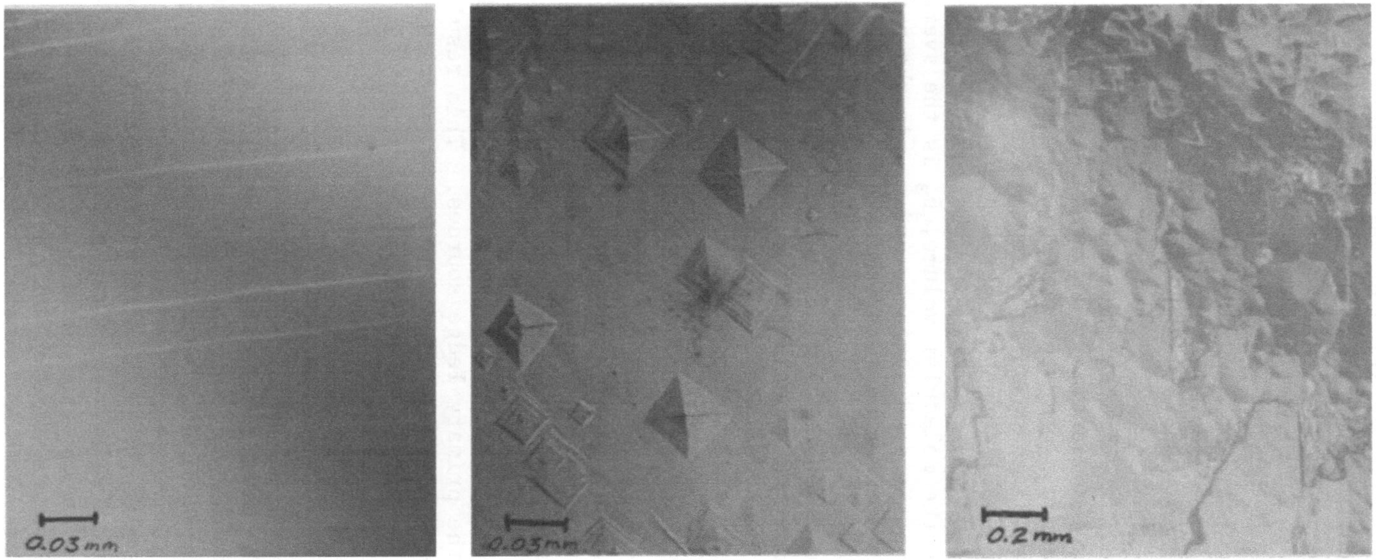

3a 3b 3c

Fig. 3. Surface changes occurring during vaporization of KCl (100) cleavage face. (a) Freshly cleaved potassium chloride crystal. (b) After vaporizing 2 hours at 418°C showing approximately square, pointed and flat-bottomed pits. (c) After vaporizing 32 hours at ∿ 420°C. The photographs are of different surfaces.

fluoride. Reported vaporization coefficients of KI [37], KCl [37, 38] also decrease with increasing temperature. Somorjai and Lester suggest that the disparate NaCl result may be due to impurity divalent cations in the crystals of Rothberg et al [36].

Surface changes accompanying vaporization of sodium chloride single crystals have been studied in a number of laboratories [37,39-42]. Our observations of vaporizing KCl (100) planes (Fig. 3) at 650-750°K using optical and interference microscopy are qualitatively similar to those Budke [39] describes for NaCl. An initially smooth surface develops a moderate density, $\sim 10^5$ cm^{-2}, of approximately square, pointed and flat-bottomed, thermal pits, many of which contain easily resolvable terraces. At the same time the surface develops very shallow valleys (which can only be seen with interference contrast) which give the unpitted surface the appearance of a gently rolling plain. As vaporization continues the square, thermal pits normally disappear while the rounded, hilly background bacomes more pronounced. The surface ultimately looks like a group of intersecting, shallow, flared, conical depressions. As with sodium chloride [40, 42], initial square pits on KCl do not correlate with dislocations revealed by chemical etching. There are many fewer thermal pits than dislocation etch pits. Polarization interferometry and interference contrast microscopy show that the slopes of most square pits are steeper in the center, as shown in Fig. 4, unlike arsenic and antimony pits. The outer sides of square pits have slopes of 2-4° with respect to the (100) surface. The perturbed area near the pit center is $\sim 4 \times 10^{-6}$ cm^2.

The observations on alkali halides have been subject to a number of different interpretations [34, 39, 40, 43]. Of these, the relation between surface energy and evaporation rate presented by Budke [39] has not been put in molecular terms. On the other hand, it is not obvious that Somorjai and Lester's [34] postulate of (a) large ledge concentrations where dislocations intersect the surface, (b) monomer and dimer formation at kinks on these ledges, (c) rapid desorption from the proximity of ledges, explains the observed surface changes. Their estimation of a short surface diffusion distance (which apparently could imply slow surface diffusion equally as well as rapid desorption) conflicts with 0.5 μm diffusion lengths derived from electron microscopic studies [40]. The accumulated kinetic results and surface observations on NaCl and KCl suggest that impurities, dislocations, and surface diffusion may all play a role in the vaporization of alkali halides, but detailed description of the mechanism appears to require further elucidation of the interrelationship between these. The difficulty in specifying a unique vaporization mechanism for the alkali halides might indicate that, in general, it may become more difficult to describe vaporization processes precisely as vaporization coeficients approach unity, and small, competing effects control vaporization kinetics.

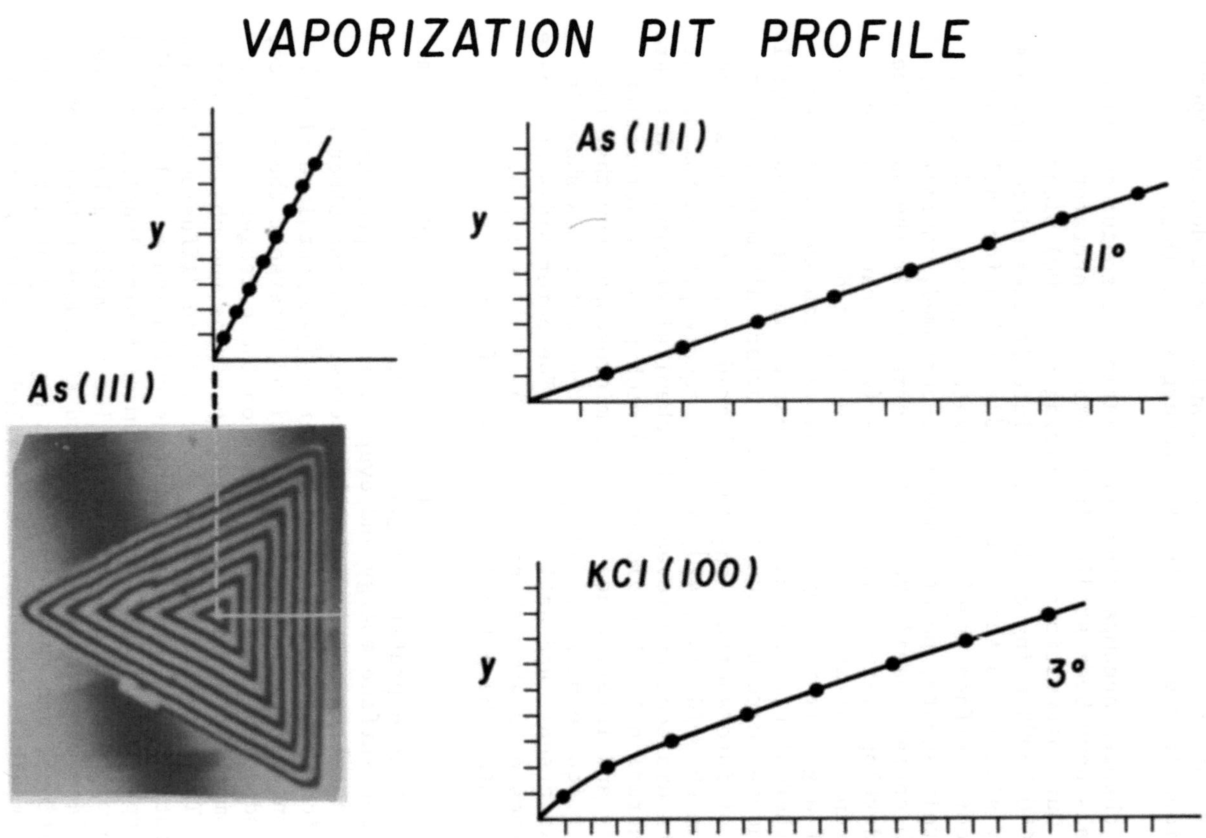

Fig. 4. Pit profiles of arsenic and potassium chloride thermal etch pits reconstructed from polarization interferometry. The arsenic pits have constant slope. Square, pointed potassium chloride pits are steeper at the pit center.

It is interesting to observe that the changing slope of thermal etch pits seen with KCl is in accord with ideas presented by Hirth and Pound [1, 2] for the shape of dislocation etch pits when surface diffusion or impurity constraints dominate vaporization kinetics. If these pits were associated with dislocations, both the steeper pit centers and the area of the perturbed region would correlate well with the reported dependence of vaporization rate upon dislocation density, which, in terms of the terrace-ledge-kink model, appears to require a larger ledge density (steeper pit) near the emergence point of the dislocation (pit center). However, association of observed pit characteristics with the reported dislocation-dependent vaporization kinetics requires a correlation between pits and dislocations which has not been established.

MODIFICATION OF VAPORIZATION RATES

It is well known that vaporization rates can be decreased by plating a non-volatile material on the surface or by adsorption of a small amount of "poison". For example, in our laboratory, gold on arsenic and oxygen on zinc have lowered measured vaporization rates. In addition, retarded vaporization can be catalyzed to bring the rate closer to that calculated from the equilibrium vapor pressure and Equation (2). This was demonstrated by Brewer and Kane [44] who increased the vaporization rate of arsenic by adding liquid thallium. Schoonmaker, Buhl, and Lemley [45] have shown that the decomposition of gallium nitride to liquid gallium and nitrogen gas is markedly enhanced in the presence of liquid gallium or indium.

A somewhat similar effect has been observed upon vaporization of arsenic single crystals in 10^{-4} to 10^{-5} torr O_2, or upon vacuum vaporization of arsenic single crystals which have been previously annealed under oxygen at similar pressures. Figure 5 shows the slow increase in measured vaporization rate (total weight loss) at constant temperature when an arsenic crystal vaporizing at the normal (steady state) rate is exposed to $\sim 8 \times 10^{-5}$ torr O_2. The rate continues to increase over a period of 10-20 hours ultimately becoming two to five times the normal free vaporization rate. The rate does not drop back to the normal free-vaporization rate if the oxygen is shut off at any stage during this process but remains high, even after vacuum vaporization for 2-3 hours. Mass spectra both during and after oxygen exposure show that the rate increase is not due to formation of $As_4O_6(g)$ or some other volatile oxide although $As_4O_6(g)$ is the thermodynamically stable species in the presence of oxygen at these pressures. However, the As_4^+ peak -- as well as fragment peaks arising from As_4 -- increases in intensity while the rate is increasing.

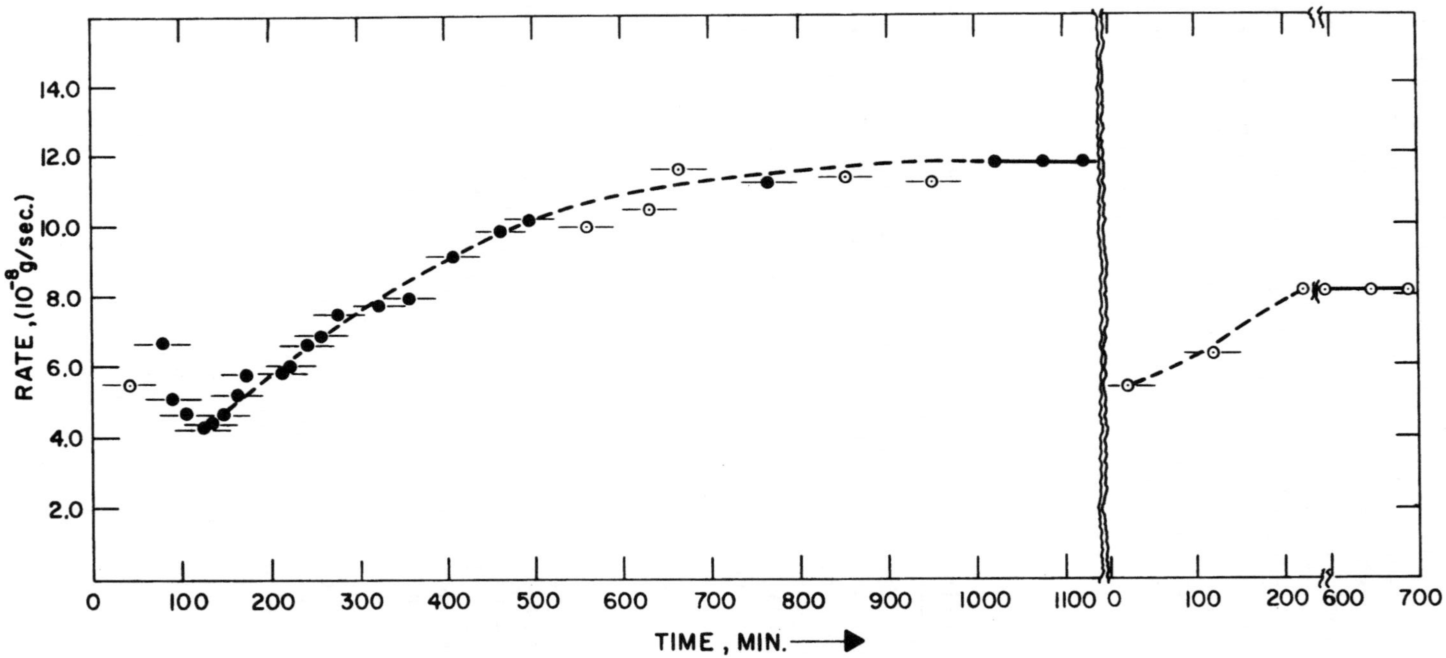

Fig. 5. Rate of weight loss of As (111) single crystal wafer under oxygen as a function of time at 559°K. The solid circles represent rates under oxygen partial pressures of 7.7 - 8.9 x 10^{-5} torr. The open circles represent rates in the absence of oxygen ($P_{O_2} < 10^{-9}$ torr). The first open circle on the left corresponds to the steady-state vaporization rate of an intersected arsenic surface before oxygen exposure. The crystal was exposed to air for 6 hours between parts I and II.

Arsenic crystals annealed under 10^{-4}-10^{-5} torr oxygen and then freely vaporized at 220-310°C behave similarly to those vaporized under oxygen. The vaporization of O_2 annealed crystals, both freshly cleaved and previously vaporized, starts out below the normal vacuum vaporization rate and increases to values 2-5 times the normal rate over a period of about ten hours. Mass spectra show only peaks from As_4 and a small background AsO^+ peak. Cleaving the surface off of such a sample caused it to behave like a freshly cleaved, oxygen-annealed sample but the ultimate rate attained, although higher than the normal vacuum vaporization rate, was lower than that of the crystal after it was originally annealed under oxygen. Although there is no measurable weight increase during O_2 annealing, neutron activation analysis indicates that oxygen-annealed crystals contain 0.01 - 0.1 atomic %O.

Arsenic crystals oxidized in nitric acid behave similarly to O_2 annealed crystals, as do crystals which have been bombarded with fast neutrons before vaporization in vacuum. Nitrogen does not affect the vaporization rate. This type of behavior is not observed with antimony where O_2 reacts to form $Sb_4O_6(g)$ under similar conditions with no unusual time-dependence. On the order of 10^{-2}-10^{-4} of the O_2 molecules colliding with antimony surfaces at 620-670°K react to form volatile Sb_4O_6 molecules at oxygen pressures of 10^{-4}-5×10^{-7} torr.

The surface of arsenic crystals annealed under oxygen are indistinguishable from fresh-cleaved surfaces before vaporization. However, after vaporization O_2-annealed crystals appear similar to arsenic vaporized in an oxygen atmosphere in that relatively steep, truncated or hexagonal, etch pits develop and the surface appears porous or amorphous, as shown in Fig. 6.

Rather tentatively, the results suggest that gaseous oxygen, chemical oxidation, or neutron bombardment catalyzes the vaporization of As_4 either by providing more suitable (kink) sites for the atomic rearrangement associated with forming tetrahedral As_4 from lattice atoms, or by lowering the activation energy required for atomic rearrangement at these (kink) sites. The long time dependence may be associated with solid-state diffusion, perhaps of vacancies formed by the introduction of oxygen, to suitable positions in the surface or lattice.

A few other observations of the effect of low pressure gases or particle irradiation upon retarded vaporization rates have been reported. Stranski and coworkers [5] report that the activation energy required to vaporize the claudetite modification of As_2O_3 is reduced by prior neutron irradiation. Bickley and Gregg [46] show that addition of 500 ppm H_2O to a nitrogen atmosphere over magnesium metal at 300°C promotes vaporization of magnesium over nitridation by weakening the nitride surface film. Under similar conditions oxygen or mechanical damage inhibits evaporation. Blank and Searcy [47] observed that introduction of $\sim 5 \times 10^{-4}$ torr nitrogen or argon reduces the sublimation rate of magnesium nitride by about a factor of one hundred.

Apparatus effects [15, 48], as well as changes in sample surface area or bulk character [14, 17, 49], can also affect measured vaporization rates.

VARIATIONS AMONG VAPORIZATION COEFFICIENTS

The mechanism described above for the vaporization of arsenic and antimony tends to support and extend earlier suggestions regarding the retarded vaporization of phosphorus and arsenic [3, 4, 44, 50-52]. The rearrangement of four nearest-neighbor atoms in the rhombohedral lattice to form a tetrahedral gaseous molecule is a very similar geometric change in arsenic and antimony. The changes in bond length and bond angle in going from the squashed pyramid

Fig. 6. Surface of arsenic (111) single crystal after vaporization under 8×10^{-5} torr oxygen for 42 hours at 558°K showing (black) porous surface around truncated pits.

configuration in the metallic lattice to the tetrahedral vapor molecule are about the same for both arsenic and antimony. In both cases, nearest neighbor internuclear distances decrease by 3%, basal internuclear distances decrease by 35%, and apex angles decrease from 97° to 60°. Yet, Table 1 shows that the vaporization coefficients, α_v, and excess activation enthalpies, $\Delta H^* - \Delta H^o$, associated with the rearrangement differ markedly.

The quantitative difference in vaporization behavior of arsenic and antimony can be rationalized as reflecting the relative thermodynamic stabilities of the gaseous tetramers. The argument presented below is, in large part, a simplified application of ideas originally proposed by Evans and Polyanyi [53] which have been extensively applied to gas-phase radical reactions [54, 55].

Consider an energy vs. reaction-coordinate diagram which represents the potential energy surface for the reaction along the minimum energy path. For the present discussion this diagram may be supposed to have a typical, simple hat shape, although it actually is expected to be appreciably more complex [4, 7]. Very roughly, the potential energy surface for these vaporization reactions might be thought of as being formed by intersection of a surface representing the potential energy for atomization of the solid with one representing atomization of the gaseous tetramer. Figure 7 very schematically illustrates the way these curves might intersect for arsenic and antimony. The energies in Fig. 7 are based upon thermodynamic data and the results in Table I, taking the zero of energy to be four gram atoms of gaseous atoms at 0°K. Figure 7 illustrates how solid and gaseous atomization surfaces which are deeper, and therefore steeper, might intersect with a somewhat higher hump. The nature and extent of the geometrical changes occurring (the "reaction coordinate") are assumed comparable, as they appear to be for arsenic and antimony. Examination of Fig. 7 also suggests that a higher hump with respect to gaseous products, $\Delta H^* - \Delta H^o$, will be formed when the equilibrium enthalpy of vaporization, ΔH^o, is small so that the minima of the intersecting solid and gaseous atomization surfaces are close in energy. A lower hump will be formed when the equilibrium vaporization enthalpy is high and the gaseous product minimum is well above that of reactant solid.

This line of reasoning suggests that very low vaporization coefficients will be associated with sublimation to form gaseous molecules which are unusually stable thermodynamically, i.e., which have large atomization energies, and with sublimation to form vapors of unusually high vapor pressure and low vaporization enthalpy. The data in Fig. 7 exemplify the application of these generalizations to arsenic and antimony. Gaseous As_4 has a higher dissociation energy (249 vs. 204) and a lower equilibrium heat of vaporization than

Sb_4 (37 vs. 51). The activation energy barrier, which is approximately proportional to T log α_v, is 9.1 kcal/mole above the enthalpy of vaporization for As_4 and 2.3 for Sb_4.

The reasoning may be phrased in somewhat more physical terms from the reverse standpoint of condensation. The vaporization (or condensation) coefficient is related to the energy required to distort the gaseous molecule into a configuration which can bond to crystal atoms (presumably at a kink site) and relax to the crystal atomic arrangement. The required distortion is more difficult if the bonding in the gaseous molecule is strong than if it is weak.

The implied relation between excess activation enthalpy, atomization energy, and equilibrium enthalpy of vaporization suggests that low vaporization coefficients are more likely to be observed with polyatomic vapor species which dominate in condensed-vapor equilibria at relatively moderate temperatures because they have small enthalpies of vaporization than with polyatomic molecules which increase in relative concentration with increasing temperature. Molecules in the latter group are often designated "high temperature molecules" [56-58].

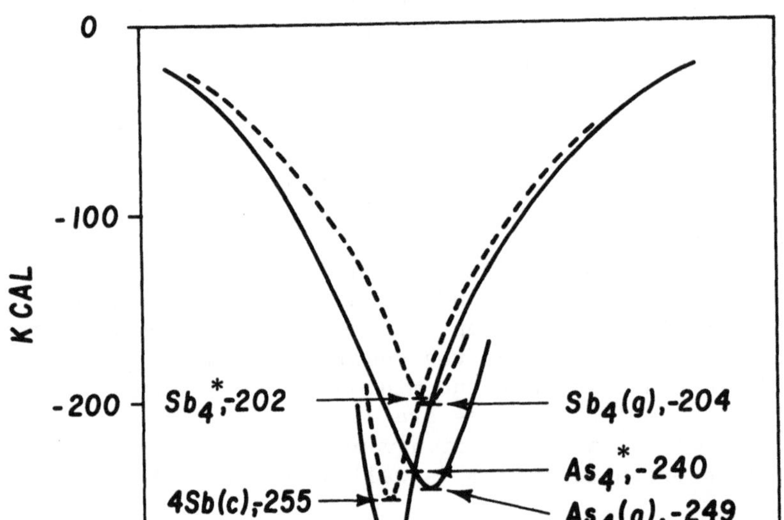

Fig. 7. Schematic potential-energy diagram illustrating how intersection of solid and gaseous atomization surfaces might determine vaporization or condensation barriers associated with molecular rearrangement. The continuous curves represent arsenic atomization potential energies; the dashed curves refer to antimony.

ACKNOWLEDGEMENT

This research was supported by the U. S. Army Research Office - Durham.

REFERENCES

1. J. P. Hirth and G. M. Pound: Progr. Materials Sci. 11 (1963).
2. J. P. Hirth: *Metal Surfaces: Structure, Energetics and Kinetics,* Proc. Joint Sem. of ASM and Met. Soc. AIME, Oct. 1962, (Amer, Soc. Metals, Cleveland, 1963), p. 199.
3. J. P. Hirth: *The Characterization of High Temperature Vapors,* edited by J. L. Margrave (John Wiley and Sons, New York, 1967) p. 453.
4. O. Knacke and I. N. Stranski: Progr. Metal. Phys. 6,181(1956).
5. W. Hirschwald and I. N. Stranski: *Condensation and Evaporation of Solids,* Proc. Intern. Symp. Dayton, Ohio, Sept. 1962, edited by E. Rutner, P. Goldfinger, and J. P. Hirth (Gordon and Breach, New York, 1964), p. 59.
6. W. G. Courtney: J. Am. Rocket Soc. 31, 751 (1961).
7. G. A. Somorjai and J. E. Lester: Progr. Solid State Chem. 4, 1 (1967).
8. G. A. Somorjai: Proc. Third International Symposium on *High Temperature Technology,* Pacific Grove, California, Sept. 1967 (International Union for Pure and Applied Chemistry, in press), p. 1.
9. G. A. Somorjai: Science 162, 755 (1968).
10. A. W. Searcy: Chapter to appear in *Chemical and Mechanical Behavior of Inorganic Materials,* (Interscience, New York), University of California, Berkeley, Lawrence Radiation Lab. Preprint UCRL-18631 (Dec. 1968).
11. B. Paul: J. Am. Rocket Soc. 32, 1321 (1962).
12. H. S. Johnson: *Gas Phase Reaction Rate Theory,* (Ronald Press Co., New York, 1966), p. 128.
13. R. C. Paule and J. L. Margrave: *The Characterization of High Temperature Vapors,* edited by J. L. Margrave (John Wiley and Sons, New York, 1967), p. 130.
14. G. M. Rosenblatt and P. K. Lee: J. Chem. Phys. 49, 2995 (1968).
15. G. M. Rosenblatt and P. K. Lee: J. Chem. Phys., in press.
16. G. M. Rosenblatt, P. K. Lee and M. B. Dowell: J. Chem. Phys. 45, 3454 (1966).
17. G. M. Rosenblatt, M. B. Dowell, P. K. Lee and H. R. O'Neal: Proc. Fourth Intl. Materials Sym., *Structure and Chem of Solid Surfaces,* Univ. California, Berkeley, June, 1968 (John Wiley and Sons, Inc. New York, 1969).
18. J. H. Wernick, J. N. Hobstetter, L. C. Lovell, and D. Dorsi: J. Appl. Phys. 29, 1013 (1959).
19. V. M. Kosevich: Kristallografiya 5, 749 (1960); Soviet Phys. Crystallography (Engl. Transl.) 5, 715 (1961).

20. M. N. Shetty and J. B.Taylor: J. Appl. Phys. 39, 3717, 5145 (1968).
M. N. Shetty, J. B. Taylor and J. G. Despault, J. Appl. Phys. 40,898 (1969).
21. W. K. Burton and N.Cabrera: Disc. Farraday Soc. 5, 33, 40 (1949).
22. F. C. Frank: Disc. Farraday Soc. 5 48 (1949).
23. W.K.Burton, N. Cabrera and F. C. Frank: Phil. Trans. Roy. Soc.(London) A243, 299 (1951).
24. M. Volmer, *Kinetic der Phasenbildung* (Steinkopff, Dresden and Leipzig, 1939).
25. W. Kossël: Nach. Ges. Wiss. Göttingen, 135 (1927); Ann. Phys. 33, 651 (1938).
26. I. N. Stranski: Z. Phys. Chem. 136, 259 (1928); (B)11, 421 (1931).
27. J. P. Hirth and G. M. Pound: Acta Met. 5, 649 (1957); J. Chem. Phys. 26, 1216 (1957); J. Phys. Chem. 64, 619 (1960).
28. J. P. Hirth: *Energetics in Metallurgical Phenomena*, Vol. II Proc.1963 Sem.,Univ. of Denver, edited by W. M. Mueller (Gordon and Breach, New York, 1965), p. 1.
29. J. R. Arthur: J. Appl.Phys. 37, 3057 (1966).
30. K. G. Robbins and J. M. Thomas: Nature 217, 1251 (1968).
31. See also B. McCarroll and G. Ehrlic in *Condensation and Evaporation of Solids*, Proc. Intern. Symp., Dayton, Ohio, Sept. 1962,edited by E.Rutner, P. Goldfinger, and J. P. Hirth (Gordon and Breach, New York, 1964), p. 521.
32. Compare reference 10 which emphasizes the importance of excess internal energy in discussions of vaporization mechanism.
33. N. Cabrera and M. M. Levine: Phil. Mag. [8] 1, 450 (1956).
34. J. E. Lester and G. A.Somorjai: J. Chem. Phys. 49, 2940 (1968).
35. R. S. Bradley and P. Volans: Proc. Roy. Soc. (London) A217, 508 (1953).
36. G. M. Rothberg, M. Eisenstadt and P. Kusch: J. Chem. Phys. 30, 517 (1959).
37. O. Knacke, R. Schmolke and I. N. Stranski: Z. Kristall. 109, 184 (1957).
38. T. E. Joyce and R. T. Grimley: J. Chem. Phys. 51, 468 (1969).
39. J. Budke: J. Am.Ceram.Soc. 51, 238 (1968); J. Appl. Phys. 40, 641 (1969).
40. H. Bethge: J. Phys. Chem. Solids, Suppl. 1, 623 (1967).
41. K. W. Keller: J. Phys. Chem. Solids, Suppl. 1, 629 (1967).
42. A. Grinberg: Phys. Status Solidi 3, 1369 (1963).
43. R. S. Bradley: Proc. Roy. Soc. (London) A217, 524 (1953).
44. L. Brewer and J. S Kane: J. Phys. Chem. 59, 105 (1955).
45. R. C. Schoonmaker, A. Buhl and J. Lemley: J. Phys. Chem. 69, 3455 (1965).
46. R. I. Bickley and S. J. Gregg; J. Chem. Soc. A1849, (1966).
47. B. A. H. Blank and A. W. Searcy: J. Phys. Chem. 72, 2241 (1968).

48. G. M. Rosenblatt: J. Phys. Chem. 71, 1327 (1967).
49. G. M. Rosenblatt: J. Electrochem. Soc. 110, 563 (1963).
50. I. N. Stranski and G. Wolff: Research 4, 15 (1951).
51. L. Pauling and M. Simonetta: J. Chem. Phys. 20, 29 (1952).
52. H. Eyring, F. M. Wanlass, and E. M. Eyring: *Condensation and Evaporation of Solids*, Proc. Intern. Symp. Dayton, Ohio, Sept.1962, edited by E. Rutner, P. Goldfinger and J. P. Hirth (Gordon and Breach, New York, (1964), pp. 3, 27.
53. M. G. Evans and M. Polanyi: Trans. Farraday Soc. 34, 11 (1938).
54. M. Boudart: *Kinetics of Chemical Processes* (Prentice-Hall, Inc., Englewood Cliffs, N. J. (1968), Chap. 8.
55. N. N. Semenov: *Some Problems in Chemical Kinetics and Reactivity*, 1, Transl.by M. Boudart (Princeton Univ. Press, Princeton,New Jersey,1958).
56. L. Brewer and A. W. Searcy: Ann. Rev. Phys. Chem. 7, 259 (1956).
57. J. Drawart and P. Goldfinger: Ann. Rev. Phys. Chem. 13, 459 (1962).
58. P. Goldfinger: Adv. High Temp. Chem. 1, 1 (1967).

KINETICS OF THE ATTACK OF REFRACTORY MATERIALS BY DISSOCIATED GASES

Daniel E. Rosner† and H. D. Allendorf

AeroChem Research Laboratories, Inc.
Subsidiary of Sybron Corp.
Princeton, New Jersey 08540

Microwave discharge/fast flow vacuum system techniques are being exploited to generate well defined streams of reactive atoms and free radicals for studies of heterogeneous reaction kinetics using resistively heated filament specimens. By these means gas/solid reactions are investigated at filament temperatures up to the sublimation threshold of refractory materials, under conditions such that the steady state reaction rates are no longer limited by (i) dissociative adsorption of diatomic gases on partially covered surfaces or (ii) reactant access and/or product escape by molecular diffusion.

We survey here the production and detection of molecular fragments and recent experimental results on the O-atom, Cl-atom, H-atom and N-atom attack of polycrystalline refractory materials forming volatile oxides, chlorides, hydrides and nitrides at temperatures up to 3000°K and at reactant pressures in the range 10^{-4} to 10^{-1} Torr. Emphasis is placed on the reactivity of tungsten, molybdenum, rhenium, graphite, boron, boron nitride and silicon carbide in dissociated gases, and simple mixtures thereof. We conclude with a discussion of mechanistic implications of our results and current extensions of the techniques to other gas/solid reactions involving free radicals (e.g., OH, F) of both fundamental and applied interest.

The step of dissociative adsorption (on either bare adsorbent sites or chemisorbed monolayers) apparently plays a major role in determining the observed kinetics of high temperature solids attacked

---

† Present address: Yale University, Department of Engineering and Applied Science, Mason Laboratory, New Haven, Connecticut 06520.

by stable *diatomic* gases† $X_2(g)$ ($X \equiv O, N, Cl, H,\ldots$). While the X_2-dissociative adsorption ("sticking") probability may be high and temperature-insensitive for densely packed planes of bare metals [5], extensive reflection can occur off clean nonmetal and semimetal surfaces, or partially covered metal surfaces, thereby setting an upper limit to the steady state reaction probability. This being the case, it is of fundamental interest to investigate the comparative behavior of solids attacked by simple gaseous atoms and molecular fragments, for which enhanced chemisorption probabilities and a simpler reaction mechanism might be anticipated. In addition to shedding light on the relative importance of dissociative adsorption as a rate-limiting step in high temperature gas-solid reactions, these data should also provide valuable information on the response of refractory solids in partially dissociated gas mixtures [6], such as now occur in aerospace and advanced technological applications. These considerations and the sparsity of meaningful kinetic data on the chemical attack of refractory solids by unstable species, motivated the long-range study outlined below. The scope of this program and its relation to other kinetic studies of atom-solid reactions can be seen from Table 1. In the present paper we discuss our experimental techniques and display experimental results for a wide variety of gas-solid systems, including several interesting systems studied in detail since the preparation of our review paper [6] for the Third International Symposium on High Temperature Technology. As in the past, emphasis will be placed on: (i) high temperature-high†† reactant pressure rate data not falsified by reactant or product diffusional limitations [30-32], or thermal accommodation effects, (ii) direct comparisons of atom-solid kinetic data with corresponding data for stable molecules attacking the same materials, using the same experimental techniques, (iii) generalizations and/or surprises linked to reactant dissociation, (iv) interaction effects associated with important mixtures of dissociated gases and (v) current studies and important questions worthy of further exploration.

EXPERIMENTAL TECHNIQUES

Atom Production

We are exploiting microwave electrical discharge techniques [33] to produce nonequilibrium but characterizable dissociated gas streams at reactant partial pressures typically of the order of 10^{-3}

† Thus, adsorbed atomic species (i.e., "adatoms") are important reaction intermediates in the attack of refractory solids by stable polyatomic gases [1-4].

†† Compared to the low pressures (typically (10^{-5}-10^{-4} Torr)) of stable molecules emphasized in mass spectrometric studies [2-4] of heated surfaces attacked by "cold" gases. Effective reactant pressures in molecular beam studies are lower still, usually being less than $\approx 10^{-6}$ Torr.

Table I. Dissociated Gas Solid Kinetics Investigations*

Atomic Gas	Solid	Investigator(s)	Reference(s)
H	C	Gould (1969)	9
	C	King and Wise (1963)	7
	C	Rosner and Allendorf (See Fig. 4)	
	C	Wood and Wise (1968)	8
	Ge	Sancier, Morrison and Wiesendanger (1966)	10
	olefins	Klein and Scheer (1962)	11
O	B	Rosner and Allendorf (1967)	6
	C	Streznewski and Turkevich (1959)	12
	C	McTaggart (1967)	13
	C	Rosner and Allendorf (1965, 1967, 1968)	14-16
	Co,Ni	Dickens, Heckingbottom and Linnett (1969)	17
	Cu	Dravnieks (1950)	18
	Ge,Si	Madix, Melin and Susu (1969)	19
	Mo	Rosner and Allendorf (1964, 1967)	20, 21
	olefins	Hughes, Scheer and Klein (1966)	22
	Pt	Fryberg (1956, 1965)	23
	Re	Rosner and Allendorf (1968)	24
	S	Cramarossa, Molinari and Roio (1968)	25
	SiC	Rosner and Allendorf (1969)	26
	W	Rosner and Allendorf (1967)	21
N	C	Goldstein (1964)	27
	SiC	Rosner and Allendorf (1969)	26
Cl	B	Rosner and Allendorf (1968)	28
	Mo	Rosner and Allendorf 1965, 1967)	29
	W	Rosner and Allendorf (1967)	6
F	W, Mo, C	Rosner and Allendorf (in progress)	--

* Reactions leading to condensed product films or volatile reaction products. Studies of surface catalyzed atom recombination are not included here.

to 5 x 10^{-2} Torr. A carrier gas (usually argon) is provided such that the total pressure level is near 1 Torr. As indicated schematically in Fig. 1, the usual configuration [6] consists of a Vycor discharge tube - Pyrex vacuum flow system coupled to mechanical pumps (not shown) which maintain flow velocities high enough to render negligible homogeneous and heterogeneous atom loss processes in the (≈28 cm) distance between the discharge cavities and the specimen. A significant advantage of this technique is the capability of producing a stream at near-ambient temperatures containing free atoms at concentrations that would ordinarily (at equilibrium) require temperatures well above 2000°K. Most experiments discussed below were carried out with stream temperatures, velocities and flow rates of the order of 300-350°K, 10^4 cm/ sec. and 20 cm^3 (STP)$(sec)^{-1}$, respectively.

The "twin discharge" arrangement (cf. Fig. 1) allows kinetic studies of high temperature surfaces simultaneously exposed to atomic gas mixtures (e.g., halogen + oxygen) of current theoretical and applied interest. Each microwave discharge cavity is supplied by an independent 2450 MHz, 125 w. Raytheon microwave power source--and there are no metal electrodes in contact with the gas stream upstream of the specimen.

Two general methods are used to generate the atoms or free radicals of interest. The first, which may be regarded as "direct", is to simply produce the atom X(g) (X=O, Cl, N,...) by microwave discharging the parent diatomic gas X_2(g), either in the presence or

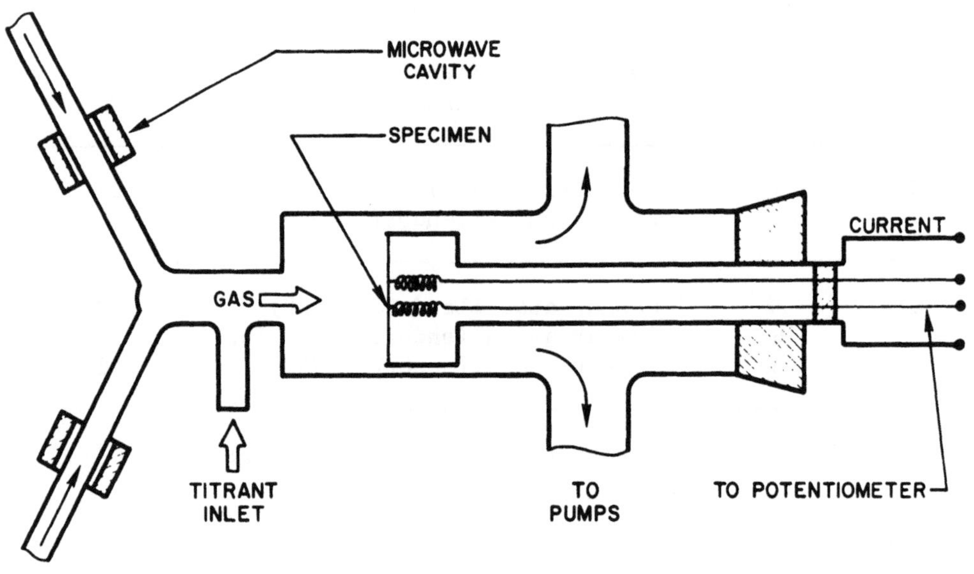

Fig. 1. Filament oxidation apparatus (Schematic)

absence of a carrier gas. This method is usually most convenient, provided the reactivity of the parent diatomic gas is negligible compared to that of the atom at the surface temperatures and reactant pressures of interest (see below). Indirect methods can also be used to generate atoms or free radicals, two examples being the quantitative production of O-atoms (in the absence of O_2) by the rapid bimolecular atom-molecular reaction [34]:

$$N + NO \rightarrow N_2 + O \qquad (1)$$

and the quantitative production of OH-radicals from the rapid bimolecular atom-molecule reaction [35]:

$$H + NO_2 \rightarrow NO + OH \qquad (2)$$

In these cases NO(g) or NO_2(g) are added immediately downstream of the microwave discharge cavity used to generate the "reactants" N-atoms and H-atoms, respectively. Clearly, care must be taken to ensure that (i) adequate time is available for these reactions to go to completion upstream of the specimen, (ii) side reactions do not occur which introduce new reactive atoms or free radicals, and (iii) stable molecules produced in these reactions (e.g., N_2, NO) do not participate in the atom or free-radical heterogeneous reaction of interest. Further details and references to the generation of atoms and free radicals using microwave discharges†, and their application to specific heterogeneous reactions may be found in our earlier papers [6, 14, 16, 21, 26, 28, 29].

Detection of Atomic Species

As noted below, we are interested in comparing observed absolute reaction rates (e.g., substrate atom removal flux) to the arrival rate of the attacking gas. To calculate the latter quantity we therefore require the absolute atom concentration, (X), in the gas mixture at the sample location. Two distinct methods have been used to accomplish this, viz. "chemical titration" and "calorimetric" techniques. Chemical titration techniques are based on light-emitting chemical reactions which ensue upon the addition of selected gases to the atom-containing stream immediately downstream of the electrical discharge region. Convenient titration reactions are available for determining absolute concentrations of O-atoms [33], N-atoms [37], and H-atoms [38]. In contrast, calorimetric techniques

† Whenthe parent diatomic gas has a sufficiently low bond energy, furnace techniques become a feasible alternative. An example is the case of fluorine ($D_o^o \approx 32$ Kcal/mole) [36] presently under study in the authors' laboratory. In other configurations (particularly static or slow-flow systems) surface catalyzed dissociation reactions (e.g., of hydrogen or oxygen) on refractory filaments have been used successfully.

are based on measurements of the heat of surface-catalyzed atom recombination reactions [39, 40]. While quite general, this method can be made absolute only in cases where (i) energy contributions from other (vibrationally or electronically) excited species can be ruled out, (ii) molecules formed from recombining atoms are fully equilibrated with the catalyst surface at the time of desorption, (iii) the "catalyst" surface is not continuously chemically attacked, and, (iv) the recombination coefficient, γ, is known, or the catalyst coil is arranged to recombine *all* atoms in the incoming stream. As described in Ref. 29, our absolute chlorine atom concentrations have been estimated in this way, using 420°K nickel filaments as the recombination catalyst. However, *relative* atom concentrations in this system are conveniently determined using the light emission accompanying gas phase atom recombination, which yields an intensity approximately proportional to the square of the atom concentration [41].

Assessment of Contributions from Excited Atoms/Molecules

While ground state atomic species are usually the main chemically reactive component of microwave-discharged diatomic gases, it is desirable to experimentally rule out non-negligible contributions to the observed reactivity due to the possible presence of electronically or vibrationally excited species (e.g., X_2^*). For example, in specific cases we have shown that the observed reaction probabilities are invariant under deliberate changes in (i) the mode of O-atom generation [21] (e.g., direct vs. indirect) or (ii) $(X_2^*)/(X)$ ratio caused by passage of the discharge "products" through selective catalyst beds [26] or tubes [21]. In the future we hope to deliberately study the surface reactivity of a well-defined excited molecule; however, in none of the cases encountered thus far has the reactivity of the gas mixture produced by eliminating the atom population shown any significant enhancement over a stream of un-discharged gas containing the same concentration of ground state diatomic molecules.

Materials

Table 2 identifies the polycrystalline (or vitreous) filament materials studied thus far. With regard to the gases, we have used extra-dry grade (Linde) oxygen, high purity grade (Linde) nitrogen, high purity grade (Matheson) chlorine, high purity grade (Matheson) hydrogen and prepurified grade (Linde) argon without further treatment.

Table II. Filament Specimens Studied in Present Program

Material	Supplier	Density[a]	Ref.[b]	Remarks
Molybdenum	Westinghouse, Inc.	10	21	High recrystallization temperature, drawn wire.
Tungsten	Westinghouse, Inc.	19	21	High recrystallization temperature drawn wire.
Rhenium	Chase Brass and Copper Co.	20	24	
Isotropic Graphite	Speer Carbon Co.	1.73	14,15	Electrographitic extruded rod.
Pyrolytic Graphite		2.1	16	Vapor deposited from ethane on tungsten substrate[b].
Boron	Texaco Experiment, Inc.	2.3	28	Vitreous coating vapor deposited on 0.5 mil diam tungsten substrate.
Boron Nitride		2.1	c	Vapor deposited from $NH_3 + BF_3$ on tungsten substrate.
Silicon Carbide	Texaco Experiment, Inc.	3.1	26	Crystalline coating vapor deposited on 0.5 mil. diam tungsten substrate.

[a] Estimated nominal density, g/ cm^3, at elevated temperature
[b] Consult reference(s) cited for further details on material characterization or deposition.
[c] This work

Reaction Rate

Our specimens are resistively heated filaments, about 3 cm long and from 0.102 mm (4 mil) to 0.508 mm (20 mil) in outer diameter. Owing to the volatility of the reaction products at the temperatures and pressures considered here, a steady state rate of specimen diameter reduction is achieved. Moreover, owing to the density of the specimens or their chemical nature, permeation or dissolution of reactants *into* the specimens can apparently be ignored under our present conditions. Accordingly two alternative techniques can be used to follow the time-independent reaction rate, depending on the homogeneity of the material. For the refractory metals and isotropic graphite, reaction rates are inferred directly from continuous measurements (at constant temperature) of the electrical resistance-time curve of the filament. The latter is determined by combining the current measurement with the voltage drop across the central 0.55 cm of the specimen (obtained using spring loaded contacts, cf Fig. 1). For "composite" specimens (e.g., pyrolytic graphite silicon carbide, boron or boron nitride deposited on a core of tungsten), the reaction rate is calculated from microscope determinations of the reduction in outer filament diameter after known exposure times. In most cases the fractional change in specimen diameter was small, and in no case was it allowed to exceed 25 percent. Departures from circularity in the cross section reacted specimens were small enough to render any errors due to this cause negligible. During an experiment the filament is maintained at constant brightness temperature by altering the voltage in accord with an optical pyrometer or infrared detector output. For homogeneous specimens whose electrical resistivity is a known, monotonic function of temperature, the true surface temperature is inferred from the electrical resistance and initial diameter. For composite specimens, or materials with unfavorable resistance-temperature characteristics, the true surface temperature is determined from the optical pyrometer reading, as corrected using (i) the reported spectral emittance-temperature relation for the specimen, and/or (ii) fixed point temperature measurements using a series of commercially available pyrometric compounds.

Absence of Diffusion Limitations

Particularly for efficient atom-solid reactions, diffusional limitations on the rate of reactant supply and/or product escape can falsify kinetic data [30-32]. Accordingly, we have paid considerable attention to eliminating such physical phenomena. As discussed in Ref. 21, this has been accomplished by combining the conditions of high gas flow rate, small specimen size and reduced total pressure. Thus, reported reaction rates and sublimation rates are not influenced by either a change in (i) the total flow rate of gas over the specimen, or (ii) the identity of the carrier gas (e.g., substitution of $N_2(g)$ or He(g) for Ar(g)). As a corollary to these findings, we conclude that the present pressure levels are not so low as to introduce systematic differences between the translational temperature of the incident reactant gas molecules and the true temperature of the surface. Thus, while only the filament is heated, we believe the present data apply to gas-solid reactions with *gas* temperatures at the reported filament temperatures.

EXPERIMENTAL RESULTS AND DISCUSSION

For purposes of comparing our reaction rate data on a wide class of gas-solid systems, and as an aid in understanding microscopic events at the interface, it is useful to express experimental results in terms of dimensionless reaction probabilities. This can be done, without making any assumptions as to the product gas species distribution, by introducing a *reaction probability* ε, defined as the ratio of the flux of adsorbent atoms† (regardless of their chemical state of aggregation) away from the surface to the collision flux (Hertz-Knudsen arrival rate) of reactant incident upon the surface. As noted above, the adsorbent element flux is obtained from the observed rate of filament radius change, the density of the adsorbent and its molecular weight.†† Also shown on many of the Arrhenius diagrams below is the predicted high temperature asymptote to ε caused by simple sublimation (vaporization) - - a process which would, of course, occur in the absence of an attacking gas. These values are based on reported Langmuir-pressures inferred from vacuum vaporization rates [42].

A synopsis of our results for the oxidation and chlorination of the refractory metals Mo(s), W(s) and Re(s) is given in Fig. 2, panels a, b, c, d. These are all Arrhenius (log ε vs. 1/T) diagrams to the same scale, and pertain to comparable reactant pressures (the number in the parenthesis is the reactant pressure in Torr). In each case one notes (i) order of magnitude enhancements in reaction probability due to dissociation§ at the lower filament temperatures, with remarkably high ε-values for the X-atom reactions even at temperatures well below 1200°K, (ii) a tendency for the O and O_2 reaction probabilities to approach each other as the temperature level is increased, with both probabilities attaining values of $\approx 10^{-1}$ or larger, (iii) a negligible sublimation contribution to adsorbent atom removal over most of the temperature range shown. For those systems

† For well-defined refractory *compounds* the steady-state adsorbent *element* fluxes are simply related by over-all stoichiometry. Thus, for SiC(s), the reported ε's can be based on either Si-atom flux, or C-atom flux, since they are equal.

†† It should be noted that if the adsorbent element flux locally varies as the nth power of the reactant partial pressure (where n is the apparent reaction order) then (since arrival rates are linear in reactant pressure) the probability ε reported herein will be proportional to the n-1 power of reactant pressure (hence, independent of reactant pressure for a first order heterogeneous reaction).

§ This itself reveals that the reported reaction rates are not externally diffusion limited, for if they were, then the enhancement ratio in reaction probability due to X_2 dissociation could not exceed $(D_{X-Ar}/D_{X_2-Ar})^m/(2)^{\frac{1}{2}}$, where $2/3 \leq m \leq 1$ and the D's are the respective Fick interdiffusion coefficients with respect to argon. Available diffusivity data [43, 44] reveal that this ratio would not be very different from unity.

characterized by relatively weak adsorbate-adsorbent binding (e.g., O-Re and Cl-Mo) one notes that ε passes through a maximum at some intermediate temperature level, above which adsorbent atom removal rates fall until sublimation becomes appreciable. In this temperature interval of $d\varepsilon/dT<0$ (or $E<0$) it is likely that X-atom desorption is appreciable, with the filaments acting as effective surface catalysts for $X_2(g)$ dissociation. Not shown are reactant pressure dependence data which, in the range of $d\varepsilon/dT>0$, typically reveal fractional order behavior for the reactant $X_2(g)$ (X=O,Cl) but first

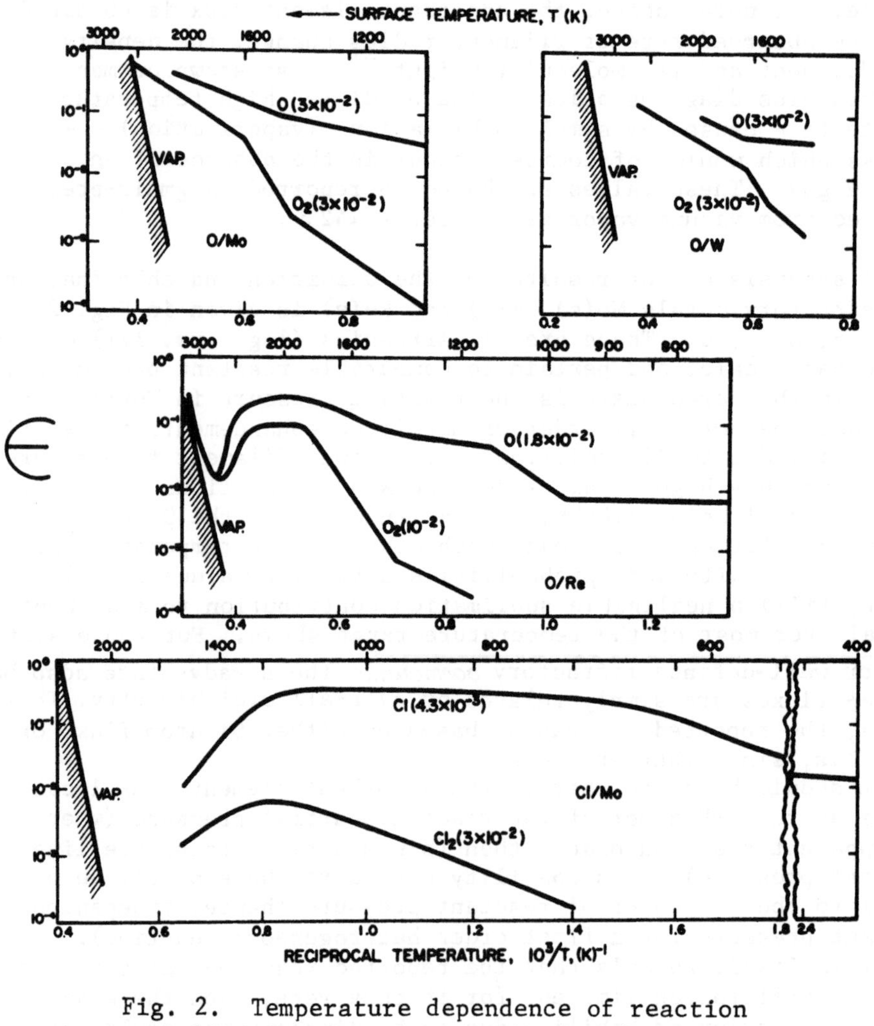

Fig. 2. Temperature dependence of reaction probabilities for the attack of refractory metals by atoms and molecules; a) molybdenum/ oxygen b) tungsten/oxygen c) rhenium/oxygen d) molybdenum/ chlorine.

order behavior for the corresponding X-atom reaction.

A synopsis of our data on the important refractory nonmetals SiC(s), C(s), B(s) and BN(s)† is given in Fig. 3, panels a, b, c, d, e). Here, one notes the following new features: (i) even the

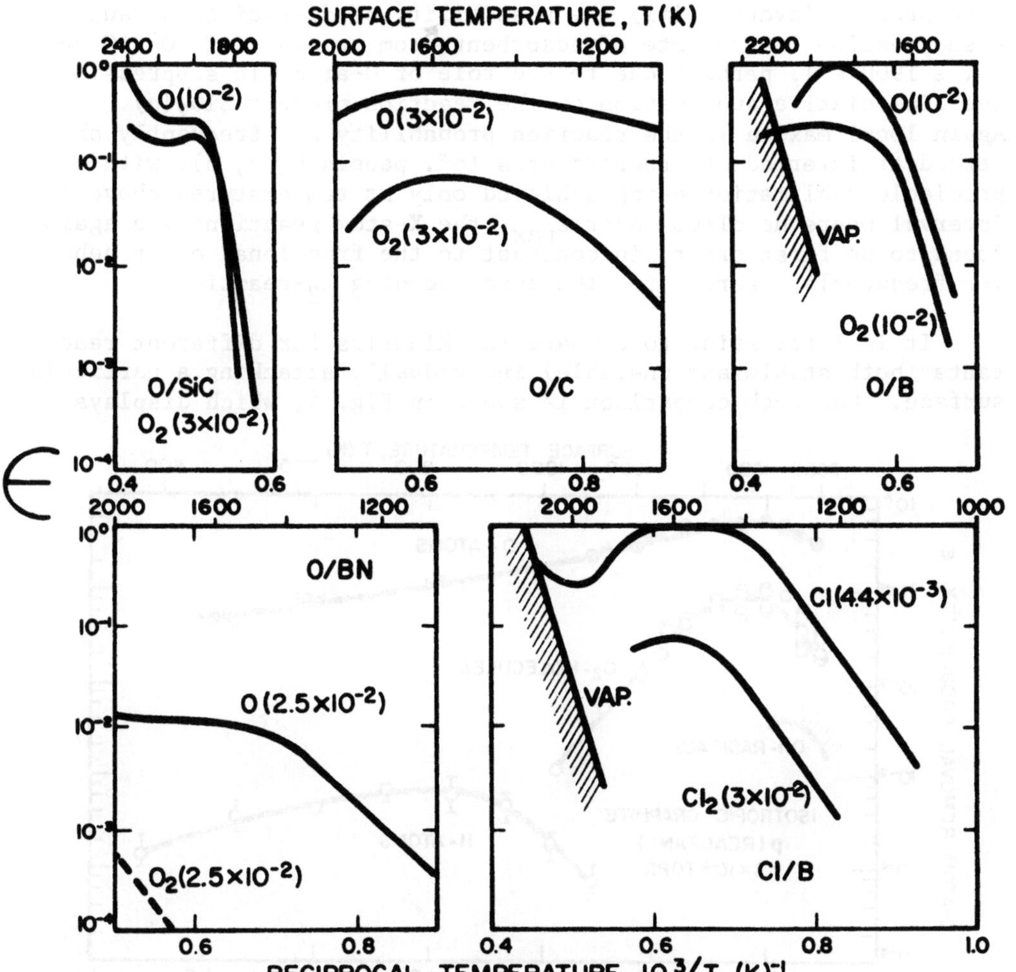

Fig. 3. Temperature dependence of reaction probabilities for the attack of refractory nonmetals by atoms and molecules; a) silicon carbide/oxygen b) graphite/oxygen c) boron/ oxygen d) boron/chlorine e) boron nitride/ oxygen.

† Results shown here for BN(s) have not been previously published, and should be considered preliminary to a more detailed account of the kinetic behavior of this system.

reaction probability for X-atom attack can be quite low ($\varepsilon<10^{-2}$) and sharply temperature dependent, especially for gas-solid systems capable of forming semi-protective product layers (e.g., O-SiC, O-B), (ii) a tendency for the dissociation enhancement in reaction probability to occur only at high temperatures (cf. O-SiC, O-B), when the respective surfaces are presumably denuded of semi-protective product layers; (iii) the possibility of dissociation causing a suppression in the rate of adsorbent atom removal (cf. O-SiC below $\approx$ 1900°K†), perhaps due to the role of O-atoms in suppressing the dissociative sublimation of the reaction product $SiO_2(c)$. Again local maxima in the reaction probability are frequently observed at intermediate temperatures (cf. panels b, c, d), with appreciable sublimation being achieved only at temperatures above an interval where $d\varepsilon/dT<0$. Near ε_{max}, the X-atom reactions are again found to be first order, in contrast to the fractional order behavior frequently observed for the corresponding X_2-reaction.

It is interesting to compare the kinetics for different reactants (both stable and unstable) individually attacking a particular surface. One such comparison is shown in Fig. 4, which displays

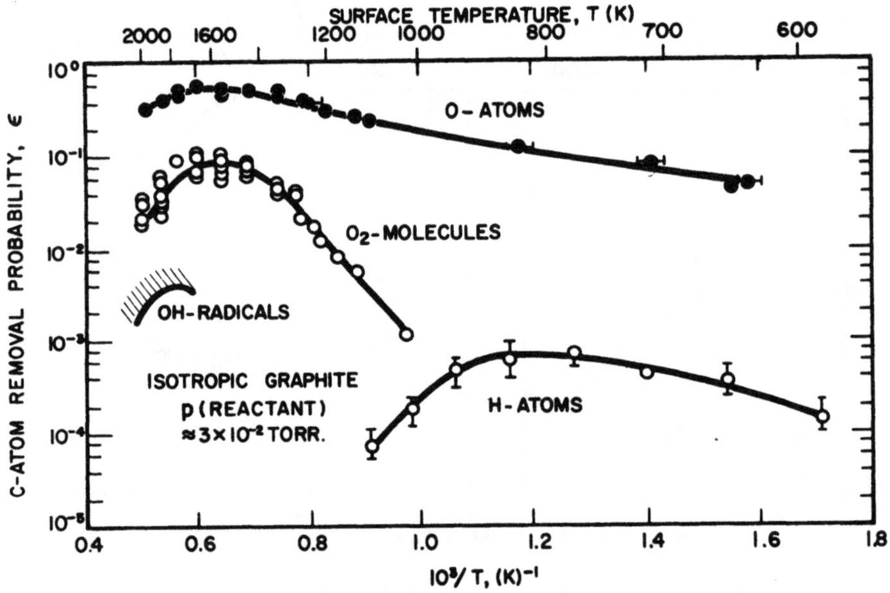

Fig. 4. Temperature dependence of the attack of isotropic graphite by O, O_2, H and OH.

† We have found that reactant dissociation markedly alters the temperature-reactant pressure locus delineating bare surface ("active") behavior from semi-protective ("passive") behavior [26]. This effect could prove quite important with regard to the use of silicide coatings in partially dissociated oxygen (or air) environments.

reaction probabilities for the attack of isotropic graphite by O(g), O_2(g), H(g) or OH(g).† These results demonstrate quite clearly that: (i) maximum reaction probabilities for atoms or free-radicals need not be large in absolute value (cf., e.g., $\varepsilon_{max} \approx 10^{-3}$ for H-atoms), nor large compared to "stable" molecules (cf. ε_{max} for O_2(g) and OH(g)), (ii) the temperature level at which ε_{max} is achieved is specific to the gas/solid system and indicative of the strength of adsorbate/adsorbent binding (cf. e.g., results for O and O_2 ($\approx$ 1600°K) with those for H-atoms ($\approx$ 800°K) at reactant pressures of about 3 x 10^{-2} Torr). As a result, it appears that less than one in a million H-atom strikes would lead to C-atom removal from 1600°K graphite, whereas essentially every other O-atom strike is successful ($\varepsilon \approx 1/2$) under identical conditions.†† However, as discussed below, one cannot conclude that these results would apply equally well when a graphite surface is *simultaneously* attacked by a gas mixture containing each of these reactants.

Indeed, rather interesting "coupling" phenomena can occur when a surface is chemically attacked by a mixture of reactants. Earlier, we reported on the enhanced rate of removal of the element tungsten in oxygen + chlorine mixtures-- a phenomenon attributed to the formation of volatile tungsten oxychloride product molecules [46, 47]. Two further examples are given below, both involving O + N mixtures (of potential interest in understanding ablation processes in dissociated air). Fig. 5 shows (i) the results of exposing 1870°K silicon carbide to O + N mixtures [48] at a constant total reactant pressure ($P_O + P_N = 1.1 \times 10^{-2}$ Torr) and (ii) corresponding results "expected" (cf. dashed curve) based on our earlier data in O-atoms alone§ (assuming "additivity" of the O-atom and N-atom "contributions"). Despite considerable scatter, it is clear that the reaction rates observed in, say, equimolar O + N mixtures are *lower* than the additivity "prediction" by more than one order of magnitude. As reported in Ref. 48, similar experiments above $\approx$ 2000°K did not show this strong coupling behavior. A consideration of the temperature dependence for this reaction [48], and the corresponding O-atom and N-atom reactions on SiC(s) [26], suggests that the observed behavior

† Our results for H(g) and OH(g) are presented here for the first time. In the latter case, extensive corrections for simultaneous contributions due to the presence of other reactive gases precluded the inference of precise values of ε. Our O and O_2 data have been reported previously [6,16] except for new points obtained below 1100°K.

†† Our measurements of the product distribution for the C(s) + O(g) reaction [45] demonstrate that CO(g) is the dominant reaction product near ε_{max}.

§ We have found [26] that the high temperature SiC(s) + N(g) atom reaction is remarkably efficient and probably constitutes the most efficient heterogeneous N-atom reaction yet discovered. In contrast, N_2(g)/SiC(s) reaction probability is lower by considerably more than two orders of magnitude.

in O + N mixtures is due to the *in situ* formation of microscope silicon oxynitride surface films which are apparently more stable (less volatile) in this environment than SiO_2(c) or Si_3N_4(c) surface phases. Another indication that the reactivity of atomic gas mixtures can bear no obvious relation to available data for simpler heterogeneous systems may be found in our experiments on the effects

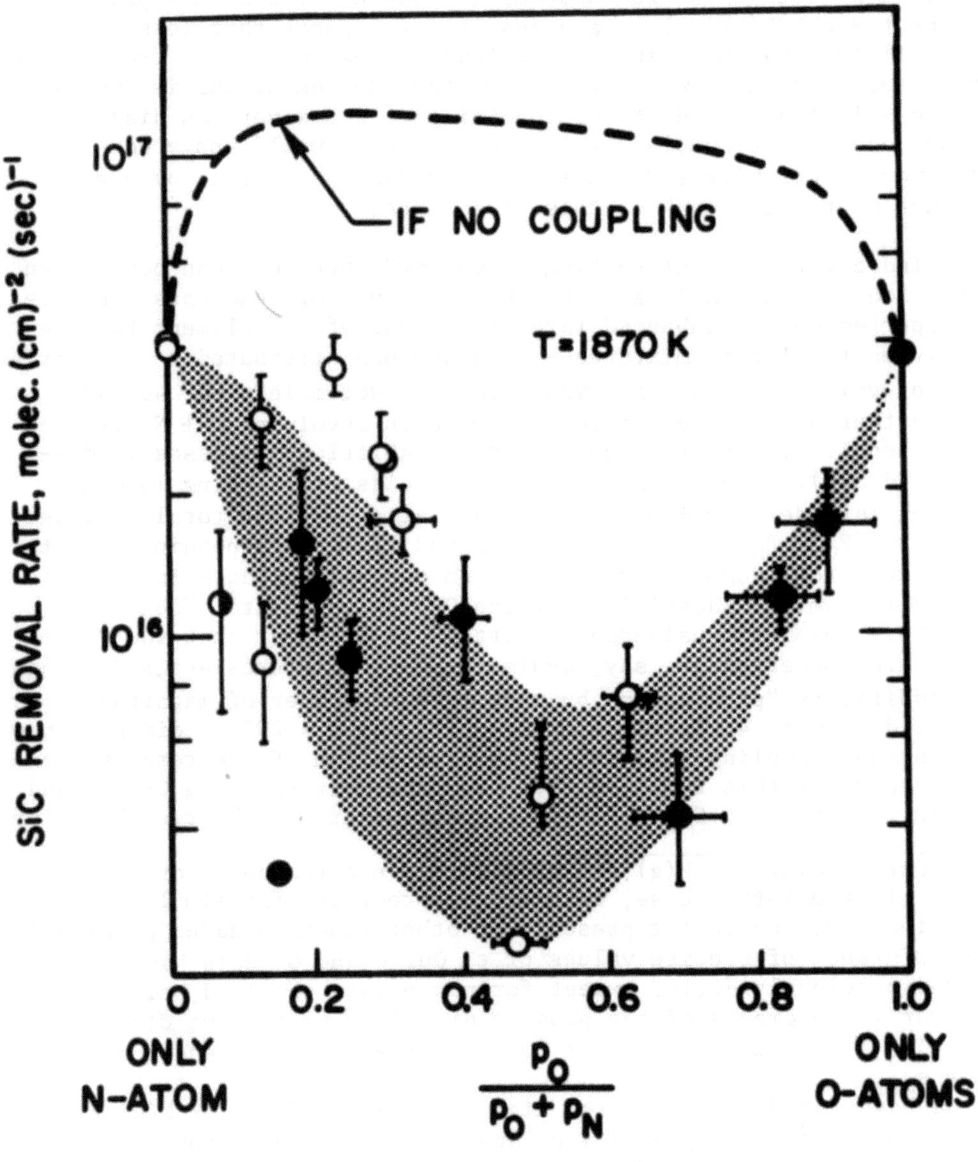

Fig. 5. Chemical attack of silicon carbide in O + N mixtures.

of N-atom addition on the B(s) + O(g) reaction (cf. Fig. 6). While N_2-addition was found to have no appreciable effect on the B(s) + O(g) reaction, N-atom addition (at constant p_O) markedly suppressed the steady-state rate of boron atom removal. For a mixture comprised of only 1 N-atom to every 10 O-atoms, the boron atom removal rates at 1645 °K are already lower than expected for the O-atom attack of boron nitride (cf. Figs. 3c, e)--and slight increases in the (N)/(O) ratio further accentuate this effect. In such cases one must conclude that the quasi-steady surface "seen" by an incident oxygen atom is not at all similar to boron nitride, but rather one from which boron atom removal (via O-atom adsorption/bond formation/desorption) is considerably more difficult. Alternative explanations based on surface catalyzed NO production/desorption appear unlikely in view of the negligible effect of N-atoms on the O-atom attack of molybdenum or tungsten (cf. Ref. 21).

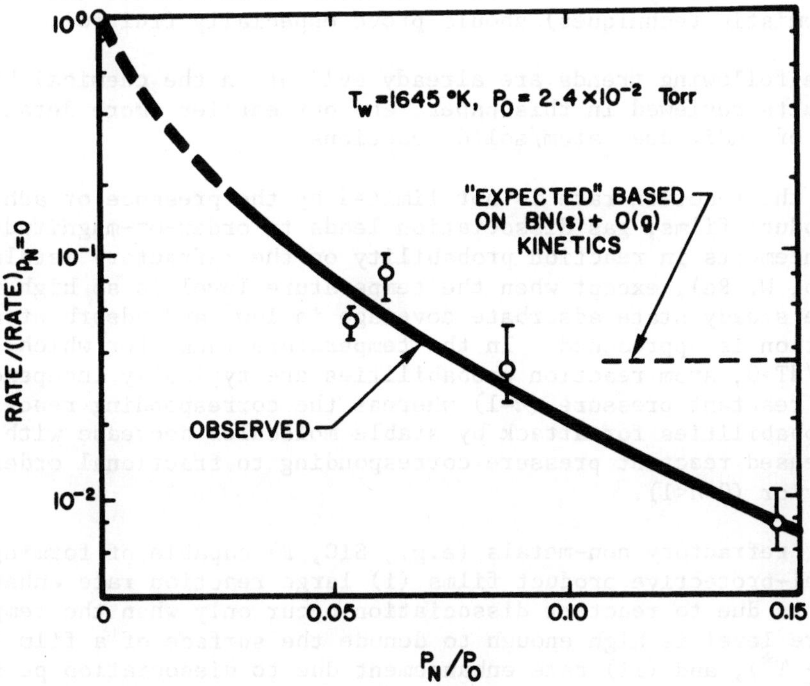

Fig. 6. Chemical attack of boron in O + N mixtures.

CONCLUSIONS

We have described a program of comparative kinetic studies involving pre-dissociated gases and solids being carried out in order to: (i) deepen our understanding of the role of dissociative adsorption and surface interaction of adsorbed atomic reaction intermediates in the chemical attack of refractory solids by stable oxidizing gases and (ii) considerably broaden the base and nature of well-defined kinetic information on important gas/solid systems at elevated temperatures. Attention has been focused on the steady-state reactivity of polycrystalline refractory solids forming volatile reaction products both in simple dissociated gas streams (containing either O, H, Cl atoms or OH radicals) and in mixed atomic systems (e.g., O + N). Being a relatively new and productive combination of experimental techniques our initial emphasis has been on surveying a broad class of gas-solid systems and environmental conditions to establish general patterns of behavior and identify significant exceptional behavior. In addition to the immediate utility of these results, and the interesting kinetic questions they raise (and sometimes answer!) we also hope the present results will help define those particular systems for which detailed measurements (including more atomistic techniques) should prove especially fruitful.

The following trends are already evident in the chemical kinetic results reviewed in this paper, and our earlier (more detailed) account of individual atom/solid reactions:

1. If the reaction rate is not limited by the presence of adherent product films, gas dissociation leads to order-of-magnitude enhancements in reaction probability on the refractory metals (Mo, W, Re), except when the temperature level is so high that the steady state adsorbate coverage is low[†] and adsorbent sublimation is approached. In the temperature range for which $d\varepsilon/dT>0$, atom reaction probabilities are typically independent of reactant pressure ($n=1$) whereas the corresponding reaction probabilities for attack by stable molecules decrease with increased reactant pressure corresponding to fractional order behavior ($0<n<1$).

2. For refractory non-metals (e.g., SiC, B) capable of forming semi-protective product films (i) large reaction rate enhancements due to reactant dissociation occur only when the temperature level is high enough to denude the surface of a film (i.e., $T \geq T^*$), and (ii) rate enhancement due to dissociation persists at ε_{max}. At lower temperatures ($T < T^*$) reactant dissociation leads to minor changes in reaction rate (e.g., O-B), and even rate suppressions[††] (e.g., O-SiC). In this regime both reaction probabilities (i.e., for X and X_2) are typically low, markedly

† Indeed, a neglible dissociation effect was reported for the oxidation of high temperature tungsten at mass spectrometer ion source pressures.

†† Due apparently to the influence of O-atom impingement on the kinetics of dissociative sublimation of SiO_2(c).

temperature dependent, and drop sharply with increasing reactant pressure ($n \leq 0$). Oxygen dissociation significantly enlarges the reactant pressure-temperature domain characterized by the steady-state presence of ("passivating") silica films on silicon carbide.

3. Even on unprotected, high temperature solid surfaces maximum absolute reaction probabilities for atom or free radical attack can be quite small ($\leq$ 1/100) as in the case of the attack of graphite by H atoms or OH radicals. For a particular gas-solid system the temperature level at which ε peaks is indicative of the strength of adsorbate/adsorbent binding (cf. H-C, O-C, O-Re) and is not significantly altered by reactant dissociation (cf. the Arrhenius diagrams for graphite oxidation, rhenium oxidation,molybdenum chlorination, and boron chlorination).

4. Despite the fact that our previous work deals only with polycrystalline filaments, our results on two microstructurally different forms of graphite [16] (e.g., isotropic and pyrolytic forms in oxygen) suggest that reactant dissociation diminishes kinetic sensitivity to microstructural details and crystal orientation.

5. In mixed atomic adsorbate systems capable of forming ternary or more complex compounds with the adsorbent, unusual kinetic behavior should be anticipated at temperature levels such that complex product molecules (or phases) are thermodynamically stable compared to simpler product molecule alternatives. This phenomenon has been observed in the systems: W(s) + O(g) + Cl(g), and SiC(s) + O(g) + N(g), and may also account for the behavior of the B(s) + O(g) + N(g) system presented above (cf. Fig. 6).

Based on available evidence we believe the enhanced reactivity of atoms observed under the present environmental conditions is due to (i) increased sticking probability on partially covered adsorbent surfaces, leading to higher steady-state adatom populations, and, hence, higher steady state reaction rates, and/or (ii) Rideal type reactivity when atoms from the gas phase are incident upon adatoms. Further measurements will be required to determine which of these contributions predominates in particular atom-solid systems, and to rationalize such findings in terms of known chemical properties of the adsorbate and adsorbent. Initial indications are that the enhanced sticking probability mechanism accounts for many of our results on the nonmetals (e.g., graphite oxidation,† boron chlorination), whereas a Rideal-type mechanism which anticipates [21] a close

† In this system, product distribution measurements [45], (e.g., $(CO)/(CO_2)$ ratios near ε_{max}) indicate that oxygen dissociation only slightly increased the tendency to produce the more highly oxygenated product (CO_2).

relationship between the reaction probability ε and the atom recombination probability γ, may be appropriate to the refractory metals at high coverages. We hope that experiments now in progress on reaction product distributions, atom recombination coefficients, and for additional atom/solid systems (including fluorination reactions) will further clarify these important mechanistic questions.

ACKNOWLEDGMENTS

It is a pleasure to acknowledge the advice of Dr. A. Fontijn in the areas of excited species generation, detection and reactivity, the helpful comments of Dr. S. C. Kurzius and other members of the laboratory staff, the assistance of A. D. Freda and A. Marsh in executing the experiments. The senior author wishes to thank Dr. J. Berkowitz-Mattuck (A.D. Little), Dr. J. McKinley (NBS), and Dr. G. C. Frybert (NASA) for their helpful correspondence and interest in phases of this work. We are especially indebted to both the U.S. Air Force Office of Scientific Research (Propulsion Division) and the Army Research Office - Durham for their support under Contracts AF 49 (638)1637 and DA 31-124-ARO-429, respectively.

REFERENCES

1. J. A. Becker, E. J. Becker and R. G. Brandes: J. Appl. Phys. 32, 411 (1961).
2. P. O. Schissel and O. C. Trulson: J. Chem. Phys. 43, 737 (1965).
3. J. B. Berkowitz, A. Büchler, J. L. Engelke and S. N. Goldstein: J. Chem. Phys. 39, 2722 (1963).
4a. J. D. McKinley: J. Chem. Phys. 45, 1690 (1966).
4b. J. D. McKinley: J. Chem. Phys. 40, 120 (1964).
5a. D. O. Hayward and B. M. W. Trapnell: *Chemisorption* (Butterworths, London, 1964).
5b. P. Kofstad: *High Temperature Oxidation of Metals* (J. Wiley, New York, 1966).
5c. G. C. Bond: *Catalysis by Metals* (Academic Press, London, 1962).
6. D. E. Rosner and H. D. Allendorf: AeroChem TP-162, September 1967; Paper No. 48, *Proc. Third International Symposium on High Temperature Technology*, (Butterworths, London, in press).
7. A. B. King and H. Wise: J. Phys. Chem. 67, 1163 (1963).
8. B. J. Wood and H. Wise: Project SQUID Technical Report SRI 26-PU, September 1968.
9. R. K. Gould: Ph.D. Dissertation, Department of Physics, University of Wisconsin (1969).
10. K. M. Sancier, S. R. Morrison and H. V. D. Wiesendanger: Catalysis 5, 361 (1966).
11. R. Klein and M. D. Scheer: J. Phys. Chem. 66, 2677 (1962).

12. J. Streznewski and I. Turkevich: Proc. *Third Symposium on Carbon* (Pergamon Press, New York, 1959), pp. 273-278.
13. F. K. McTaggart: *Plasma Chemistry in Electrical Discharges* (Elsevier Press, Amsterdam, 1967).
14. D. E. Rosner and H. D. Allendorf: Carbon 3, 153 (1965).
15. D. E. Rosner and H. D. Allendorf: AIAA J. 3, 1522 (1965).
16. D. E. Rosner and H. D. Allendorf: AIAA J. 6, 650 (1968).
17. P. G. Dickens, R. Heckingbottom and J. W. Linnett: Trans. Faraday Soc. (in press).
18. A. Dravnieks: J. Am. Chem. Soc. 72, 3761 (1950).
19. R. J. Madix, G. Melin and A. Susu: Presented at the Second International Symposium on High and Intermediate Energy Molecular Beams, Cannes, France, July 1969; see also: R. J. Madix and A. A. Susu: (in press)
20. D. E. Rosner and H. D. Allendorf: J. Chem. Phys. 40, 3441 (1964).
21. D. E. Rosner and H. D. Allendorf: J. Electrochem. Soc. 114, 305 (1967).
22a. A. N. Hughes, M. D. Scheer and R. Klein: J. Phys. Chem. 70, 798 (1966).
22b. R. Klein and M. D. Scheer: J. Phys. Chem. 72, 616 (1968).
22c. M. D. Scheer and R. Klein: Science (AAAS) 144, 1214 (1964).
23a. G. C. Fryberg and H. M. Murphy: J. Chem. Phys. 24, 175 (1956).
23b. G. C. Fryberg: J. Chem. Phys. 42, 4051 (1965).
24. D. E. Rosner and H. D. Allendorf: J. Chem. Phys. 49, 5553 (1968).
25. F. Cramarossa, E. Molinari and B. Roio: J. Phys. Chem. 72, 84 (1968).
26. D. E. Rosner and H. D. Allendorf: (in preparation).
27. H. W. Goldstein: J. Phys. Chem. 68 39 (1964).
28. D. E. Rosner: J. Phys. Chem. 72, 4159 (1968).
29. D. E. Rosner and H. D. Allendorf: J. Phys. Chem. 69, 4290 (1965).
30. D. E. Rosner: AIAA J. 2, 593 (1964).
31. D. E. Rosner: Int. J. Heat Mass Transfer 9, 1233 (1966).
32. D. R. Rosner: *Eleventh Symposium (International) on Combustion* (The Combustion Institute, Pittsburgh, 1967), pp. 181-196.
33a. See, e.g., F. Kaufman: *Progress in Reaction Kinetics*, ed. G. Porter (Pergamon Press, New York, 1961).
33b. B. A. Thrush: Science 156, 470 (1967).
34. J. E. Morgan, L. Elias and H. I. Schiff: J. Chem. Phys. 33, 930 (1960).
35. F. Kaufman and F. B. DelGreco: J. Chem. Phys. 35, 1895 (1961).
36. V. H. Dibeler, J. A. Walker and K. W. McCulloh: J. Chem. Phys. 50, 4592 (1969).
37. A. N. Wright and C. A. Winkler: *Active Nitrogen* (Academic Press, New York, 1968).

38. L. F. Phillips and H. I. Schiff: J. Chem. Phys. 37, 1233 (1962).
39. L. Elias, E. A. Ogryzlo and H. I. Schiff: Can. J. Chem. 37, 1680 (1959).
40. D. E. Rosner: American Rocket Soc. J. 32 1065 (1962).
41. M. A. A. Clyne and D. H. Stedman: Trans. Faraday Soc. 64, 1816 (1968).
42. G. A. Somorjai and J. E. Lester: *Progress in Solid State Chemistry*, ed. H. Reiss (Pergamon Press, Oxford, 1967), pp. 1-52.
43a. J. E. Morgan and H. I. Schiff: Can. J. Chem. 42, 2300 (1964).
43b. H. Wise and B. J. Wood: *Advances in Atomic and Molecular Physics*, eds. D. R. Bates and I. Estermann, Vol. 3 (Academic Press, New York, 1967), pp. 291-353.
44. R. M. Fristrom and A. A. Westenberg: *Flame Structure* (McGraw-Hill, New York, 1965).
45. H. D. Allendorf and D. E. Rosner: AeroChem TP-195R, December 1968 (Carbon, in press).
46. D. E. Rosner and H. D. Allendorf: AIAA J. 5, 1489 (1967).
47. J. D. McKinley: *Proc. VI International Symposium on the Reactivity of Solids*, eds. J. W. Mitchell, R. C. Devries, R. W. Roberts and P. Cannon (Wiley-Interscience, New York, 1969), pp. 345-351.
48. D. E. Rosner and H. D. Allendorf: AeroChem TP-208, June 1969 (AIAA J., in press).

DISCUSSION OF PAPER

C. B. Alcock: When reaction probability for atomic oxygen is considerably less than unity, have you been able to discover the fate of the unused atom flux. If there is any significant amount of recombination on the surface of the reacting solid, the surface temperature might be higher than anticipated.

D. Rosner: Surface heating is not a factor because surface temperature is being monitored and controlled in both the presence and absence of atomic gas. We would be very interested, ourselves, in simultaneously determining what the atom recombination probability is, but so far we have not been successful. There are some measurements being made now which may lead to recombination probabilities or the sum of the recombination and reaction probabilities on the surface. What we have reported here, as has been indicated, is the probability that a sticking atom will lead to substrate atom removal.

J. Drowart: Is it possible that excited atomic or molecular states produced in the microwave discharge are in part responsible for the observed effects?

D. Rosner: The possibility of excited species reacting has naturally concerned us and we've done several experiments that lead us to believe that the reported results pertain to atomic reactions. One of the methods we have used is to study the reaction using two distinct means of generating the same species. For example, if one generates atoms directly, by dissociating O_2 and compares the results in reaction probabilities to atomic oxygen generated by nitrogen atom reactions,one finds that the probabilities are the same within experimental error.

THE GROWTH OF HIGHLY DEFECTIVE SCALES

C. E. Birchenall

Department of Chemical Engineering

Newark, Delaware

Wagner's theory of scale growth, with variable valences, should apply to highly defective scales as well as to nearly stoichiometric scales. Oxides with highly defective structures exhibit complex groupings of defects and ions, some resembling other phases that are stable in the same system. However, the complexes are not likely to persist unchanged through a sequence of transport steps. High vacancy densities decouple tracer atom and vacancy motions quickly, so that correlation corrections cannot be based on dilute models.

At one time, equilibria in molten silicates were forced to fit the mass action law using only concentrations by the invention of complex ions.Increasingly precise and varied data undermined this structure, excessively complicated as it was by oversimplification. The more rigorous phenomenological treatment of carefully collected data that evolved mirrors the structural complexities of the molten silicates, but has not yet produced quantitative pictures of those structures.

The beguiling success of simple defect models in explaining properties of the alkali and silver halides and the utility of Wagner's theory in explaining scale growth encourage attempts to patch up deviations from these models with defect complexes that behave as if they are in dilute solution. That is, the complexes are taken to have energies of formation that are independent of composition and the degree of order. It is too early to advocate abandoning these efforts, for many valuable insights may still come from them. However, increased emphasis on the experimental and phenomenological aspects of the subject seems to be desirable.

The recent conference [1] on Mass Transport in Oxides produced agreement on at least two points: (a) only in those oxides that are highly defective are impurity effects sufficiently unimportant that intrinsic behavior is readily observed; (b) in such highly defective oxides long-range, composition-dependent interactions among defects pose theoretical problems that have not been solved.

In connection with (a), Wagner's theory [2] is restated below in a manner that argues its applicability to the growth of highly defective [3], as well as nearly stoichiometric, ionic phases. Concerning (b), arguments are presented that defect complexes, though inevitable in highly defective compounds, seldom have kinetic identities (that is, persist unchanged through a sequence of transport steps), and their energies and structures depend on composition and order.

WAGNER'S THEORY OF SCALING

For an explicit frame of reference, consider a rigid lattice through which ions diffuse on their own sublattices or through the interstitial space independently of each other apart from satisfying electroneutrality requirements. Electrons or electron holes are much more mobile than either ion. In principle a steady state is achieved in a specimen of thickness ξ with metal M supplied to the inner face at a constant chemical potential (μ) and removed from the outer face at a lower chemical potential. The latter chemical potential is fixed by the applied pressure $p(X_2)$ of the electronegative gas, which also determines the chemical potential of X in the specimen at this face. X is removed as it arrives at the inner face at a constant lower chemical potential determined by the dissociation pressure of the compound in equilibrium with metal M.†

Steady state transport is entirely by M^{z1+} and X^{z2-} ions and electrons, e^-, (or electron holes, p^+, for which the present treatment could be adapted††). z_1 and z_2 are not constants, but are weighted averages for the participating valence states (see the appendix for justification).

† If M is in an alloy, its chemical potential is reduced and the dissociation pressure is increased. The alloy concentration in contact with a growing scale is likely to be time dependent, and the oxide may dissolve the other metallic components of the alloy.

†† If the oxide shifts from p to n at some composition within its stable phase range, local recombination sites must be able to balance the p^+ and e^- currents to maintain local equilibrium to satisfy the model.

A virtual equilibrium is assumed to exist at every point in the compound layer.

$$M = M^{z_1+} + z_1 e^- \; ; \qquad \mu_M + \mu_1 + z_1 \mu_3$$
$$X = \tfrac{1}{2} X_2 = X^{z_2-} - z_2 e^- ; \quad \mu X = \tfrac{1}{2} \mu X_2 = \mu_2 - z_2 \mu_3 \qquad (1)$$

The Gibbs-Duhem equation for a binary system at constant temperature and pressure yields

$$y_M d\mu_M + y_X d\mu_X = 0$$

where the y's are mole fractions. By stoichiometry

$$\frac{y_M}{y_X} = \frac{z_2}{z_1}$$

hence

$$d\mu_M = -\frac{z_1}{z_2} d\mu_X \qquad (2)$$

The flux J of numbers of charges associated with the ions or electrons in equivalents per unit area per unit time is given by [2]

$$J_i = -\frac{u_i n_i z_i}{NV} \left(\frac{1}{N} \frac{d\mu i}{ds} \pm z_i e \frac{d\Phi}{ds} \right) \qquad (3)$$

where + or - is selected to match the sign of the charge or the particle. u_i is the mobility, n_i is the number of i ions in volume V, N is Avogadro's number, Φ is the local electrostatic potential, and s is the distance from the inner face of the specimen measured along the diffusion direction. The driving force (in parentheses in Eq. 3) is the gradient of the electrochemical potential of the charged particle. Avoidance of space charge accumulation requires that for any area normal to the flux

$$J_1 - J_2 - J_3 = 0 \qquad (4)$$

When nearly all of the conductivity is by electron (or hole) migration the electrochemical potential of the electrons, $\zeta_3 = \mu_3 - Ne\Phi$, is nearly uniform through the conductor.

That is $\frac{d\zeta_3}{ds}$ 0, or

$$\frac{1}{N} \frac{d\mu_3}{ds} \simeq e \frac{d\Phi}{ds} \qquad (5)$$

Substituting Eqs. 1, 2, and 5 into Eq. 3 yields for the fluxes

$$J_1 = -\frac{u_1 n_1 z_1}{N^2 V}\frac{d\mu_M}{ds} = \frac{u_1 n_1 z_1^2}{N^2 V z_2}\frac{d\mu_X}{ds} \qquad (6)$$

$$J_2 = -\frac{u_2 n_2 z_2}{N^2 V}\frac{d\mu_X}{ds}$$

Obviously the approximation leading to Eq. 5 yields $J_3 = 0$, not an accurate driving force for electron flux. It is more accurate to evaluate J_3 by substituting J_1 and J_2 from Eqs. 6 into Eq. 4, because the small electrochemical potential gradient for electrons is determined by the unequal fluxes of the much less mobile ions.

Then

$$J_3 = J_1 - J_2 = \frac{1}{N^2 V}\left(\frac{u_1 n_1 z_1^2}{z_2} + u_2 n_2 z_2\right)\frac{d\mu_X}{ds} \qquad (7)$$

Recall that the tracer self-diffusion coefficient D* is given by

$$D^*_i = f_i u_i kT \qquad (8)^\dagger$$

where f_i is the correlation coefficient [4].

Growth of the layer depends on the sum of the rates of delivery of cations to the outer face and anions to the inner face where the arriving ion combines with ions in the metallic and gaseous reservoirs to form new layers of compound. These contributions are effectively independent because the mobile electrons accommodate the differences in ion mobilities. Usually one contribution is much smaller than the other and often is negligible by comparison. The total of the ion fluxes (in opposite directions) is

$$J = |J_1| + |J_2| = \frac{1}{N^2 VkT}\left(\frac{z_1^2}{z_2} n_1 \frac{D^*_1}{f_1} + n_2 z_2 \frac{D^*_2}{f_2}\right)\frac{d\mu_X}{ds} \qquad (9)$$

Stoichiometry requires that $n_1/n_2 = z_2/z_1$. Wagner's rational rate constant k_r is obtained by integrating Jds over the scale thickness ξ

---

† At least in the high defect region of the scale the correlation factor must approach unity for reasons to be given below

$$k_r = \int_0^{\xi} J ds = J_{ave.}\xi = \frac{1}{NVRT}\int_{\mu'_X}^{\mu''_X}\left(\frac{z_1 D_1^*}{f_1} + \frac{z_2 D^*_2}{f_2}\right) n_2 d\mu_X \quad (10)$$

The single prime refers to the inner face; the double prime refers to the outer face. For steady-state, diffusion-controlled growth, the variation of J_i with s/ξ, hence J_{ave}, must be independent of ξ (or of the time).

The chemical potential of the electronegative element is related to its pressure as a diatomic gas by

$$\mu_X = 1/2\ \mu X_2 = 1/2\ RT \ \ln P_{X2} \quad (11)$$

Substituting Eq. 11 into Eq. 10

$$k_r = \frac{1}{2NV}\int_{P_{X_2}(diss.)}^{P_{X_2}(appl.)}\left(\frac{z_1 D^*_1}{f_1} + \frac{z_2 D^*_2}{f_2}\right) n_2 \ d \ \ln P_{X_2} \quad (12)$$

Eq. 12 differs from the one given by Wagner only in treating n_2 as dependent on P_{X2} and inclusion of correlation factors. Usually it is a satisfactory approximation to consider the concentration of the electronegative ion to be constant over the scale thickness. Then

$$k_r = \frac{C_2}{2}\int_{P_{X_2}(diss.)}^{P_{X_2}(appl.)}\left(\frac{z_1 D^*_1}{f_1} + \frac{z_2 D^*_2}{f_2}\right) d \ \ln P_{X_2} \quad (12')$$

where $C_i = n_i/NV$

The oft-suggested equivalence of Eq. 12 to Darken's equation[†] is illusory. For metals, (a) Eq. 4 is not the necessary condition governing electron flow, (b) the stoichiometry-valence conditions

† Darken's equation can be written as

$$D = (y_2 D^*_1 + y_1 D^*_2)\ \alpha$$

where D is the chemical interdiffusion coefficient, y's are mole fractions, and α is a thermodynamic factor that depends on the variation of activity with concentration.

do not apply, (c) nor are the atom motions independent in a substitutional alloy. They compete for the same set of sites, in contrast to ionic crystals. (d) Moreover, Eq. 12, is the sum of the counter currents of the ions, not a measure of their relative motion as given by Fick's coefficient for a chemical interdiffusion process.

Himmel et al. [3] showed that Eq. 12 modified for cases in which cation diffusion is much greater than anion diffusion, was consistent with growth rate and cation diffusivity measurements for wustite, but they regarded the use of variable valence to be an approximation and the agreement to be somewhat fortuitous. Condit [5] found that his ion self-diffusion coefficients in ferrous sulfide permitted calculation of rate constants reported previously by Meussner and Birchenall. [6] †

Himmel et al. [3] pointed out that it was essential not only for the net flux of charge through any plane to be zero (Eq. 4) but also for the fluxes of each kind of atom to be constants through the thickness of the scale. The atom fluxes are proportional to the molar (i.e. gram ion) fluxes $j_1 = J_1/z_1$ and $j_2 = J_2/z_2$. Using the relations already given, for cation flux only, through every cross-sectional plane

$$j_1 = \frac{1}{2\xi} \int_{P_{X_2}(\text{diss.})}^{P_{X_2}(\text{appl.})} C_2 \frac{D^*_1}{f_1} \, d \ln P_{X_2} \tag{13}$$

which is to be compared with the flux in equivalents at a particular plane.

$$J_1 = \frac{C_2}{2} z_1 \frac{D^*_1}{f_1} \frac{d \ln P_{X_2}}{ds} \tag{6'}$$

The compatibility of these conditions is seen if z_1 is taken to be constant and Eq. 6' is integrated from s=o to s=ξ. However, neither of these conditions, electroneutrality and mass conservation, depends in itself on the near stoichiometry of the growing phase. A growing scale must satisfy these conditions in any case by appropriate internal adjustments of the valences, mobilities and gradients. Consequently Wagner's equation is as valid for highly defective scales as for nearly stoichiometric scales.

† New measurements of iron self-diffusion coefficients in $Fe_{1-\delta}S$ by Hobbins [7] showing a marked decrease near stoichiometric composition should have little effect on the values reported at high sulfur pressures and should improve agreement at low sulfur pressures.

At any point

$$D^*_1 = \frac{\int_{P_{X_2}(\text{diss.})}^{P_{X_2}(\text{appl.})} C_2 \frac{D^*_1}{f_1} \, d \ln P_{X_2}}{\xi f_1 C_2 \frac{d \ln P_{X_2}}{ds}} \tag{14}$$

or, if C_2 is nearly constant,

$$D^*_1 = \frac{\int_{P_{X_2}(\text{diss.})}^{P_{X_2}(\text{appl.})} \frac{D^*_1}{f_1} \, d \ln P_{X_2}}{\xi f_1 \frac{d \ln P_{X_2}}{ds}} \tag{14'}$$

COMPLEXES AND CLUSTERS IN HIGHLY DEFECTIVE OXIDES

The temptation to invoke complexes in highly defective oxides has been stimulated by the diffraction studies of Roth [8] on quenched wüstite, showing the presence of cations in tetrahedral interstitial sites with two neighboring octahedral cation sites vacant. This configuration is like a bit of the spinel structure and may account for the easy nucleation of magnetite when oxygen-rich wüstite is quenched.

Anderson [9] has pointed to the need to enrich statistical treatments of defect equilibria in ionic crystals with microdomains of ordered defects to account for both thermodynamic and structural observations on a wide variety of oxides and sulfides. Libowitz [10] combined these notions into an ingenious description of the dissociation pressures of wüstite.

J. B. Wagner [11] followed Landler and Komarek [11] in the use of correlation coefficients for an ideal face-centered cubic lattice [4] with low vacancy density in discussing recent studies on chemical interdiffusion in wüstite. The following simple calculation illustrates the improbability that defect clusters in such highly defective oxides can be followed unambiguously through a series of diffusive displacements. Also, owing to the high vacancy density, a tracer atom and a vacancy with which it exchanges are likely to have their motions uncorrelated quickly by the intervention of other vacancies.

In $M_{1-\delta}X$, if vacancies were randomly distributed and interstitial sites were not occupied to a comparable extent, the probability that any particular site is vacant would be δ . An atom that has just jumped into a vacant site not only has that vacancy as a nearest neighbor (excluding anion sites which are not considered), but also may have other of the remaining C-1 equivalent sites vacant. C is the number of nearest cation neighbor sites. The probability that there are r additional vacancies in the C-1 sites is [12]:

$$P_r = \frac{(C-1)\ (C-2)\ .\ .\ .\ .\ .\ (C-r)}{r!}\ \delta^r\ (1-\delta)^{C-r-1} \qquad (15)$$

The total number of vacant neighbor sites is r+1 = n. For r = 0, hence n = 1, the probability that all C-1 sites are full must be found, which is $(1-\delta)^{C-1}$. Table 1 gives the result of the calculation, where larger vacancy concentrations were not considered when the probability dropped to or below 0.005. C was taken to be 12. P_1 corresponds to n = 2, and so on.

Table 1.

The probability that a cation that has just replaced a vacancy has n nearest vacant sites; for randum vacancy distribution, cation deficiency δ, and 12 nearest neighbor sites.

δ/n	1	2	3†	4	5	6	7	f_{est}
0.01	0.895	0.099	0.005					0.80
0.05	0.569	0.329	0.087	0.014	0.001			0.84
0.10	0.314	0.384	0.213	0.071	0.016	0.002		0.90
0.15	0.167	0.325	0.286	0.152	0.054	0.013	0.002	0.98

† Three vacancies in twelve sites corresponds to the composition $Fe_9O_{12} \equiv Fe_3O_4$.

Note that only one sixth of the cations that have just made a diffusion jump in a δ = 0.15 compound would not find at least a second vacancy on a nearest cation site and more than half would have at least three vacancies in the first cation coordination shell. Directional correlation of the next ion jump to the one that brought it to this position must be very low. For this extreme case it is surely better to take f equal to unity than to 0.78, which applies to a cubic close-packed cation arrangement. f_{est}. in the last column of Table 1 is only slightly better than rough guess about the increasing randomness of jumping of a tracer atom as δ increases. It is hardly worthwhile to do a better calculation because the initial assumption of random vacancy distribution is not likely to apply to a real case. Nevertheless it is clear that correlation coefficients based on dilute defect models do not apply to highly defective systems.

In strictly a configurational sense, Roth's vacancy-interstitial complex needs only a single excess vacancy; see Fig. 1-a (although the energy of the interstitial ion is decreased if it has additional vacant cation sites near it). The simplest complex (Fig. 1-b) is equivalent to a cation half way between one equilibrium site and its neighbor along a normal diffusion path in the NaCl structure. If this configuration has a modest degree of stability, as it appears to have, the mid-point of the diffusion path is a dimple in the energy surface. The saddle points must lie symmetrically at the points where the jumping cation must squeeze through the anion plane.

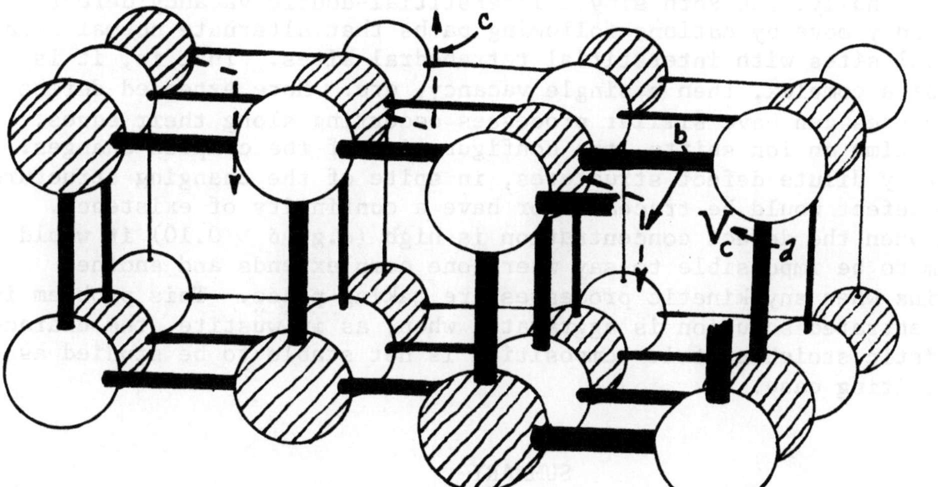

Fig. 1. A portion of the NaCl structure showing (a) a single cation vacancy, (b) the displacement of a neighboring cation that results in Roth's complex, and (c) the propagation of similar displacements down a <110> row of cations to produce an extended complex resembling an element of the spinel structure.

The vacancy pair in Roth's complex defines a <110> vector in the wüstite structure (dashed diagonal in top face in Fig. 1). If the first cation is displaced to a tetrahedral site below the vector, (Fig. 1-b), the next cation can be displaced to the interstitial site above (Fig. 1-c). From the single original vacancy a line defect of considerable length might form which is a configurational element of the spinel structure. This point is belabored to show that big defect complexes may not require high local vacancy concentrations. However, a high local vacancy concentration should greatly increase the variety of complexes that are energetically probable. Moreover, the energy of a particular local configuration must depend on the degree of disorder around it, which in turn must depend upon both concentration and temperature.

Recent measurements of cation order in the spinel Mg Fe_2O_2 by Faller [13] have shown a slightly modified Bragg-Williams relation for order versus temperature, implying order-dependent ordering energy. What is true in this respect of Fe^{3+} and Mg^{2+} should be equally true of Fe^{2+}, Fe^{3+} and cation vacancies. The difficulty in retaining high temperature order by quenching casts doubt on earlier spinel ordering studies that were not carried out at the equilibrium temperature.

Finally, the Roth single interstitial-double vacancy defect can only move by cations following paths that alternate normal octahedral sites with interstitial tetrahedral sites. That is, it is first a complex, then a single vacancy, etc. More extended defect complexes can have similar processes occurring along their edges. Each time an ion shifts, the configuration of the complex changes. In very dilute defect structures, in spite of the changing structure, the defect would be traceable or have a continuity of existence. But when the defect concentration is high (e.g. $\delta \sim 0.10$) it would seem to be impossible to say where one complex ends and another begins when any kinetic processes are taking place. This problem in concentrated solution is aggravated when, as in wüstite, the characteristic stoichiometric composition is not stable to be studied as a limiting case.

SUMMARY

Although a detailed structural description of a highly defective oxide appears to be out of immediate reach, as does any statistical treatment of the equilibrium and kinetic properties that does not greatly simplify the model, Wagner's theory of scale growth appears to be applicable to highly defective oxides as well as nearly stoichiometric oxides. Its success lies in its phenomenological character insofar as the detailed defect structure of the oxide is concerned.

REFERENCES

1. *Mass Transport in Oxides,* October 22-25, 1967, Nat. Bur. Stds. Special Publication 296, 1968. J. B. Wachtman, Jr. and A. D. Franklin, eds. See especially pp. vi, 209-211, 214-217.

2. C. Wagner, in *Atom Movements,* Amer. Soc. for Metals, Cleveland, 1951, pp. 153ff.

3. L. Himmel, R. F. Mehl and C. E. Birchenall, Trans. AIME 197 , 827 (1953).

4. K. Compaan and T. Haven, Trans. Farad. Soc. 52, 786 (1956); *Ibid.* 54, 1498 (1958).

5. R. H. Condit, Ph.D. Dissertation, Princeton University, 1960, p. 67

6. R. A. Meussner and C. E. Birchenall, Corrosion 13, 677t (1957).

7. R. R. Hobbins, Jr., Ph.D. Dissertation, University of Delaware, 1969, to be published.

8. W. L. Roth, Acta Cryst. 13, 140 (1960).

9. J. S. Anderson, in *Nonstoichiometric Compounds,* Am. Chem. Soc., Washington, 1963, pp. 1ff.

10. G. G. Libowitz, in Ref. 1, pp. 109ff.

11. J. B. Wagner, Jr., in Ref. 1, pp. 65ff. and P.F.J. Landler and K. L. Komarek, Trans. AIME 236, 138 (1966).

12. H. Margenau and G. M. Murphy, *Mathematics of Physics and Chemistry,* D. Van Nostrand Co., 1956, p. 432.

13. J. G. Faller, Ph.D. Dissertation, University of Delaware, 1967, to be published.

APPENDIX

Average Mobilities when Valence is Variable

If the cation valence is variable it is necessary to show that an average, but concentration dependent, mobility can be associated with each concentration. As an explicit example, consider oxide M_2O, for which local association-dissociation equilibria can be written:

$$M = M^+ + e^- \qquad \mu_0 = \mu_1 + \mu_3$$

$$M^+ = M^2 + e^- \qquad \mu_1 = \mu_2 + \mu_3$$

$$\text{Sum:} \quad M = M^{2+} + 2e^- \qquad \mu_0 = \mu_2 + 2\mu_3$$

$$\vdots \qquad\qquad \vdots$$

When electron mobility is much greater than that for any of the atomic or ionic species, Eq. 5 enables the velocity v_1 of the charged (as well as neutral) particles to be written in terms of the driving force $d\mu_0/ds$. Thus

$$V_0 = -\frac{u_0}{N}\frac{d\mu_0}{ds}$$

$$V_1 = -\frac{u_1}{N}\left(\frac{d\mu_1}{ds} + e\frac{d\Phi_3}{ds}\right) = -\frac{u_1}{N}\frac{d\mu_0}{ds}$$

$$v_2 = -\frac{u_2}{N}\left(\frac{d\mu_2}{ds} + 2e\frac{d\Phi}{ds}\right) = -\frac{u_2}{N}\frac{d\mu_0}{ds}$$

and so on, until all valence states are accounted. The total flux of the metallic component is

$$j = \sum_{k=0,1,2,...} C_K v_K = -\frac{1}{N}\frac{d\mu_0}{ds}\sum_{k=0,1,2,...} u_k C_k$$

If $\overline{C}$ is the composition based on a total assay of the metallic component

$$\overline{C} = \sum_{k=0,1,2,....} C_k$$

Defining $\overline{u}\,\overline{C} = \Sigma\, u_k C_k$,

$$\overline{u} = \frac{\Sigma\, u_k C_k}{\Sigma C_k} = \Sigma\, x_k u_k$$

where

$$x_k = C_k / \Sigma\, C_k$$

that is, x is the mole fraction.

Hence $\overline{u}$ is simply the arithmetic mean for the contributing species and

$$j = -\frac{\overline{u}\,\overline{C}}{N}\,\frac{d\mu_O}{ds}$$

Thus the u_i in the development of Wagner's equation for oxides of variable valence are $\overline{u}$'s as defined above. The shift in valence states, the variation of $\overline{u}$ with $\overline{C}$ or μ_O, must take place in such a manner that uniform flux of each kind of atomic particle is maintained while the electroneutrality condition is satisfied simultaneously.

DISCUSSION OF PAPER

C. B. Alcock Since the author suggests that the point defect should be replaced by a larger structural unit, does he consider that it would be useful to proceed by analogy with the structure of the next higher oxide in making models of defects?

C. E. Birchenall If the higher oxide has a similar structure, ion displacements corresponding to the structural differences are likely candidates for defect models. It was by this procedure that I constructed the extended wüstite defect in Fig. 1. However, when the two oxides have larger differences in structure the comparison may not be as useful as direct application of imagination to the parent structure.

P. E. Childs I would like to point out that much structural work has been done on wüstite since Roth's [1] work, both on quenched samples and at high temperatures. All these studies contain Roth's conclusion that some iron is present in interstitial positions in wüstite, but suggest that the defect clusters are far more complex than Roth suggested. Koch and Cohen [2], using x-rays, found evidence in quenched wüstite for large clusters which were ordered on a superlattice and did not resemble the magnetite structure. The complexity of the defect structure has also been shown by diffuse neutron scattering on quenched samples. [3] Only limited data are available at high temperatures, but both Koch and Cohen [2] and Maier [4] found diffuse scattering from samples at equilibrium at high temperatures. The use of the simple two vacancy-interstitial' model for wüstite thus seems to have been overworked, and account should be taken of more recent work.

1. W. L. Roth, Acta Cryst., 13, 140 (1960).
2. F. Koch and J. B. Cohen, Acta Cryst., B25, 275 (1969).
3. P. E. Childs, D. Phil Thesis, Oxford (1967).
4. J. Maier, Mem. Sci. Rev. Met.

C. E. Birchenall I agree that the Roth model is overworked. I tried to emphasize the dangers of attributing defect structures determined on quenched specimens to the growing phase at high temperature. Wüstite is more hazardous than most oxides because of its tendency to decompose at low temperatures. However, the defect equilibria are temperature-dependent even in stable oxides.

The additional references you cite should be helpful to anyone venturing into this field.

F. S. Pettit In the derivation of equations, J appears to have been taken as constant. Is J taken as constant and can J be assumed to be constant for the growth of an oxide with large deviations from stoichiometry?

C. E. Birchenall (partly communicated) The molar flux j_i is assumed to be constant at each cross section to conserve matter. However, the equivalent flux J_i may vary from one cross section to another as the effective valence of the flowing ions changes.

Because of this pertinent question, I have added a clarifying statement to Eq. 10 which seems to me to preserve the Wagner form of the derivation with the introduction of little error. However

J_1 and J_2 are not exactly uniform through the scale thickness.

A more accurate calculation of the growth rate constant follows, where k is in cm^2 per second. The growth rate is the sum of the thickness added by the flux of cations to the outer surface and the flux of anions to the inner surface. Using the molar fluxes which are independent of position in the scale, V_1'' is the volume increase for a gram mole of cations at the composition of the outer surface and V_2' is the volume increase for a gram mole of anions at the inner surface. The growth rate is

$$\frac{d\xi}{dt} = \frac{k}{\xi} = | j_1 V_1'' | + | j_2 V_2' |$$

$$= \frac{1}{RT} \left(\frac{z_1 c_1}{z_2} \frac{D^*_1}{f_1} V_1'' + c_2 \frac{D^*_2}{f_2} V_2' \right) \frac{d\mu_x}{ds}$$

$$k \int_0^1 d\,(s/\xi) = k = \int_{\mu_x'}^{\mu_x''} c_2 \left(\frac{D^*_1 V_1}{f_1} + \frac{D^*_2 V_2'}{f_2} \right) d\mu_x$$

Let me emphasize that k is the normal rate constant, not the rational rate constant.

CHEMICAL DIFFUSION IN WÜSTITE AND CHROMIUM-DOPED MANGANOUS OXIDE

Peter E. Childs and J. Bruce Wagner, Jr.

Department of Materials Science, Norwestern University

Evanston, Illinois

The chemical diffusion coefficient, $\tilde{D}$, is the proportionality constant in Fick's second law when a concentration (activity) gradient is propagated through the sample. By following the rate of equilibration, for example, the rate of uptake of oxygen within a single phase (intraphase kinetics) $\tilde{D}$ may be determined. It is shown that $\tilde{D}$ is a function of composition for compounds exhibiting finite deviations from stoichiometry. When sufficiently small increments of composition are chosen for equilibration, $\tilde{D}$ may be considered essentially constant. Hence the dependence of $\tilde{D}$ on composition may be obtained by measuring $\tilde{D}$ over a number of closely spaced composition increments. In such cases, it is shown that $\tilde{D}$ is inversely proportional to a function of the concentration of the mobile defects, (cation vacancies in MnO and in FeO). Chromium-doped manganous oxide and ferrous oxide (wüstite) are treated in detail as examples, and literature data are reviewed. The available theories relating the chemical diffusion coefficient to the self diffusion coefficients are reviewed, and criteria are presented for distinguishing between surface-controlled kinetics and bulk-diffusion-controlled kinetics.

INTRODUCTION

In recent years many studies have been made of the intraphase equibration kinetics in oxide systems [1-21]. Such studies involve measurements of the rates of oxidation and reduction within a bivariant, single-phase region. In such systems the composition of the solid phase is uniquely determined by the oxygen pressure and the temperature. Quantitative information on the equilibration behavior of important metallurgical systems in reactive atmospheres is thus obtained. It also enables self-diffusion coefficients, obtained usually from tracer diffusion studies, and deviation from stoichiometry to be correlated with equilibration data, providing that a suitable phenomenological theory is available. Oxides are particularly inviting for study because of their practical importance, the wide occurrence of nonstoichiometric metal-oxygen systems and the availability of convenient methods for controlling oxygen pressures. Wüstite, $Fe_{1-x}O$, nonstoichiometric ferrous oxide, and manganous oxide, $Mn_{1-x}O$, are notable examples and have been extensively studied. Both compounds exist over a wide range of composition at high temperatures, and the phase diagrams of the $Fe_{1-x}O$ [22-25] and $Mn_{1-x}O$ [26-28] systems are fairly well established. They provide convenient model systems for studying the effect of varying concentrations of cation vacancies on high temperature properties. Many other nonstoichiometric oxide systems have been studied and Table I collects the available data in the literature on the metal-deficient systems. Wagner [29] and O'Keefe [30,31] have also discussed chemical diffusion in oxides. Measurements have also been made on oxides where oxygen is the more mobile species, e.g. Nb_2O_5, [13-14] UO_2 [21], $SrTiO_3$ [19], $CaTiO_3$ [18] and rare earth oxides [15,16]. In ZnO the movement of Zn interstitials has been followed by a conductivity technique [20]. In Table I, comparable values of the chemical diffusion coefficient, $\tilde{D}$, and the cation tracer diffusion coefficient, D_M^T, are quoted where available. The important point to note is that the effective chemical diffusion coefficient obtained from equilibration studies is always 10^2 - 10^3 times larger than the self-diffusion coefficients. Any interpretative theory must explain the magnitude of this difference. Only in the case of $Fe_{1-x}O$, $Mn_{1-x}O$, $SrTiO_{3-x}$ and $CaTiO_{3-x}$ has the effect of composition been studied to any extent.

When the oxygen pressure at the surface of an oxide is changed to another value lying within the phase field at the temperature, net oxidation or reduction occurs. Ions move in the solid phase to regain a uniform chemical potential throughout the sample. In the metal-deficient oxides considered in the present paper, metal ions and electron holes have been shown to be the more mobile species, the oxygen lattice remaining intact and relatively immobile. This has been shown in $Fe_{1-x}O$ [12] and $Mn_{1-x}O$ [33] only by marker studies so far, but where the oxygen diffusion coefficients have been measured the ratio, D_O/D_M, is very low, e.g. $<10^{-2}$ in NiO [34] and $< 10^{-4}$ in CoO [35]. Consequently when the oxygen potential is increased at the surface of such an oxide, oxidation occurs, and metal ions move outwards to the surface of the sample. The reverse occurs for reduction. There is also an opposite flow of metal vacancies and electron holes in order to maintain charge and site balance. When intraphase kinetic studies are performed, the only phase-boundary reaction is at the solid-gas boundary. Then the rate determining process may be either the bulk diffusion of metal ions towards or away from the surface, in oxides exhibiting predominantly electronic conduction, or the oxygen-exchange reaction at the surface. The diffusion process is an example of chemical diffusion since we have mass transport under the action of a chemical potential gradient (see next section). A similar diffusion situation occurs when an oxide layer is formed on a metal during oxidation. During the oxidation of a metal, the kinetics may initially be non-parabolic for the formation of thin oxide layers, but as the layer thickens, chemical diffusion across the oxide layer becomes rate-determining [36,37] An additional solid-solid boundary is present in this case. Moreover, if more than one oxide phase is stable more complex systems are possible, e.g. Fe-FeO - Fe_3O_4 - Fe_2O_3, where the slowest transport process is rate-determining. Consequently gross oxidation or reduction of an oxide to another phase may be determined by the chemical diffusion coefficient in that phase [38]. Another experimental arrangement by which chemical diffusion in oxides may be studied is the analog of the classical metal interdiffusion experiment. Two oxides are placed together and annealed e.g. FeO - MgO [39], FeO - MnO [40], NiO - MnO [41], and a Matano interface develops. By determining the concentration profiles of the two species through the interface chemical diffusion coefficients can be obtained. These various methods for obtaining information on chemical diffusion processes are discussed later in this contribution.

Intraphase equilibration studies offer the most precise way of studying chemical diffusion in oxides which exist over an accessible range of pressures. The experimental variables can be better controlled, the complication of phase-boundary reactions is reduced, and the composition dependence of the effective chemical diffusion coefficient can be studied conveniently in such measurements.

The main question arising from such equilibration experiments is the significance of these measured rate constants, and the effective diffusion coefficients derived from them. We must ask whether we are dealing primarily with a surface reaction rather than a bulk diffusion process. Ideally the selection and control of the experimental variables and a careful analysis of the resulting equilibration kinetics will enable this question to be answered. In the next section the available theories relating chemical and self-diffusion coefficients are reviewed and discussed. The nature of the surface reaction on oxides and the criteria for distinguishing between surface and bulk diffusion control are elaborated in a later section. The current understanding of the high-temperature oxidation-reduction reactions of wüstite and manganous oxide are discussed in the last section, mainly in the light of work done in this laboratory. The two systems will be used as examples to illustrate the oxidation and reduction behaviour of oxides of variable composition at high temperature.

RELATIONS BETWEEN THE CHEMICAL AND SELF-DIFFUSION COEFFICIENTS

First the diffusion coefficients used in this paper will be defined. The self-diffusion coefficient of the ith species is denoted

TABLE I

Chemical Diffusion and Cation Tracer Diffusion Coefficients for Metal-Deficient Oxides

Compound	$\tilde{D}$ (cm^2/sec)		Conditions	Method	Ref.	D_M^T (Cm^2/sec)	Conditions	Ref.
MnO	0.3–0.8x10^{-6} at 1031°C	Ox	CO_2–CO sc/	Cond.	1	4x10^{-11} to 3x10^{-9}	CO_2CO sc	51
	7.5–1.6x10^{-6} at 1031°C	Red				at 1032°C across	RA	
		(Region B*)	pc†			the phase field		
	3.0x10^{-6} at 1030°C	Ox						
	1.2x10^{-6} at 1030°C	Red				1x10^{-11} to 3x10^{-11}	H_2O–H_2 sc/	42
		(Region A*)				at 1000°C		
	9.5–0.2x10^{-6} at 1300°C	Redox	CO_2–CO sc	TG	2		pc†† RA	
	3.0–0.3x10^{-6} at 1000°C	Redox						
		(Region B*)	CO_2–CO pc†	Cond.	a	$10^{-24} < p_{O_2} < 10^{-13}$		
MnO-Cr	3.0–0.14x10^{-6} at 1037°C	Redox						
		(Region B*)	CO_2–CO sc	Cond	a			
FeO	1.4–0.4x10^{-6} at 1000°C	Red	CO_2–CO pc†	TG	3	4.7–9.6x10^{-8} at 983°C	H_2O–H_2 pc	43
	1.1–0.4x10^{-7} at 1000°C	Redox	CO_2–CO pc†	TG	4	across the phase field	SA	
	5.9–3.0.10^{-6} at 1003°C	Red	H_2O–H_2 pc††	TG	5	7x10^{-8} at O/Fe	H_2O–H_2 pc	44
	7.8–1.7x10^{-7} at 1100°C	Red	CO_2–CO pc†	TG	6	1.087 at 983°C		
	4.1–1.6x10^{-7} at 1055°C	Redox	CO_2–CO pc†	TG	7	4.0–7.4x10^{-8}	H_2O–H_2	45
	3.5x10^{-7} at 1060°C, O/Fe 1.086–1.118	Ox	CO_2–CO pc††	Cond.	8	across the phase field	RA	
	5.0–20x10^{-6} at 1062°C	Redox	CO_2–CO pc†	TG	9	1.03–2.35x10^{-7} at 1100°C across the phase field	CO_2–CO p pc†† SA	46

FeO-Cr	1.0×10^{-6} at $1000^{o}C$, O/Fe 1.125-1.05	Red	CO_2-CO 0.67wt% Cr	pc†	TG	3			
	$4.6-2.9\times10^{-6}$ at $1100^{o}C$	Red	CO_2-CO 0.88wt% Cr	pc†	TG	6			
	$2.6-1.6\times10^{-6}$ at $1100^{o}C$	Red	CO_2-CO 1.70wt% Cr	pc†	TG	6			
FeO-Ti	$3.4-28\times10^{-6}$ at $1100^{o}C$	Red	CO_2-CO 0.5 wt% Ti	pc†	TG	6			
FeO-Li	6.4×10^{-6} at $1100^{o}C$	Red	CO_2-CO N_{Li} 10^{-3}	pc†	TG	6			
FeO-Mg	$7.9-3.1\times10^{-7}$ at $988^{o}C$	Ox	CO_2-CO 18 at% Mg	pc††	Cond.	8			
	$12.0-0.6\times10^{-7}$ at $994^{o}C$	Ox	CO_2-CO 31.65at% Mg	pc††	Cond.	8			
CoO	3.2×10^{-7} at $1000^{o}C$ for $3\times10^{-6}p_{O_2}$ 0.51 atm	Ox	Ar-O_2	sc	Cond.	10	$0.9-2.6\times10^{-9}$ at $1000^{o}C$ $4.7\times10^{-2}<p_{O_2}<1$ atm	pc	44
NiO	1.3×10^{-7} at $1000^{o}C$ for $3.6\times10^{-6}p_{O_2}$ 1 atm	Ox	Ar-O_2	sc	Cond.	10	0.55×10^{-11}	pc/sc	47
							0.999×10^{-11}	pc/sc	48
							2.49×10^{-11}	sc	49
	$1.9-4.7\times10^{-5}$ at $1005^{o}C$ 10^{-6} to 1 atm p_{O_2}	Ox	Ar-O_2	pc††	Cond.	12	All at $1000^{o}C$ and $p_{O_2}=0.21$ atm.		
Cu_2O	3×10^{-4} at $1000^{o}C$	Ox		sc	Cond.	19	3.2×10^{-8} at $1000^{o}C$, $p_{O_2}=$ 1 atm	pc	50

Key:
- a P.E. Childs this work
- † as-grown samples from the metal
- †† dense, shaped polycrystal samples
- sc denotes single crystal samples
- pc denotes polycrystal samples
- RA residual activity method

Key:
- SA Surface activity method
- S Sectioning method
- Cond. conductivity method for following kinetics
- TG thermogramimetric method for following kinetics
- * terminology after Hed and Tannhauser [26]

by D_i, and is the diffusivity due to thermal excitation in a region of constant chemical potential. Under the influence of a chemical potential gradient, the diffusivity increases, and the intrinsic diffusion coefficient, D_i^*, is defined by:

$$D_i^* = D_i \, [\partial \ln a_i / \partial \ln N_i] \tag{1}$$

where a_i and N_i are the activity and mole fraction of the ith species and the term $[\partial \ln a_i/\partial \ln N_i]$, is known as the thermodynamic factor. In a region of constant chemical potential this term is unity. For every flux of atoms moving by a defect mechanism, there is an equal and opposite flux of defects. In the case of a vacancy mechanism, for example, a vacancy diffusion coefficient, D_{V_i}, may be defined [52] as:

$$D_{V_i} = D_i \cdot \frac{n_i}{n_{V_i}} = D_i \cdot \frac{1 - x}{x} \tag{2}$$

for a compound M_{1-x}, where n_i, n_{V_i} are numbers of ith ions and i^{th} vacancies. There is some confusion of definition at this point, and D_{V_i} is often given as [30]:

$$D_{V_i} = \frac{D_i}{N_{V_i}} , \tag{3}$$

where N_{V_i} is the mole fraction of vacancies. N_{V_i} is often taken as being equal to x, but it is not a very clearly defined concept. If half the sites are vacant $D_{V_i} = D_i$, and thus Eq. (2) seems to be valid. As $x \to 0$, Eq. (2) reduces to Eq. (3), which is thus applicable at low vacancy concentrations.

Self-diffusion coefficients are usually determined by observing the movement of labelled atoms under a uniform chemical potential, and the tracer diffusion coefficient, D_i^T, is given by

$$D_i^T = D_i f , \tag{4}$$

where f is the correlation factor and depends on the diffusion mechanism and the lattice type. The vacancy correlation factor for a dilute solution of vacancies in a f.c.c. lattice is 0.78146. The last definition we require is that of the chemical or inter-diffusion coefficient. This is defined as the proportionality constant in Fick's law for the case when mass transport occurs under a chemical potential gradient and both species move.

$$\tilde{J} = J_1 - J_2 = -\tilde{D} \cdot \partial\tilde{c}/\partial x, \tag{5}$$

where the J's are the fluxes of the two species through a given plane normal to the diffusion direction and $\tilde{c}$ is the absolute value of the excess concentration of one component. It may be considered as the flow of a deviation from stoichiometry through the crystal restoring a uniform composition. The inter-relation of $\tilde{D}$, D_i and the deviation from stoichiometry is one of the desired aims of this study and various expressions relating these terms are described below.

Wagner's Equation [53]

This was derived originally from a consideration of tarnishing reactions and related the rational rate constant, k_r, for tarnishing with chemical diffusion of the metal and non-metal species. It applied to an electronic semiconductor ($t_e \simeq 1$) with a small deviation from stoichiometry.

$$\tilde{J} = \frac{k_r}{\Delta x} = \left\{ \frac{300}{96{,}500} \cdot \frac{(t_M + t_X)\, t_e \sigma}{N_e} \cdot \frac{1}{|z_M|} \cdot \frac{\partial \mu_M}{\partial \tilde{c}} \right\} \cdot \frac{\partial \tilde{c}}{\partial x} \cdot \tag{6}$$

Here t_M, t_X, t_e are the transport numbers of the metal, the non-metal, and the electronic carrier, $\tilde{c}$ is the absolute excess concentration of one species in equivalents per cm^3, x is the distance variable, μ_M is the chemical potential of the metal, and z_M the metal valence. Using the Nernst-Einstein equation, which applies to dilute solutions of defects, $\tilde{D}$ can be related to the self-diffusion coefficients. For metal deficient oxides $D_M >> D_O$ and the final expression becomes:

$$\tilde{D} = (1 + |z_M|)\, D_M \cdot \frac{N_M}{N_{V_M}}, \tag{7}$$

using Eq. (2) to relate D_{V_M} and D_M, or as given by Wagner [54]:

$$\tilde{D} = \frac{3(1 - x)}{x} \cdot \frac{D_M^T}{f} \tag{8}$$

for a binary, bi-bivalent electronic semiconductor, $M_{1-x}X$. The factor, 3, in Eq. (8) is due to the diffusion potential, $(1 + z_M)$. An equation similar to Eq. (7) may also be derived by considering the fluxes in the solid oxide [55]. It can be shown that using the effective charge on the vacancy, m, instead of z_M:

$$\tilde{D} = (1 + m) \cdot D_{V_M} \tag{9}$$

This is Wagner's equation using the effective charge of the vacancy, m, instead of the cationic charge, z_M, and it shows the dependence of D on the vacancy diffusivity.

Brebrick's Equation [56]

A univalent ionic semiconductor with a composition not far removed from stoichiometry was considered. His derivation is:

$$\tilde{D} = k_X \cdot \lambda^2 \left\{1 + \left(\frac{k_M - 1}{kX}\right) \cdot \frac{n_{V_M}}{n_{V_M} + n_{V_M}}\right\} \cdot \left\{1 + \left(\frac{\delta^2 + 4K_s}{\delta^2 + 4K_i}\right)^{\frac{1}{2}}\right\} \quad (10)$$

Here k_X and k_M are related to D_X and D_M by concentration-dependent Nernst-Einstein expressions, for example:

$$D_M = k_M \cdot \lambda^2 \frac{n_{V_M}}{S}, \quad (11)$$

where λ is the mean ion jump distance, k_M is the cation jump frequency, n_{V_M} is the number of cation vacancy sites, and S is the total number of sites on a given sublattice at a given vacancy concentration. In Eq. (10), n_{V_X} is the number of anion vacancy sites, δ is the deviation from stoichiometry, K_S is the Schottky constant and K_i the intrinsic constant.

When the concentration of cation vacancies is much greater than of the anion vacancies $n_{V_M}/(n_{V_M} + n_{V_X}) \to 1$, and the deviation from stoichiometry is large enough that $\delta^2 >> 4K_s$, $4K_i$, Eq. (10) reduces to:

$$\tilde{D} = 2 \cdot D_M \cdot \frac{S}{n_{V_M}}, \quad (12)$$

with a similar expression holding when $n_{V_X} >> n_{V_M}$. This again returns us to Wagner's equation, using Eq. (3) for D_{V_M}, this time for a univalent compound, i.e. $(1 + z_M) = 2$. Brebrick's equation, however, assumes complete ionization of the vacancies, and its application at low defect concentrations requires a knowledge of the Schottky and intrinsic constants, and the ion jump parameters.

Darken's Equation [57]

Darken derived an equation describing interdiffusion in binary alloys from empirical considerations:

$$\tilde{D} = N_X \cdot D_M^* + N_M \cdot D_X^* \tag{13}$$

Using Eq.(10) and noting that:

$$\frac{\partial \ln a_M}{\partial \ln N_M} = \frac{\partial \ln a_X}{\partial \ln N_X} = \frac{\partial \ln a_i}{\partial \ln N_i} \tag{14}$$

for a binary compound, we find:

$$\tilde{D} = (N_X \cdot D_M + N_M \cdot D_X) \cdot \frac{\partial \ln a_i}{\partial \ln N_i} \tag{15}$$

and for $D_M >> D_X$ this reduces to:

$$\tilde{D} = N_X \cdot D_M \cdot \frac{\partial \ln a_i}{\partial \ln N_i} \tag{16}$$

where the thermodynamic factor, ($\partial \ln a_i / \partial \ln N_i$), contains the driving force for mass transport evaluated at a composition N_X. The thermodynamic factor may be obtained from experimental activity versus composition data, but it is also possible to derive an analytical expression for it if $\ln a_M$ and $\ln N_M$ can be expressed in a suitable form. The addition or subtraction of oxygen to a metal-deficient oxide, MO_{1+y}, with cation vacancies is usually represented by the equation:

$$\frac{1}{2} O_2 \rightleftarrows O_O + V_M^{m'} + m\oplus , \tag{17}$$

where the Kroger-Vink notation is used and $V_M^{m'}$ represents a cation vacancy with an effective charge of m- with respect to the metal sublattice, and $\oplus$ represents an electron hole. The vacancy and hole concentrations are related by the equilibrium constant for this reaction:

$$K_d = \frac{[V_M^{m'}] [\oplus]^m}{p_{O_2}^{1/2}} \tag{18}$$

and by the electroneutrality condition:

$$m [V_M^{m'}] = [\oplus] \tag{19}$$

from Eqs. (18) and (19) we can write:

$$[V_M^{m'}] = y = A \cdot p_{O_2}^{\frac{1}{2(m+1)}} \tag{20}$$

where A is a constant. A similar equation can be written for ⊕ with the same pressure dependence and a different constant.

Since $N_M = 1/\ (2 + y)$ and $a_O = (p_{O_2}/p^o_{O_2})^{\frac{1}{2}}$ we can derive an expression for the thermodynamic factor (T.F.) for an oxide MO_{1+y} [58] using Eq. (20):

$$\frac{\partial \ln a_M}{\partial \ln N_M} = - \frac{N_O}{N_M} \cdot \frac{\partial \ln a_O}{\partial \ln N_M} = (m+1) \frac{(2 + y)}{y} \cdot (1 + y)$$

or for the formulation $M_{1-x}O$ we have:

$$\text{T.F.} = \frac{(m + 1) \quad (2 - x)}{(1 - x) \cdot x} \tag{21}$$

Since $N_O = 1/(2-x)$ in this case, we can write Eq. (16) as:

$$\tilde{D} = (m + 1) \ D_M \cdot \frac{1}{(1 - x) \ x}, \tag{22}$$

an expression which is remarkably close to Wagner's equation.

Manning's Expression [59]

From considerations of the atomistic mechanism of vacancy diffusion, Manning obtained expressions for the intrinsic diffusion coefficients in metals which allow for a net vacancy flow. This gives the Darken expression with an additional term, for example [60]:

$$D_i^* = D_i \left(\frac{\partial \ln a_i}{\partial \ln N_i}\right) \cdot s \tag{23}$$

where

$$s = 1 + \left(\frac{\partial N_M/\partial x}{\partial N_X/\partial x}\right) \cdot \frac{2N_i (D_M - D_X)}{M(N_M D_M + N_X D_X)}, \tag{24}$$

where m is a constant related to the correlation factor and is 7.15 for f.c.c. crystals. For interdiffusion the Eq. (15) becomes:

$$\tilde{D} = (N_M D_X + N_X D_M) \cdot \left\{ \frac{\partial \ln a_i}{\partial \ln N_i} \right\} \cdot s' \tag{25}$$

where s′ is given by:

$$s' = 1 + \frac{2N_M N_X (D_M - D_X)^2}{m(N_M D_X + N_X D_M)(N_M D_M + N_X D_X)} \tag{26}$$

and for $D_M >> D_X$, s′ reduces to 1.28. Thus:

$$\tilde{D} = 1.28\, N_X D_M \cdot \left(\frac{\partial \ln a_i}{\partial \ln N_i} \right) \tag{27}$$

Price's Equation [1]

Price made the assumption that the measured chemical diffusion coefficient in intraphase equilibration studies was an intrinsic cation vacancy diffusion coefficient, $D^*_{V_M}$. From expressions for the thermodynamic factor and mole fractions of vacancies and cations, he obtained the expression:

$$\tilde{D} = D^*_{V_M} = D_M (m + 1) \cdot \frac{(1 - x)}{x^2} \tag{28}$$

for $M_{1-x}O$. It can be seen by comparison with Wagner's Eq. (8) or Eq. (9), that Eq. (28) merely contains an extra factor, the inverse of the vacancy fraction. Consequently $\tilde{D}$ is inversely proportional to the defect concentration in the oxides considered here (FeO and MnO).

In Fig. 1 the variation of the expressions, Eqs. (9), (27) and (28) is shown as a function of the vacancy fraction, x, in an oxide $M_{1-x}O$. The similarity of Darken's and Wagner's equation does not allow them to be distinguished on this plot; Price's equation has a different slope and is several orders of magnitude higher.

METHODS OF DETERMINING CHEMICAL DIFFUSION COEFFICIENTS IN OXIDES

The various methods which yield information on chemical diffusion processes in oxide systems are discussed in terms of interphase

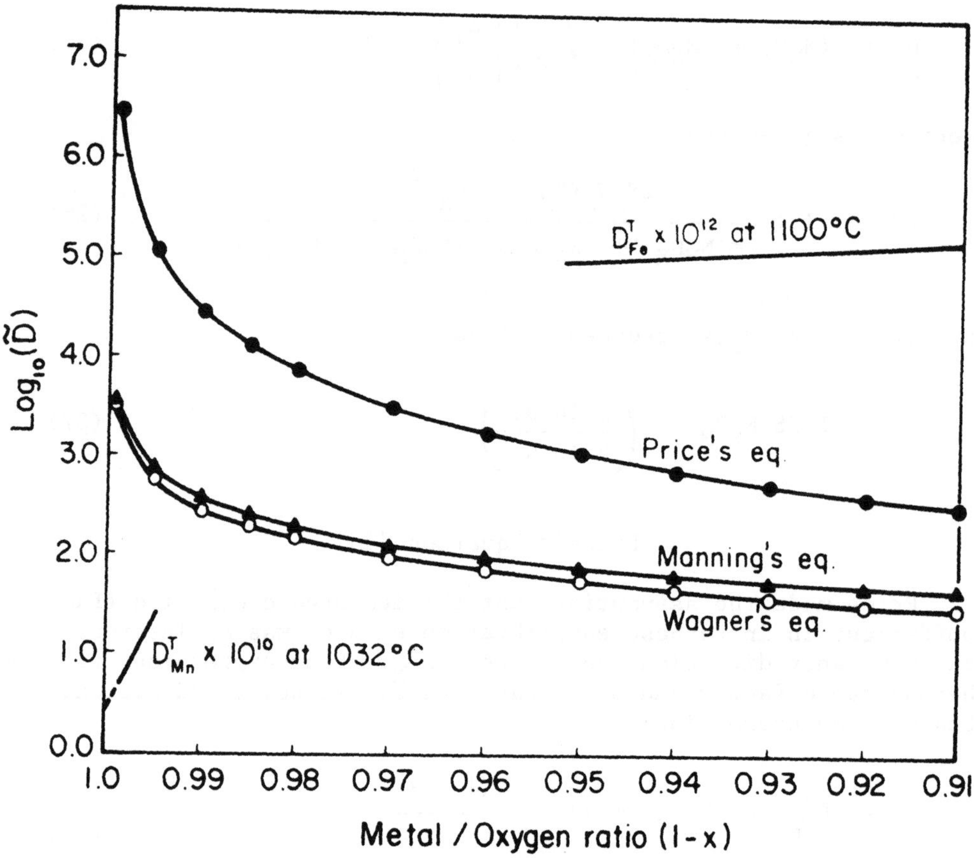

Fig. 1. The chemical diffusion coefficient, $\tilde{D}$, as a function of the metal-oxygen ratio, (1 - x), calculated from Eqs. (9), (27) and (28), with $D_M = 1.0$ and m = 2. The curves show the expected concentration dependence of $\tilde{D}$ apart from the variation of the self-diffusion coefficient. Darken's Eq. (22) is not shown as it lies too close to the curve of Wagner's equation. $D^T_{Fe} \times 10^{12}$ at 1100°C from Hembree and Wagner [46] and $D^T_{Mn} \times 10^{10}$ at 1032°C from Price and Wagner [51] are also shown over the appropriate composition ranges.

and intraphase kinetics. Table II is a schematic representation of the various diffusion couples involved in these methods.

Interphase Kinetics

This describes the reaction between two solid phases across a solid-solid boundary, where chemical diffusion of material across one oxide phase is rate-determining. Three types of experiments may be distinguished.

1. Oxidation of one phase to the next highest oxide phase, by controlling the temperature and oxygen pressure. This includes the isothermal parabolic oxidation of a metal [63, 61, 62], e.g. Mn → MnO [33] and the oxidation of an oxide to the next highest oxide, e.g. FeO → Fe_3O_4 [38]. Reduction experiments are not usually diffusion controlled through the bulk solid because a porous second phase is formed, through which the reducing gases can move to the interface. Parabolic growth of the outer phase (phase 2) during oxidation of the substrate (phase 1) indicates diffusion control, whereas linear growth indicates reaction at one of the phase boundaries is rate-controlling. For meaningful results the oxide formed must be compact and non-porous, and in good contact with the starting phase. The parabolic rate constant, k_p, may be obtained by measuring the weight gain/unit area ($\Delta m/A$) for thin plate samples or the change in oxide thickness, and is given by

$$k_p = \frac{\Delta m}{A}^2 \cdot \frac{1}{t} = \frac{4D}{\pi} (\Delta c)^2 \quad , \tag{29}$$

where Δc is the total change in anion concentration across the growing phase. The parabolic and rational rate constants (Eq. 6) are related by:

$$k_r = z_0^2 \cdot \tilde{V} \cdot k_p/2 \cdot M_0^2 \quad , \tag{30}$$

where $\tilde{V}$ is the equivalent volume of the oxide, M_0 and z_0 are the atomic weight and valence of oxygen respectively. The chemical diffusion coefficient, $\tilde{D}$, thus obtained is an average over the composition gradient in the oxide layer. One surface of the oxide is always at the composition corresponding to the solid-solid boundary, and at the gas-solid surface the composition can be varied up to the upper phase boundary of the outer phase (phase 2). For metal-deficient transition metal monoxides growing on the corresponding metals, Pettit [63] has shown that measurements of k_p at various oxygen pressures of metal oxidation enables the cation diffusion coefficient and oxygen activity across the oxide phase to be related.

TABLE II

Schematic Representation of Methods of Determing
Chemical (Interdiffusion) Coefficients

a) Interphase kinetics

i) Movement of metal ions

<—————— ——————>

Oxygen	Oxide (i)	Metal or oxide (i-1)	Oxide (i)	Oxygen
(p_{O_2},T)	Phase 2	Phase 1	Phase 2	(p_{O_2},T)

Boundary activities (left to right): a_0'', $a_{0_{12}}$, $a_{0_{12}}$, a_0''

ii) Movement of metal ions

<—————— ——————>

Oxygen	Oxide (i+1)	Oxide (i)	Metal or oxide (i-1)	Oxide (i)	Oxide (i+1)	Oxygen
(p_{O_2},T)	Phase 3	Phase 2	Phase 1	Phase 2	Phase 3	(P_{O_2},T)

Boundary activities (left to right): a_0'', $a_{0_{32}}$, $a_{0_{21}}$, $a_{0_{21}}$, $a_{0_{32}}$, a_0''

iii) Interdiffusion of ions

<——————> <——————>

Oxygen	Oxide 2	Oxide 1	Oxide 2	Oxygen
(P_{O_2},T)	Phase 2	Phase 1	Phase 2	(P_{O_2},T)

Boundary activities (left to right): a_0'', $a_{0_{21}} \sim a_0''$, $a_{0_{21}} \sim a_0''$, a_0''

b) Intraphase equilibration kinetics

Oxygen	<——> Metal ions <——>	Oxygen
(P_{O_2},T)	Oxide phase	(P_{O_2},T)
	a_0'' $a_0 \lessgtr a_0''$ a_0''	

2. Multiphase systems produced by oxidation of a metal, or lower oxide, in a high oxygen pressure so that a series of oxide layers are formed, the outer one being in equilibrium with the gas phase. For example, the oxidation of iron in 1 atm. oxygen gives Fe - FeO - Fe_3O_4 - Fe_2O_3 above 570°C [38]. If diffusion in any layer is rate-controlling the thickness of the layer is proportional to the square root of time. In the case above the formation of the FeO layer from Fe and Fe_3O_4, and of the Fe_3O_4 layer from FeO and Fe_2O_3 were found to be diffusion controlled. Wagner [64] has recently considered the interpretation of the parabolic rate constants obtained from such multiphase diffusion couples, and has derived general equations for the chemical diffusion coefficient, $\tilde{D}$. In the formation of oxide phase i from oxide phase (i - 1) and a gas phase with an oxygen pressure corresponding to the coexistence of phases i and i + 1, we have:

$$\int_{N_2(i)'}^{N_2^{(i)''}} \tilde{D} \cdot dN_2 = \frac{(N_2^{(i)} - N_2^{(i-1)''})(1 - N_2^{(i)})}{(1 - N_2^{(i-1)''})} \cdot k_{III}^{i}, \tag{31}$$

and for phases (i-1 and (i+1) coexisting with i we have:

$$\int_{N_2^{(i)'}}^{N_2^{(i)''}} \tilde{D} \cdot dN_2 = \frac{(N_2^{(i)} - N_2^{(i-1)''})(N_2^{(i+1)'} - N_2^{(i)})}{(N_2^{(i+1)'} - N_2^{(i-1)''})} \cdot k_{II}^{i} \tag{32}$$

In Eqs. (31) and (32) k_{III}^{i} and k_{II}^{i} are the corresponding parabolic rate constants of formation of phase i under these conditions given by Eqs. (33) and (34); $N_2^{(i)} = 1/2\,(N_2^{(i)'} + N_2^{(i)''})$, the average mole fraction of component 2 in phase i; $N_2^{(i)'}$ is the mole fraction of component 2 in phase i at the i/(i-1) boundary, and $N_2^{(i)''}$ the equivalent mole fraction at the i/(i+1) boundary. Also:

$$k_{II}^{i} = \left[\frac{\Delta x^{(i)^2}}{2t}\right]_{i-1,\ i+1} \tag{33}$$

$$k_{III}^{i} = \left[\frac{\Delta x^{(i)^2}}{2t}\right]_{i-1,\ O_2(i+1)} \tag{34}$$

It is clear that $\tilde{D}$ appears behind the integral sign in Eq. (32). Nevertheless, only an average value of $\tilde{D}$ across the oxide phase has been obtained by this method so far.

3.) Interdiffusion of two oxide phases is the analog of the classical metal diffusion couples. A diffusion couple may be formed of two oxides which have the same structure and are miscible, for example $Fe_{1-x}O$ - MgO [39] or $Fe_{1-x}O$ - $Mn_{1-x}O$ [40]. On annealing the couple, counterdiffusion of the cations and oxygen will occur, preserving site and charge balance and a Matano interface develops for both cation species [30]. Wagner [64] has given a rigorous definition of the interdiffusion coefficient as:

$$\tilde{D} = V_M \cdot \frac{(N_2 J_1 - N_1 J_2)}{\partial N_2/\partial x} , \tag{35}$$

where J_1, J_2 and N_1, N_2 are the fluxes and mole fractions of the two components, V_m is the molar volume and x is the distance parameter. For a dilute solution of component 2 in component 1 (or vice versa) it can be assumed that V_m is constant and $N_1 \simeq 1$, giving

$$\tilde{D} = - \frac{J_2 \cdot V_m}{\partial N_2/\partial x} \quad \text{if} \quad N_2 << 1 \tag{36}$$

In an interdiffusion experiment, determining the concentration profile of either cation enables the average interdiffusion coefficient of the two cations to be evaluated. The usual expression for a Matano interface analysis is [65]:

$$\tilde{D} = \frac{1}{2t} \cdot \frac{\partial x}{\partial c} \cdot \int_c^o x \, dc, \quad \text{and} \quad \int_o^1 x \, dc = 0, \tag{37}$$

where x is the distance from the origin of the Matano interface and c the concentration of the interdiffusing species. With oxides of variable composition, interdiffusion will require the creation or annihilation of vacancies, changing V_m. The vacancy profile may also have to be considered in such a system. A concentration dependent $\tilde{D}$ is shown by an asymmetric penetration profile, which may have a sharp cut-off. No measurements have been made with a couple composed of the same oxide at different compositions, where labelled atoms could be used to follow the interpenetration. Measurements have been made on incremental (Fe,Mn) O couples differing in manganese concentrations [40]. Control of the experimental variables is difficult when the oxygen activity in the two oxides is different, and the meaning of interdiffusion coefficients obtained under such non-equilibrium conditions is uncertain.

Intraphase Kinetics

Where an oxide can exist over a range of composition, oxygen may be added to or removed from the oxide within the single-phase region. No solid-solid boundaries are present, and it may be considered as interdiffusion of oxygen with the oxide [30] resulting in a change of stoichiometry. In metal-dificient oxides, such as considered in this paper, the cations are the more mobile species, and a change in oxygen pressure produces a vacancy excess or deficit at the surface, as indicated by Eq. (17). These then diffuse into or out of the solid to regain thermodynamic equilibrium. For metal-deficient oxides, we are thus considering a chemical diffusion process characterised by the diffusivity of cation vacancies (see previous section). The vacancy and oxygen excess concentrations are always equal at a given point in the crystal. If the oxide is compact and non-porous we can consider that we have diffusion from a constant source into a finite solid. In solving Fick's second law for this system, we assume the following boundary conditions for the vacancy concentrations, c_V, for a thin plate of thickness 2ℓ:

$c_V = c_{V_f}$ for $x = \pm\ \ell$ at $t > 0$ (instantaneous surface equilibration)

$c_V = c_{V_i}$ for $-\ell \leq x \leq +\ell$ at $t = 0$ (uniform initial composition)

$c_V = c_{V_f}$ for $-\ell \leq x \leq \ell$ at $t = \infty$ (uniform final composition)

Here t is the time variable in the equilibration, x is the distance parameter, and subscripts i and f indicate the initial and final states. We also assume that the chemical diffusion coefficient, $\tilde{D}$, is independent of composition, that diffusion is normal to the surface, that there is no change in cross-section, and that no phase change or precipitation occurs in the reaction. The resulting solution for $c_V(x, t)$ is [66]:

$$c_V(x,t) = c_{V_f} - (c_{V_f} - c_{V_i}) \frac{4}{\pi} \sum_{n=0}^{\infty} \frac{(-1)^n}{2n+1} \exp \{- \tilde{D} \cdot (2n+1)^2 . \pi^2 t/4\ell^2) \cdot \cos \frac{(2n+1)\pi x}{2\ell} \} \cdot \quad (38)$$

This solution is most conveniently given in terms of either the remaining fractional total composition change, Σ, or the fractional change in total composition, Q.

$$\Sigma = 1 - Q = 1 - \frac{\text{total change in concentration of diffusing species at time } t}{\text{total change in concentration of diffusing species at time } \infty}$$

$$= 1 - \frac{\Sigma \left[c_V(x,\ t)\right] - c_{V_i}}{c_{V_f} - c_{V_i}} \tag{39}$$

$$\text{i.e.}\quad Q = \frac{\left[\frac{1}{2\ell} \int_o^{\ell/2} c_V(x,\ t) \cdot dx\right] - c_{V_i}}{c_{V_f} - c_{V_i}} \tag{40}$$

$$= 1 - \sum_{n=0}^{\infty} \frac{8}{(2n+1)^2 \pi^2} \cdot \exp \{ \tilde{D} \cdot (2n+1)^2 \cdot \pi^2 \cdot t/4\ell^2 \}$$

For long times this series for a thin plate is usually truncated to give the familiar analytical expression for $\tilde{D}$:

$$\sum = 1 - Q = \frac{8}{\pi^2} \cdot \exp (- \tilde{D} \cdot \pi^2 \cdot t/4\ell^2) \tag{41}$$

which is valid for $Dt/\ell^2 \geq 0.30$, or $(1 - Q) \leq 0.7$.

At short times we can consider that we have diffusion from a plane source into a semi-infinite solid, and this yields the solution [66]:

$$1 - Q = 1 - 2 \frac{(Dt)^{\frac{1}{2}}}{\ell} \left\{ \frac{1}{\pi^{\frac{1}{2}}} + 2 \sum_{n=1}^{\infty} (-1)^n \left[\frac{1}{\pi^{\frac{1}{2}}} \exp \left(- \frac{n^2 \ell^2}{Dt} \right) - \frac{n\ell}{\sqrt{Dt}} \cdot \operatorname{erfc} \left(\frac{n\ell}{\sqrt{Dt}} \right) \right] \right\}$$

$$\simeq 1 - 2 \sqrt{\frac{Dt}{\pi \ell^2}} \tag{42}$$

the latter being an analytical solution useful at short times. Alternatively at the beginning of the reaction, a parabolic rate law analysis could be used in a weight gain experiment, giving from Eq. (29):

$$\tilde{D} = \frac{\Delta m}{A}^2 \cdot \frac{\pi}{4t\ (\Delta c)^2} \tag{43}$$

where $\tilde{D}$ is averaged over the composition interval Δc.

Q is most conveniently determined by thermogravimetric measurements, since the weight change of the sample gives the total concentration change at a given time, i.e.

$$Q = \frac{\Delta w_t}{\Delta w_f} = \frac{w_t - w_i}{w_f - w_i}, \qquad (44)$$

where w_i, w_f and w_t are the weights of the sample at t = 0, ∞ and t, respectively. However, we may use any property which is determined by the total change in the excess oxygen content at a given time to follow the course of the reaction. For small changes in composition, electronic conductivity measurements provide a more sensitive way of following the reaction. The conductivity must be proportional to the excess oxygen concentration, and the mobility of the carriers and the degree of ionization must not change during the reaction. Campbell, et al [19] and Price [1] have discussed the use of conductivity measurements in this way. If the condition holds that $\sigma \propto \Sigma\, c_V(x, t)$ we can write:

$$Q = \frac{\Delta\sigma_t}{\Delta\sigma_f} = \frac{\sigma_t - \sigma_i}{\sigma_f \quad \sigma_i}, \qquad (45)$$

where σ_i, σ_f and σ_t are the electronic conductivities at t = 0, ∞ and t, respectively. Eq. (43) may be used in Eq. (40), when the conductivity is proportional to the vacancy concentration. In the n-type region of MnO the electrons dominate the conductivity, and consequently this is inversely proportional to the vacancy concentration. Price [1] has derived an expression to allow $\tilde{D}$ to be derived from conductivity measurements in the n-type region, where:

$$\sum = 1 - Q = 1 - \frac{\Delta\sigma_t}{\Delta\sigma_\infty} = \frac{Z}{Z-1}\left[1 - \frac{1}{1 + (Z-1)\frac{4}{\pi}\cdot\exp(-\pi^2\cdot\tilde{D}\cdot t/4\ell^2)}\right] \qquad (46)$$

where $Z = \left[P_{O_2i}/P_{O_2f}\right]^{1/[2(m+1)]}$, the ratio of c_{V_i} to c_{V_f}. This is a truncated expression valid only at long times. For a given $\tilde{D}$ it gives the same equilibration time as Eq. (41), but the equilibration curve shows less curvature. In the four-probe conductivity determination the actual conductivity measured is that of the central portion of the sample between the potential probes. In this case we may need to include diffusion from the side-edges, but for a sample where the voltage probes are more than a diffusion length in from the end-edges, diffusion through the end-edges can be neglected. Price [1] derived an equation valid at long times allowing for this. It is:

$$\sum = 1 - Q = \left(1 - 0.274\,\frac{2\ell}{b}\right) \cdot \frac{8}{\pi^2} \exp\left(-\,\frac{\pi^2\,\tilde{D}\,t}{4\ell^2}\right) \qquad (47)$$

where b is the width of the sample. If $b >> 2\ell$ then this reduces to Eq. (41) for a thin plate.

One of the main assumptions in deriving these solutions is the assumption of instantaneous surface equilibration. This implies that the oxygen pressure change over the sample is a step-function and that the surface reaction is much faster than the diffusion process. If this is not so Eq. (38) should be modified, the term $(c_{V_f} - c_{V_i})$ being replaced by $(c_{V_s}(t) - c_{V_i})$, where the $c_{V_s}(t)$ denotes the surface concentration as a function of time. If we can represent $c_V(t)$ by a suitable function we still have an expression for the concentration profile which involved $\tilde{D}$. In practice the oxygen concentration change at the surface can probably be represented by an exponential function [66]:

$$c_{V_s}(t) = (1 - e^{-\beta t}).\,(c_{V_f} - c_{V_i}) + c_i$$

$$\text{i.e. } \beta \to \infty,\ c_{V_s}(t) \to c_{V_f}\ ;\quad t \to 0\ ,\quad c_{V_s}(t) \to c_{V_i} \qquad (48)$$

Eq. (38) can be solved explicitly for this surface concentration giving:

$$\sum = 1 - Q = \exp(-\beta t) \cdot \left(\frac{\tilde{D}}{\beta\ell^2}\right)^{1/2} .\tan\left(\frac{D}{\beta\ell^2}\right)^{1/2} + + \frac{8}{\pi^2} \sum_{n=0}^{\infty} \frac{\exp\left(-\pi^2\tilde{D}t(2n+1)^2/4\ell^2\right)}{(2n+1)^2\left[1 - (2n+1)^2\,\{\tilde{D}\pi^2/4\beta\ell^2\}\right]} \qquad (49)$$

When $\beta \to \infty$ this gives us Eq. (40), and Eq. (48) is valid providing $\beta \neq [\tilde{D}\,(2n+1)^2\pi^2]\,/4\ell^2$.

If a surface reaction is partly rate-controlling, Eq. (48) is probably not the correct expression to represent the surface concentration, and some other expression for $c_{V_s}(t)$ is required. Landler and Komarek [5] suggested that the H_2O-H_2 reduction of wüstite was a case of combined control and gave a linear expression for the surface reaction. Their equation is:

$$\frac{dm}{dt} = k_s \cdot A \cdot (c_{Fe_s} - c_{Fe_f}) \quad \text{gm/sec} \qquad (50)$$

where A is the surface area of the sample, k_s the surface rate constant in oxygen atoms/sec/cm^2/unit deviation of iron concentration in wüstite from equilibrium in gm. Fe/cc^3. Iron ions were considered as diffusing into the sample from the surface. Fick's second law can be solved for a linear surface reaction giving the following expression:

$$\sum = 1 - Q = \sum_{1}^{\infty} 2 \exp\left(- \frac{4\tilde{D}t}{2} \cdot \beta_n^2\right) \cdot \left[\beta^2 + \left(\frac{2\beta_n^2 \cdot \tilde{D} \cdot M_O}{k_s \ell M_{Fe}}\right) + \frac{2\beta_n^2 \tilde{D} M_O}{k_s \ell M_{Fe}}\right] \quad (51)$$

where $\cos \beta_n = \beta_n / 2D\ M_O / k_s \ell\ M_{Fe}$, and M_O M_{Fe} are the atomic weights of oxygen and iron respectively. $\tilde{D}$ cannot be obtained analytically from expressions like Eqs. (23) or (21), and Landler and Komarek [5] compared the experimental and calculated curves to determine the necessary β_n values.

The nature of the surface reaction is discussed in the next section, and there it will be seen that some information on the surface reaction of oxides is available from thin foil studies. The oxygen exchange process on the surface is not altered by changing the sample thickness, but its relative importance with respect to the diffusion process is.

One of the assumptions made in solving Fick's second law for bulk diffusion is a concentration independent $\tilde{D}$. This is not true in the systems considered here, but for small composition intervals $\tilde{D}$ is approximately constant. Analysing the last part of the equilibration curve using Eq. (41) gives a $\tilde{D}$ which is appropriate to the final composition. The analysis of the initial part of the curve should give a $\tilde{D}$ near the initial composition, and will also be complicated by the surface reaction. The exact composition dependence of $\tilde{D}$ may be lost if too large a composition interval is used, and it is desirable to use as small an interval as possible.

THE SURFACE REACTION

The only phase boundary present in intraphase kinetic studies is the gas-solid boundary, and reaction at this boundary provides an alternative rate-controlling process to bulk diffusion. The surface reaction is always present, but it becomes important only when its rate is comparable with or less than the diffusion rate. Under these conditions, the experimental equilibration data can only be interpreted if the relative magnitudes of the two rate constants can be ascertained. In this section the nature of the surface reaction on wüstite and manganous oxide will be described. The factors affecting the rates of the surface reaction and the bulk diffusion process will be discussed, in order to find criteria to distinguish between them.

The Nature of the Surface Reaction on Oxides

Obviously the surface reaction will depend on the atmosphere used to control the oxygen pressure over the sample. The oxygen-exchange process with the surface will be different in CO_2-CO, H_2O-H_2 and Ar-O_2 mixtures, and in general the surface reaction rates will be different. The compositions of wüstite and manganous oxide are usually controlled by using CO_2-CO or H_2O-H_2 mixtures. In this laboratory only CO_2-CO and Ar-O_2 mixtures have been used in equilibration studies. In addition to considering the process by which oxygen is added to or removed from the surface, we may also have to consider the effect of H or C dissolving in the sample [67-69]. The physical nature of the surface may also affect the surface reaction, for example the surface roughness and/or the crystallographic orientation.

The oxygen uptake by a sample may be determined by any of the following steps:

i) mass transport in the gas phase

ii) dissociation of CO_2 into CO and adsorbed oxygen atoms/ions at the oxide/gas interface, perhaps via an activated complex *or* dissociation of H_2O at the surface in a similar manner.

iii) incorporation of oxygen into the oxide at the oxide-gas interface.

iv) migration of metal ions (vacancies) and electrons through the oxide to the surface.

Step (i) may be a diffusion controlled process. However, the activation energy for such a process is usually low, 10 - 15 Kcal per mole as compared to the activation energy for solid state diffusion, 20 - 100 Kcal per mole. Moreover, a large change in the linear flow rate of the gases over the sample will alter the kinetics if (i) is important. Steps (ii) and (iii) give rise to surface control proper, and step (iv) represents diffusion control. One of the main problems in investigating the surface reaction is that of determining the mechanism of step (ii).

The overall reactions for the addition of oxygen at the surface are:

$$CO_{2_g} \rightleftarrows CO_g + O_O + m\oplus + V_M^{m'} \qquad (52)$$

and

$$H_2O_g \rightleftarrows H_{2_g} + O_O + m \oplus + V_M^{m'} \tag{53}$$

but these may represent several successive reaction steps. In order to derive a rate expression, we need to know the rate determining step(s) and the nature of the absorbed species. The derivation of the surface rate expression is considered in the Appendix, and alternative expressions of it are given there. Using Wagner's [70,71] expression for the rate of oxygen addition, in atoms per second, we have:

$$\frac{dn}{dt} = k' \, P_{CO_2} - k'' \, P_{CO} \, . \tag{54}$$

Here k',k'' are the rate constants for addition and subtraction of oxygen. Eq. (54) may be transformed into the expression:

$$\frac{dn}{dt} = k' \, (a_{O_s}) \cdot P_{CO_{2_f}} \cdot a_{O_s}\left(\frac{1}{a_{O_s}} - \frac{1}{a_{O_f}}\right), \tag{55}$$

where $k' \, (a_{O_s})$ is the rate constant at a surface oxygen activity a_{O_s}, and $P_{CO_{2f}}$ and a_{O_f} are the equilibrium CO_2 pressures and oxygen activity above the oxide surface. At the beginning of the reaction where $a_{O_s} \simeq a_{O_i}$ we obtain:

$$\frac{dn}{dt} = k'(a_{O_i}) \cdot P_{CO_{2_f}} \cdot a_{O_i}\left(\frac{1}{a_{O_i}} - \frac{1}{a_{O_f}}\right) \tag{56}$$

where a_{O_i} is the initial oxygen activity in the compound or at the metal/oxide interface in the case of metal oxidation. An equivalent expression to Eq. (55) may be obtained in H_2O-H_2 mixtures [72], substituting $P_{H_2O_f}$ for $P_{CO_{2_f}}$ and using a different rate constant.

Recently some intraphase kinetic measurements have been made on very thin wüstite samples (10 - 15 μ). Grabke [73] studied the exchange of $^{14}CO_2$ with the wüstite surface as a function of CO_2/CO ratio and temperature,and found the empirical expression for the oxygen exchange surface rate constant to be:

$$k' \, (a_O) = \text{const.} \left(\frac{P_{CO_2}}{P_{CO}}\right)^{-u} \tag{57}$$

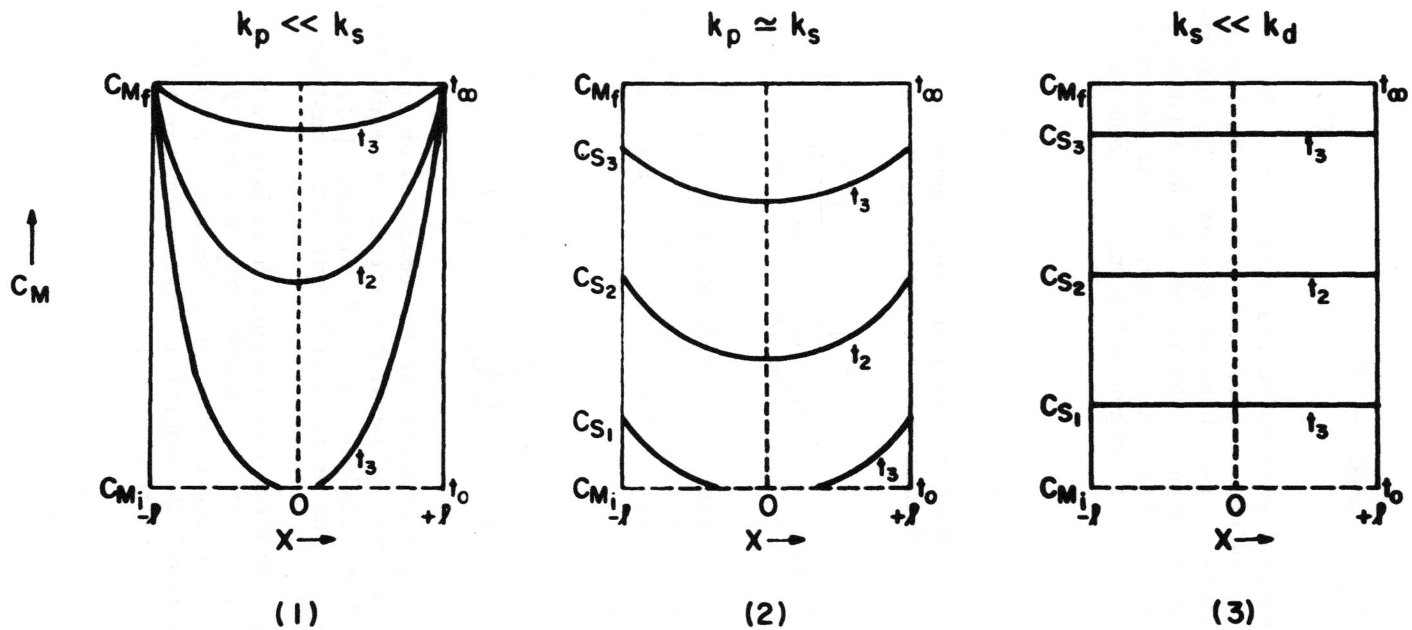

Fig.2. Schematic illustration of the concentration profiles in an equilibrating metal-deficient oxide, for various times (t) after the start of the reaction. c_{M_i} and c_{M_f} are the initial and final metal concentrations at the surface, at times $t = t_o$ and $t = t_\infty$ respectively. c_{S_1}, c_{S_2}, c_{S_3} are the metal concentrations at the surface for a finite surface reaction for times t_1, t_2 and t_3 respectively, where $t_o < t_1 < t_2 < t_3 < t_\infty$. The three cases illustrated are for different relative magnitudes of k_s, the surface rate constant, and k_p, the parabolic rate constant. The sample thickness is 2 ℓ and x is the distance parameter in the diffusion direction.

where the const. = 1.2×10^{-2} exp (-34100/RT) and u = 0.6 at 983°C. The velocity of oxygen transfer is proportional to the partial pressure of CO_2 and depends on the oxygen activity at the oxide surface, i.e. the CO_2/CO ratio, and was different on other oxides. The rate is strongly temperature dependent, and the dependence on oxygen activity increases with increasing temperature. Grabke suggests that the reaction:

$$CO_{2_g} + 2e^- \rightleftarrows CO_g + O^{2-}_{ad} \tag{58}$$

may explain the observed dependence. Stotz has studied the oxygen uptake of wüstite foils in CO_2-CO and H_2O-H_2 [74] and of Cu_2O and NiO in oxygen [75], by following equilibration kinetics with conductivity measurements. At 900°C the initial equilibrium rate for wüstite is essentially that of the reactions:

$$CO_{2_g} \rightleftarrows CO_g + O_{ad} \tag{59}$$

$$H_2O_g \rightleftarrows H_{2_g} + O_{ad}$$

no such measurements have yet been reported on manganous oxide.

All equilibration reactions will be surface controlled in the initial stages, because the diffusion rate → ∞ as the diffusion length → 0. We would like to know at what sample thickness diffusion control takes over and in what range of thicknesses mixed control is to be expected. In the oxidation of iron, the kinetics are found to change from linear to parabolic in both CO_2-CO [36] and H_2O-H_2 [72] mixtures. In CO_2-CO the critical oxide thickness for diffusion control to become important seems to be $> 10^{-2}$ cm. In H_2O-H_2 the surface reaction is faster, and the changeover to diffusion control occurs at smaller oxide thicknesses. The oxidation of thick Mn plates in CO_2-CO (∿ 1000 μ) shows only parabolic kinetics indicating that decomposition of CO_2 is faster on MnO or that diffusion is slower in MnO than in FeO. The kinetics of Mn oxidation or MnO equilibration in H_2O-H_2 mixtures have not been measured. Wagner [76] has discussed the change from linear to parabolic kinetics in the oxidation of metals to a metal-deficient oxide.

Distinguishing Between Surface Reaction and Bulk Diffusion Control

Three limiting cases for the relative magnitudes of k_s, the surface rate constant and k_p, the diffusion rate constant, can be distinguished and are illustrated in Fig. 2.

(1) k_p << k_s bulk diffusion control
(2) k_s^p ≃ k_p^s combined bulk diffusion and surface reaction control
(3) k_s << k_p surface reaction control

For each case the metal (or vacancy) concentration profiles through the samples are markedly different, and consequently the experimental weight gain versus time curves will be different. In case (1) the initial kinetics will be parabolic and in case (3) they will be linear. Case (2) will show mixed kinetic behavior. In principle it is possible to change the equilibration kinetics exhibited by a given compound from case (1) through (3) by altering the experimental conditions, e.g. sample size or temperature. Increasing the temperature favors surface control since diffusion is more strongly temperature dependent [61], and increasing the thickness of a sample favors diffusion control. It is possible to point out some experimental criteria whereby we can distinguish between the alternative rate-controlling processes. Ideally the starting samples should be well-annealed, dense, homogeneous, smooth single crystals with a suitable geometry, e.g. a thin plate, long cylinder or a sphere. This eliminates several complicating experimental variables which are difficult to control, i.e. grain boundary and pore diffusion, edge effects, stress effects and surface irregularities.

The factors which may alter the rate-controlling mechanism or may enable us to distinguish between surface or bulk diffusion control are:

i) nature of the oxygen-determing atmosphere.
ii) analysis of the kinetics, especially in the early stages of reaction.
iii) sample size and shape.
iv) grain size or crystal orientation of the sample.
v) inert-gas dilution.
vi) variation in the gas flow-rate.
vii) sample composition/oxygen partial pressure at the surface.
viii) temperature.

The importance of these factors is discussed briefly below.

i) Suitable gas mixtures, e.g. H_2O-H_2, CO_2-CO, Ar-O_2 are used to produce atmospheres of variable oxygen pressure. The oxygen exchange process is different in each case and usually the surface rates will differ. For a given sample changing the atmosphere, but maintaining the same oxygen pressure, may change the rate thus indicating surface control. In oxidation of iron at 950°C the kinetics are linear in CO_2-CO [77] and parabolic in H_2O-H_2 [72]. The other component in the gas phase (C, H or Ar) may also dissolve in or permeate the solid, and affect the kinetics [67-69]. Meussner, et al [78] have shown that wüstite has approximately the same equilibrium properties in CO_2-CO and H_2O-H_2 mixtures, suggesting that the solubility of hydrogen or carbon is negligible compared to the defects due to non-stoichiometry. But in compounds with lower native defect concentrations such inadvertent doping may be important.

ii) Initial linear kinetics indicate phase boundary control, while parabolic kinetics indicate bulk diffusion control. In the case of mixed kinetics we can write [61]:

$$\left(\frac{\Delta m}{A}\right) \cdot \frac{1}{k_s} + \left(\frac{\Delta m}{A}\right)^2 \cdot \frac{1}{k_p} = t \tag{60}$$

or on rearranging:

$$\frac{1}{k_s} + \left(\frac{\Delta m}{A}\right) \cdot \frac{1}{k_p} = t \;/\; \left(\frac{\Delta m}{A}\right) \tag{61}$$

Plotting $(\Delta m/A)$ versus $t/(\Delta m/A)$ gives $1/k_s$ as the intercept and $1/k_p$ as the gradient, providing diffusion control is significant. If $k_s >> k_p$ this equation reduces to Eq. (29) and if $k_s << k_p$ we obtain the linear rate expression:

$$\left(\frac{\Delta m}{A}\right) \cdot \frac{1}{k_s} = t \tag{62}$$

Usually the initial part of the experimental curves is known least accurately, because of uncertainties in the zero time for the atmosphere change. Also diffusion initially will be very fast and the initial slope of the equilibration curve would reflect only the surface reaction at intermediate times the curve will evidence mixed control and at long times diffusion control. Consequently the complete composition-time curves should be considered in determining the rate-controlling mechanism.

iii) The dimensions of the sample are particularly crucial in determining the rate-controlling step, in that the mean diffusion distance ($\bar{\ell}$) determines whether diffusion or the surface reaction will be slower. Thin samples are invariably surface-controlled and enable the details of the surface reaction on a given compound to be studied [73-75]. As the thickness increases the diffusion rate decreases as $\bar{\ell}^2$ and a change over from linear to parabolic kinetics occurs [36]. Within a given region of control the time, t_Q, taken for a given fraction of the reaction to occur, Q, depends differently on the sample dimensions. For example, for a plate sample $t_Q \propto \bar{\ell}^2$ for diffusion control, and $t_Q \propto V/A$, i.e. $\propto \bar{\ell}/2$, for surface control. By changing the sample thickness, for a given set of experimental conditions, we should be able to determine the rate constants for diffusion and the surface reaction. In a weight gain experiment this would be done conveniently by starting with a thick, dense, rectangular plate and successively grinding down the faces to reduce the thickness, and measuring equilibration times at each thickness.

iv) The grain size and/or crystal orientation of the sample surface may determine the surface reaction rate. The growth of FeO [71] and MnO [33] on the metal in CO_2-CO mixtures was found to give preferentially (100) faces, presumably because these grow faster than other faces. Too little information is available on the effect of grain size or orientation on the equilibration kinetics to enable any generalization to be made. Studies of single crystal samples would enable different surface orientations to be used, but the results of such studies have yet to be reported.

v) Inert-gas dilution using argon or nitrogen offers an easy way of distinguishing surface and bulk diffusion control. If the oxygen-producing atmosphere is mixed with an inert-gas keeping the total pressure, flow-rate and oxygen potential constant, then the effective surface area for a given sample thickness will be decreased. If the surface reaction is rate-controlling the equilibration rate should decrease in proportion to the degree of dilution. In the previous section we saw the surface rate on wüstite was proportional to p_{CO_2} for a given p_{O_2} and thus the surface rate should decrease with argon dilution, and this has been observed [36]. No change in rate should be observed for a diffusion controlled process, unless the transport of oxygen to the surface were hindered by too much dilution; i.e. unless the mass-transport diffusion process in the gas phase becomes significant.

vi) A variation in the gas flow-rate, within the limits required for adequate mixing and thermal equilibration, may disturb the oxygen-exchange equilibrium at the surface and hence the reaction rate and may indicate diffusion-controlled mass transport in the gas phase. The effect of flow-rates on the equilibration kinetics should be investigated, and in comparative studies, constant total linear flow-rates should be used. Constant flow rates are also important in eliminating noise in thermogravimetric determinations [2].

vii) The rate of the surface reaction and of the diffusion process both depend on the composition of the sample, which is determined by the oxygen potential in the surrounding atmosphere. If these two dependencies are different and are known, measuring the equilibration rate as a function of composition should indicate the rate-controlling process. The rate of oxygen exchange to a wüstite surface has been shown to be proportional to $(p_{CO_2}/p_{CO})^{-u}$, where u is 0.6 at 983°C and increases with decreasing temperature [73]. The pressure dependence of the diffusion coefficient seems to be complex and can only be estimated at present from such expressions as given in the last section.

viii) Surface reaction and diffusion are both thermally activated processes and in general the experimental "activation energy" of the equilibration rate will have several contributing factors.

For certain critical sample thickness for a given sample, the rate-controlling process may be changed by altering the temperature [77]. In general low temperatures favor diffusion control, and at high temperatures the surface reaction is favored.

EXPERIMENTAL RESULTS

The equilibration kinetics of wüstite and manganous oxide have been studied in this laboratory, and the experimental details are given here. Thermogravimetric measurements were used to follow the equilibration of wüstite samples because of the wide range of composition, and hence large weight change, across the phase field, as shown in Fig. 3. At the temperatures of these measurements, $\leq$ 1200°C, manganous oxide shows a much smaller deviation from stoichiometry as is shown in Fig. 4. At 1000°C the maximum vacancy concentration is ∿ 1.2% and at 1200°C ∿ 5.2%. Consequently in this system electronic conductivity measurements have been used to study the equilibration kinetics. The common features of the two methods are first described, and then the individual features of the thermogravimetric and conductivity methods are given. Experimental results on the two systems are then presented separately.

Method

CO_2-CO mixtures were used to determine the oxygen partial pressures over the sample. Fig. 3 shows the phase diagram of wüstite plotted as temperature versus log (CO_2/CO), together with isocomposition lines. In Fig.5 the phase diagram of MnO is given as log p_{O_2} versus 1000/T, showing the region which can be covered using CO_2-CO mixtures. The lower limit in oxygen pressure is set by the decomposition of CO to carbon and this is indicated on the figures. Matheson C.P. grade CO and "Bone Dry" CO_2, and Linde high purity dry Argon (99.996%) were used as the starting materials, and each was separately purified. For CO_2 Linde molecular sieves, #3A were used to remove water before and after a copper turnings furnace at ∿ 550°C, which was used to remove oxygen from the gas. A similar purification line was used for argon except that Ascarite or Linde molecular sieve #4A was also included to remove CO_2. Linde molecular sieves, #4A were used for the CO line which removed water and CO_2, and Ascarite was also used to remove CO_2. The purified gases were monitored through calibrated capillaries, and the flow-rates adjusted by altering the pressure head across the capillaries. Dibutyl phthalate was used in the reservoirs and as the manometric fluid. Each capillary was calibrated for the particular gas by timing the motion of a soap bubble in a burette. Values of flow-rate versus pressure head were fitted to the best least squares line (usually a quadratic), and from this a set of calibration tables was generated. To cover the required range of CO_2/CO ratios several capillaries were required for

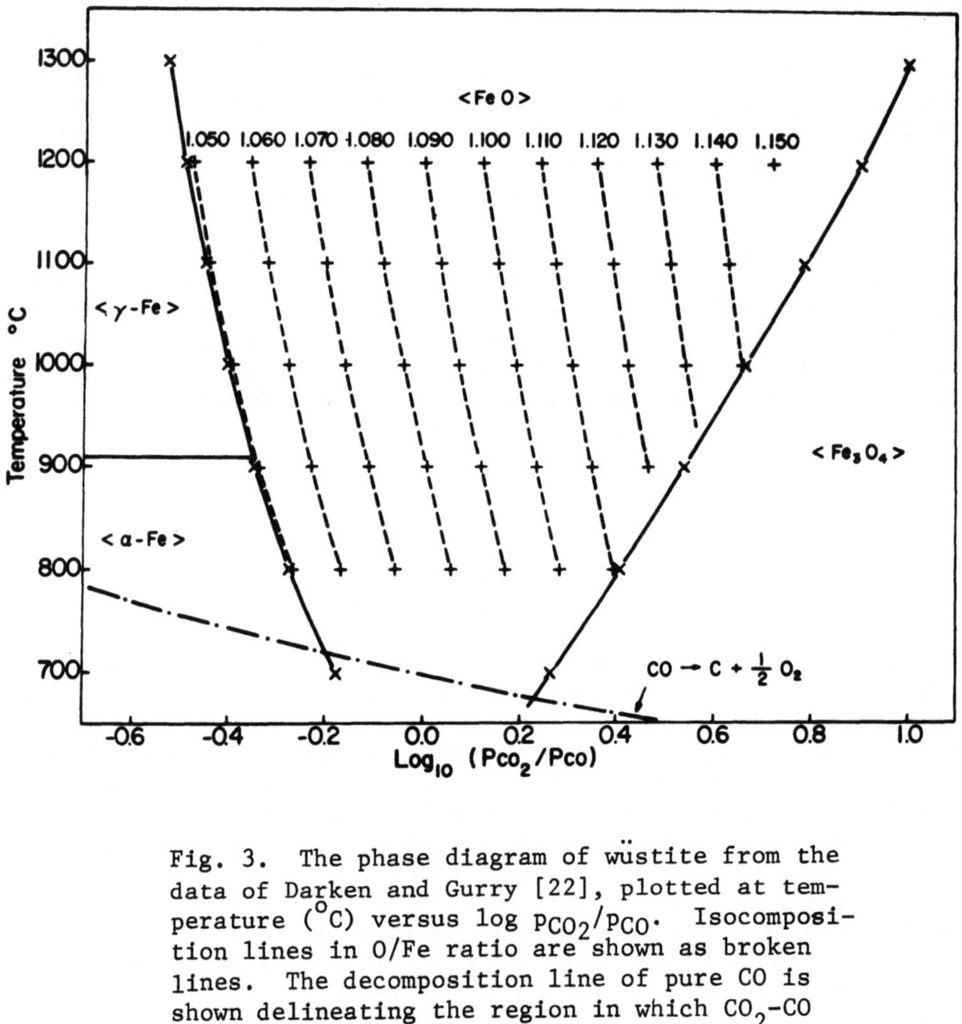

Fig. 3. The phase diagram of wüstite from the data of Darken and Gurry [22], plotted at temperature (°C) versus log p_{CO_2}/p_{CO}. Isocomposition lines in O/Fe ratio are shown as broken lines. The decomposition line of pure CO is shown delineating the region in which CO_2-CO mixtures can be used.

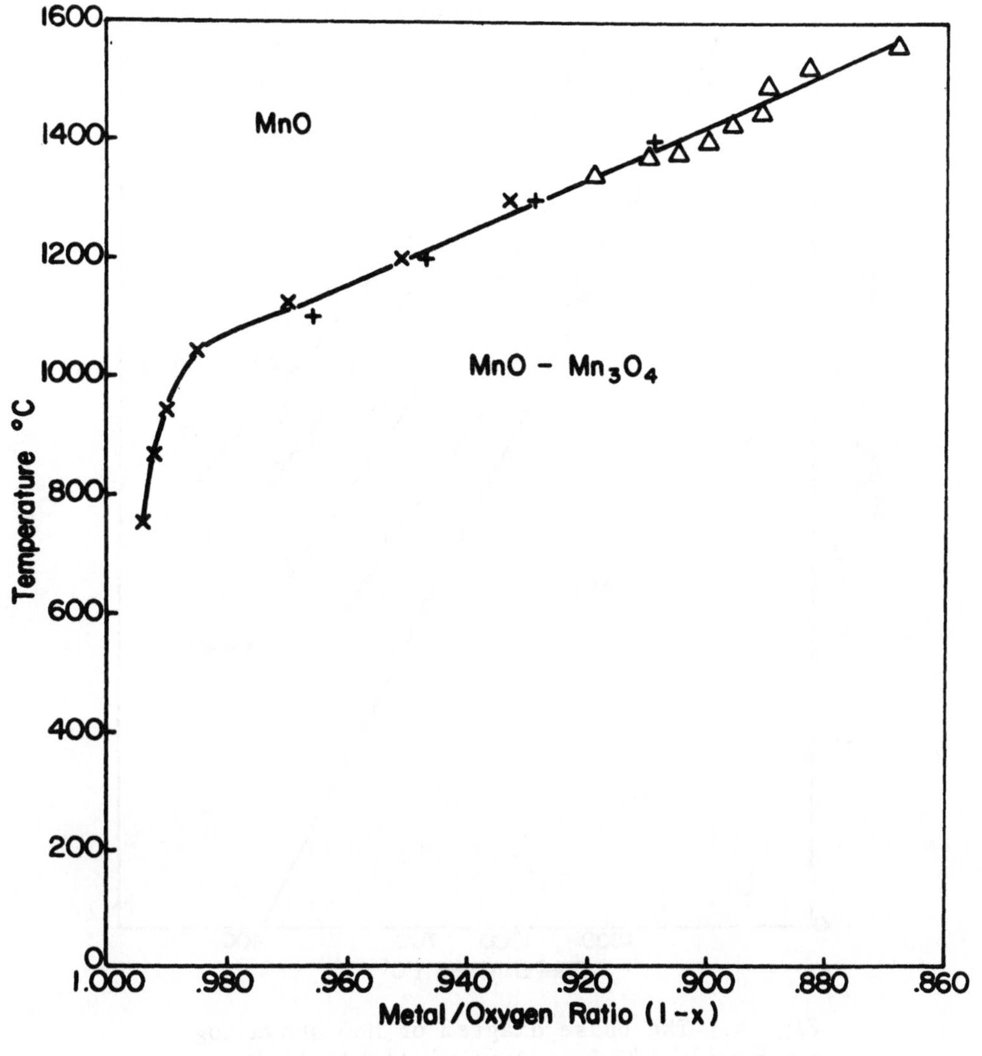

Fig. 4. The phase diagram of MnO compiled from available data, plotted as temperature (^{o}C) versus Mn/O ratio. x Fender and Riley [27], + estimated from Hed and Tannhauser [26], Δ Schmahl and Hennings [28].

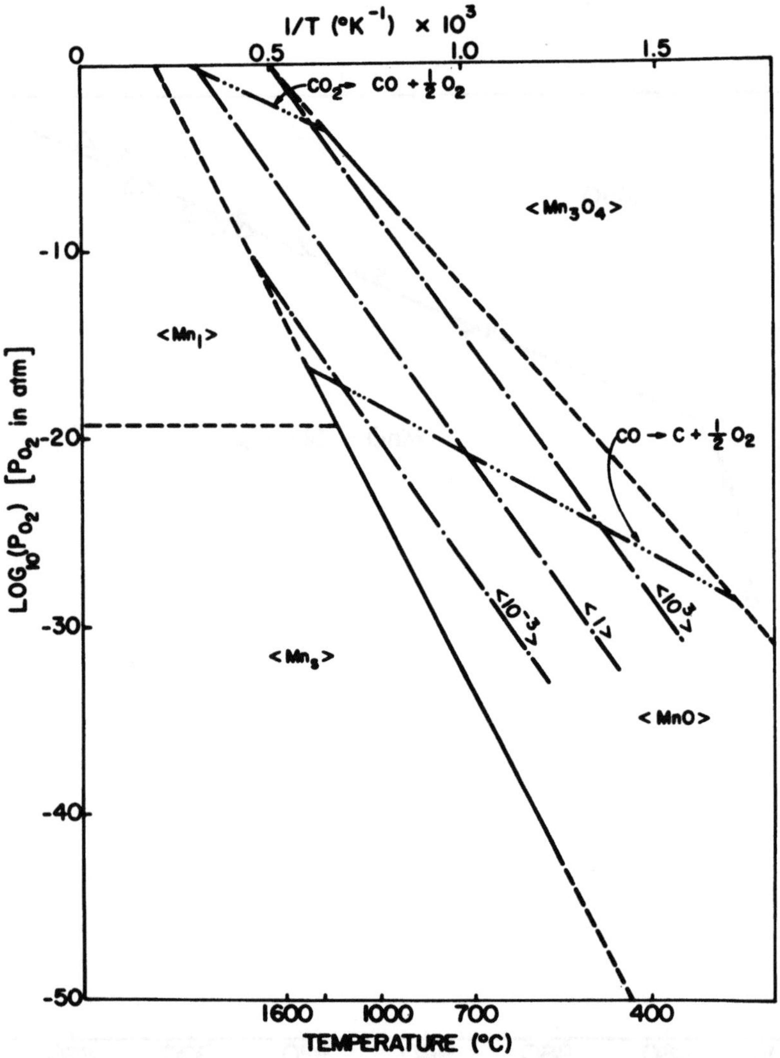

Fig. 5. The phase diagram of MnO shown log pO_2 versus 1/T from data available in the literature. Solid lines indicate experimental determinations of the phase boundaries, the broken lines are extrapolated. The pressure-temperature plots for CO_2/CO ratios of 10^3, 1 and 10^{-3} are shown (—· —), and together with the decomposition lines of pure CO_2 and pure CO (-····-); these delineate the region that can be studied with CO_2-CO mixtures.

CO_2 and CO, particularly for MnO. FeO is stable for a range of CO_2-CO ratios approximately 10:1 to 1:10, whereas MnO requires in excess of 1000:1 to 1:1000. The extremities of the MnO phase thus cannot be reached using CO_2-CO mixtures, and gas ratios of 500:1 are the best attainable experimentally. The individual flow-rates could be measured to better than 1%, giving an accuracy in the ratios of $\pm$ 1.5%. For very low CO or CO_2 flows, the accuracy is less than this. The gases were mixed by passing through a long column of glass beads and entered the furnace at the bottom. Flow-rates of around 0.9 cm/sec were maintained in the hot zone of the furnace to prevent thermal segregation of the gases [22].

Marshall resistance furnaces (maximum temperature 1200°C) were used in a vertical position. 3/4" or 1" I.D. Mullite reaction tubes sealed to Pyrex tapered joints at each end were placed in furnaces. The temperatures were controlled to $\pm$ 1°C by an Electromax controller using Pt-Pt · 10% Rh thermocouples. The furnace hot zones were about 5 cm long, with a variation of $\pm$ 0.5°C in this distance. Pt-Pt ·10% Rh thermocouples were used to measure the temperature at the sample position, with cold junctions maintained at 0°C in an ice-bath, and shielded leads to the potentiometer.

Isothermal measurements were made of the equilibration rates by following the change in weight or in the conductivity of the sample, as illustrated in Fig. 6. The samples were equilibrated at a point in the phase field determined by the CO_2/CO ratio and the temperature. In wüstite this was related to a particular O/Fe ratio, but in chromium-doped manganous oxide measurements were made as a function of p_{O_2} as the thermodynamics of this system are not yet known. When equilibrium had been reached the furnace was sealed off at the bottom with the old mixture inside, with the gas-flow bypassing it. The pressure heads were rapidly adjusted (< 1 min) to give a new CO_2-CO ratio and the gas-train purged for about 4 minutes. The new mixture was then admitted into the furnace and allowed to displace the old mixture. The zero time was taken when the first change in weight or conductivity was detected. Monitoring the sample weight or conductivity during the static period indicated that equilibrium was maintained. The property being measured was then followed as a function of time until no further change was detectable and the new equilibrium had thus been reached. The procedure could then be repeated, keeping within the phase field at all times. Increasing the oxygen pressure (CO_2/CO ratio) isothermally, increases the oxygen content of the sample and is termed an oxidation run. Similarly reducing the oxygen pressure gives a reduction run. The oxidation and reduction runs are specified by the initial and final oxygen pressures. After measuring equilibration rates in CO_2-CO mixtures,various amounts of argon were added, maintaining the same oxygen pressure, and the rates re-measured. Measurements were made with successive steps in oxygen pressure between the phase limits, and the reproducibility of the equilibrium values was good. This

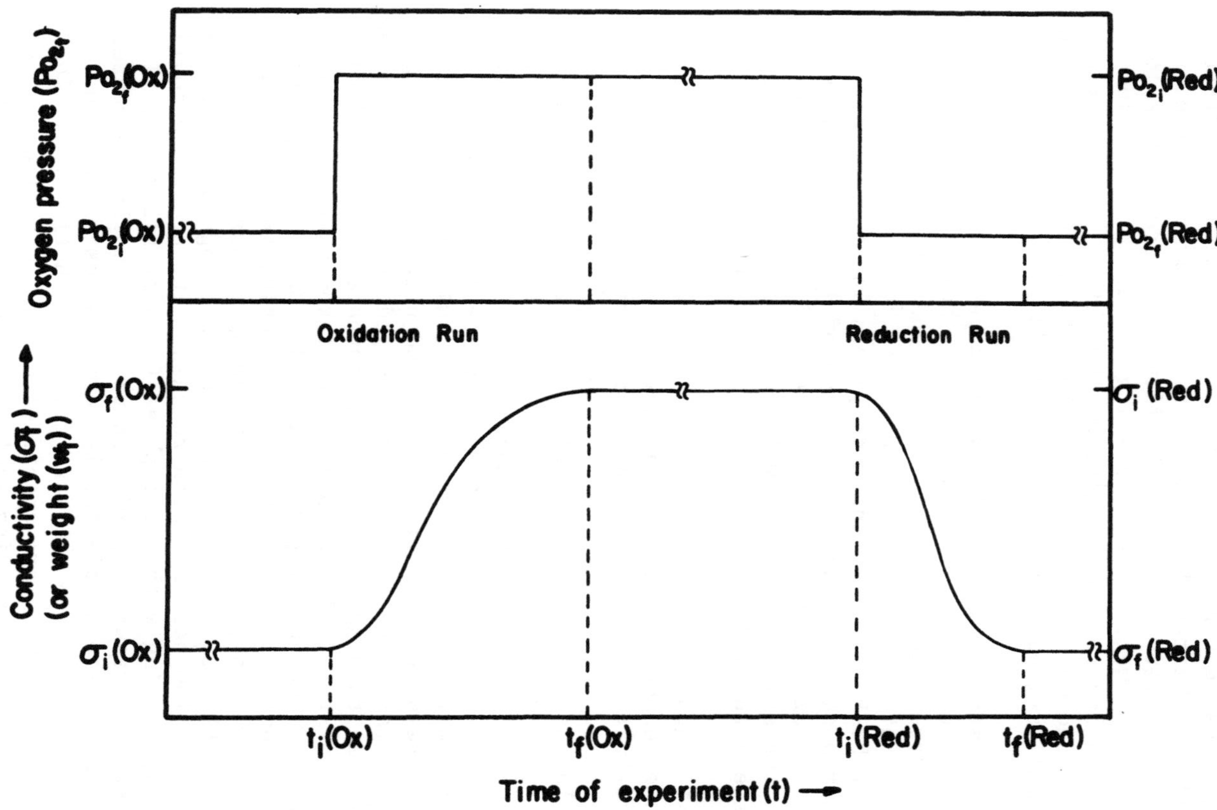

Fig. 6. Schematic representation of the interphase equilibration curves for a cation-deficient oxide,with predominantly electronic conduction. The corresponding changes in p_{O_2} and conductivity (or weight) are shown for both oxidation and reduction runs. Subscripts i and f indicate the initial and final states of the equilibration reaction.

provides an in situ check on the state of the sample. Measurments were made at different temperatures, up to 1200°C, to investigate the temperature dependence of the equilibration rates.

The experimental equilibration data, property versus time, was analysed using the previously discussed equations with programs written in Fortran IV for a CDC 6400 computer, with graphical outputs. Chemical diffusion coefficients were calculated from the curves using Eqs. (41) and (43) and compared for diluted and undiluted gas mixtures. A significant difference in the derived values obtained under these conditions, or in oxidation or reduction to the same point, was taken to indicate a significant change in equilibration time and hence a contribution of the surface reaction. In this case the value of $\tilde{D}$ obtained is an effective chemical diffusion coefficient, as it includes a contribution from the surface reaction.

The wüstite samples for the thermogravimetric studies were grown from high-purity iron* platelets, with approximate dimensions 1" x 0.5"x 0.030." Details of the spectrographic analysis of the iron are in Table III. The samples were supported by a platinum wire from a quartz fiber attached to the balance pen of an Ainsworth semi-micro automatic recording balance. The sensitivity of the balance is 0.0001 gm and the weight changes were recorded on a chart at 2 feet per hour. The samples were grown in situ and then the oxidation/reduction and dilution studies were performed isothermally. The sample thicknesses were calculated assuming growth perpendicular to the principal face of the sample, and allowing for the change in lattice parameter of wüstite with temperature and composition. This method produced polycrystalline samples of fairly small grain size, (about 2mm) but exhibiting considerable surface roughness. Studies are under way on dense, shaped wüstite samples to eliminate any uncertainty in thickness or surface condition. The equilibration curves for a given sample were very reproducible after prolonged oxidation and reduction across the phase field. There was less agreement between different samples, probably because of the uncertainties in the absolute thickness and differences in surface conditions. Zero weight was established with the suspension system alone. Equi-concentration jumps were made cross the phase field, with intervals of 0.01 in O/Fe ratio, for both oxidation and reduction in undiluted and diluted CO_2-CO mixtures.

The chromium-doped MnO samples were cleaved from a single-crystal boule supplied by Marubenii-Iida (America) inc., and grown by Nakazumi Crystals, Japan. Portions of the boule were analysed by Schwartzkopf Microanalytical Laboratory and gave an average composition of 0.95 at.% Cr over the boule. Spectrographic analysis indicated a low impurity level, and the results are given in Table III.

* The zone-refined iron was furnished by the Battelle Memorial Institute through the courtesy of the American Iron and Steel Inst.

TABLE III

a) Analysis of the High-Purity Iron Used as the Starting Material for the Wüstite Studies

Aluminus	15	Manganese	< 0.1
Antimony	< 5 ND	Molybdenum	< 5 ND
Arsenic	< 5 ND	Nickel	20
Beryllium	< 0.2	Nitrogen	< 1
Boron	< 5	Oxygen	17
Cadmium	< 5 ND	Phosphorous	9
Calcium	< 10 ND	Silicon	10
Carbon	< 10	Sulfur	12
Chromium	5	Tin	< 5 ND
Cobalt	5	Titanium	< 1 ND
Copper	7	Tungsten	< 5 ND
Hydrogen	0.3	Vanadium	< 1 ND
Lead	< 1 ND	Zinc	< 10 ND
Magnesium	< 5	Zirconium	< 1 ND

b) Semi-Quantitative Spectrographic Analysis of the MnO-Cr Single Crystal as Supplied by Nakazumi Crystals Co., Japan

Aluminum	ND	Magnesium	ND
Arsenic	ND	Manganese	PH
Antimony	ND	Mercury	X
Barium	ND	Molybdenum	ND
Boron	ND	Niobium	ND
Bismuth	ND	Nickel	ND
Cadmium	ND	Phosphorus	X
Calcium	ND	Potassium	X
Carbon	X	Sodium	ND
Chromium	ML	Sulfur	X
Cobalt	ND	Silicon	ND
Copper	ND	Silver	ND
Iron	ND	Tantalum	ND
Lead	ND	Tellurium	X
Lithium	ND	Tin	ND

P 10 - 100%
M 1 - 10%
H upper half of range
L lower half of range

Lowest detection level quoted << 0.01%, ND indicates not detected, and X indicates not tested.

The crystals cleaved along {100} planes, and Laue photographs from these surfaces indicated a good single crystal. The samples were ground to the final slab shape using 500A SiC paper. Four parallel holes, ∿ 0.015" in diameter, were drilled ultrasonically along the center-line of the sample and normal to the largest face. Platinum wires, coated thinly with platinum paste, were push-fitted into these holes as to form probes for the four-probe d.c. conductivity method. The sample dimensions and the inner probe spacing were measured to 0.002 cm using a travelling Vernier microscope. Typical sample dimensions were 8 x 4 x 1 mm^3. The probe leads were then spot-welded to the platinum wires in the conductivity rig, so the sample was about 1 mm away from the head of a Pt-Pt · 10% Rh thermocouple. The contacts were found to be ohmic and stable for long periods in the furnace over a range of temperatures. Compared with wrapped Pt leads these probes are much more stable, give better contact, the spacing does not change and the spacing can be measured more accurately. All electrical leads from the sample were shielded, and thermal-free copper switches were used in the circuitry. The current was provided by a 12 V battery, reduced through a variable resistor to give a current of 3-4 mA in the sample circuit. The voltage drop across the inner sample probes were compared with that across a standard 10 Ω resistance. The potential drop was measured with the current flowing in both directions to eliminate thermal e.m.f.'s in the sample. Initially a Leeds and Northrup K3 potentiometer with a d.c. null detector was used to follow the kinetics, but the response time of this system is slow. It was replaced by a Doric integrating microvoltmeter, a six-range instrument with digital readout, microvolt sensitivity and fast reading time. Much faster kinetics could be followed with this instrument and a conductivity measurement took about 15 sec, including monitoring the temperature. Readings were taken at uniform time intervals, the smallest being a minute, and were recorded manually. This limits the shortest equilibration times that can be studied ∿ 10-20 minutes. When the two measuring systems were intercompared they agreed to within a few μV. Equilibrium conductivities were reproducible over long periods of time in the furnace, though after repeated cycling the values decreased slightly (less than 6% in the worst case). On removal the samples were found to be roughened and polycrystalline on the surface and the dimensions were found to be slightly changed. The sample mounted in the rig was inserted fairly slowly into the furnace, to minimize thermal shock, and in a fast flow of CO_2 to avoid oxidation. The atmosphere was then changed to a CO_2/CO ratio within the p-type region. After equilibrium was attained measurements were made in the initial linear p-type region. In order to do this the log σ - log p_{O_2} diagram had to be determined for these samples [79]. The linear p-type region for the chromium doped MnO extends over two orders of magnitude in oxygen pressure. This region was sub-divided into eight approximately equal intervals for the experiments.

The solubility of chromium in wüstite has been measured across the phase field [80], but only limited data is available on MnO-Cr near the p-n transition [81]. These data would indicate that below 1200°C our sample would be saturated with chromium, and would thus be two-phase. The lack of hysteresis effects with temperature and

p_{O_2} in our samples suggest they were single-phase. An x-ray powder photograph of a ground sample of the as-received boules gave a lattice parameter of ∿ 4.45 Å, compared with ∿ 4.45 Å for a similar undoped MnO sample, with no signs of a second phase. Our sample thus seems to be single phase, but a detailed thermogravimetric study of Cr solubility in MnO is needed. A polycrystalline MnO sample was also grown by oxidizing a Mn plate wrapped with four Pt leads. The ends were ground down to reveal a void similar in size to the original Mn plate, thus confirming that cations are the mobile species. The probe spacing and cross-sectional area of this sample were measured with a travelling microscope, and then the equilibration kinetics were determined.

Equilibration Kinetics of Wüstite

More studies have been made of the intraphase equilibration kinetics of wüstite than of any other system, as can be seen in Table I. A marked variation is observed by different workers in the magnitude of the chemical diffusion coefficients derived using Eq. (40). This is illustrated more explicitly in Figs. 7 and 8, which show available measurements of $\tilde{D}_{FeO}$ as a function of oxygen/metal ratio at ∿ 1000°C and ∿ 1050°C. Most workers have explained the experimental equilibration curves as due to diffusion control, except for Landler and Komarek [5] who decided that mixed kinetics were present in H_2O-H_2 mixtures. All other work has been done in CO_2-CO mixtures, and thus the latter are directly comparable. In order to make meaningful comparisons we also need especially to consider the dimensions and the nature of the samples used. Table IV collects all the important experimental details.

Information is also available on the reaction of CO_2-CO mixtures with wüstite surfaces (see previous section), and the temperature and oxygen pressure dependence of the surface reaction are known [73]. It appears from iron oxidation studies that the decomposition of H_2O-H_2 mixtures is faster than that of CO_2-CO mixtures on wüstite. Consequently for a given sample thickness and experimental temperature, diffusion control should be more important in H_2O-H_2 mixtures than in CO_2-CO mixtures. Stotz [74] showed that the initial equilibration rates, measured by the relaxation in conductivity, were essentially that of the surface reaction in both CO_2-CO and H_2O-H_2 mixtures. The lower equilibration rates observed by Fujii and Meussner [4] probably represent surface control in thin samples.

Several factors may be important in producing the significant differences in the magnitude and pressure dependence observed where fairly thick wüstite samples were used, and where consequently diffusion control might be expected. The thickness of the sample determines the equilibration time for a thin plate sample (see Eq. (41)), but this expression is only valid if the length and width of the

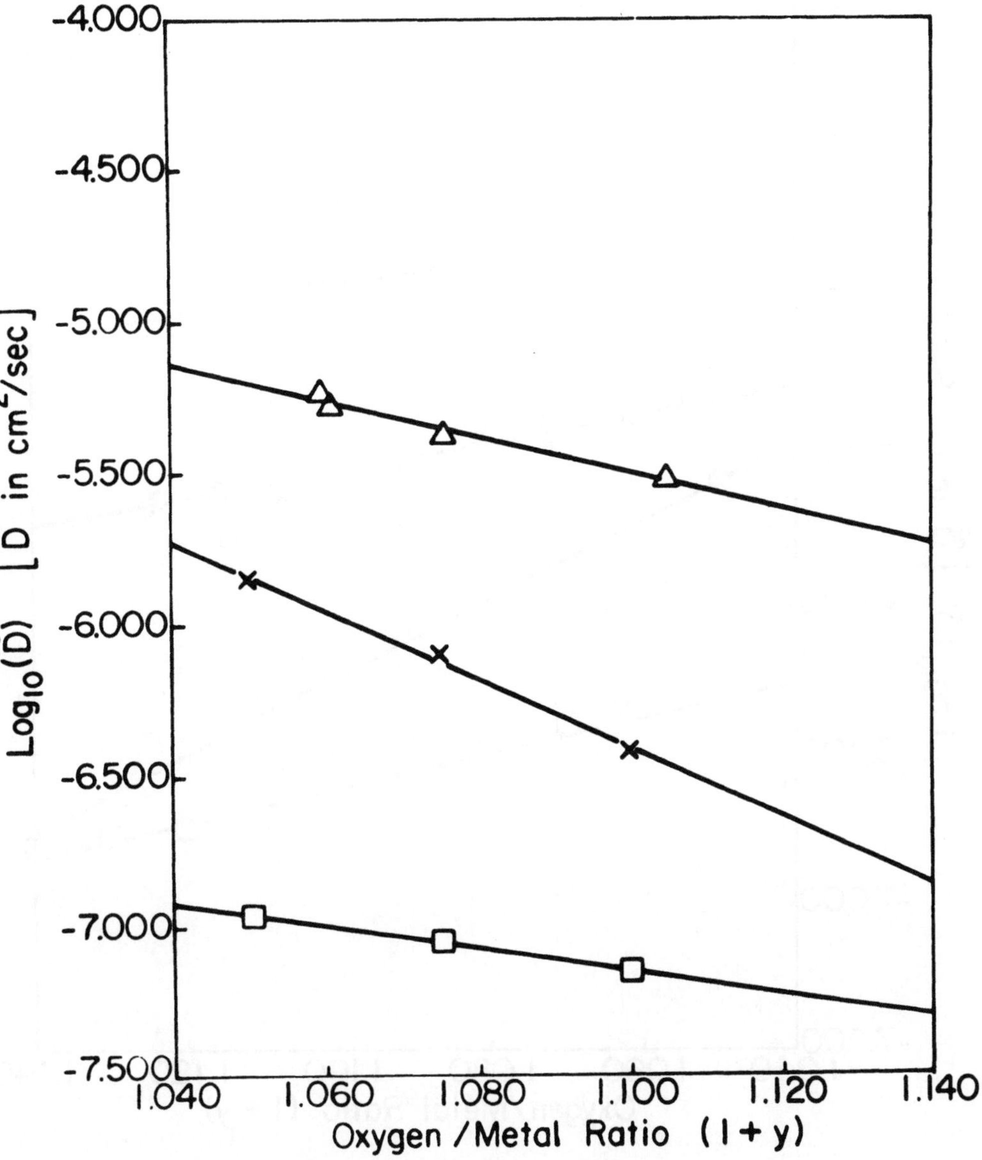

Fig. 7. Literature data on the chemical diffusion coefficient, $\tilde{D}$, of wüstite calculated by means of Eq. (41), as a function of O/Fe ratio, around 1000°C. Δ Landler and Komarek [5] at ∿ 1003°C in H_2O-H_2 mixtures; x Levin and Wagner [3] at 1000°C in CO_2-CO mixtures; □ Fujii and Meussner [4] at 1000°C in CO_2-CO mixtures. Only reduction runs are shown.

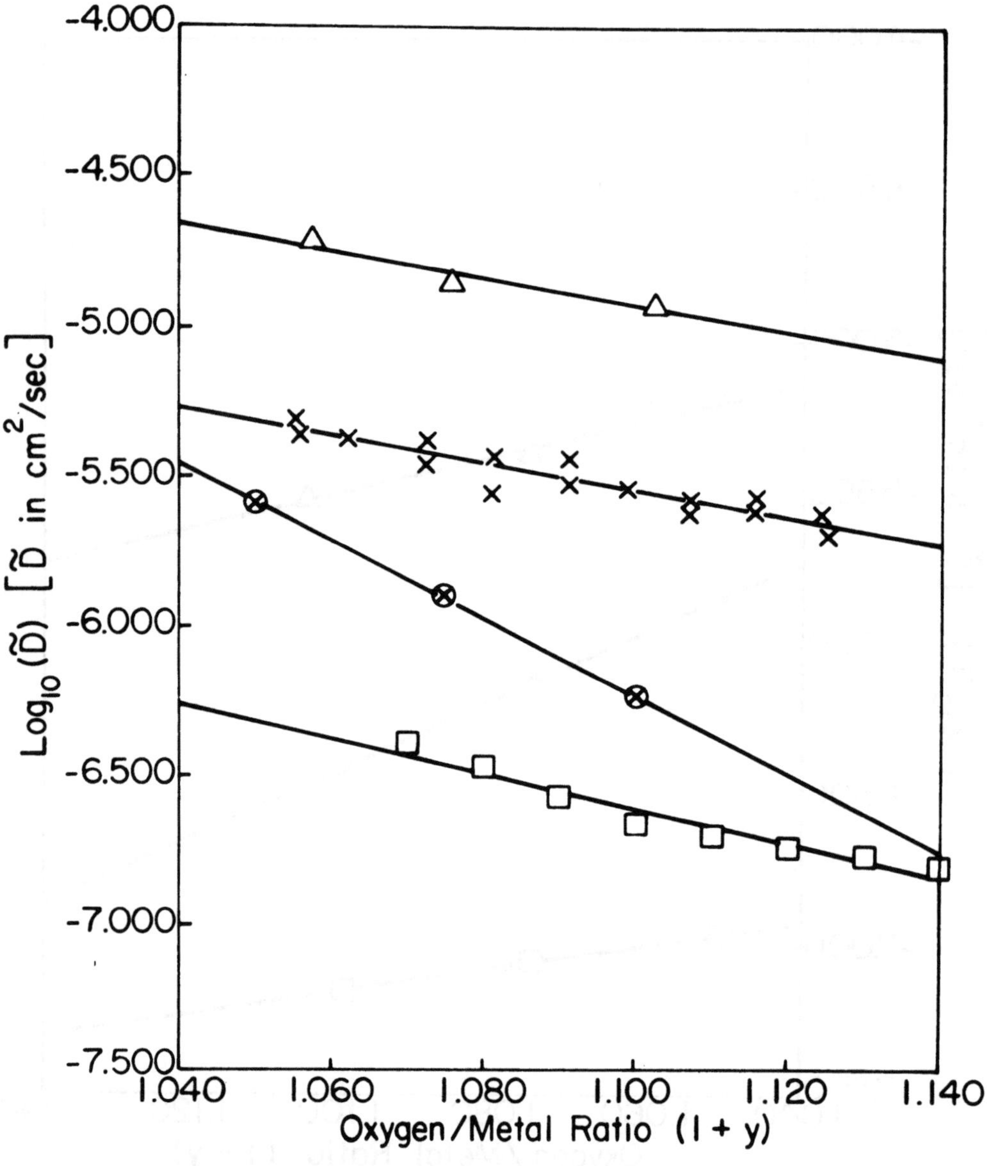

Fig. 8. Literature data on the chemical diffusion coefficient, $\tilde{D}$, of wüstite calculated by means of Eq.(41) as a function of O/Fe ratio, around 1050°C. Δ Landler and Komarek [5] at 1050°C in H_2O-H_2 mixtures; x Laub [9] at 1062°C in CO_2-CO mixtures; ⊗ Levin and Wagner [3] at 1050°C in CO_2-CO mixtures; □ Campbell at 1055°C in CO_2-CO mixtures. Only reduction runs are shown.

TABLE IV
Comparison of Wüstite Samples Used in Determining Equilibrium Kinetics

Reference	Sample Dimensions	A/ℓ(mm)	Grain Size	Method of Preparation	Atmosphere	
R.Levin and Wagner[3]	∿13x7.5x2.5 mm^3	39	∿3mm	Grown from iron plates (also Cr-doped)	CO_2+ CO	Red
Landler and Komerek[5]	14.68x10.41x5.02mm^3	30.4	5mm	Ground from solidified wüstite	H_2O-H_2	Red
Fujii and Meussner[4]	0.15mm thick foils	---	---	Grown from iron foil	CO_2-CO	Redox
Stotz[4]	15 μ foils	---	---	Grown from iron foil	CO_2-CO/H_2O-H_2	Ox
Hembree[6]	∿13x6.5x1.2mm^3	∿70	---	Grown from iron plates (also Cr, Ti, Li dopants)	CO_2-CO	Red
Campbell[7]	∿2.0mm thick plates	---	several mm	Grown from iron plates	CO_2-CO	Redox
H. Levin[8]	10x2x1mm^3	20	∿ Single Crystal	Ground from solidified wüstite (also Fe-Mg-O)	CO_2-CO	Ox
Laub[9]	∿26x13x1.5mm	225	∿ 2mm	Grown from iron plates	CO_2-CO + Ar	Redox

sample are much greater than the thickness. If diffusion in from the edges of the sample affects an appreciable volume of the sample, then the thin plate solution will not be valid. The equilibration curves change in shape as the edges become important, and an exact analysis of the diffusion profiles is formidable. The solution for a brick-shaped sample (three-dimensional diffusion) usually given [29] is merely the product of three one-dimensional solutions. It does not allow for the overlapping of diffusion profiles from adjacent edges, but does give some idea of the effect of the other dimensions on the equilibration kinetics. In Fig. 9 the curves of (1 - Q) versus time for two sample sizes are shown, comparing the one-dimensional and three-dimensional solutions. As the thickness increases for a given sample area the two curves diverge. In general if we analyze the kinetics by Eq. (40) when a thin plate expression is not valid, the derived diffusion coefficient will be larger than the actual value. None of the samples described in Table IV approach the thin-plate condition very closely, but those used recently by Laub [9], in unpublished work, have the highest area to thickness ratio. A correction for edge effects will alter the published values of the chemical diffusion coefficient and may explain many of the discrepancies. The samples used by Landler and Komarek [5] have the lowest area to thickness ratio, and they allowed partially for this by using a "mean" diffusion length in Eq. (40). Their analysis was also based on the shape of the equilibration curves. These are markedly affected by edge effects, and this may also affect their interpretation.

The physical character of the sample, for example surface condition,density, orientation of the crystallites and grain size, may also affect the equilibration rates. Ideally measurements should be made on single crystal samples, with various surface orientations. For wüstite,dense, solidified samples of large grain size are the closest this can be approached. Most work has been done on as-grown samples, grown in situ from iron plates in a controlled atmosphere. The resulting samples are somewhat irregular in shape, with rounded edges, have uneven surfaces and variable grain size. The importance of these various factors is difficult to determine, and an average sample thickness is assumed. This inaccuracy in defining the actual diffusion distance is probably the other major source of error in the absolute magnitudes of reported data. Thermal expansion and lattice parameter changes in wüstite can be allowed for, and providing the overall dimensions do not change with time, meaningful comparisons can be made on the sample. For example oxidation versus reduction, or diluted versus undiluted runs can best be compared on the same sample,eliminating the absolute differences between samples.

Since the derived diffusion coefficients are strongly dependent on the oxygen pressure, small composition steps within the phase field are required to isolate the dependence. Levin and Wagner [3] used very large composition intervals, and in Figs. 7 and 8 their

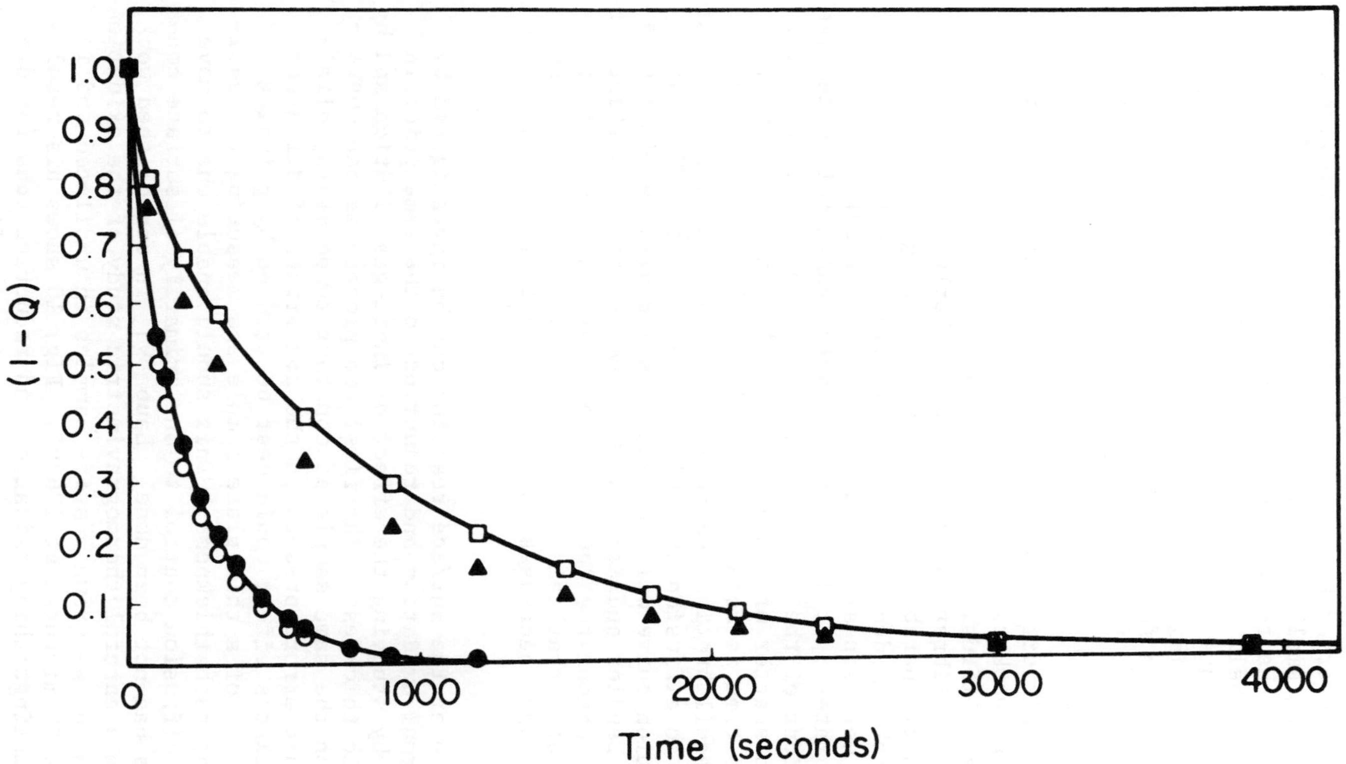

Fig. 9. Plot of (1 - Q) versus time (secs) for two different sample sizes, utilizing the full thin plate solution [Eq. (40)] and the full brick solution [29]. Filled and open circles: sample 2.6 x 1.3 x 0.15 cm^3 for the thin plate and brick solutions respectively; open squares and filled triangles: sample 2.6 x 1.3 x 0.3 cm^3 for the thin plate and brick solutions respectively. Calculated assuming $\tilde{D} = 10^{-6}$ cm^2/sec and not a function of composition.

results show the greatest pressure dependence. At a given temperature the gradients observed by other workers are quite similar and were determined using small composition intervals. Most of the results illustrated in the two figures were obtained in reduction experiments. Only Campbell [7] and Laub [9] have done both oxidation and reduction experiments within the phase field. The oxygen pressure dependencies are similar, but the results of the two workers differ in magnitude by a factor of ten. As-grown samples were used in both cases, and the difference is probably due to the two factors detailed above.

Assuming only bulk diffusion control, there does seem to be some consistency between workers, allowing for an uncertainty in the absolute magnitudes. Diffusion control has been assumed because of the shape of the equilibration curves, particularly at long times, and the approximately parabolic initial weight changes. However, Landler and Komarek [5] decided that the initial weight decreases in H_2O-H_2 reduction were linear, and derived a surface rate constant from the initial rates. It does not seem very meaningful to analyse the initial portions of the kinetic curves obtained gravimetrically in this way. The scatter of the data is usually large and the uncertainty in zero time is more important initially. Also we would expect all such equilibration reactions to be surface controlled initially, changing to diffusion control at longer times. The shape of the equilibration curve will reflect this and also the influence of edge effects. Landler and Komarek [5] also assumed a constant surface reaction throughout the equilibration, but the later work of Grabke [73] indicates that this is not true. The surface rate for the addition of oxygen decreases as the O/Fe ratio increases across the phase field.

The importance of the surface reaction can be investigated by comparing isothermal oxidation and reduction to the same point in the phase field, by studying the effect of inert-gas dilution and by varying the sample thickness. The first two procedures can conveniently be done on the same sample at constant temperature, eliminating the absolute errors present in the comparison of different samples. The thickness effect could best be studied by grinding down the large faces of a thin-plate sample and measuring the equilibration rates at each thickness. This should enable one to move from a region of diffusion control to one primarily of surface control. So far this has not been done. Laub [9], in unpublished work, has recently done a careful thermogravimetric study of the oxidation-reduction kinetics of as-grown wüstite samples in undiluted and in argon-diluted CO_2-CO mixtures at 1062°C. Fig. 10 shows his results on one sample, distinguishing oxidation and reduction runs for diluted and undiluted CO_2-CO mixtures. Reduction is slightly slower than oxidation in both cases, but they are essentially the same within the experimental error. There is a significant reduction in the equilibration rate when 50% argon dilution is used, indicating

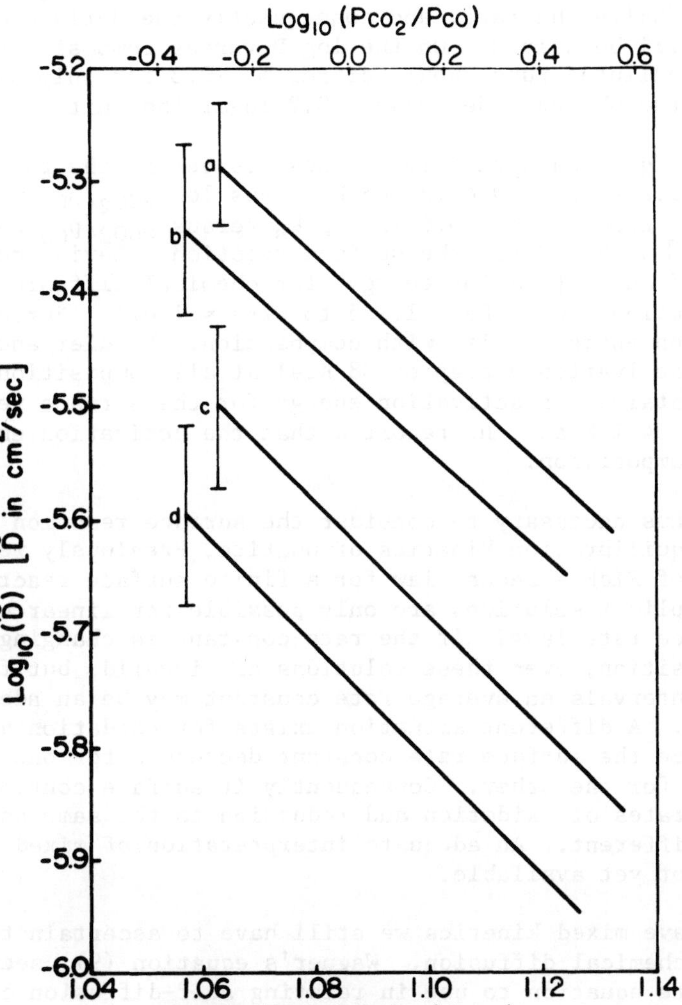

Fig. 10. The effect of argon dilution on the equilibration kinetics of wüstite at 1062°C from Laub [9]. Oxidation and reduction values of the chemical diffusion coefficient, $\tilde{D}$, derived by Eq. (41) for undiluted and 50% argon diluted CO_2-CO mixtures are plotted versus O/Fe ratio. The error bars represent the 95% confidence limit. a and b: oxidation and reduction for 50% argon diluted mixtures.

that a surface reaction is important. If the surface rate is proportional to p_{CO_2} (see Eq. (55)), then diluting with 50% argon would approximately halve the rate, and consequently the derived diffusion coefficient will be halved. On the log $\tilde{D}$ versus composition plot the diluted and undiluted runs should differ $\dot{\sim}$ -0.3 for only surface control, and the observed decrease $\sim$ 0.2 logarithm units.

Another significant feature of these curves is the p_{O_2} dependence, where the gradient of $\sim$ -0.4 in log $\tilde{D}$ versus log p_{CO_2}/p_{CO} may be compared with the gradient of -0.6 in log k_s versus p_{CO_2}/p_{CO} obtained by Grabke [73] at 983°C for the surface reaction. Levin and Wagner [3] also found an "activation energy" for chemical diffusion of $\sim$ 32 Kcal for reduction from O/Fe = 1.125 to O/Fe = 1.075. Furthermore this activation energy varied with composition. Landler and Komarek [5] found an activation energy of 38 Kcal at all compositions. Grabke [73] obtained an activation energy for the surface reaction at CO_2/CO = 1 of 34.1 Kcal. He reported that the activation energy varied with composition.

It is thus necessary to consider the surface reaction in interpreting the equilibration kinetics of wüstite. Previously we discussed the solution of Fick's second law for a finite surface reaction, and noted that explicit solutions are only possible for linear or exponential surface rate laws. If the rate constant is changing with the surface composition, even these solutions are invalid, but for small composition intervals an average rate constant may be an adequate approximation. A different situation exists for oxidation and reduction runs,since the surface rate constant decreases for one process and increases for the other. Consequently if surface control is present, the rates of oxidation and reduction to the same point will probably be different. An adequate interpretation of mixed control kinetics is not yet available.

If we have mixed kinetics we still have to ascertain the contribution of chemical diffusion. Wagner's equation (9), seems to be the appropriate equation to use in relating self-diffusion coefficients and the chemical diffusion coefficient. From Eq. (9) we have

$$\tilde{D} = (m + 1) \cdot D_{V_M} = (m + 1) \cdot D_M \cdot \frac{1 - x}{x}$$

$$= (m + 1) \cdot \frac{D_M^T}{f} \cdot \frac{(1 - x)}{x} \qquad (63)$$

For a given degree of ionization and constant correlation factor, $\tilde{D}$ will be reduced by increasing vacancy concentration, and increased by increasing D_{Fe}^T. In Fig. 1 we saw the shape of the function $(m + 1)\ (1 - x)/x$, and in Fig. 11 the available data on D_{Fe}^T are shown.

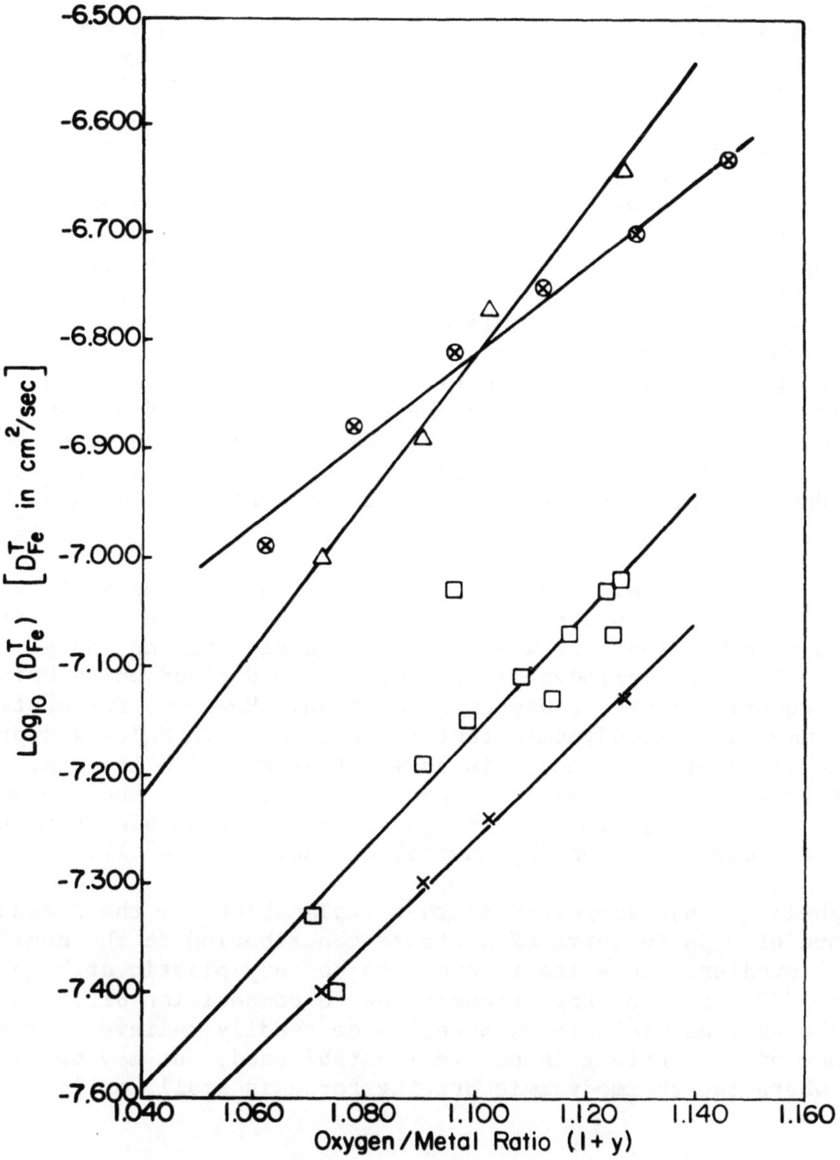

Fig. 11. Tracer diffusion coefficients, D^T_{Fe}, in cm^2/sec versus the O/Fe ratio in wüstite. Δ and x: Desmarescaux and Lacombe [45] interpolated from their data at 1100°C and 983°C; ⊗ Hembree and Wagner [46] at 1100°C; □ Himmel, et al. [43] at 983°C. The solid lines are least mean squares lines through the interpolated points for Desmarescaux and Lacombe [45].

From the available data D^T_{Fe} and $1/x$ seem approximately to cancel each other, as shown in Fig. 12 where calculated $\tilde{D}$'s are given at 1100°C from Hembree and Wagner's D^T_{Fe} data [46]. The magnitude of the calculated $\tilde{D}$ is close to that determined experimentally assuming instantaneous surface equilibration. The assumption of a constant correlation factor has been questioned by Campbell [7] and by Birchenall in this conference. Values of f are usually given for an ideal lattice and when the vacancy concentration is high the derivation is no longer correct. It seems likely that f increases with increasing vacancy concentration, and this would tend to reduce $\tilde{D}$. If $\tilde{D}$ were obtained free from the complications of the surface reaction, it could be used to calculate f, but the discrepancy seen in Fig. 11 seems too large to be explained by variation in m or f. One of the problems in this comparison is the lack of accurate, comprehensive tracer diffusion measurements. The reported pressure dependencies of D^T_{Fe} vary markedly, and D^T_{Fe} has only been measured at a few temperatures. Only a limited comparison of $\tilde{D}_{FeO}$ and D_{Fe} can thus be done, assuming Eq. (4) and (2) to be consistent with each other.

Several workers have used Darken's equation as a basis of comparison, though this is strictly applicable only to the interdiffusion of a binary alloy. It was shown previously that an analytic expression for the thermodynamic factor can be derived which brings Darken's equation close to Wagner's equation. However, for wüstite the experimental thermodynamic factor, $(\partial \ln a_O/\partial \ln N_O)$, is approximately constant at ~ 60, and this does not agree with the analytical expression (21). Using the experimental values of the thermodynamic factor the agreement with $\tilde{D}_{exp}$ is also reasonable in magnitude but differs in oxygen dependence, as shown in Fig. 12.

Campbell [7] has suggested another explanation for the pressure dependence of $\tilde{D}_{FeO}$ in terms of a stress contribution to the chemical potential gradient. Wüstite is known to be very plastic at high temperatures [82] and thus any stresses due to composition differences across the samples would be expected to be readily relieved. The importance of this effect is not well established, but may be important where the thermodynamic driving force is small, as is wüstite.

The effect of metal dopants on the equilibration kinetics of wüstite has been studied to a limited extent, but the overall picture is still not clear. Both Levin and Wagner [3] and Hembree [6] investigated the effect of chromium doping in wüstite, and found a slight reduction in the derived chemical diffusion coefficient, compared with their own results on undoped wüstite. Titanium also reduced the equilibration rate and lithium increased it slightly [6]. These results are in accord with the effect of these dopants on the equilibrium vacancy concentration in wüstite, and the observation that the derived $\tilde{D}$ in wüstite decreases with increasing vacancy concentration.

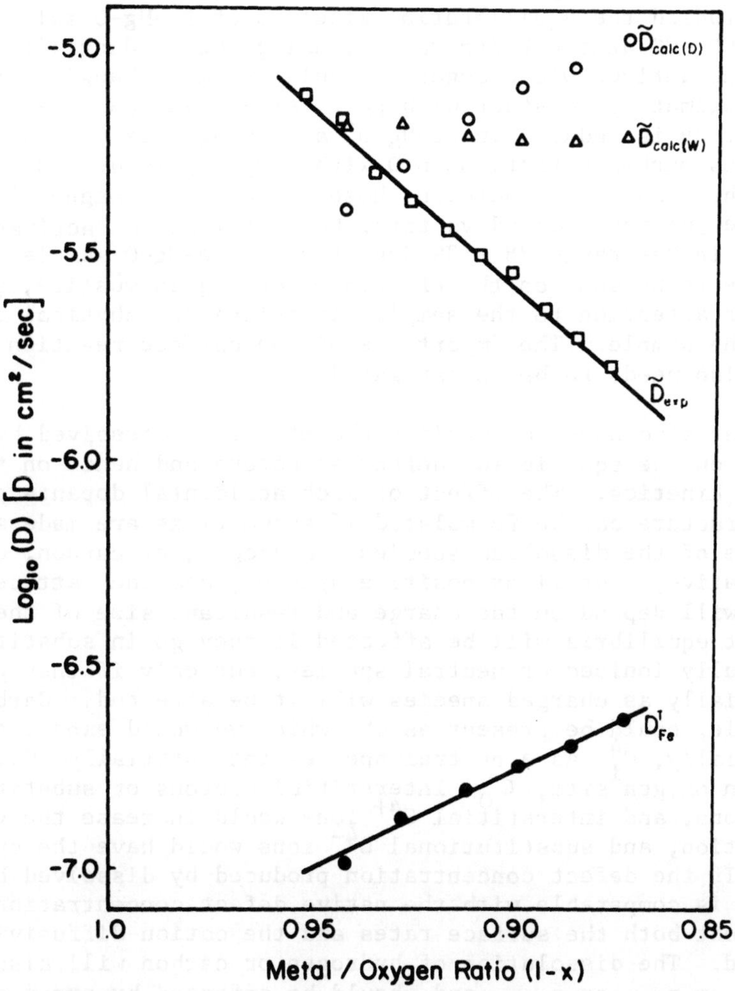

Fig. 12. Comparison of experimental and calculated values of the chemical diffusion coefficient, $\tilde{D}$, of wüstite at 1100°C as a function of Fe/O ratio. Open squares are $\tilde{D}_{exp}$ from Hembree [6], filled circles are D^T_{Fe} from Hembree and Wagner [46], both in cm^2/sec. The solid lines are the least mean squares lines for these data. Open triangles are $\tilde{D}$ calculated from Wagner's equation, (9) and the open circles are $\tilde{D}$ calculated from Darken's equation, (22), using experimental values for the thermodynamic factor.

The small differences observed, however, are within the experimental errors present in comparing different samples. Recently H. Levin [8] has studied the equilibration kinetics of Fe-Mg-O solid solutions in CO_2-CO at two Fe/Mg ratios, using the conductivity technique. The derived $\tilde{D}$'s changed irregularly with changing p_{O_2}, and were approximately constant with p_{O_2}. At a given p_{O_2} the vacancy concentration is reduced in Fe-Mg-O, and changes less with p_{O_2}, so that a less marked variation in $\tilde{D}$ with oxygen pressure is to be expected. The results lay between those of Levin and Wagner [3] and Hembre [6] for undoped wustite. Levin [8] found "activation energies" in the range 28 - 39 Kcal for his Fe-Mg-O samples. More work needs to be done on the effects of doping in wüstite, paying particular attention to the sample dimensions and physical constitution of the sample. The importance of the surface reaction on doped samples also needs to be investigated.

We may also have to consider the effect of dissolved hydrogen or carbon on the equilibrium defect structure and hence on the equilibration kinetics. The effect of such accidental dopants on the defect structure can be formulated if assumptions are made about the nature of the dissolved species. Hydrogen, or carbon, could go in as negative, neutral or positive species, and the lattice site occupied will depend on the charge and resultant size of the ion. The defect equilibria will be affected if they go in substitutionally as fully ionised or neutral species, but only if they go in interstitially as charged species will it be affected. Carbon, for example, could be present as C^{4+} which we would expect to find interstitially, C_i^4 as a neutral species interstitially, C_i, or as C^{4-} on an oxygen site, C''_O Interstitial protons or substitutional hydride ions, and interstitial C^{4+} ions would increase the vacancy concentration, and substitutional C^{4-} ions would have the opposite effect. If the defect concentration produced by dissolved hydrogen or carbon is comparable with the native defect concentration, we would expect both the surface rates and the cation diffusivity to be altered. The dissolution of hydrogen or carbon will also be dependent on p_{H_2O} or p_{CO_2}, and should be affected by argon dilution.

No clear evidence has been presented for the appreciable solution of hydrogen or carbon in wüstite. Carbon migrates through wüstite layers on iron [83], and we suspect that if carbon (and hydrogen) are soluble that they go in and migrate as neutral, interstitial species. The ion formation energies and sizes favor this, and in this case the equilibrium defect structure and the kinetics should be unaltered. An unambiguous answer requires an analysis for hydrogen or carbon in annealed samples, and if they are found to be present, the location and charge of the dissolved species should also be determined.

Equilibration Kinetics of Manganous Oxide

Only a limited amount of work has been done on the diffusion properties of manganous oxide. Three studies have been made of the equilibrium kinetics of MnO, all using single crystal samples about 1 mm thick with (100) faces. The kinetics of undoped MnO equilibrating in CO_2-CO mixtures has been studied by Price [1] using the conductivity technique between 854°C and 1147°C, and by Bransky and Tallan [2] using the thermogravimetric technique up to 1400°C. Some equilibration data has also been collected on as-grown polycrystalline MnO by Price and in the present work. In this paper results on chromium-doped manganous oxide single crystals are also reported, using the conductivity technique between 1037°C and 1206°C in CO_2-CO mixtures. A summary plot of the available chemical diffusion data in MnO is given in Fig. 13. By following conductivity changes the equilibration kinetics can be followed in both the p-type and n-type regions of MnO [84, 85], though appreciable weight changes only occur in the p-type region [85, 26]. Fig. 13 shows that the chemical diffusion coefficients derived using Eq. (12) and (17) decrease across the phase field, that is with increasing oxygen pressure. The slope of the lines increases with increasing temperature. The results obtained by Price [1] between 854°C and 1037°C lay essentially on the same line, and this behavior was also observed for Cr-doped MnO at 1037°C and 1100°C. Above 1100°C the chemical diffusion coefficient increases with temperature, but no meaningful activation energy can be derived. The expressions for $\tilde{D}$ given previously are complex functions of p_{O_2}, and the data were represented on a log-log plot only for convenience. The available data of Bransky and Tallan [2] at 1300°C seem to curve over on this plot. The 95% experimental confidence limit shown for Cr-doped MnO shows the reproducibility of the data, and these errors are less than in Price's work. The apparently greater diffusion coefficients obtained from Cr-doped MnO is thus probably not a significant fact, since we would expect $\tilde{D}$ to be smaller for Cr-doped MnO. Price [1] extrapolated his results into the n-type region. Equilibration is very much faster in this region where the vacancy concentration changes are small, and the few points obtained in the n-type region for Cr-doped MnO indicate that $\tilde{D}$ is approximately independent of p_{O_2} in this system. This may indicate a change in the rate-determining process.

Price [1] also suggested, from a comparison of equilibration times, that surface control was of some importance in MnO equilibrations. Specifically he stated that oxidation runs in the p-type region were diffusion controlled, and reduction runs were surface controlled. In the n-type region he suggested that both oxidation and reduction were diffusion controlled. In Cr-doped MnO no significant difference was observed in oxidation and reduction runs to the same oxygen pressure in the temperature range 1037-1206°C. A detailed study of the effect of argon dilution at 1206°C indicated that surface control was unimportant at this temperature for the Cr-doped MnO. The results are shown in Fig. 14. The diffusion coefficients obtained for

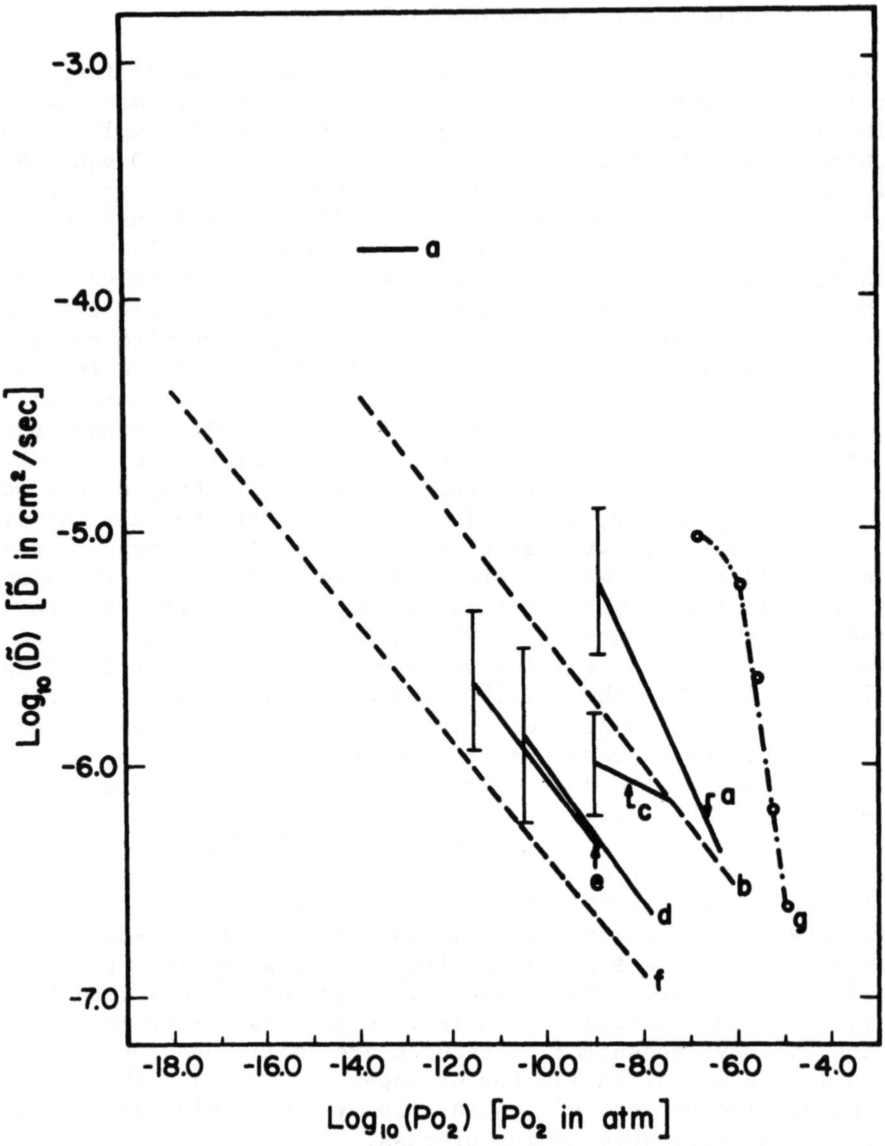

Fig. 13. Comparison of the available chemical diffusion coefficients, $\tilde{D}$, in cm^2/sec for MnO as a function of the oxygen pressure, in atm. Solid lines: present work on MnO-Cr, (error bars indicating the 95% confidence limits): a at 1200°C; c at 1149°C; d at 1100°C; e at 1037°C. Broken lines: data of Price [1] on single crystal MnO: b at 1147°C; f at 854 to 1031°C. Line g, (o--o): points of Bransky and Tallan [2] at 1300°C on single crystal MnO.

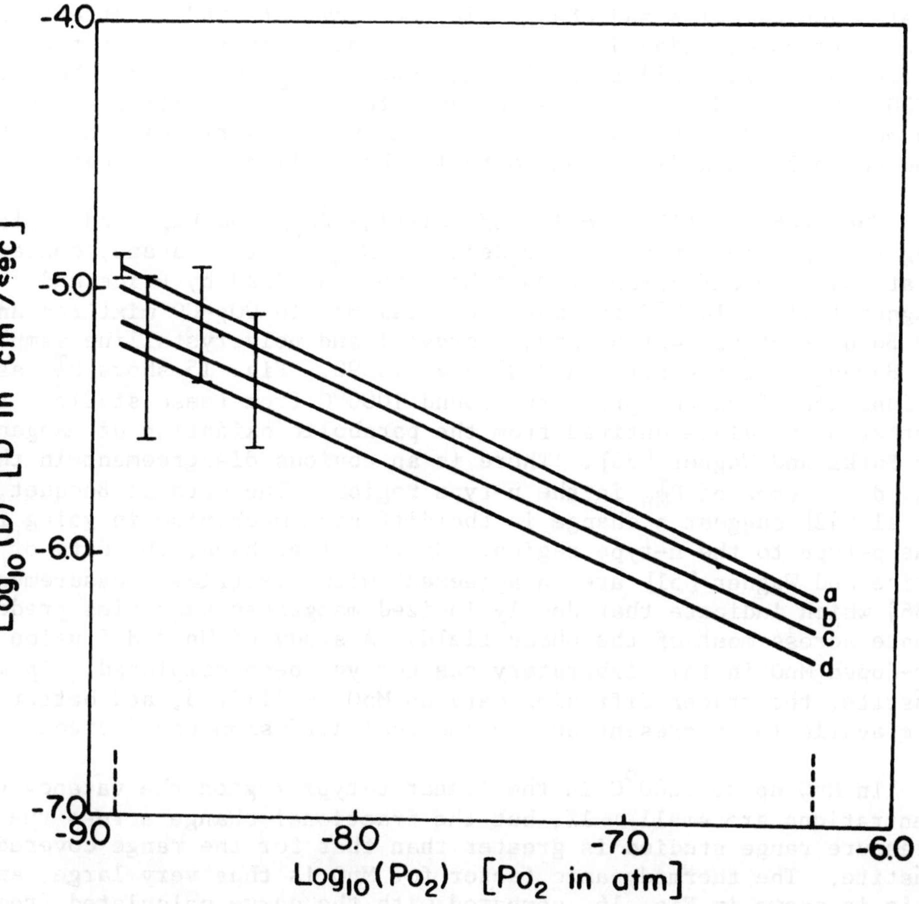

Fig. 14. The effect of argon dilution on the equilibration kinetics of MnO-0.95 at.% Cr single crystals at 1206°C. The chemical diffusion coefficient, $\tilde{D}$, from Eq. (47) in cm^2/sec is plotted versus the oxygen pressure, in atm. The solid lines are the least mean squares lines of the experimental data over the indicated pressure range, with the error bars indicating the 95% confidence limit. a: sample #1 at 1204°C in undiluted CO_2-CO mixtures; b: sample #2 at 1200°C in CO_2-CO mixtures containing 33% argon; c: sample #2 at 1206°C in 50% argon diluted CO_2-CO mixtures; d: sample #2 at 1206°C in undiluted CO_2-CO mixtures.

two different argon dilutions lie within the experimental scatter of the data for undiluted CO_2-CO mixtures. These experiments were only performed in the p-type region. Bransky and Tallan [2] also found that argon dilution and changes in the flow rate had no effect on the kinetics of undoped MnO. This evidence, together with the observed parabolic oxidation of manganese in CO_2-CO mixtures between 1000° and 1200°C [33], seems to indicate that the surface reaction is much faster on MnO surfaces than on wüstite surfaces, and that consequently bulk diffusion controls the equilibration kinetics.

The problem still remains of relating $\tilde{D}_{MnO}$ and D_{Mn}, and explaining the apparent inverse dependence of $\tilde{D}_{MnO}$ on the vacancy concentration. The diffusion of Mn^{54} has been studied by Price and Wagner [51] at 1032°C in single crystal MnO in CO_2-CO mixtures and by Bocquet, et al. [42] on single crystal and polycrystalline samples in H_2O-H_2 mixtures between 900° and 1150°C. Fig. 15 shows D^T_{Mn} as a function of oxygen pressure around 1000°C from these studies, together with values derived from the parabolic oxidation of manganese by Fueki and Wagner [33]. There is an obvious disagreement in the p_{O_2} dependence of D^T_{Mn} in the n-type region. The data of Bocquet, et al [42] suggest a change in the diffusion mechanism in going from the p-type to the n-type region. On the other hand, the data of Price and Wagner [51] are in agreement with electrical measurements [86] which indicate that doubly ionized manganese vacancies predominate across most of the phase field. A study of Mn^{54} diffusion in Cr-doped MnO in this laboratory has not yet been completed. As with wüstite, the tracer diffusion data on MnO is limited, and better data are available at present on the chemical diffusion coefficients.

In MnO up to 1200°C in the linear p-type region the vacancy concentrations are small, $\sim$ 1%, but the fractional change across the oxy-pressure range studied is greater than that for the range covered in wüstite. The thermodynamic factor for MnO is thus very large, and this is shown in Fig. 16, compared with the curve calculated from Eq. (21) for m = 2. This term changes by a factor of $\sim$ 20, between 2000 and 100, across the region of the phase field studied. The chemical potential gradient which provides the driving force for chemical diffusion is thus very high, especially at the low vacancy concentrations. Both Darken's and Wagner's equations give quite good agreement in magnitude and seem to produce the observed p_{O_2} dependence of $\tilde{D}_{MnO}$ at the lowest p_{O_2}'s studied, as is shown in Fig. 17.

The small changes in composition mean that the volume changes on oxidation or reduction are small. The change in lattice parameter with oxygen pressure has been neglected at the temperatures and pressure ranges of these measurements. A stress contribution to the chemical potential will also be small compared to the thermodynamic contribution.

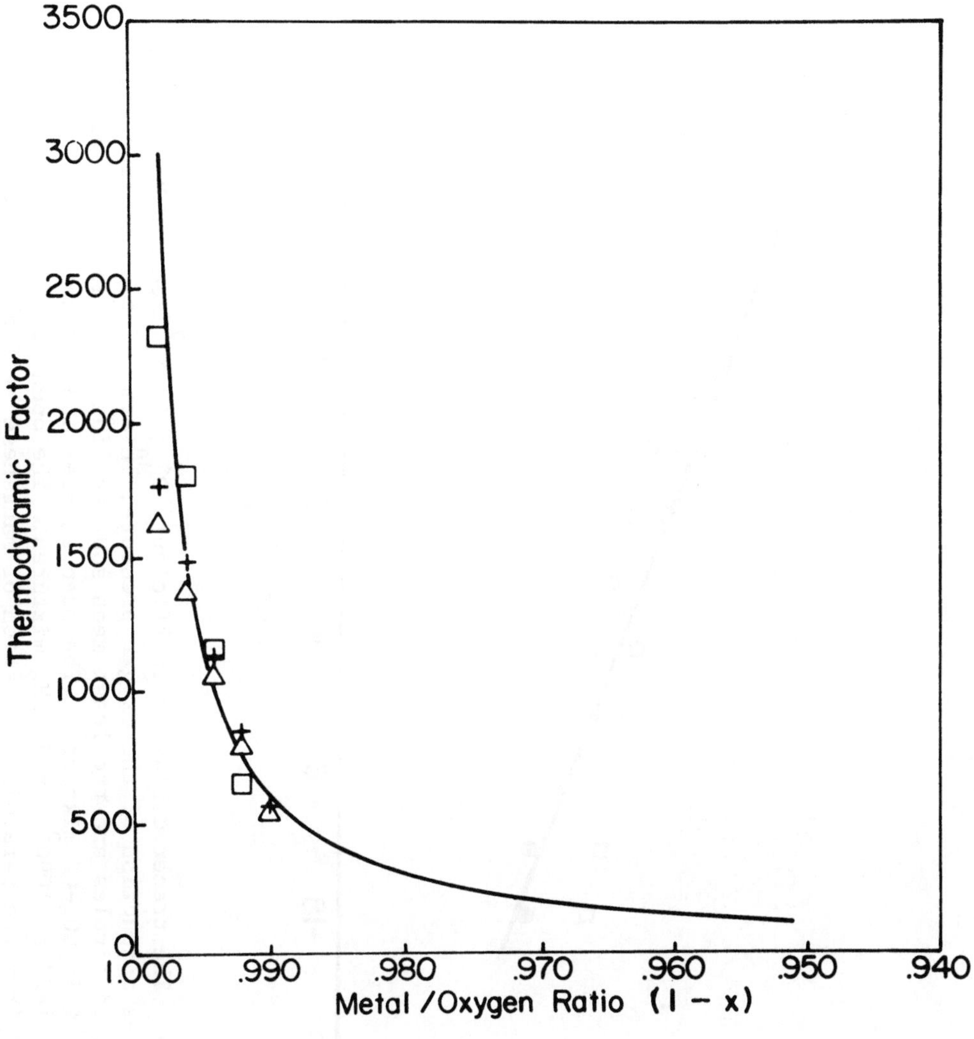

Fig. 15. The thermodynamic factor of MnO. The curve calculated from Eq. (21) for m = 2 is compared with experimental values taken from the data of Fender and Riley [27],as a function of the Mn/O ratio. □, + and Δ are experimental values determined at 1000, 1050 and 1100°C.

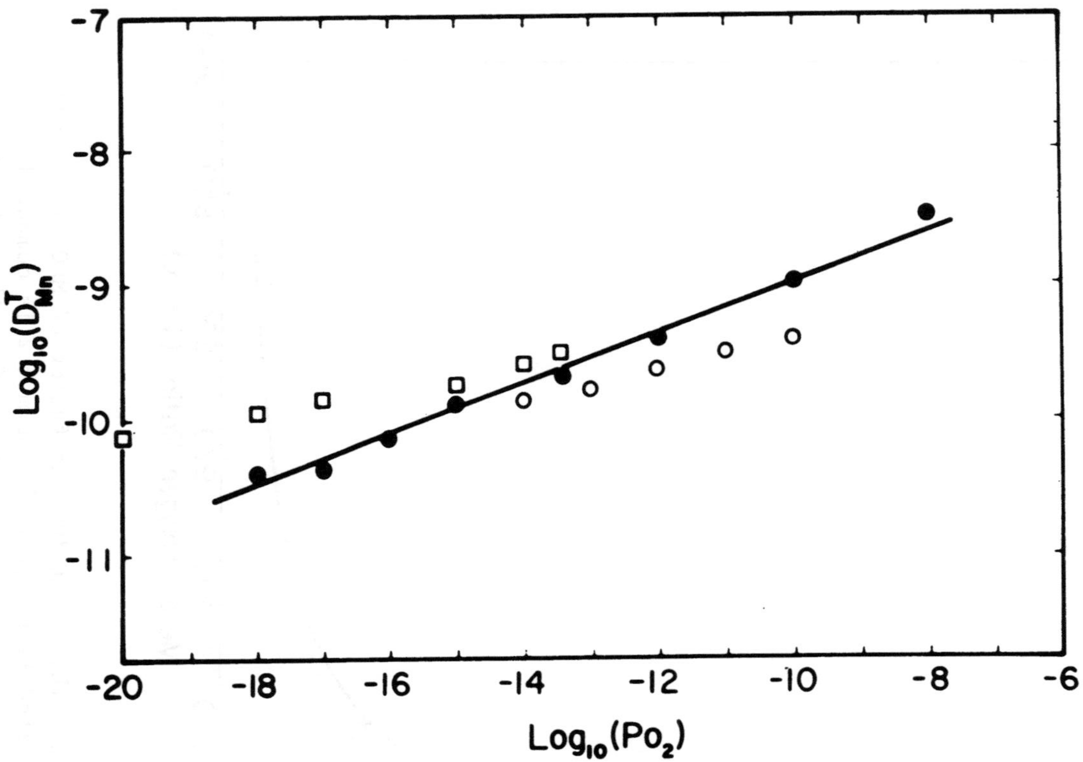

Fig. 16. The tracer diffusion coefficient, D^T_{Mn}, in cm^2/sec for MnO around 1000°C as a function of the oxygen pressure, in atm. The solid line and the filled circles are the least mean squares line and points from price at 1031°C in CO_2-CO mixtures. The open squares are the points of Bocquet, et. al. [42] at 1000°C in H_2O/H_2 mixtures. The open circles are the values derived from the parabolic oxidation of manganese by Fueki and Wagner [33] at 1000°C.

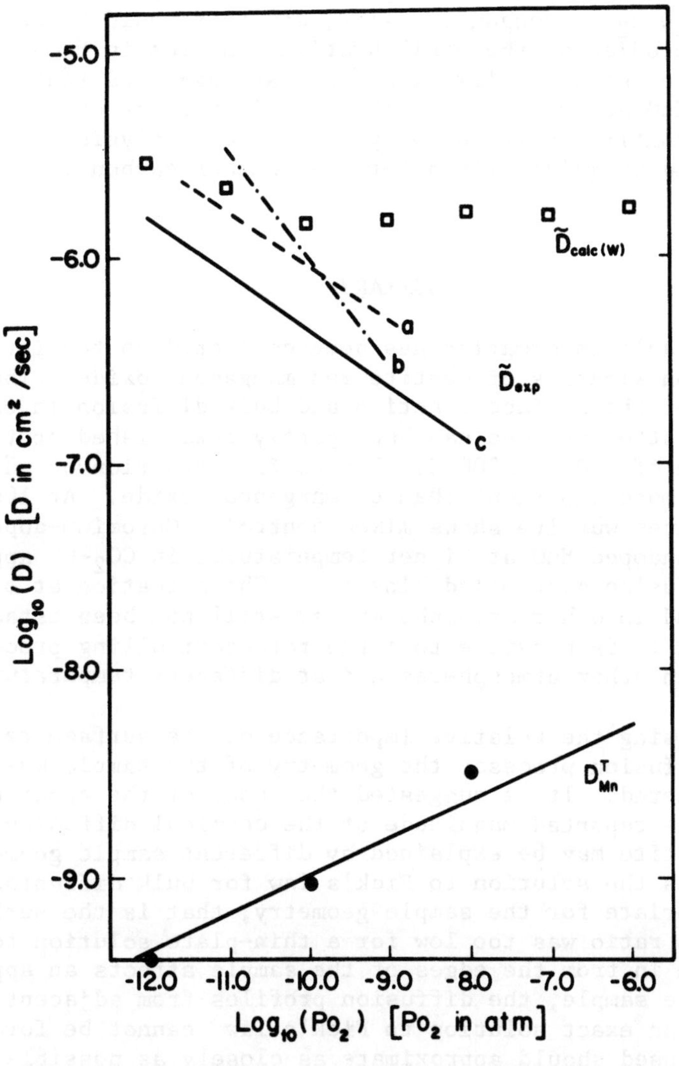

Fig. 17. Chemical diffusion coefficient, $\tilde{D}$, in cm^2/sec for MnO as a function of the oxygen pressure, in atm. Open squares: calculated using Wagner's equation (9), with m = 2 and thermodynamic data from Fender and Riley [27], and D_{Mn}^T from Price [1] at 1032°C. a is the experimental $\tilde{D}$ for MnO-Cr at 1037°C; b for polycrystal MnO at 1000°C; c for single crystal MnO at 1031°C from Price [1]. Filled circles: values of D_{Mn}^T from Price and Wagner [51] together with the least mean squares line through all their data.

No studies have yet been reported of the surface reaction of CO_2-CO or H_2O-H_2 on manganous oxide, and this would be desirable. Comparable studies of the equilibration kinetics in H_2O-H_2 mixtures would also be useful. Simkovich [68] has suggested that carbon is soluble in MnO and goes in as C_i^{4+}. No evidence was obtained in these studies to confirm this, and only a chemical analysis on annealed samples could unequivocally determine whether carbon is in fact present.

SUMMARY

Considerable information has been collected on the intraphase equilibration kinetics of wüstite and manganous oxide. The relative importance of the surface reaction and bulk diffusion in controlling the equilibration process has been partly established at temperatures in the range of 800 to 1300°C. The surface reaction on wüstite appears to be more important than on manganous oxide. At 1062°C in CO_2-CO mixtures wüstite shows mixed control. Chromium-doped MnO at 1206°C and undoped MnO at higher temperatures in CO_2-CO appear to exhibit diffusion controlled kinetics. The situation at other temperatures and in other atmospheres has still not been established clearly, and it is possible that the rate-controlling process will be changed in other atmospheres and at different temperatures.

In assessing the relative importance of the surface reaction and the bulk diffusion process, the geometry of the sample must be carefully considered. It is suggested that most of the apparent discrepancies in the reported magnitude of the chemical diffusion coefficient in wüstite may be explained by different sample geometries. In many cases the solution to Fick's law for bulk diffusion used was inappropriate for the sample geometry, that is the surface area to thickness ratio was too low for a thin-plate solution to be used. If diffusion in from the edges of the sample affects an appreciable volume of the sample, the diffusion profiles from adjacent faces will overlap, and an exact solution to Fick's law cannot be formulated. The samples used should approximate as closely as possible a one-dimensional diffusion system, and the best geometries for this are the thin plate, the cylinder or the sphere. The thin-plate sample is usually the most convenient to use and one criterion of such a sample is the face area to thickness ratio (A/ℓ), as was illustrated in Table IV. This ratio is related to the volume of the sample which experiences one-dimensional diffusion, and obviously this volume should be maximized. For example in a plate 20 x 20 x 1 mm^3, 81% of the total volume experiences one-dimensional diffusion, and A/ℓ = 400 mm. In order that 90% of the volume should experience one-dimensional diffusion, the thin-plate sample dimensions should be at least 40 x 40 x 1 mm^3, A/ℓ = 3200 mm, far larger than any samples that have been used so far. A cylindrical sample offers an easier way of increasing the volume affected by one-dimensional diffusion, while still retaining a reasonable sample size. For example a cylinder 20 mm long by 2 mm in diameter has only 10% of the total volume affected by diffusion from the end faces. We suggest that particular attention should be directed to the choice of the sample dimensions and geometry in future studies.

The chemical diffusion coefficients obtained in the present study are 10 -100 times greater than the corresponding tracer diffusion coefficients. This difference arises because the chemical diffusivity is essentially the diffusivity of the vacancies in ionic compounds with predominantly electronic conduction, where vacancies are the majority lattice defects, as in FeO and MnO. Consequently the ratio of $\tilde{D}$ to D_M^T is inversely proportional to the vacancy concentration. Since D_M^T is found to be proportional to the vacancy concentration the chemical diffusion coefficient, $\tilde{D}$, should be aproximately independent of composition. This is not observed experimentally, and in the two systems considered here, FeO and MnO, $\tilde{D}$ appears to decrease with increasing vacancy concentration. Part of this decrease may be due to the surface reaction contribution in wüstite, but this is not the case in MnO. This composition dependence thus has to be explained for both systems.

Similar behavior is shown by other systems. Grace and co-workers have studied the equilibration kinetics of $SrTiO_3$ [17] and $CaTiO_3$ [18] in H_2O-H_2 mixtures. Anion vacancies are the predominant defects, and argon dilution studies indicated the kinetics were diffusion controlled. $\tilde{D}$ decreased with increasing vacancy concentration, that is with decreasing H_2O/H_2 ratio. A recent study of cubic cuprous sulfide [90], using a polarization cell to study chemical diffusion, also gives a $\tilde{D}$ which decreases with increasing defect concentration. Studies of the parabolic growth of $Fe_{1-x}S$ on iron [91] gave values of D_{Fe}, which when related to $\tilde{D}_{FeS}$ by Darken's equation gave a function which increased and then decreased with increasing S/Fe ratio. This result may be significant, but the intraphase equilibration kinetics should be investigated to determine $\tilde{D}$ directly.

Sufficient experimental evidence is on hand to suggest that $\tilde{D}$ decreases with increasing deviation from stoichiometry in compounds which have vacancies as the majority defect. Using the conventional relations given in this paper, these results would yield calculated self-diffusion coefficients which decrease with increasing vacancy concentration. This is not found in self-diffusion coefficients derived from tracer diffusion studies or from the parabolic oxidation of the metal. In the growth of a metal-deficient oxide on a metal, the parabolic rate constants obtained in various oxygen pressures invariably increase with increasing oxygen pressure in the gas phase. This increase is explained by the expression:

$$k_p = \text{const.} \left[a''_O - a'_O \right] = \text{const.} \left[p''^{\frac{1}{2}}_{O_2} - p'^{\frac{1}{2}}_{O_2} \right], \quad (64)$$

where the double primes refer to the outer gas atmosphere in equilibrium with the oxide surface, and the single prime to the effective oxygen pressure or activity at the metal-metal oxide interface. Such a relation will also apply to the intraphase oxidation of the same oxide, where the single prime then refers to the initial oxygen activity in the oxide, which at the start of the oxidation is unaffected in the center of the sample. In this case similar relations should apply to reduction within the phase. If the inner oxygen activity is kept fixed at the metal-oxide boundary, or by performing experiments with different outer oxygen pressures for the same initial composition in the intraphase experiments, then k_p will increase as p''_{O_2} increases. This behavior is found for the oxidation of Mn [33], and in the intraphase equilibration of FeO [3], k_p increases for reduction as the size of the composition interval increases. Unfortunately similar data on the variation of k_p with increasing composition increments, starting from the same initial composition, are not available for the intraphase oxidation of FeO or MnO. When equal composition increments are taken across the phase field in wüstite, in either reduction or oxidation, the initial parabolic kinetics and the analysis of the last part of the kinetics using a thin plate solution, both yield a $\tilde{D}$ which decreases with vacancy concentration. This difference is related to the size of the composition interval used, and only when small increments are used across the phase field does the composition dependence of $\tilde{D}$ become apparent. Wagner's [64] formulation of the parabolic rate constants, given in Eqs. (31) and (32), shows this, in that only the integral ($\tilde{D}\, dN_2$) between N_2'' and N_2' can be obtained from parabolic oxidation studies. This integral will increase as ($N_2'' - N_2'$) increases, whatever the composition dependence of $\tilde{D}$. Unless this dependence is known only an apparent $\tilde{D}$ can be obtained, and this will increase with increasing oxygen gradient across the oxide. When the same gradient in oxygen activity is present, but at a different average oxygen activity, i.e. at various points in the phase field, the dependence of $\tilde{D}$ on composition becomes apparent. The discrepancy between k_p determinations and intraphase measurements of $\tilde{D}$ thus seems to lie in a faulty interpretation of k_p, when the oxygen activity gradient across the oxide is successively increased.

The decrease of the chemical diffusion coefficient with increasing vacancy concentration and its relation to the tracer diffusion coefficient still have to be explained. The relation of $\tilde{D}$ to the vacancy diffusivity, as in Eq. (9) seems to be correct, so that D_{V_M} does not appear to be related to D^T_{Mn} by Eqs. (2) and (4). Changes in the correlation factor, f, or the diffusion potential term, (1 + m), do not seem to be able to account for the observed discrepancy. D_{V_M} may be related to a vacancy mobility by the Nernst-Einstein equation and this mobility is usually assumed to be constant with composition, and at low concentrations this is probably true [91].

If the vacancy diffusivity is inversely proportional to the vacancy concentration, the observed composition dependence of $\tilde{D}$ would be explained. The mobility of metal atoms is related to the concentration of vacancies by an equation such as [56]:

$$D_M = k_M \cdot \lambda^2 \cdot x , \qquad (65)$$

where k_M, and λ are the ion jump parameters and x is the vacancy fraction $M_{1-x}X$. Thus increasing the vacancy concentration makes it easier for the metal atoms to move. If a similar expression is true for vacancies then they should be less mobile the fewer occupied sites there are. That is the vacancy diffusivity should be proportional to (1 - x), the metal fraction, and for a constant degree of ionization, D_{V_M} will decrease as the vacancy concentration increases. But this is only Eq. (2) in a different guise, and we note that the Nernst-Einstein formulation is only applicable to dilute solutions of charged species. When interactions become significant the effective mobility decreases, in this case the mobility of the vacancies. In MnO it appears that Wagner's treatment can explain the observed $\tilde{D}$ at the lowest vacancy concentration, but after about 0.5% vacancies, Wagner's equation seems to be inapplicable. Introducing an activity coefficient for the vacancies instead of the concentrations, i.e. using the Debye-Hückel theory, may be the solution to this problem. However, it may be that experimental measurements of $\tilde{D}$ offer the only way of determining the vacancy mobility. In FeO the electronic conductivity is observed to become invariant with p_{O_2} at the highest electron hole concentrations, indicating a decrease in electron hole mobility. There is evidence for vacancy clustering also in FeO within the phase field [92], and it may be that we are observing the effect of this on the vacancy diffusivity. In MnO the conductivity appears to behave as expected by the mass action interpretation of the defect structure. But here the vacancy concentration is much lower and the Nernst-Einstein relations seem to apply for part of the composition range studied.

It should be emphasized that the "activation energies" obtained from a plot of log $\tilde{D}$ versus 1/T are composite quantities. They contain both the energy of formation or destruction of the defects and the energy of migration of the defects, but not in a simple combination. It is only a fortuitous set of circumstances which results in linear plots of log $\tilde{D}$ versus 1/T. In general such plots will be non-linear and in cases such as MnO, the formation and migration energies change in such a way as to yield virtually zero "activation energy" between 850° and 1100°C. Above these temperatures $\tilde{D}$ increases with temperature.

Most workers have been content with an agreement in magnitude between experimental and calculated chemical diffusion coefficients, and in quoting an "activation energy." We believe we have shown that the situation is not that simple if the compositional dependence of $\tilde{D}$ is taken into consideration, and this dependency may contain the most significant information. It means that in appreciably nonstoichiometric systems accurate equilibration kinetics cannot be calculated from self-diffusion coefficients, unless the variation of vacancy mobility with composition is known.

We would like to make some suggestions for further work in this area. The effect of surface orientation should be studied, as well as extending the studies on the effect of argon dilution and sample thickness on the equilibration kinetics over a wide range of temperature in FeO and MnO. When mixed control is present, some effective method is required to separate the two contributions to the equilibration rate. Detailed complementary measurements of cation and anion self-diffusion coefficients are required, because the lack of such data is a major drawback in understanding these diffusion processes. A large number of systems are available where $\tilde{D}$ could be measured as a function of composition either by the thermogravimetric or the conductivity technique. Some work has been done on UO_{2-x} [21] and on carbides [93] and should be extended, but other systems such as $Fe_{1-x}S$ or CeO_{2-x} should be investigated. This would test whether $\tilde{D}$ is always inversely proportional to the majority defect concentration in appreciably nonstoichiometric systems. Careful attention must be made to defining the sample geometry and to sample characterization. This applies to studies of the surface reaction as well as the chemical diffusion process. Most studies of the catalytic activity of surfaces have been carried out under very badly defined catalyst surface conditions, and the surface reaction can best be investigated at thermodynamic equilibrium, where chemical diffusion does not occur. The exchange mothod developed by Grabke [73] seems to offer a good way of doing this. Much more information could be obtained from relaxation measurements on thin foils [74], [75], in particular to studying the way the surface concentration changes through the course of the equilibration.

In this review we have tried to indicate the wealth of useful information that may be obtained from a quantitative study of intraphase equilibration kinetics, both on the surface reactions and chemical diffusivity of nonstoichiometric systems. Moreover we have pointed out the experimental details necessary to obtain relatively unambiguous rate constants. Finally we have shown the need, not only for additional experimental data, but also for further theoretical considerations of the relation between the chemical diffusion coefficient and the self-diffusion coefficients obtained from tracer experiments.

APPENDIX

Expressions for the Rate of the Surface Reaction on Wüstite

Wagner [70] proposed that the rate of decomposition of CO_2-CO mixtures on wüstite could be represented by the equation:

$$\frac{dn}{dt} = k'(a_O)\, p_{CO_{2_f}} - k''(a_O)\, p_{CO_f}, \tag{A-1}$$

where dn/dt is the rate of addition of oxygen atoms to the surface, $k'(a_O)$ and $k''(a_O)$ are the rate constants for the forward and backward reactions respectively and depend on the oxygen activity at the surface, a_O, and $p(CO_2)_f$, $(p(CO)_f$ are the partial pressures of CO_2 and CO in the gas phase. We shall show here the various equivalent expressions for the surface rate that can be derived from this equation, and then show that an equivalent expression can be obtained by formulating an activated complex mechanism for the decomposition. Subscripts i, s and f indicate the initial, surface and final conditions, i.e. at $t = t_o$, t and t_∞ respectively for the oxide.

Initially dn/dt = 0, and thus:

$$\frac{k'(a_{O_i})}{k''(a_{O_i})} = \frac{p_{CO_i}}{p_{CO_{2_i}}} \tag{A-2}$$

whence from Eq. (A-1) we have:

$$\frac{dn}{dt} = k'(a_{O_i})\left(p_{CO_{2_f}} - \frac{p_{CO_{2_i}}}{p_{CO_i}} \cdot p_{CO_f}\right) \tag{A-3}$$

at the start of the reaction as dn/dt → 0. For the growth of FeO on Fe the initial conditions are fixed by the Fe/FeO equilibrium, i.e.

$$\frac{p_{CO_{2_i}}}{p_{CO_i}} = K'_i \tag{A-4}$$

and also:

$$p_{CO_{2_f}} = N_{CO_{2_f}} \cdot p' \quad \text{and} \quad p_{CO_f} = N_{CO_f} \cdot p' \qquad \text{(A-5)}$$

where $N(CO)_{2_f}$ and N_{CO_f} are the mole fractions of CO_2 and CO in the gas phase and $p' = p(CO_2)_f + p_{CO_f}$. From this the familiar equation is obtained [71]:

$$\frac{dn}{dt} = k'(a_{O_i}) \cdot p'(1 + K'_i)(N_{CO_{2_f}} - N_{CO_{2_i}}) \qquad \text{(A-6)}$$

For a given CO_2/CO ratio over the metal the initial oxidation rate will depend on p' and will thus be affected by argon dilution. The rate constant obtained is that for the surface oxygen activity corresponding to the wüstite composition in equilibrium with iron at a particular temperature. An alternative formulation of (A-6) is the following:

$$\frac{dn}{dt} = k'(a_{O_i}) \cdot p_{CO_{2_f}} \left(1 - \frac{p_{CO_{2_i}}}{p_{CO_i}} \cdot \frac{p_{CO_f}}{p_{CO_{2_f}}}\right) \qquad \text{(A-7)}$$

Since for CO_2-CO mixtures we have the equilibrium constant:

$$K_p = \frac{p_{CO}\, p_{O_2}^{1/2}}{p_{CO_2}} \qquad \text{(A-8)}$$

we can represent the ratios p_{CO_2}/p_{CO} in terms of the corresponding oxygen pressures, i.e.

$$\frac{dn}{dt} = k'(a_{O_i})\, p_{CO_{2_f}} \left(1 - \frac{p_{O_{2_i}}^{1/2}}{p_{O_{2_f}}^{1/2}}\right) \qquad \text{(A-9)}$$

An oxygen activity may now be defined by the equation:

$$a_O = \left(\frac{p_{O_2}}{p_{O_2}^{o}}\right)^{1/2} \qquad \text{(A-10)}$$

where $p_{O_2}^{o}$ is the oxygen pressure in the standard state. Thus Eq. (A-10) gives us in Eq. (A-9):

$$\frac{dn}{dt} = k'(a_{O_i}) \cdot p_{CO_{2_f}} \left(1 - \frac{a_{O_i}}{a_{O_f}}\right) \qquad \text{(A-11)}$$

or

$$\frac{dn}{dt} = k'(a_{O_i}) \frac{p_{CO_{2_f}}}{a_{O_f}} (a_{O_f} - a_{O_i}) \qquad \text{(A-12)}$$

From Eqs. (A-8) and (A-10) we have at a given temperature:

$$\frac{p_{CO_{2_f}}}{a_{O_f}} = \frac{p_{CO_f}}{K_p} \cdot (p^o_{O_2})^{1/2} = p_{CO_f} \cdot K'_p \qquad \text{(A-13)}$$

where $K'_p = (p^o_{O_2})^{\frac{1}{2}}/K_p$, and is constant. Thus Eq. (A-12) reduces to:

$$\frac{dn}{dt} = k'(a_{O_i}) \cdot p_{CO_f} \cdot K'_p (a_{O_f} - a_{O_i}) \qquad \text{(A-14)}$$

or by taking a_{O_i} outside in Eq. (A-11) we have:

$$\frac{dn}{dt} = k'(a_{O_i}) \cdot p_{CO_{2_f}} \cdot a_{O_i} \left(\frac{1}{a_{O_i}} - \frac{1}{a_{O_f}}\right) \qquad \text{(A-15)}$$

The Eqs. (A-6), (A-9), (A-11), (A-14) and (A-15) only apply to the initial stages of the oxidation of either a metal to an oxide or the intraphase oxidation of an oxide to another composition within the phase field.

We now have to ask what happens past the initial stages of the reaction. The oxygen activity at the surface changes between a_{O_i} and a_{O_f} and at any time it has a value a_{O_s}, in the case of a finite surface reaction. The rate constants k' and k'' vary throughout the reaction as a function of the surface activity of oxygen, a_{O_s}. Initially from Eq. (A-1) the rate of addition of oxygen is a maximum, and Eq. (A-3) holds. At t_∞, i.e. after equilibration with the gas phase, the rate of addition of oxygen is zero and:

$$\frac{k''(a_{0_f})}{k'(a_{0_f})} = \frac{p_{CO_{2_f}}}{p_{CO_f}}, \quad \frac{dn}{dt} = 0 . \tag{A-16}$$

But using Eqs. (A-8) and (A-10) we can write this ratio of rate constants in terms of the oxygen activity at the surface, i.e.

$$\frac{k''(a_{0_f})}{k'(a_{0_f})} = a_{0_f} \cdot K'_p \tag{A-17}$$

and

$$\frac{k''(a_{0_i})}{k'(a_{0_i})} = a_{0_i} \cdot K'_p \tag{A-18}$$

It seems reasonable at intermediate times to relate the ratio $k''(a_{0_s})/k'(a_{0_s})$ to the surface oxygen activity, a_{0_s}:

$$\frac{k''(a_{0_s})}{k'(a_{0_s})} = a_{0_s} \cdot K'_p \tag{A-19}$$

Thus Eq. (A-1) gives us

$$\frac{dn}{dt} = k'(a_{0_s})(p_{CO_{2_f}} - K'_p \cdot a_{0_s} \cdot p_{CO_f}) \tag{A-20}$$

and a series of relations can be derived as above, for example:

$$\frac{dn}{dt} = k'(a_{0_s}) \cdot K'_p \cdot p_{CO_f}(a_{0_f} - a_{0_s}) \tag{A-21}$$

Grabke [73] obtained an empirical relation for the rate of the forward reaction, $k'(a_0)$, at various temperatures, finding that:

$$k'(a_0) = \text{const.} \left(\frac{p_{CO_2}}{p_{CO}}\right)^{-u} = \text{const.}\ a_0^{-u} K_p'^{-u} \tag{A-22}$$

Substituting (A-22) in (A-21) gives us:

$$\frac{dn}{dt} = k'(p_{CO_2}/p_{CO} = 1) \cdot K_p'^{(1-u)} \cdot p_{CO_f} \cdot a_{O_s}^{-u} (a_{O_f} - a_{O_s}) \qquad \text{(A-23)}$$

The variation of the rate of addition of oxygen is thus a complicated function of the surface oxygen activity and at present there seems to be no way of analyzing this except in the initial stages of the reaction. This expression was given by Grabke [73], and is also similar to the equation derived by Wagner [36, 37] to explain the transition from linear to parabolic kinetics in metal oxidation:

$$\frac{dn}{dt} \propto p_{CO_{2_f}} \left[a_{O_s}^{-2/3} - \frac{a_{O_2}^{1/3}}{p_{CO_{2_f}}/p_{CO_f}} \right] \qquad \text{(A-24)}$$

We shall now show that equivalent expressions for the surface rate on wüstite can be derived assuming that CO_2 (or H_2O) is decomposed on the surface. The decomposition of molecular oxygen, present in the gas phase,is less plausible because of its low concentration and the higher dissociation energy of O_2 versus CO_2 (H_2O). All the gaseous species will be adsorbed to some extent on the surface and these will be denoted by a subscript s. CO_{2_s} and CO_s are assumed to be in equilibrium with the gas phase. For a finite surface reaction the oxygen activity at the surface is not that of oxygen in the gas phase,i.e. the surface and the gas phase are not in equilibrium with respect to the oxygen activity. If we consider an adsorbed CO_2 molecule decomposing via an activated transition state on the oxide surface giving adsorbed oxygen, then we combine the general features of the mechanisms suggested by Hauffe [61], by Wagner [70] and by Turkdogen,et al [72]. i.e.

$$CO_{2_g} \rightleftarrows CO_{2_s} \rightleftarrows CO_{2_s}^{\dagger} \rightleftarrows CO_g + O_s \, , \qquad \text{(A-25)}$$

where subscript g denotes the gas phase and superscript † indicates an activated intermediate.

For the forward reaction we can write an equilibrium constant:

$$K^{\dagger} = \frac{a^{\dagger}}{a_{CO_{2_s}}} = \frac{a^{\dagger}}{p_{CO_{2_g}}} \qquad \text{(A-26)}$$

where $a^\dagger$ is the surface activity of the complex. Denoting the fractional surface coverage of the ith species by θ_i, we have:

$$\theta_{CO_2} + \theta_\dagger + \theta_O + \theta_{CO} + \theta_\square = 1, \qquad \text{(A-27)}$$

where $\theta_\square$ is the fraction of unoccupied sites, and $\theta_\dagger$ is the fraction of sites occupied by the activated complex. The activity of a species on the surface can be written as [87]:

$$a_i = k_i \cdot \frac{\theta i}{\theta} \qquad \text{(A-28)}$$

where k_i is a proportionality constant. We can substitute for $a^\dagger$ in Eq. (A-26) and using Eyring's rate expression [88] for the forward reaction we obtain:

$$\text{Rate of forward reaction} = C \cdot \theta_\dagger = C \cdot \frac{K\dagger}{k^\dagger} \cdot \theta_\square \cdot p_{CO_2} \qquad \text{(A-29)}$$

where C = RT/h · Γ, the universal specific rate constant times Γ, the number of sites/ unit area.

Similarly for the backward reaction we can write:

$$K^{\dagger'} = \frac{a^\dagger}{p_{CO} \cdot a_{O_s}} \qquad \text{(A-30)}$$

and consequently

$$\text{Rate of backward reaction} = C \cdot \theta_\dagger = C \cdot \frac{K^{\dagger'}}{k^\dagger} \cdot \theta_\square \cdot p_{CO} \cdot a_{O_s} \qquad \text{(A-31)}$$

The net reaction rate is given by subtracting Eqs. (A-29) and (A-31) which gives us:

$$\frac{dn}{dt} = \frac{C}{k^\dagger} \cdot \theta_\square \left(K^\dagger \cdot p_{CO_2} - K^{\dagger'} \cdot p_{CO} \cdot a_{O_s} \right) \quad \text{atom O/cm}^2\text{/sec} \qquad \text{(A-32)}$$

which is the form expected if the formation of O_s is the rate-determining process. If oxygen is the major absorbed species at the surface we can use an approximate expression for $\theta_\square$ [89], i.e. as $\theta_O \rightarrow 1$, $\theta_\square \rightarrow 0$ and all other $\theta_i \rightarrow 0$, so that:

$$a_{O_s} = \frac{k_O}{\theta_\square} \cdot \theta_O \rightarrow \frac{k_O}{\theta_\square} \qquad \text{(A-33)}$$

We can thus substitute for $\theta_\square$ in Eq. (A-31) giving:

$$\frac{dn}{dt} = \frac{C}{k^\dagger} \cdot k_0 \cdot \left(K^\dagger \frac{p_{CO_2}}{a_{0_s}} - K^{\dagger'} p_{CO} \right) \qquad \text{(A-34)}$$

As the final equilibrium is reached $dn/dt \to 0$, and consequently:

$$K^\dagger \cdot \frac{p_{CO_{2_f}}}{a_{0_f}} = K^{\dagger'} \cdot p_{CO_f} \qquad \text{(A-35)}$$

But since CO_2 and CO were assumed always to be in equilibrium with the surface, then $p_{CO_2} = p_{CO2_f}$ and $p_{CO} = p_{CO_f}$ in Eq. (A-34), and we obtain:

$$\frac{dn}{dt} = \frac{C}{k^\dagger} \cdot k_0 \cdot K^\dagger \left(\frac{p_{CO_{2_f}}}{a_{0_s}} - \frac{p_{CO_{2_f}}}{a_{0_f}} \right)$$

$$= \frac{C}{k^\dagger} \cdot k_0 \cdot K^\dagger \cdot p_{CO_{2_f}} \left(\frac{1}{a_{0_s}} - \frac{1}{a_{0_f}} \right) \qquad \text{(A-36)}$$

From (A-33) we have:

$$\frac{k_0}{k^\dagger} = \frac{a_{0_s}}{a^\dagger} \qquad \text{(A-37)}$$

and thus Eq. (A-35) becomes:

$$\frac{dn}{dt} = \frac{C}{a^\dagger} \cdot K^\dagger \cdot p_{CO_{2_f}} \, a_{0_s} \left(\frac{1}{a_{0_s}} - \frac{1}{a_{0_f}} \right), \qquad \text{(A-38)}$$

which is identical in form to Eq. (A-15), where:

$$k' (a_{0_s}) = \frac{C \cdot K^\dagger}{a^\dagger} = C' \qquad \text{(A-39)}$$

A similar expression can be derived for the decomposition of H_2O-H_2 mixtures on wüstite as given by Turkdogan et al [72]:

$$\frac{dn}{dt} = C'' \cdot p_{H_2O_f} \cdot a_{O_s} \left(\frac{1}{a_{O_s}} - \frac{1}{a_{O_s}} \right), \qquad \text{(A-40)}$$

where only the constant term, C'', is changed.

All these expressions have been given in terms of the oxygen activity at the surface, since oxygen is being added in the process. From Eq. (20) we see that the vacancy concentration in the oxides considered here is also dependent on the oxygen pressure, that is on the oxygen activity at any particular point. The oxygen excess and vacancy excess at any point are equal. Since we have from Eq. (17):

$$\left[V_M^{m'}\right] = y = C_V \cdot \text{const.} = A \cdot p_{O_2}^{\frac{1}{2(m+1)}} \qquad \text{(A-41)}$$

we can also write:

$$a_O = C_V^{m+1} \cdot A' = y^{m+1} \cdot A'' \qquad \text{(A-42)}$$

where A' is a constant given by const. $/A^{m+1} \cdot p_{O_2}^{o\frac{1}{2}}$. Hence we may write (A-38) as:

$$\frac{dn}{dt} = C' \cdot p_{CO_{2f}} \cdot y_s^{m+1} \left(\frac{1}{y_s^{m+1}} - \frac{1}{y_f^{m+1}} \right) \qquad \text{(A-43)}$$

That is we can relate the rate of addition of oxygen to the vacancy concentration at the surface, or to the oxygen excess at the surface, compared with the equilibrium values. This assumes that the thermodynamics of defect formation are not changed at the surface, that is that local equilibrium prevails.

ACKNOWLEDGEMENT

This work was supported by the Advanced Research Projects Agency through a grant from the Materials Research Center of Northwestern University and by the American Iron and Steel Institute.

REFERENCES

1. J. B. Price, Ph.D. Thesis, Northwestern University (1968).
2. I. Bransky and N. M. Tallan, to be published in Vacuum Microbalance Technique 8, (1969).
3. R. L. Levin and J. B. Wagner, Jr., Trans AIME 233, 159 (1965).
4. C. T. Fujii and R. A. Meussner, NRL Report, 27 (1965).
5. P. F. J. Landler and K. L. Komarek, Trans. AIME 236, 138 (1966).
6. P. L. Hembree, M. S. Thesis, Norwestern University (1966).
7. R. H. Campbell, Ph. D. Thesis, Arizona State University (1968).

8. H. A.Levin, Ph.D. Thesis, Norwestern University (1969).
9. L. W. Laub, unpublished Ph.D. research, Northwestern University (1969).
10. J. B.Price and J.B. Wagner, Jr., Z. Phys. Chemie N.F. 49, 257 (1966).
11. G.A.Slack, private communication to S. P. Mitoff, J. Chem Phys. 35, 882 (1961)
12. R. Morlotti, Z.Naturforsch., 24a, 441 (1969).
13. W. H. Chen and R. A. Swalin, J. Phys. Chem. Solids 27, 57 (1966),
14. J. S. Sheasby, W. W. Smeltzer and A. E. Jenkins, J. Electrochem. Soc. 115, 338 (1968).
15. C. D. Wirkus, M.F.Berard and D. R. Wilder, J. Am. Ceram. Soc. 50, 113 (1967).
16. M. F. Berard, C.D.Wirkus and D. R. Wilder, J. Am. Ceram. Soc. 51, 643 (1968).
17. L. C. Walters and R. E. Grace, J. Phys. Chem. Solids 28, 245 (1967).
18. W.L. George and R. E. Grace, J. Phys. Chem. Solids 30, 889 (1969).
19. R. H.Campbell, W. J. Kass and M. O'Keefe, *Mass Transport in Oxides*, NBS Special Publication No. 296, 173 (1968).
20. D. G. Thomas and J. J. Lander, J. Chem. Phys. 25, 1136 (1956).
21. K. W.Lay, to be published.
22. L. S. Darken and R.W. Curry, J. Am. Chem. Soc. 67, 1398 (1945).
23. P. Vallet and P.Raccah, Mem. Sci. Rev. Met. 62, 1 (1965).
24. R. L. Ackermann and R. W. Sandford, Jr., ANL-7250 (1965).
25. B. E.F.Fender and F. Riley, J. Phys. Chem. Solids 30, 793 (1969).
26. A.Z.Hed and D.S.Tannhauser, J. Electrochem. Soc. 114, 314 (1967).
27. B. E. F. Fender and F. Riley, Proc. Int'l. Conf. *Extended Defects in Non-Metallic Solids*, (1969), to be published.
28. N. G. Schmahl and D.F. K. Hennings, Arch Eisenhüttw. 40,395 (1969).
29. J. B. Wagner, Jr., *Mass Transport in Oxides*, NBS Special Publication No.296, 65 (1968).
30. M. O'Keefe, Proc. Int'l. Conf., *Sintering and Related Phenomena* G. C. Kuczynski Ed., Gordon and Breach (N.Y.), 57 (1967).
31. M. O'Keefe, Proc.Int'l. Conf., *Extended Defects in Non-Metallic Solids*, (1969), to be published.
32. L.B.Pfeil, J.Iron & Steel Inst. (London) 119, 501 (1929).
33. K. Fueki and J. B. Wagner, Jr., J. Electrochemical Soc. 112, 970 (1965).
34. M. O'Keefe and W. J. Moore, J. Phys. Chem., 65, 1438 (1961).
35. J. B. Holt, Proc. Brit. Ceram. Soc., 9, 157 (1967; W. K. Chen and R. A. Jackson, J.Phys. Chem. Solids 30, 1309 (1969).
36. F. S. Pettit and J.B. Wagner, Jr. Acta Met. 12, 35 (1964).
37. C.Wagner, Ber. Bunsenges. physik Chem. 72, 778 (1968).
38. M.H.Davies, M. T. Simnad, C. E. Birchenall, Trans. AIME 191, 889 (1951); ibid. 197, 1250 (1953).

39. B. J.Wuensch and T.Vasilos, J. Chem. Phys. 36, 2917 (1962); E.B. Rigby and I. B. Cutler, J. Am. Soc. 48, 95 (1965).
40. N. de Libanati, J. Philibert and J. Manenc. Mem. Sci. Rev. Met. 63, 127 (1966).
41. H. Schmalzried and J. B. Holt, Z. Physik. Chem. (Frankfurt) 60, 220 (1968).
42. J. P.Bocquet, M. Kawahara and P. Lacombe, Comptes Rendus (Paris), 265C,1318 (1967).
43. L. Himmel, R. F. Mehl and C.E. Birchenall, Trans. AIME 197, 827 (1953).
44. R. E. Carter and F.D. Richardson, Trans. AIME 200, 1244 (1954).
45. P. Desmarescaux and P. Lacombe, Mem. Sci. Rev. Met. 60, 899 (1963);
P. Desmarescaux, J. P. Bocquet, and P. Lacombe, Bull. Soc. Chim.France 15, 1106 (1965).
46. P. L. Hembree and J. B. Wagner, Jr. Trans. AIME 245, 1547 (1969).
47. R.Lindner and A.Akerström, Z. Phys. Chem. 6, 162 (1956); Discuss. Faraday Soc. 23, 133 (1957).
48. M. T.Shim and W. J. Moore, J. Chem. Phys. 26,802 (1957).
49. J. S. Choi and W. T. Moore, J. Phys. Chem. 65, 1308 (1962).
50. W. J. Moore and B. Selikson, J. Chem. Phys. 19, 1539 (1951) and 20, 927 (1952)
51. J. B. Price and J. B. Wagner, Jr. to be published in J. Electrochem. Soc.
52. N. F.Mott and R. W. Gurney, *Electronic Processes in Ionic Crystals*, Oxford University Press, p. 34 (1953).
53. C. Wagner, Z.Physik. Chem. B21, 25 (1933) and ibid, B32, 447 (1936).
54. C. Wagner, A.S.M. Symposium, *Atom Movements*, Cleveland, p. 153 (1951).
C. Wagner, Z. Physik Chem. B22, 447 (1936).
55. C. Wagner, J.Phys. Chem.Solids 29, 1925 (1968).
56. R. F. Brebrick, J.Appl. Phys. 30, 911 (1959).
57. L. S. Darken, Trans. AIME 175, 184 (1948).
58. P. E.Childs, to be published.
59. J. R. Manning, Phys. Rev. 139A, 126 (1965); J. R. Manning, Acta Met. 15, 817 (1967).
60. N.L. Peterson, *Solid State Physics*, 409 (1969).
61. K. Hauffe, *Oxidation of Metals*, Plenum Press, New York (1965).
62. P. Kofstad, *High-Temperature Oxidation of Metals*, Wiley, New York (1966); J.Bénard ed., *L'Oxydation des Métaux*, Vols. I and II, Gauthier-Villars, Paris (1962 and 1964).
63. F.S. Pettit, J. Electrochem. Soc. 112, 1249 (1966).
64. C.Wagner, Acta Met. 17, 99 (1969).
65. C. Matano, Japan J.Phys. 8, 109 (1933); W. Jost, *Diffusion in Solids, Liquids and Gases*, 3rd Edition, Academic Press, New York, p. 31 (1960).
66. J. Crank, *The Mathematics of Diffusion*, Oxford University Press, Oxford (1956).

67. S. Stotz and C.Wagner, Ber. Bunsenges physik Chem., 70, 781 (1966).
68. G. Simkovich, Private communication to J. B. Wagner, Jr. (1969).
69. S. Zelouf and G. Simkovich, Trans. AIME 245, 875 (1969).
70. C. Wagner, Mimeographed notes, Department of Metallurgy, Massachusetts Institute of Technology, (Spring 1955).
71. F. Pettit, R. Yinger and J. B. Wagner, Jr., Acta Met. 8, 617 (1960).
72. E. T.Turkdogan,W. M. McKewan and L. Zwell, J. Phys. Chem. 69, 327 (1965).
73. H. J. Grabke, Ber.Bunsenges, physik. Chem. 69, 49 (1965).
74. S. Stotz, Ber. Bunsenges. physik. Chem. 70, 37 (1966).
75. S. Stotz, Ber. Bunsenges. physik. Chem. 70, 769 (1966).
76. C.Wagner, Ber. Bunsenges. physik. Chem. 70, 776 (1966).
77. K. Hauffe and H.Pfeiffer, Z. Metallk. 44. 27 (1953).
78. R. A.Meussner, L. E. Richards and C. T. Fujii, Rept. N.R.L. Prog, 26 (1965).
79. P. E.Childs and J. B. Wagner, Jr., to be published.
80. C. T. Fujii and R. A. Meussner,Trans. AIME 242, 1259 (1968).
81. M. O'Keefe and M.Valigi, to be published.
82. G. Vagnard and J.Manenc, Mem. Sci. Rev. Met. 61, 767 (1964); B. Ilschner, B. Reppich and E. Riecke, Discussions Faraday Soc. 38, 243 (1964); B. Reppich, Phys. Stat. Solidi 20, 69 (1967).
83. C. T.Fujii and R. A.Meussner, J. Electrochem. Soc. 111, 1216 (1964). and Rept. N.R.L. Prog. 30 (Dec. 1963).
84. A. Z.Hed and D.S. Tannhauser, J. Chem. Phys. 47, 2090 (1967); F.Marion and A. Duguesnoy, Rev. Hautes. Temp. et Refracts. 2, 201 (1965); H. L. McKinzie, Ph.D. Thesis, Arizona State University (1968).
85. N. G. Eror, PhD.Thesis Northwestern University (1965).
86. P. Nagels and M. Denayer, Solid State Comms. 5, 193 (1967); H.J. DeWit and C. Crevecoeut, Phys. Letters, 25A, 393 (1967); M. Guishi, N. M. Tallan and D. S. Tannhauser, Solid State Comms. 6, 135 (1968).
87. E. T. Turkdogan and P. Grieveson, J. Electrochem. Soc. 114, 59 (1967).
88. S. Glasstone, K. J. Laidler and H. Eyring, *The Theory of Rate Processes*, McGraw-Hill, New York (1941).
89. E. T. Turkdogan, Trans. AIME 242, 1665 (1968); W. L. Worrell and E.T.Turkdogan, Trans. AIME, 242, 1673 (1968).
90. K.Weiss, Ber. Bunsenges. physik. Chem. 73, 338 (1969).
91. J.B.Holt, H. G. Sockel, H. Schmalzried, J. Am. Ceram. Soc. 52, 375 (1969).
92. W. R. Roth, Acta Cryst. 13, 140 (1960);F. Koch and J. B. Cohen, Acta Cryst. B25, 275 (1969); P. E. Childs, D. Phil. Thesis, Oxford University (1967).
93. T. C. Wallace in *Fundamentals of Refractory Compounds*, Plenum Press, New York, p. 133 (1968).

DISCUSSION OF PAPER

G. Simkovich Was there a difference in $\tilde{D}$ in MnO between the C. Wagner & Darken equations and in what direction did each go with an increase in cation vacancies?

J. B. Wagner: The equations of C. Wagner and of Darken both yielded virtually the same values of the chemical diffusion coefficient for the intraphase equilibration of undoped manganous oxide. In this case, calculations were performed using the tracer diffusion data of J. B. Price and the thermodynamic factor obtained from the data of Fender and Riley. The chemical diffusion coefficient so calculated was almost independent of composition whereas the experimental determination yielded values which decreased with increasing oxygen content and hence increasing cation vacancy concentration. It should be pointed out that the calculated values of the chemical diffusion coefficient represent a balancing of the self or tracer diffusion term and the thermodynamic factor in the equations under consideration. The self or tracer diffusion coefficient increases with increasing oxygen content corresponding to increasing cation vacancy concentration while the thermodynamic factor decreases with increasing oxygen content. For undoped manganous oxide at 1032°C, these two factors just balance one another in the equations of C. Wagner and of Darken and the *calculated* chemical diffusion coefficient is virtually independent of composition.

Finally, I point out that neither the equations of C. Wagner nor of Darken are being tested by the experiments cited which involve compounds exhibiting wide deviations from stoichiometry. The equation of Darken was derived specifically for interdiffusion of metals while that of C. Wagner was derived for ionic crystals under the assumption of non-interacting particles-- a condition which is not met in manganous or ferrous oxide.

OXYGEN TRANSPORT PROPERTIES OF RARE EARTH OXIDES

LeRoy Eyring

Department of Chemistry

Arizona State University

The usefulness of the rare earth oxides as model systems for studies which include material transport is pointed out. The results so far obtained in kinetics studies of these systems are sketched and the whole placed against the emerging understanding of nonmetallic solids having extended defects.

Whatever may be said of present understanding of material transport mechanisms in oxides with low concentrations of defects (and this has recently been reviewed by Birchenall[1]) one must admit that next to nothing is known of these processes in oxides having extended defects. The beginnings of a foundation are being laid for such an understanding in several systems. The results on the rare earth oxides are to be summarized very briefly here.

There is much to recommend the rare earth oxides as the choice for such comprehensive studies. The higher oxides of cerium, praseodymium and terbium, especially praseodymium, exhibit complicated phase relationships including ordered intermediate compounds belonging to an homologous series having the formula $R_n O_{2n-2}$ (n = 4, 7,9, 10, 11, 12 and ∞ for Pr) each of which has a small but measureable range of composition. In addition there are, in the praseodymium oxide system, two genuinely non-stoichiometric regions of wide composition range separated by a narrow miscibility gap [2]. The structures of all these are related to fluorite and bixbyite [3]. There exists also polymorphism in certain of the line phases. Furthermore the praseodymium oxide system and certain regions of the terbium oxide system can be studied at temperatures and oxygen pressures where unusual furnace and crucible materials and gas-buffer systems with their contamination possibilities are not required. Add to this the feasibility of subtle modifications of the system by the admixture of the chemically and structurally related other earths and actinide elements and one sees the practically unlimited possibilities of obtaining experimental information necessary for a comprehensive understanding of solid oxides. The actinide element oxides are analogous in a limited way yielding a set of oxides with related structures encompassing nearly one third of the known elements.

Startling advances into understanding grossly defective specimens have recently resulted from observations using the high resolution electron microscope. It is now clear that in the quenched samples which have been studied concentrations of random point defects are not high. It is likely that the same is true at moderately high temperatures as well. Grossly defective material restructures itself to economize on free energy and new structural features appear. The most studied so far are those oxides adjusting to oxygen deficiency by undergoing crystallographic shear in one dimension or forming block structure with crystallographic shear in two dimensions. In addition there are compounds where the clustering of defects occurs, such as in the Wüstite phase, FeO_{1-x}. As more is learned, other types of structural defects will appear and more suitable classifications will be made.

The premonitory effects of melting and hybrid crystal formation described by Ubbelohde [4], microheterogeneity discussed by Ariya [5], microdomains discussed by J. S. Anderson [6] and microdomains and pseudo-phase formation discussed by Hyde et al, [2], are less direct manifestations of extended defects. It has become clear that the extension of the treatment of materials with point defects to those with extended defects simply taking into account pairwise interactions will not suffice. Rather, new theories based on the new structural, thermodynamic and kinetic information must be devised and developed.

Studies of the kinetics of oxygen transport and phase reactions have been begun on the sesquioxides and other phases of narrow composition range for their own intrinsic interest and as a foundation for studies of nonstoichiometric and other non-classically behaving substances. It shall be the intent of this paper to sketch briefly what has been learned in recent years of the rate of oxygen transport and phase reactions in the rare earth oxides and to place this information against the larger study of the chemistry of materials having extended defects.

A SUMMARY OF RECENT TRANSPORT STUDIES

Some of the more important parameters describing the transport behavior of rare earth oxides together with references to the original experiments are given in Tables I and II.

Heterogeneous reactions have been studied and the reaction mechanism suggested by analysis of the rate of reaction. In these studies a model is assumed and the mathematical consequences determined. A satisfactory fit between the data and the calculated behavior suggests a plausible model.

Phase Boundary Controlled Reactions

If the sample consists of spheres whose reaction rate is controlled by a phase boundary retreating into the sphere at a uniform rate a plot of the cube root of the unreacted remainder against time will be a straight line. In such a reaction nucleation of the new phase must be rapid over the entire surface yielding a spherical phase boundary. Excellent agreement between this model and the reduction rate of PrO_2, $PrO_{1.78}$ and the oxidation rate of $PrO_{1.83}$ has been observed (Table I) [7, 8].

Diffusion Controlled Reactions

A diffusion mechanism including phase boundary movement has been proposed [9] for the oxidation of substoichiometric rare earth sesquioxides although rather than the phase boundary formation itself being rate controlling it is chemical diffusion through the reaction product. If on the other hand the rate controlling step is simply diffusion into the sphere along a radial concentration gradient a plot of the logarithm of the amount of unreacted sample against time gives a curve which falls off rapidly at first while a uniform gradient is being established then becomes linear. Such a mechanism apparently obtains in the reduction of $PrO_{1.71}$. In the case of the reduction of $TbO_{1.82}$ and $TbO_{1.71}$ only the linear behavior at longer times is observed. (Table I) [7]. The chemical diffusion coefficient may be calculated from these experiments if one knows the radius of the particles being studied. In a few cases the rate of oxidation of the solid [10] or liquid [11] rare earth metals have yielded chemical diffusion information from the Wagner equation.

Self Diffusion Studies

More recently studies of self-diffusion of oxygen in rare earth oxides has been undertaken (Table II) employing heterogeneous isotope exchange [12, 13]. The results are treated to yield surface and diffusion parameters.

A sectioning technique was used to obtain diffusion and surface parameters for calcia stabilized zirconia [14]. These results are included in Table II for comparison with the data from the rare earth oxides.

TABLE I - HETEROGENEOUS REACTIONS

Reaction		Rate Controlling Mechanism	Activation Energy (Kcal)	Temperature Range	Oxygen Pressure	D_o ($cm^2/sec \times 10^4$)	D ($1000^oK \times 10^{10}$)	Ref.
$PrO_{2.00}$	(a)	Phase Boundary	50	289-347	vac			8
$PrO_{1.83}$	(b)	" "	26.9	265-305	1 atm			8
$PrO_{1.78}$	(c)	" "	49.6	407-493	vac			7
$PrO_{1.71}$	(d)	Diffusion	42	700-970	vac			7
$*TbO_{1.82}$	(e)	"	41.5	383-471	vac			7
$*TbO_{1.71}$	(f)	"	∿ 40	570-643	vac			7
Sc_2O_3	(g)	Diffusion with a moving boundary	38.30	1106-1298	air	7.72	2.27	9
Y_2O_3	(g)	" "	19.58	1064-1241	air	0.0606	27.6	9
Dy_2O_3	(g)	" "	26.24	1087-1235	air	0.163	5.27	9
Ho_2O_3	(g)	" "	40.53	1050-1274	air	71.8	8.26	9
Er_2O_3	(g)	" "	30.12	1060-1292	air	1.31	0.958	9
Tm_2O_3	(g)	" "	45.56	1015-1235	air	114.	1.97	9
Lu_2O_3	(g)	" "	29.76	1020-1297	air	1.88	1.58	9
Y_2O_3	(h)	Metal Tarnishing	44.50	900-1400	air	14.3	0.37	10
R_2O_3	(h)	" "					∿100(1600°C)	11

*Behavior deviates from diffusion-like mechanism in the early stages of reaction (modified diffusion behavior).

(a) $PrO_{2.00} \rightarrow PrO_{1.83} + 0.085\ O_2$

(b) $PrO_{1.83} + 0.085\ O_2 \rightarrow PrO_{2.00}$

(c) $PrO_{1.78} \rightarrow PrO_{1.71} + 0.035\ O_2$

(d) $PrO_{1.71} \rightarrow PrO_{1.50} + 0.105\ O_2$

(e) $TbO_{1.82} \rightarrow TbO_{1.71} + 0.055\ O_2$

(f) $TbO_{1.71} \rightarrow TbO_{1.50} + 0.105\ O_2$

(g) $RO_{1.50-\delta} + \delta/2\ O_2 = RO_{1.50}$

(h) $R + 3/2\ O_2 = RO_{1.50}$

TABLE II

Oxygen Self-Diffusion	Activation Energy (Kcal)	Temperature Range	D_o (cm^2/sec)	D (1000°K)	Ref.
$Zr_{0.858}Ca_{0.142}O_{1.858}$	31.20	781-1097	1.8×10^{-2}	8.1×10^{-8}	14
Nd_2O_3(99.999%)(a)	31	700-1000	1.3×10^{-4}	7.7×10^{-10}	12
Sm_2O_3(99.9%)(b)	23.5	700-950	9.2×10^{-6}	8.96×10^{-10}	12
Sm_2O_3(99.998%) (b)	21.3	700-1000	6.0×10^{-6}	1.39×10^{-9}	12
Er_2O_3(99.999%)				$\sim 10^{-14}$	12
ι-Pr_7O_{12}(low temp) (c)	8.21	725-930	9×10^{-8}	3.58×10^{-9}	13
ι-Pr_7O_{12}(high temp)(d)	19.05	725-930	2.9×10^{-8}	1.65×10^{-9}	13

(a) Very small oxygen pressure dependence.
(b) Virtually no oxygen pressure dependence.
(c) $D = 1.6 \times 10^{-2} \exp(-27{,}000/RT) f_{II}^{\frac{1}{2}}$. f equals the atom fraction of excess oxygen.
(d) $D = 1.3 \times 10^{-3} \exp(-24{,}000/RT) f_{I}^{\frac{1}{2}}$.

DISCUSSION

The superficial results of the experiments performed so far confirm that oxygen diffusion is relatively rapid being only one or two orders of magnitude smaller than in calcia stabilized zirconia. This is to be contrasted with the cation diffusion which is negligible at the temperatures of these experiments.

The absence of a pressure dependence of the self-diffusion rate in the sesquioxides was surprising and the unusual pressure dependence of self-diffusion for the $Pr_7O_{12+\delta}$ may ultimately be understood in terms of a complex diffusion mechanism.

It comes as no surprise that results obtained from chemical and self-diffusion measurements are not the same. At present there are insufficient data to make any attempt at correlation meaningful.

Although all the data are for compounds which are said to be line phases they all have a measureable range of composition hence their behavior must be considered in terms of complex defect interaction and are therefore the foundation for studies of grossly nonstoichiometric oxides. When rationalized in these terms they shall make their peculiar contribution to full understanding.

The mechanism of formation of transition metal oxides and oxide fluorides which have regularly spaced crystallographic shear planes as structural features have received some attention. B. G. Hyde [15] discusses three mechanisms in the light of recent high resolution electron microscopic studies on the NbO_2F, V_2O_5, MoO_3 and TiO_2 systems. It is apparent from these studies that the nucleation of partial dislocation loops is well advanced at very low levels of nonstoichiometry and that these loops grow longitudinally as they serve as a sink for the low concentration of random vacancies formed in the host lattice as oxygen is removed. However, this mechanism, proposed by Anderson and Hyde [16, 17], cannot provide for the lateral motion of the planes of crystallographic shear needed to secure their order in the final structures. In at least one case this mechanism is seen to operate in conjunction with a cooperative mechanism of cation diffusion by which an entire shear plane may move laterally (perhaps by means of steps or screw dislocations) as had been proposed by Andersson and Wadsley [18].

O'Keeffe [19] has recently reviewed the problems associated with attempting a mechanistic description of diffusion or phase reaction in materials having extended defects. Some simple models are discussed including diffusion of vacancies with nearest neighbor repulsion on a single cubic lattice. The fluorite-related oxides are given as an example.

The consequences of the simple cubic treatment is that the conductivity is drastically reduced when certain restrictions on vacancy environments are taken into account. In the case of no nearest neighbors, a maximum conductivity is reached at a concentration of vacancies of 0.10 (compared to 0.50 for random vacancies, or corresponding higher values at higher temperatures).

O'Keeffe observes that in $Zr_{1-x}Ca_xO_{2-x}$ the conductivity goes through a maximum at $x \simeq 0.12$ at 1000°C and at higher x at higher temperatures. The activation energy for conductivity also increases with x [20]. Also discussed are some problems in the interpretation of atom transport measurements in Wüstite where the extended defects are known to be clusters of rather large size.

Of course the nonstoichiometric rare earth oxides are not shear structures or clusters of the type discussed above. The structural feature which has been proposed to account for the homologous series and range of composition is a string of oxygen vacancies in characteristic crystallographic directions [21, 22]. The detailed mechanism of the formation of this structural feature or the movement of oxygen in structures containing it have not yet been seriously proposed.

It is clear that a satisfactory atomic mechanism of diffusion in such materials as the rare earth oxides must take into account the nature of the defects and expecially their interaction. These may or may not correlate with the structural entity considered as the extended defect but the final results must be consistant with the structural and thermodynamic properties

ACKNOWLEDGEMENT

The United States Atomic Energy Commission has supported much of the work reported here; we are grateful to them.

REFERENCES

1. C. E.Birchenall in *Mass Transport in Oxides*, J. B. Wachtman, Jr. and A. D. Franklin, Eds., N.B.S. Special Publication, 296 (1968) p. 119.
2. B. G. Hyde, D. J. M. Bevan and L. Eyring, Trans. Roy. Soc., London, Ser. A No 1106, 259, 583 (1966).
3. J. O. Sawyer, B. G. Hyde and L. Eyring, Bull. Soc. Chim. Fr., 1190 (1965).
4. A. R. Ubbelohde, J. Chim. Phys. 62, 33 (1966).
5. S. M. Ariya and Yu. G. Popov, J. Gen. Chem., Moscow 32, 2077 (1962).
6. J. S. Anderson, Proc. Chem. Soc., 166 (1964).
7. U. E.Kuntz and L. Eyring, *Kinetics of High Temperature Processes*, W. D. Kingery, Ed., John Wiley and Sons, Inc., New York, (1965) p. 50.
8. B. G.Hyde, E. E. Garver, U. E. Kuntz and L. Eyring, J. Phys. Chem.,69, 1667 (1965).
9. M. F.Berard, C. D. Wirkus, and D. R. Wilder, J. Am. Ceram. Soc., 51, 643 (1968).
10. H. J.Borchardt,J. Inorg. Nucl. Chem., 26, 711 (1964).
11. J. Brett and L. Seigle, J. Less-Common Metals 10, 22 (1966).
12. G. D.Stone, G. R.Weber and L. Eyring, *Mass Transport in Oxides*, J. B. Wachtman, Jr., and A. D. Franklin, Eds., N.B.S. Special Publication 296 (1968) p. 179.
13. G. R. Weber and L. Eyring, Proceedings, International Colloquy of the C.N.R.S. on the Rare Earth Elements, Paris-Grenoble, May 5-10, 1969, in press.
14. L.A.Simpson and R. E. Carter, J. Am. Ceram. Soc. 49, 139 (1966).
15. B. G.Hyde and L. A. Bursill, *Chemistry of Extended Defects in Non-Metallic Solids*, L. Eyring and M. O'Keeffe, Eds., North-Holland, Amsterdam, 1969, in press.
16. J. S. Anderson and B. G.Hyde, Bull. Soc. Chim. Fr., 1215 (1965).
17. J. S. Anderson and B. G. Hyde, J. Phys. Chem. Solids, 28, 1393 (1967).
18. S. Andersson and A. D. Wadsley, Nature 211, 581 (1966).
19. M. O'Keeffe, *Chemistry of Extended Defects in Non-Metallic Solids*, L. Eyring and M. O'Keeffe, Eds., North-Holland, Amsterdam, 1969, in press.
20. R. E.Carter and W. L. Roth, in *Electromotive Force Measurements*

in High Temperature Systems, C. B. Alcock, Ed., American Elsevier, New York.

21. B. G. Hyde and L.Eyring, *Rare Earth Research III*, L. Eyring, Ed., Gordon and Breach, New York, (1965) p. 623.
22. B. G.Hyde, D. J. M. Bevan and L. Eyring, *International Conference on Electron Diffraction and Crystal Defects*, Melbourne, II C-4,(1965), published and distributed for the Australian Academy of Science by Pergamon Press (1966).

THE ROLE OF CARBON IN REACTIONS IN THE U-C-O-N SYSTEM

T. B. Lindemer, J. M. Leitnaker, and K. E. Spear

Metals and Ceramics Division, Oak Ridge National Lab.

Oak Ridge, Tennessee

Carbon plays a vital role in reactions in the U-C-O-N system. This has been demonstrated by extensive studies of the kinetics and thermodynamics of reactions in the system from 1400 to 2000°C, including those of UO_2 with carbon, UC with nitrogen, and UC_2 with nitrogen. Carbon can act either as a reductant or can control the reaction rate. It can control the rate of reactions either by diffusion processes, by the distribution of free carbon within the solid reactants, by the formation of compounds more rich in carbon, or by the formation of a non-equilibrium type of free carbon. In all cases, the role of carbon is dominant.

INTRODUCTION

The major component of future high-performance nuclear fuels will be one or more of the compounds or solid solutions in the U-C-O-N system. There are two major areas of technical interest in this system: the fuel cycle and the chemical stability. Several typical chemical reactions have been studied between 1400 and 2000°C to determine basic kinetic and thermodynamic properties: the reaction of UO_2 with carbon, UC with nitrogen, UN with graphite, and UC_2 with nitrogen.

Carbon plays an important role in the kinetics of these reactions. It acts as a reductant with UO_2. It controls the reaction rate in UC_2-N_2 reactions by a diffusion process and by its initial distribution as free carbon. Its rejection in the UC-N_2 reaction leads to the formation of UC_2. Additionally, carbon can precipitate in a non-equilibrium form of free carbon. This discussion will review several studies that permit these conclusions.

EXPERIMENTAL TECHNIQUES

The general experimental methods used in the kinetic studies are described below, while techniques particular to a specific study are described in the discussion of that study. In many of the studies, the uranium compound was available as fully dense spheres with diameters of 500 μm. These were reacted for different times, and the outer diameters of the sperically symmetrical phases involved in the reaction were determined on equatorial cross sections of 25 spheres at each time by a metallographic method [1]. The principal data obtained were r_o, the radius of the sphere at zero time, and r, the radius of the reactant core at time t. These measurements were used in models that exist in the literature for reactions in solid spheres. The model for a reaction controlled by diffusion through the product layer is [2]:

$$r_o^2 \left[1 - \left(\frac{r}{r_o}\right)^2 \left(3 - 2\frac{r}{r_o}\right)\right] = k_D t, \quad (1)$$

in which k_D is in units of square centimeters per second. A model for control by a mechanism at the interface between the core and the product takes the form [3]:

$$r_o - r = k_I t, \quad (2)$$

in which k_I is in units of centimeters per second. A model for control at the surface is [4]:

$$r_o \left[1 - \left(\frac{r}{r_o}\right)^3\right] = 3\, k_S t, \quad (3)$$

in which k_S is in units of centimeters per second. A plot of the left-hand side of Eqs. (1), (2), and (3), versus time for a series of experiments at a given temperature and incremental time intervals will often exhibit linearity for the left-hand-side function for one of the models; the plots for the other two models will be nonlinear. These characteristics provide one means for eliminating certain mechanisms from consideration as rate controlling.

Other techniques were used to obtain kinetic information from the thermodynamic studies. The intermediate stages of a reaction that eventually attained equilibrium were followed by the techniques of chemical analysis and Debye-Scherrer x-ray powder diffraction. In one case, the thermodynamic information developed from these studies could be used to show the existence of a non-equilibrium but kinetically stable form of carbon.

UO_2-C REACTION

The reaction of UO_2 spheres and graphite was studied between 1400 and 1750°C [5]. This reaction is important because UO_2 is a starting material commonly used for making other nuclear fuels and because carbon is generally used as a reductant in the process. The UO_2 and graphite were reacted in a rotating tungsten crucible to maintain dynamic contact between graphite and the spheres. The reaction yielded two quite different morphologies, depending on initial heating conditions. These results are illustrated in Fig. 1. The configuration in Fig. 1(b), in which the UO_2 maintained contact with the surface of the sphere until the reaction was nearly complete. was obtained by heating the reactants to temperature at a rate greater than 50°C/min. The other configuration, which has a spherically symmetrical layer of UC_2, was obtained by a heating rate of 25°C/min or less. The initial heating rate apparently affects the initial nucleation and growth of the UC_2 on the surface of the sphere. The conversion rate for the nonuniform reaction was 2 to 5 times faster than that for the uniform reaction, a fact that has potential importance in the design of a process for the production of nuclear fuel.

The overall chemical conversion must occur by some sequences of mechanisms, all of which are either diffusion processes or interface reactions. There are illustrated schematically in Fig. 2. An example of an interface reaction at the surface of the sphere is the solution of graphite into UC_2. This is followed by diffusion of the carbon through the UC_2 to the interface between UC_2 and UO_2, where reaction produces UC_2 from UO_2. In the symmetrical reaction, the oxygen released by the reaction is postulated to go into solid solution in the UC_2, where it diffuses to the surface and reacts with graphite to form CO.

The mechanism controlling the symmetrical configuration of Fig. 2 was deduced by comparing the results from the symmetrical and nonsymmetrical reactions. This comparison indicated that the diffusion of oxygen through the UC_2 was the rate-controlling step [5].

The cause of the faster reaction in the nonsymmetrical configuration was also deduced [5]. The oxygen released at the interface between UO_2 and UC_2 diffuses through the UO_2 neck to the surface of the sphere, where it reacts with carbon to form CO. This process must be faster than the diffusion of oxygen through the UC_2, and the conversion is therefore more rapid. The effect of grain size is not believed to change the rate-controlling steps observed under the conditions presented here.

Quantitative data was obtained for the uniform reaction by the metallographic technique described earlier. The average measurements for r and r_o at each time were used in the model for control

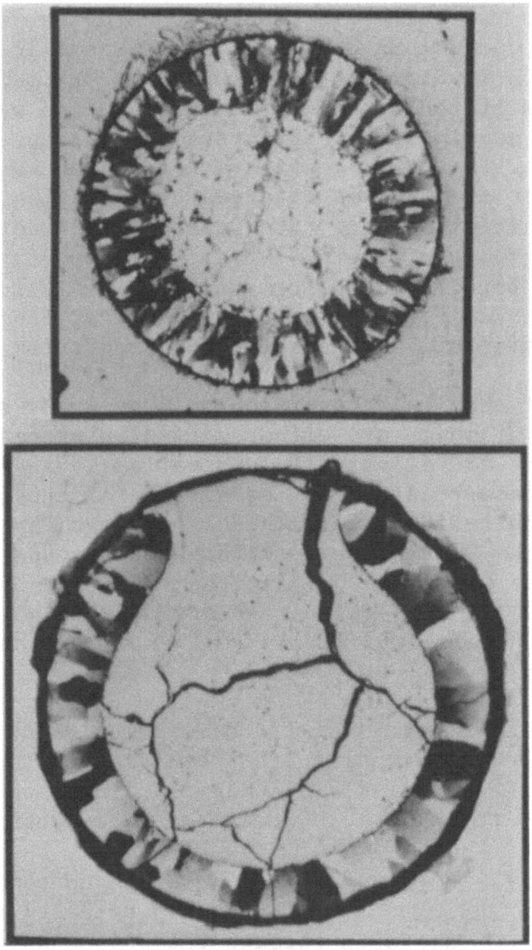

Fig. 1. Symmetrical (a) and Nonsymmetrical (b) Configurations Observed in the Reaction of UO_2 and Carbon. (a) 1700°C, 60 min. (b) 1750°C, 90 min. Cracks result from quenching at the end of the experiment. UO_2 is the core phase, UC_2 is the outer phase, and some graphite particles adhere to the outer surface. 150x. Etchant: 1:1:1 solution of water, acetic, and nitric acids.

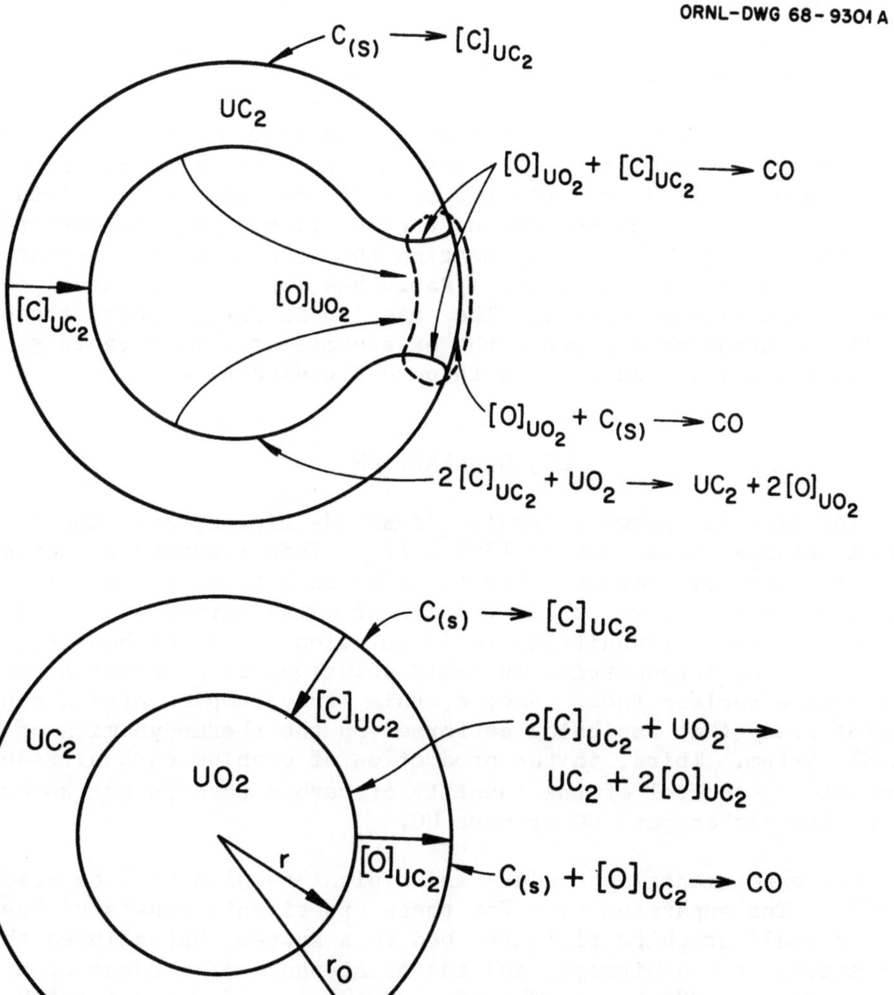

Fig. 2. Schematic Representation of Symmetrical (Lower) and Nonsymmetrical (Upper) Configurations Observed in the Reaction of the UO_2 and Carbon. The proposed mechanisms occurring during the reaction are shown.

by diffusion in spheres. The slope, k_D, of the plot at each temperature was fitted to the Arrhenius relation to obtain the activation energy, 90 kcal/mol.

The methods and results of this investigation compare most closely with those of Ainsley et al. [6]. They studied the reaction of pelletized powder mixtures of UO_2 and carbon under vacuum from 1300 to 1550°C and concluded that the diffusion of either carbon or oxygen was rate controlling. They obtained an activation energy of 92 kcal/mol for the first-order rate constant.

In summary, potentially useful information has been found in the study of the reaction of UO_2 and carbon. For example, the conversion process is 2 to 5 times faster if contact can be maintained between the UO_2 and carbon during the reaction. For microspheres, this could be accomplished by heating the reactants to temperatures at a minimum rate of 50°C/min. (Rapid heating to temperature seems to be a determining factor.) The reaction is faster under these conditions, apparently because the oxygen migrates more rapidly through the oxide than it does through the dicarbide.

UC_2-N_2 REACTION

The reaction between uranium dicarbide microspheres and nitrogen was studied from 1500 to 1700°C [7]. This reaction was considered for several reasons. First, it belongs to a general class of reactions that involve the conversion of metal carbides with nitrogen to produce a carbonitride solid solution. This carbonitride, with plutonium carbonitride in solid solution, is a potential high-performance nuclear fuel. Second, this study complemented a concurrent study that was being performed on the thermodynamics of the U(C,N) system. Third, in the production of uranium carbonitride from oxides, control of the quantity of carbon used is not as critical as for either pure UN or pure UC.

The experimental procedure and typical results will be described briefly. The apparatus used for these experiments consisted basically of a small graphite fluidized bed in a system that allowed the pre-heating of the nitrogen and the maintenance of a constant nitrogen pressure of 380 torr. The UC_2 spheres were dropped into the bed at zero time; about 200 to 1000 spheres were removed with a carbon bucket after each of eight equal-time increments. These spheres were examined by the technique described earlier [1] to obtain the conversion as a function of time.

A representative particle observed for each of the different times is shown in Fig. 3. The core consists of uranium dicarbide, the intermediate layer is uranium carbonitride, and the outer layer and dark precipitates are free carbon. The precipitates formed on

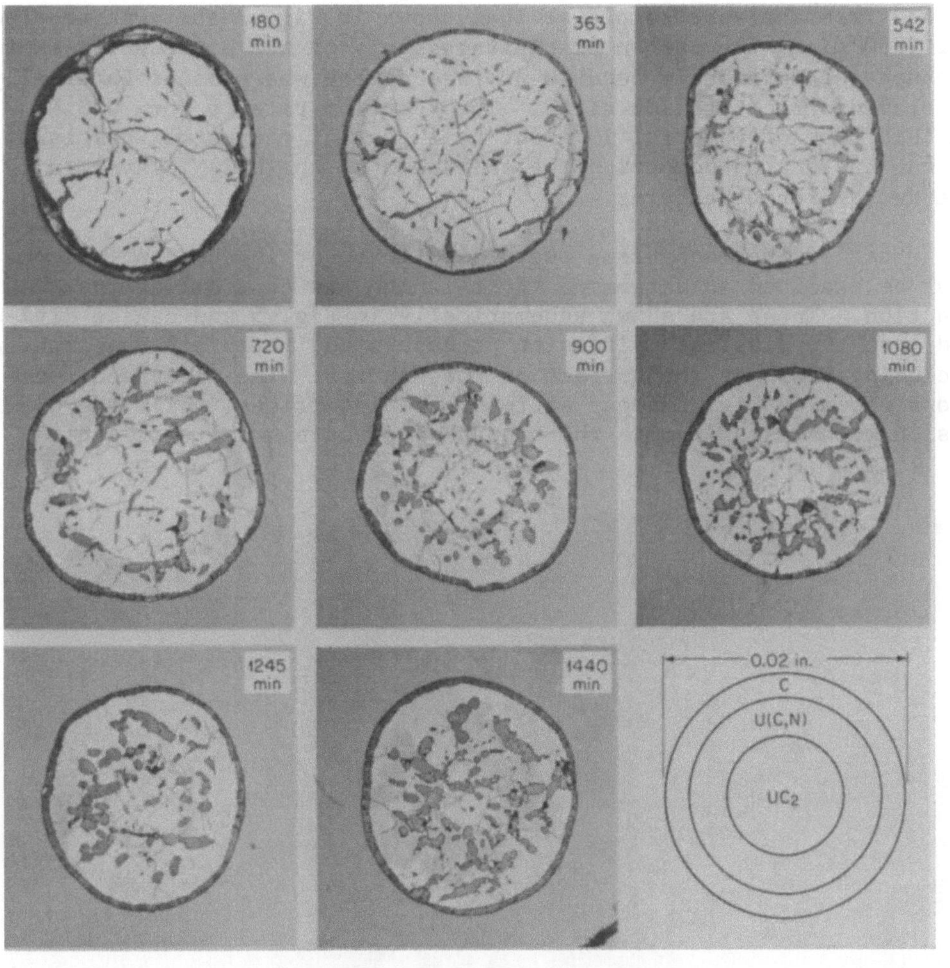

Fig. 3. Near-Equatorial cross section of approximately 500-µm-OD UC_2 spheres reacted with nitrogen at 1700°C for the indicated times. The outer layer and internal precipitates are free carbon, the core is UC_2, and the intermediate layer is U(C,N). 150x. Etchant: 1:1:1 solution of water, acetic, and nitric acids.

free carbon that was present in the original UC_2 microspheres to the extent of about 1 vol %.

A particle was occasionally observed that contained little or no internal free carbon, as shown in Fig. 4. This sphere was obtained from the same sample as that shown in Fig. 3 that was reacted for 1080 min. The carbon precipitates only on the outer surface of the sphere, apparently because no free carbon was present internally to provide precipitation sites. The reaction rate to form U(C,N) in this type of material is about half that of the normal particle, as can be seen by comparing the thickness of the U(C,N) layer in the two different spheres.

Each of the mechanisms postulated to occur in the conversion must be examined to determine if it is the one controlling the reaction. These are shown schematically in Fig. 5 and can be divided into two classes: diffusion processes and interface reactions, including those at the surface of the spheres. The interface reactions shown in Fig. 5 occur at two-phase interfaces, while the diffusion mechanisms involve the transport of carbon and nitrogen in

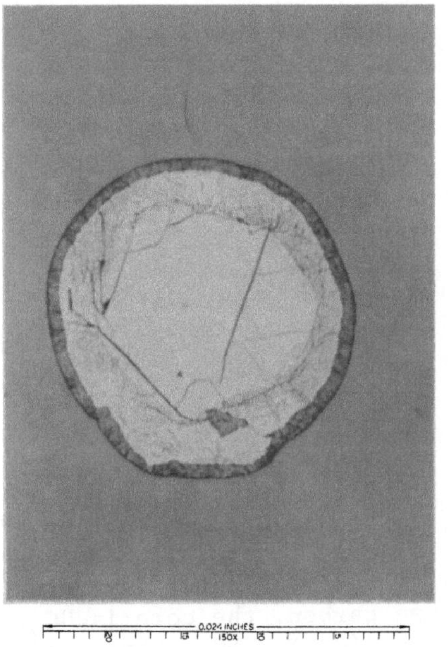

Fig. 4. Cross section of UC_2 sphere that initially contained almost no free carbon. This sphere was from the same sample as that in Fig. 3 that was reacted for 1080 min. Note the smaller amount of U(C,N) present. 150x. Etchant: 1:1:1 solution of water, acetic, and nitric acids.

the system. The interfacial mechanisms are proposed to be the following:*

A. N_2 to nitrogen in the outer layer of free carbon (nitrogen carbon).
B. $Nitrogen_{carbon}$ to $[N]_{U(C,N)}$ at the interface between U(C,N) and the outer layer of free carbon.
C. $Nitrogen_{carbon}$ to $[N]_{U(C,N)}$ at all interfaces between free carbon and both U(C,N) and UC_2.
D. [C] to $C_{(free)}$ at all interfaces between free carbon and both U(C,N) and UC_2.
E. $[N]_{U(C,N)}$ reacting with UC_2 to form U(C,N) and [C] at the interface between U(C,N) and UC_2.

The diffusion mechanisms are postulated to be these:

F. [C] diffusing from the interface between UC_2 and U(C,N) through the U(C,N) or UC_2 to adjacent free carbon sites.
G. $[N]_{U(C,N)}$ diffusing from the sphere surface through the U(C,N) to the interface between UC_2 and U(C,N).
H. $[N]_{U(C,N)}$ diffusing to the interface between UC_2 and U(C,N) from adjacent interfaces of free carbon and U(C,N).
I Nitrogen diffusing through the thin outer layer of free carbon.

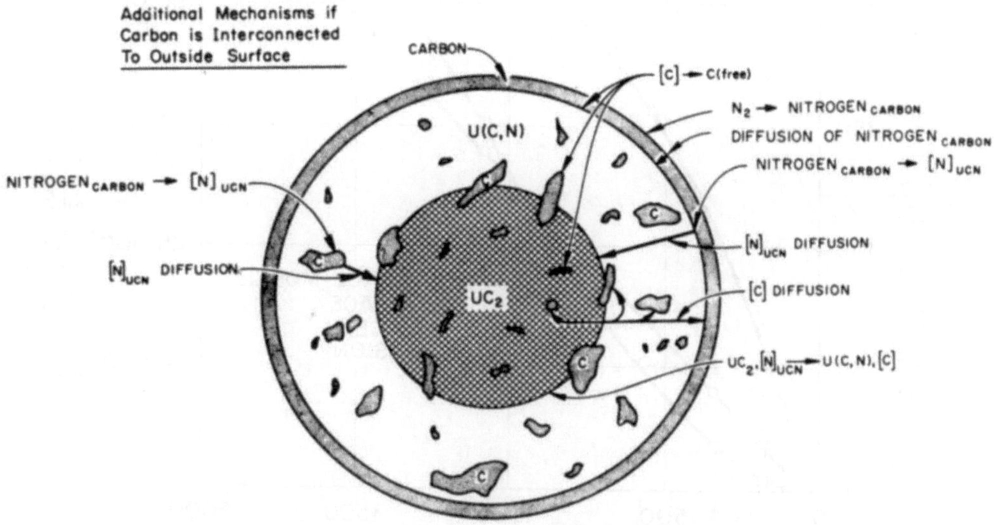

Fig. 5. Schematic representation of the phases present in the reaction of UC_2 and nitrogen. The proposed mechanisms occurring during the reaction are also shown.

---

* Brachets indicate a species in solid solution.

Mechanisms C and H are rather unique in that one of them may be rate controlling if, and only if, the free carbon served as an interconnected path for N_2 to diffuse from the surface to the interior of the sphere.

The kinetic models used for the analysis of the results for conversion versus time permitted the elimination of many of these possible mechanisms as the ones controlling the conversion process. Plots of the functions of Eqs. (1), (2) and (3) are shown in Fig. 6, which indicates that the mechanism controlling the reaction occurs at the interface between U(C,N) and UC_2. Only mechanisms E, F, and H can occur in this region. It was concluded that mechanism F, the diffusion of [C] from the interface between UC_2 and U(C,N) to the free carbon sites closest to the UC_2-U(C,N) interface, was controlling during the conversion of UC_2 [7]. For the spheres shown in Fig. 3, these sites are at or near the interface. For spheres lacking the internal sites, such as shown in Fig. 4, the nearest free carbon site is the surface of the sphere.

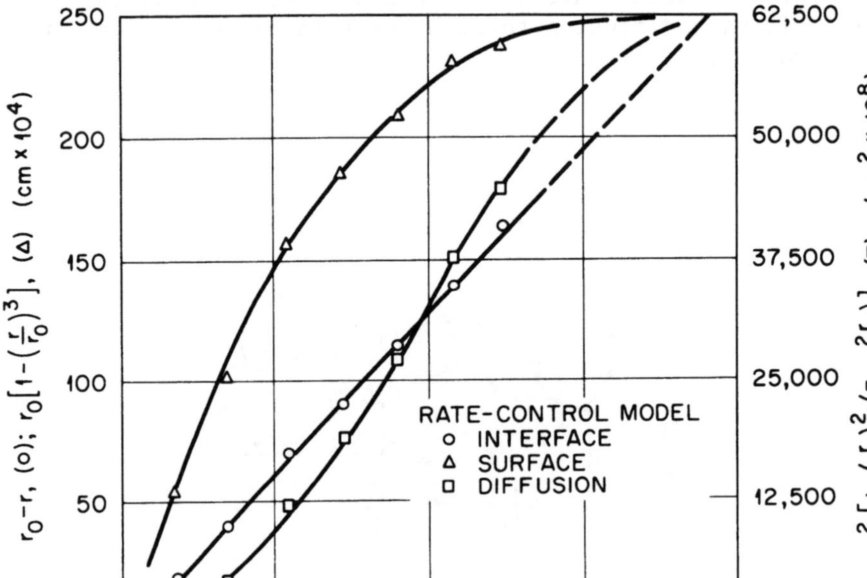

Fig. 6. The time- dependent plots of the function in Eqs. (1), (2), and (3) for the reaction at 1700°C of Nitrogen at 380 Torr with UC_2 containing free carbon (Run A5164-54).

The uniform internal distribution of free carbon in the microsphere therefore affected the reaction rate [7]. This distribution maintains an approximately constant carbon-to-carbon separation during the entire conversion process. This permits an approximately constant flux of carbon via Fick's first law. The flux, in cubic centimeters of carbon per square centimeter of interface between U(C,N) and UC_2 per unit of time, limits the rate of conversion of UC_2 to the region near the interface of UC_2 and U(C,N). The reaction would, therefore, follow Eq. (2) even though the diffusion of carbon was controlling the reaction. The activation energy for the diffusion of carbon was obtained for this system from the Arrhenius relation and the least-square slope, k_I, of plots such as that shown in Fig. 6. The slopes at 1500, 1600 and 1700°C and a N_2 pressure of 380 torr were fitted to the Arrhenius equation to give

$$k_I = 95.6 \exp [-78{,}000/RT] \text{ cm/s.} \qquad (4)$$

Carbon is consequently involved in two important effects on the reaction of UC_2 and N_2. As has been shown, one is control of the reaction by the diffusion of carbon; the second effect results from the number and distribution of sites available for the precipitation of carbon. Increasing the number of sites was observed to result in an increase in the reaction rate.

UC-N_2 REACTION

Several observations were made about the kinetics of the reaction between UC and nitrogen during a thermodynamic study of the UC-UN solid solution from 1500 to 1900°C [8]. The -325 mesh, arc-melted UC was exposed to a given pressure of nitrogen in a graphite furnace and then analyzed by chemical and Debye-Scherrer techniques. The results of this study are understood more readily by reference to the U-C-N quasi-equilibrium diagram of Fig. 7. (U_2C_3 is not shown because the kinetics for its formation are unfavorable in these experimental situations.)

The reaction proceeded in at least three stages. In the first and most rapid stage, the UC reacted to form the high-carbon U(C,N) solution that is in equilibrium with UC_2. The carbon rejected when the U(C,N) was formed from UC reacted with the adjacent UC to form UC_2, as shown in Fig.8,but formed little or no free carbon. The second stage began after the UC had reacted to form the U(C,N) and UC_2. The UC_2 then reacted with nitrogen to form additional high-carbon U(C,N) and free carbon. The third stage of the reaction began after the UC_2 had been converted. The U(C,N) reacted with nitrogen to reject more free carbon and slowly formed the nitrogen-rich U(C,N) composition in equilibrium with carbon and nitrogen. This third stage appeared to be controlled by the diffusion of the nonmetal atoms; the x-ray results, which are representative of the surface of the material, always indicated more combined nitrogen in the U(C,N) than did the results of the chemical analysis.

These results seem to fit the pattern established in the previous experiments. Again, a primary factor in the reaction appears to be the precipitation of free carbon. It is considerably slower than the formation of the UC_2 in the first stage of the reaction. The stepwise nature of the reaction is also revealing. It is apparent that the usual practice of writing a chemical equation for the overall reaction, $UC + x/2\ N_2 \rightarrow U(C_{1-x}N_x) + xC$, reveals little about the way the reaction actually takes place or about the intermediate phases that may form.

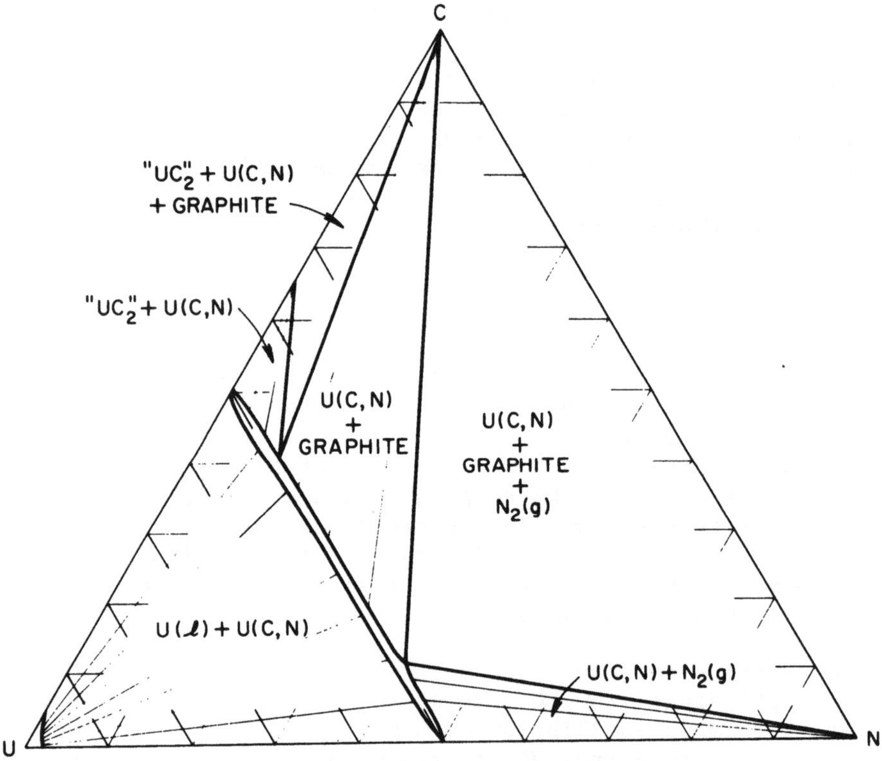

Fig. 7. The U-C-N quasi-equilibrium diagram at 1700°C and a pressure of 0.5 atm. U_2C_3 is not shown because its formation is kinetically difficult in the present experiments.

ACTIVE CARBON

Another interesting phenomenon is the form of the free carbon that precipitates in some of these reactions between the carbide and nitrogen. The carbon can appear in a form that is chemically more active than graphite. Whether it does or not depends on the reactants and the reaction conditions. Active carbon has been observed in both the U-C-N and V-C-N systems.

The existence of active carbon was postulated from the results of a thermodynamic study of two reactions in the U-C-N system [9]. The starting materials in the first reaction were UC and N_2. These reacted to produce the U(C,N) solid solution and free carbon. The second reaction involved UN and graphite and produced U(C,N) and N_2. The reactions were performed under identical conditions of temperature and nitrogen pressure and, therefore, would be expected to give the same thermodynamic information at equilibrium.

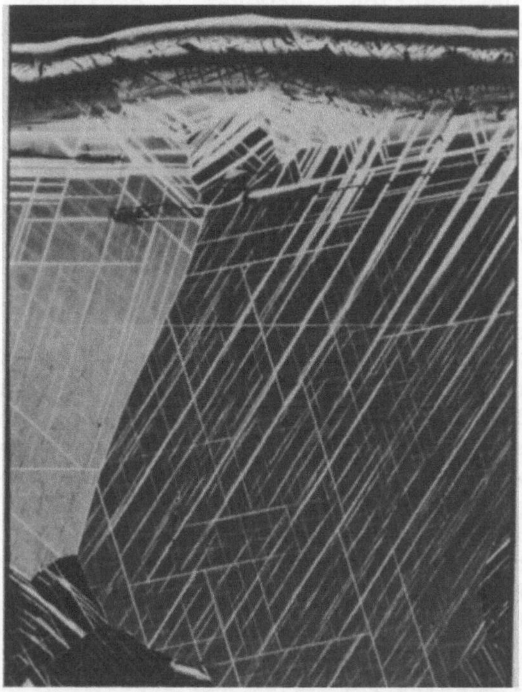

Fig. 8. The Microstructure observed furing the first stage of the reaction of UC and nitrogen. The outer phase is U(C,N), the irregular phase layer is a UC_2-U(C,N) mixture, and the inner phase is essentially UC. Etchant: 1:1:1 solution of water, acetic acids. 160x.

The results of the two reactions were not always identical, however. The data for the equilibrium reaction

$$UC_{(s)} + 0.5\ N_{2(g)} \rightleftarrows UN_{(s)} + C_{(s)}, \qquad (5)$$

were used to evaluate the function J^o_{298}, where

$$J^o_{298} = T\left[-R\,\ln\left(\frac{X_{UN}}{X_{UC}\ P^{0.5}_{N_2}}\right) - \Delta\left(\frac{G^o_T - H^o_o}{T}\right)\right] \qquad (6)$$

This function* is plotted in Fig. 9 versus the mole fraction of UN.

The results at 1700, 1800, and 1900°C for the reactions UC plus N_2 and UN plus graphite all give statistically the same results for the equilibrium reaction. However, a deviation was observed at 1500°C. While the reaction of UN and graphite gives the same results as those at higher temperatures, the results for UC and N_2 give a statistically significant, lower value. At 1500°C, the carbon produced from the reaction of UC and N_2 appears to be in a state that is more active than graphite. This more active form of carbon causes the observed difference of about 500 cal in J^o_{298}. This difference is approximately the enthalpy of the reaction of graphite to form active carbon.

The evidence for active carbon has been substantiated by studies of vanadium carbonitride solutions [10]. The reactions studied here were the same as in the U-C-N system, except that vanadium was substituted for uranium. The reactions were carried out simultaneously in a twin crucible made from graphite. The principal analytical result obtained was the lattice parameter of the carbonitride solid solution, as shown in Fig. 10. Note that if the carbon that precipitates from the reaction of VC and N_2 is more active than graphite, then thermodynamic considerations predict that the carbonitride produced will be richer in carbon than the carbonitride from the reaction of VN and graphite. Figure 10 shows that the reaction of VC plus N_2 always resulted in a lattice parameter for the carbonitride higher than that of the reaction of VN plus graphite. The observed difference in the lattice parameter was experimentally shown not to

---

* If the UC-UN solid solution were ideal and if the activity of carbon were unity, the quantity J^o_{298} would be identical to the heat of reaction, ΔH^o_{298}, for Reaction (5). Such a quantity serves as a useful means of comparing results at different temperatures since $RT\ln\gamma$ at a particular composition will change slowly with temperature and will not change at all for an ideal or regular solution.

result from errors in determinations of the lattice parameter, the kinetics of the addition or removal of carbon or nitrogen, differences in impurity levels, or temperature gradients across the twin crucible. The conclusion is that the precipitated free carbon is in a more active state than graphite.

These observations are supported by other experiments. Lewis, Frisch, and Margrave [11] observed a difference of 220 cal/mol in the combustion of pyrolytic carbon annealed at 2200 and 3600°C. Browning and Emmett [12] made equilibrium measurements in the system carbon-methane-hydrogen and found that carbon obtained from the decomposition of Fe_3C had a free energy about 300 cal/mol higher than that of graphite.

The effect of active carbon could be important in other systems.

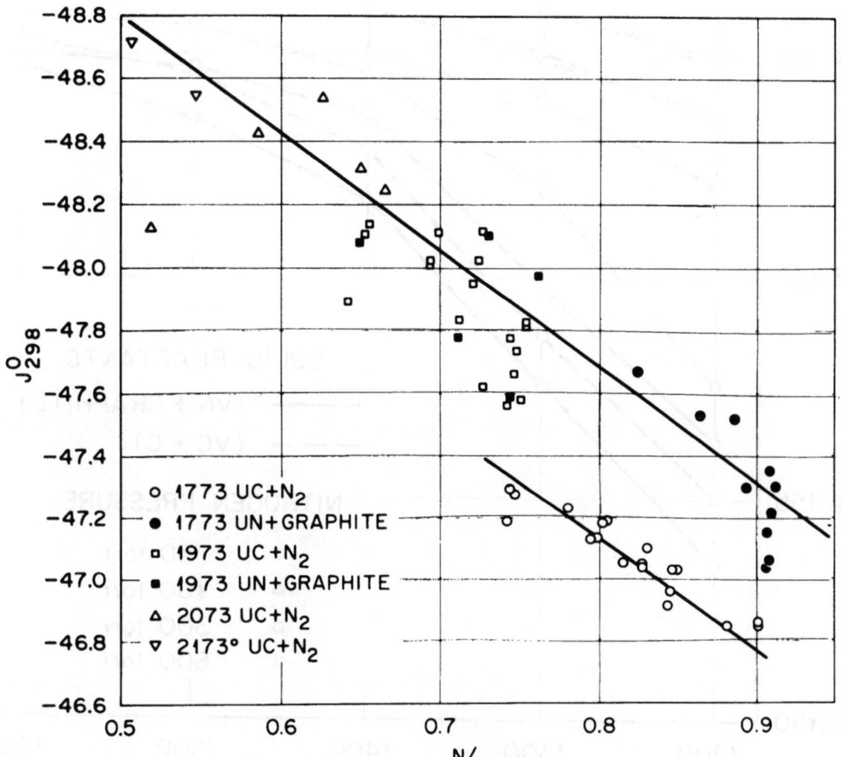

Fig. 9. The quantity J^o_{298} [Eq. (6)] as a function of the ratio of N:(C+N) for the specified reactions in the U-C-N system.

In the study of metal carbide vaporization reactions that produce free carbon, the enthalpy of formation calculated by assuming the free carbon to be graphite may be as much as 0.5 kcal/mol of carbon too negative unless the carbon could precipitate on preexisting graphite particles. In a peritectoid decomposition of solids that gives free carbon as a product, the decomposition temperature would be lowered if active carbon were precipitated.

CONCLUSIONS

The results of the studies reviewed in this paper permit the following general conclusions about specific reactions in the U-C-O-N system between 1400 and 2000°C:

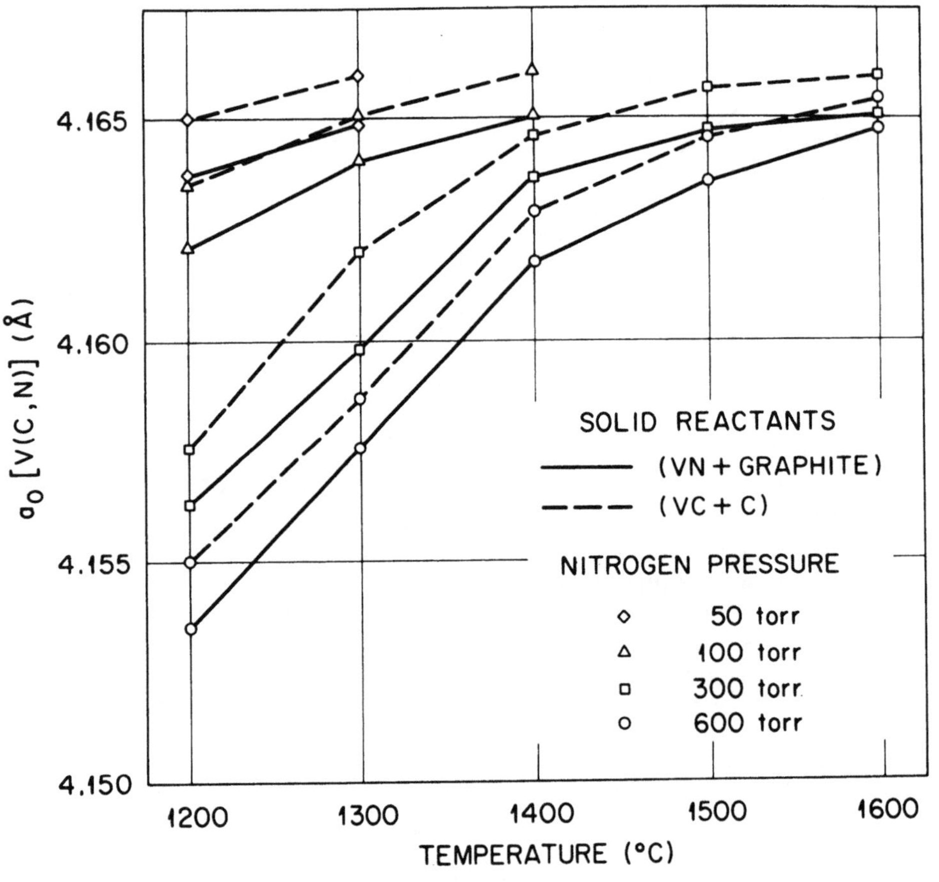

Fig. 10. The lattice parameter of V(C,N) obtained by reacting either VN and graphite or VC and nitrogen at the indicated temperatures and pressures.

1. The primary rate-controlling mechanism in the reaction of UO_2 with graphite, UC with N_2, and UC_2 with N_2 was the diffusion of a non-metal atom through one of the uranium compounds in the system.

2. Carbon has a significant role in these reactions. The observed effects are listed below.

(a) Diffusion of carbon controls during the conversion of UC_2 by N_2 to U(C,N).
(b) An initial distribution of free carbon in the UC_2 increased the rate of the reaction of UC_2 with N_2.
(c) Carbon rejected in the first stage of the reaction of UC and N_2 combines with unreacted UC to form UC_2; free carbon appears only in the later stages of this reaction.
(d) Below about 1600°C, the free carbon produced by these reactions can be thermodynamically more active than graphite. This behavior was shown to affect the results of thermodynamic and lattice parameter studies.

3. The physical configuration of the components in a system can have significant effects on the reaction rate. The following effects were noted:
(a) In the reaction of graphite with UO_2 spheres, the reaction rate was 2 to 5 times faster when the UO_2 maintained contact with the surface of the sphere.
(b) In the reaction of UC_2 with N_2, the initial distribution of free carbon in the UC_2 was observed to affect the rate by as much as a factor of 20.

4. The reaction to the final equilibrium product may occur in several stages, as was observed in the reaction of UC with N_2. In this reaction, these stages involved the appearance and subsequent reaction of intermediate phases or compositions.

ACKNOWLEDGMENT

This research was sponsored by the U. S. Atomic Energy Commission under contract with the Union Carbide Corporation.

REFERENCES

1. T. B. Lindemer and M. D. Allen, ASM (Am. Soc. Metals) Trans. Quart. 62(2), 547-48 (1969).
2. R.E.Carter, J. Chem. Phys. 35(3), 1137-38 (1961).
3. W. M. McKewan, Trans. Met. Soc. AIME 218, 2-6 (1960).
4. S. Yagi and D. Kunii, *Fifth Symposium on Combustion*, pp. 231-44, Reinhold Publishing Corp., New York, 1955.

5. T. B. Lindemer, M.D. Allen, and J. M. Leitnaker, J. Am. Ceram. Soc. 52(5), 233-37 (1969).
6. R. Ainsley, B. R. Harder, N. Hodge, R. G. Sowden, D. B. White, and D. C. Wood, *New Nuclear Materials Including Non-metallic Fuels, Proc. Conf. Prague,* pp. 349-376, International Atomic Energy Agency, Vienna, 1963.
7. T. B. Lindemer, M. D. Allen, and J. M. Leitnaker, to be submitted to the J. Am. Ceram. Soc.
8. J. M. Leitnaker, T. B. Lindemer, and C. M. Fitzpatrick, to be submitted to the J. Am. Ceram. Soc.
9. J. M. Leitnaker, *Thermodynamics of Nuclear Materials, 1967,* pp. 317-330, International Atomic Energy Agency, Vienna, 1968.
10. K. E. Spear and J. M. Leitnaker, J.Am. Ceram. Soc. 52(5), 257-62 (1969).
11. D. C. Lewis, M. A. Frisch, and J. L. Margrave, Carbon 2, 431-32 (1965).
12. L. C. Browning and P. H. Emmett, J. Am. Chem. Soc. 73, 581-88 (1951).

DISCUSSION OF PAPER

R. Benz: The reaction $2UC_2 = U_2C_3 + C$ has been reported to occur at temperatures of about 1500°C and below. In your work with U-C-O and U-C-N phases, was any observation made that might indicate a dependence of the lower limiting temperature of stability of the tetragonal UC_2 phase on impurity content?

The U_2C_3 phase is not shown in the illustrated isotherm of the U-C-N ternary phase diagram. Was this phase seen in the reaction products of your experiments.

T. B. Lindemer: We have no direct evidence on the effect of nitrogen on the stability of UC_2, however, indirect evidence indicates that UC_2 can dissolve about 1-2 at.% of nitrogen.

In a dynamic experiment of the type we have carried out the U_2C_3 phase appears to be very difficult to form, and we almost never observed it.

KINETICS OF THE OXIDATION OF LIQUID COPPER AND THE EFFECTS OF INTERFACIAL SILICA

C. G. Glen and F. D. Richardson

Nuffield Research Group in Extraction Metallurgy

Imperial College, London

It has been found that the oxidation of levitated droplets of liquid copper by mixtures of oxygen and nitrogen is impeded by the presence of silica films on the metal surface; these can be formed when as little as 2 ppm of silica is present in the metal. When these films are removed by evaporation or by HF, the rates of oxidation are much increased and the reaction becomes controlled by mass transfer in the gas phase at all temperatures between 1100° and 1600°C. The silica films do not impede vaporisation of copper.

The spreading of silica on liquid copper has been studied in relation to its effect on the kinetics of oxidation, and the effects of adding CaO, Al_2O_3 and ZrO_2 have been investigated.

Measurements made with nickel droplets exposed to mixtures of N_2+O_2 at temperatures of 1500° and 1700°C suggest that the 4 ppm silicon present in the specpure metal had no effect on the oxidation rate because the silica evaporated, but that added silica powder retarded oxidation.

INTRODUCTION

In view of the possibilities which are now emerging, of refining metals at high speeds by dispersing them as sprays, a study has been made of reactions between drops of metal and flowing gases.

The drops have been treated and levitated simultaneously by RF induction. Cold reacting gases have been passed around the drops so as to keep their temperatures at appropriate values, and the kinetics of reaction have been studied under these conditions.

EXPERIMENTAL

Apparatus

For most of the experiments the drops of metal (0.25 to 0.50 gm) were levitated in vertical pyrex or quartz tubes 9 and 12 mm internal diameter. Gas flow was downwards and the detailed arrangements were very similar to those used previously by Toop and Richardson [1]. A typical arrangement is shown in Fig. 1. For experiments in which oxide powders were injected into the gas stream, the apparatus was similar to that used by Distin et al [2], the gas inlet tube being 12 mm in diameter. The levitation coil was energised from a 450 kc 15 kva Radyne generator. At the end of each experiment a copper quenching block was pushed up the reaction tube and brought into contact with the levitated specimen. The generator was then switched off.

Materials

Cylinder gases were used. All were dried with silica gel and magnesium perchlorate. Oxygen was removed from nitrogen and argon

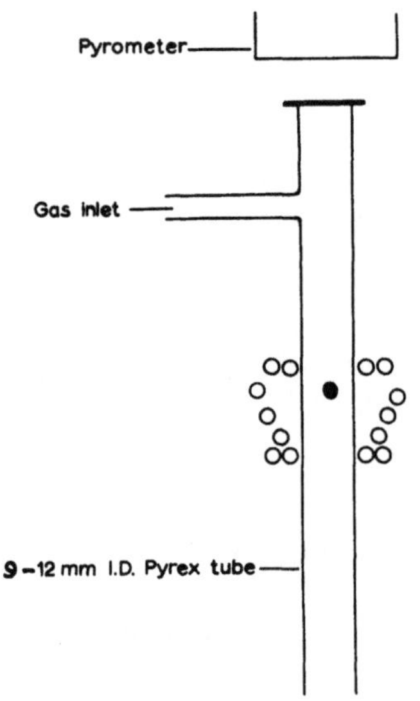

Fig. 1. Typical levitation apparatus

by passing the gases through copper gauze at 500°C. Mixtures were made by means of capillary flowmeters and checked by continuous emf analysis in the case of mixtures of oxygen and nitrogen.

The metals were of Specpure standard.

Temperature Measurement

It is not possible to probe levitated drops with thermocouples so temperatures were measured by radiation pyrometry. Continuous records were obtained from a Land partial radiation pyrometer (QSH 100/5) which was coupled to a potentiometric recorder. It worked on a wavelength of 0.9μ and was modified to make the main lens aperture 0.5 in. in diameter; when focussed at 5 in from the front lens, the minimum target was 0.06 in. in diameter. For the wavelengths on which this pyrometer operated, special emissivity measurements had to be made.

Analysis

Oxygen in the metals was determined by reduction in hydrogen, the loss in weight being measured [1].

Procedure

The normal procedure was to levitate the drop and bring to temperature in N_2 flowing at the same rate as the gases used for the subsequent reaction. The reacting gas mixture was then switched into the gas inlet of Fig. 1 and the starting time noted. The experiment was terminated by switching off the power to the generator and catching the sample in a copper mould or on a quenching block

RESULTS

Emissivities

The emissivities were measured on samples of 100 gm of metal held in an alumina crucible 2 cm in diameter, with a reentrant bottom for accommodating a Pt-Pt.Rh thermocouple. The crucible was filled to the brim and heated by the same RF generator as used for the levitation work. The copper and nickel, containing various concentrations of oxygen,were held in a stream of purified argon and the Land radiation pyrometer temperatures were obtained as a function of thermocouple temperatures. The results obtained are shown in Figs. 2 and 3. The emissivities are plotted in Fig. 4; the

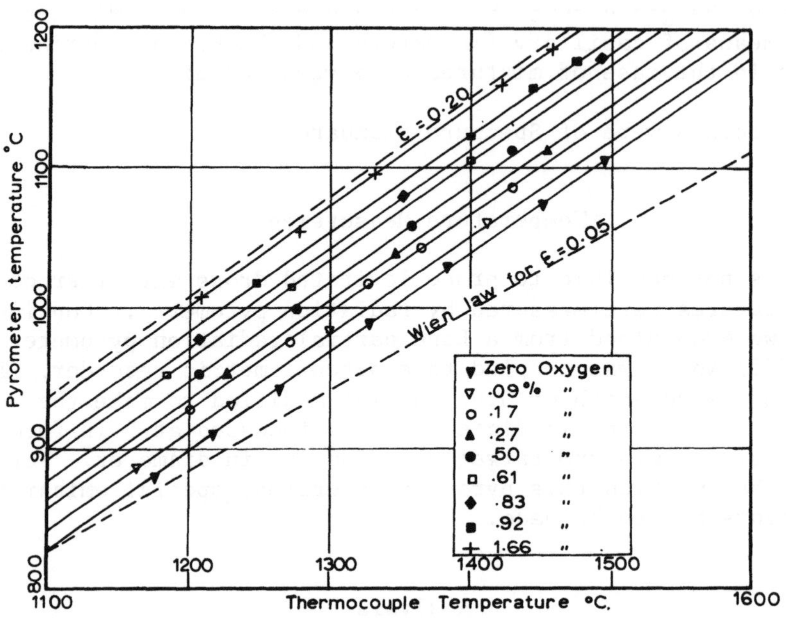

Fig. 2. Radiation pyrometer versus thermocouple temperatures for Cu [O] in argon.

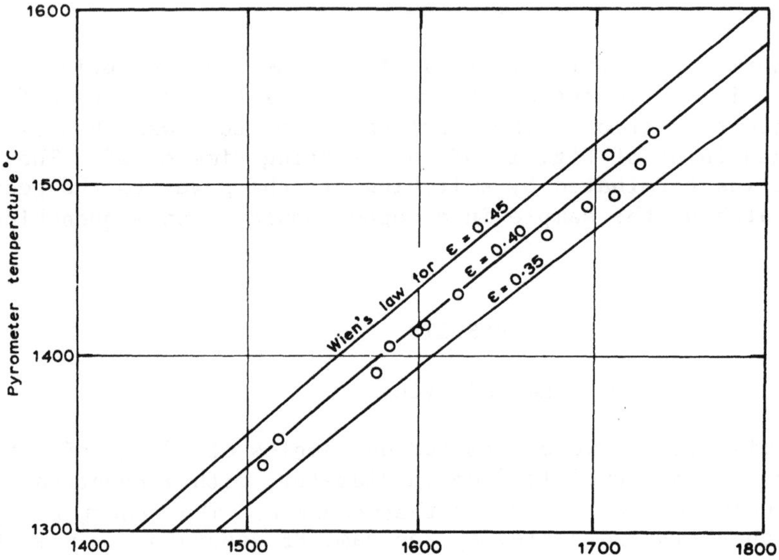

Fig. 3. Radiation pyrometer versus thermocouple temperatures for Ni in argon.

increase caused by oxygen is similar to that observed by Toop and Richardson [1] for 6650Å and greater. It is known from earlier work [1] that the emissivity of nickel is not significantly affected by oxygen at concentrations of less than 1 wt %.

Reaction between Copper and Oxygen

Although specpure copper was used and this contains less than 10 ppm of impurities, it was discovered during the course of the kinetic studies that surface films on the copper seriously retarded the reactions.

It took some time before it was realised that such films were intervening and that they were composed entirely of silica. Silicon and/or silica are present at less than 2 ppm in specpure copper, but when the metal is melted, any silicon present can react with oxygen in the metal or in the surrounding gas and form a coherent surface film at all temperatures. Alternatively any silica present forms a similar film. This film can be removed by evaporation in nitrogen in about 10 minutes at 1500°C and in shorter times at higher temperatures. At lower temperatures this silicon or silica can be removed

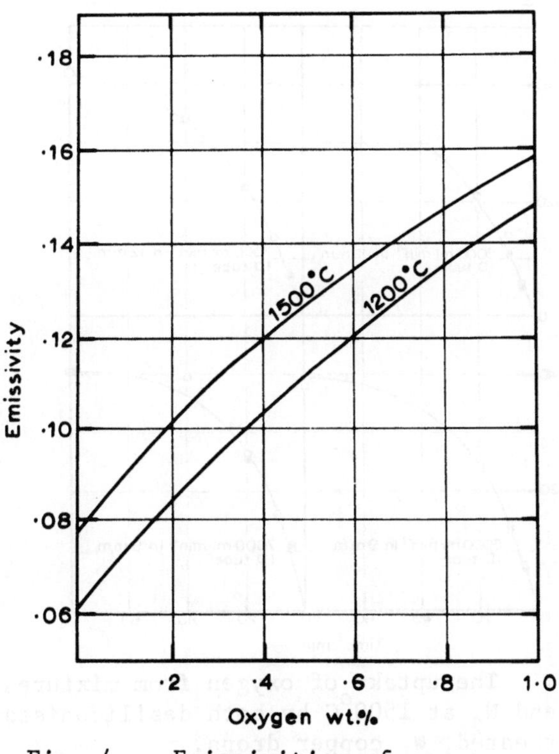

Fig. 4. Emissivities of copper + oxygen solutions, at 0.9μ.

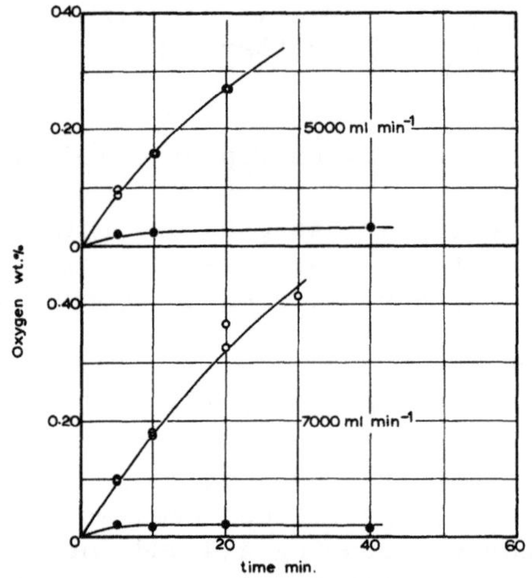

Fig. 5. The uptake of oxygen from mixtures of O_2 + N_2 at 1400^oC, by both desiliconised, O and untreated, ●, copper drops, in 9 mm tube with pO_2 = 1.11×10^{-4} atm.

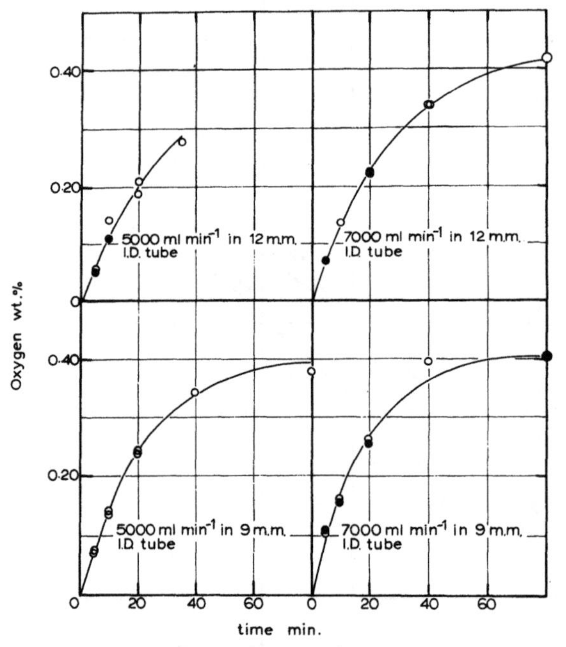

Fig. 6. The uptake of oxygen from mixtures of O_2 and N_2 at 1500^oC by both desiliconised, O and untreated, ●, copper drops.
pO_2 = 1.11×10^{-4} atm.

by heating the drop in nitrogen or argon so that the film forms, quenching, treating with HF, drying and relevitating. The film cannot be removed by nitric or hydrochloric acid, but it can be coagulated by blowing about 1 mgm of powdered lime on to the drop when levitated.

Typical results obtained at 1400°C and 1500°C are shown in Figs. 5 and 6† for drops weighing 0.25 gm, from which the silica had been previously removed by HF and for drops for which this was not done. By means of a careful study of the relationship between power to the coil, oxygen content of the metal, emissivity and pyrometer signal, a method was found controlling the power so as to keep the metal drop at constant temperature in spite of the marked effect of the oxygen on the coupling.

Vaporisation of Copper Metal

The rates of vaporisation from desiliconised drops of copper weighing 0.25 gm were measured in purified nitrogen $P(O_2 \sim 10^{-6}$ atm) at 1400°, 1500° and 1600°C at various flow rates. (The values were obtained by weighing drops before and after the experiments and the results are shown in Fig. 7. Rates below 75 x 10^{-2}cm min were obtained with 12 mm tube, those above with 9 mm. Some similar experiments were made with uncleaned drops and these showed the silica film had no significant effect.

Experiments on Silica Films

It was observed visually that on quenching a specpure drop, the crown of the drop had a different texture from the remainder. By means of an electron probe scan it was possible to find silicon in this region over areas up to 100 microns wide. If these areas were removed by polishing, no silicon was found 1 mm below. Similarly no silica could be found after etching the entire sample in HF after levitation after 10 min. in the liquid state in nitrogen.

All the silicon was removed from a number of specpure copper drops by levitation and treatment with HF. These drops were then remelted in an alumina crucible and some samples withdrawn for levitation tests in O_2+N_2 mixtures, to confirm that silica was not reintroduced by this procedure. To the remainder of the desiliconised metal, silicon was added to give 20 ppm. Samples of 0.25 gm were then levitated in O_2+N_2 ($P(O_2) = 1.4 \times 10^{-2}$ atm) flowing at 2500 ml min^{-1} at 1400 and 1500°C. The uptake of oxygen was again slow, unless these samples were quenched, treated with HF and then further oxidised. The results are shown in Fig. 8. In this case the blocking action of silica persists at 1520°, because of the greater

† All flow rates stated are STP rates except in Figs. 13, 14 and 15 for which flow rates at room temperature are used.

quantity present.

Experiments were then made with an apparatus similar to that used by Distin et al [2] to establish whether silica powder injected into the gas stream and so carried to the surface of a cleaned drop levitated in nitrogen would cause retardation of oxygen transfer from N_2+O_2 mixtures.

About 1-3 mgm of powder (about 50 micron) was taken up by each 0.25 gm drop which was then exposed to N_2+O_2 (P(O_2 = 1.1 x10^{-4} atm) flowing at 5000 ml min^{-1} for 10 minutes. In the absence of the silica addition a drop in this apparatus, takes up 0.2 to 0.3% oxygen at 1100° to 1500°C in the ten minutes. The results obtained with the silica additions are shown in Fig. 9. At temperatures of 1300°C and above, the rate of transfer was reduced by 50 to 80% depending on the temperature. It thus appeared that the silica was able to spread over the surfaces above about 1250°C. In experiments in which the silica was added after oxidation had proceeded for five minutes, it was found that the spreading and subsequent retardation of oxidation was unaffected by the presence of the surface active

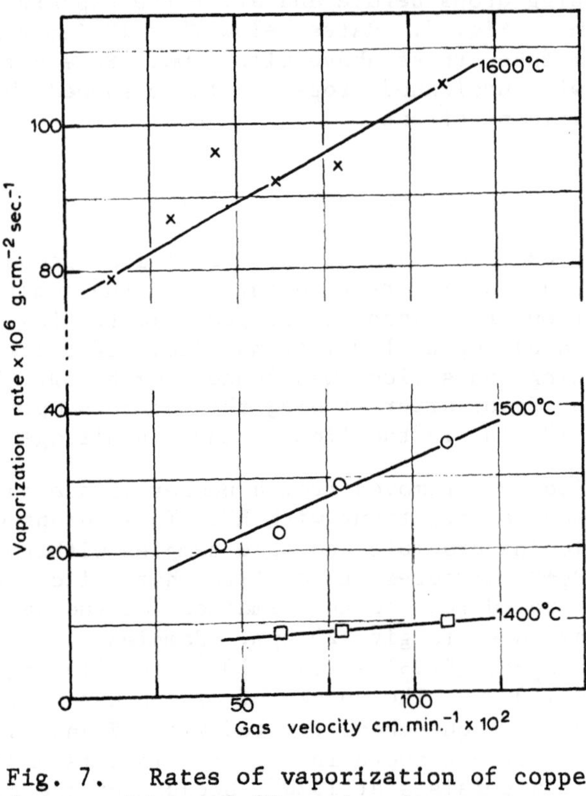

Fig. 7. Rates of vaporization of copper from drops weighing 0.25 gm in nitrogen.
Flow rates are for gas at STP.

oxygen. The results are shown in Fig. 10.

When CaO and MgO powders were added to levitated drops containing 2 ppm of silicon, the silicon immediately ceased to impede oxygen transfer. When CaO and MgO were added to cleaned drops, they had no effect on oxidation.

When Al_2O_3 and ZrO_2 powders were added to drops containing 2 ppm of silicon, the blocking action of the silicon was not affected. Neither oxide had any effect on cleaned drops.

Some experiments were done with specpure nickel exposed to nitrogen containing oxygen at 10^{-4} atm at 1500°C, 1600°, and 1700°C. Unfortunately the limitations of levitating and heating simultaneously, prevented experiments being made at lower temperatures. It was found that the reactions proceeded at the same speed whether the

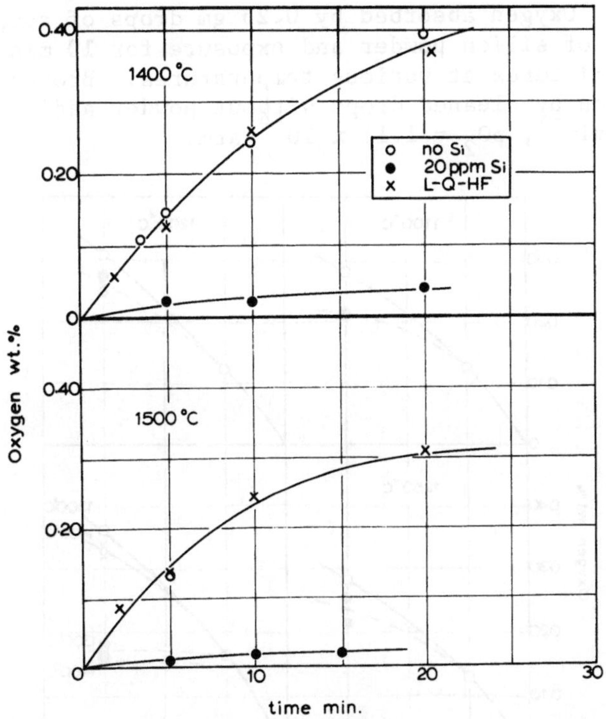

Fig. 8. Effect of added silicon on uptake of oxygen for $N_2 + O_2$ mixtures ($pO_2 = 1.4x10^{-4}$ atm.) at 1400° and 1500°C. Results from remelted de-siliconised metal O; from desiliconised metal after addition of 20 ppm silicon ●; after cleaning drops containing 20 ppm silicon X.

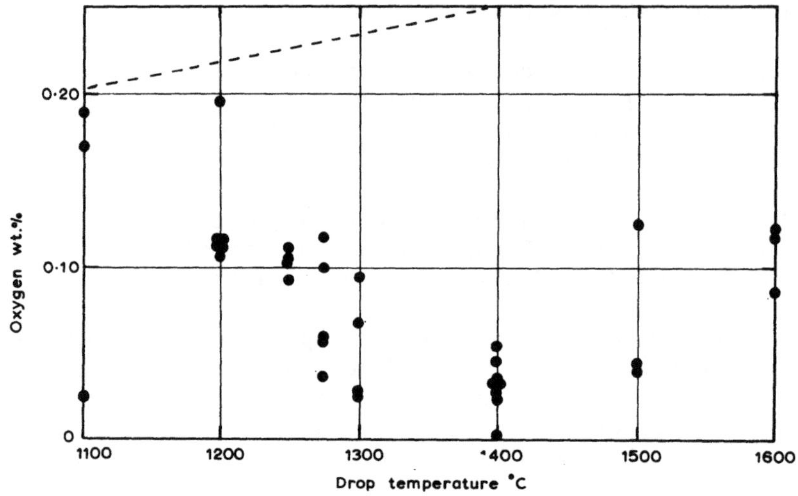

Fig. 9. Oxygen absorbed by 0.25 gm drops of copper after addition of silica powder and exposure for 10 min to $N_2 + O_2$ mixtures at various temperatures. Broken line shows absorption by cleaned drops without powder additions. 5000 ml min^{-1}, $pO_2 = 1.11 \times 10^{-4}$ atm.

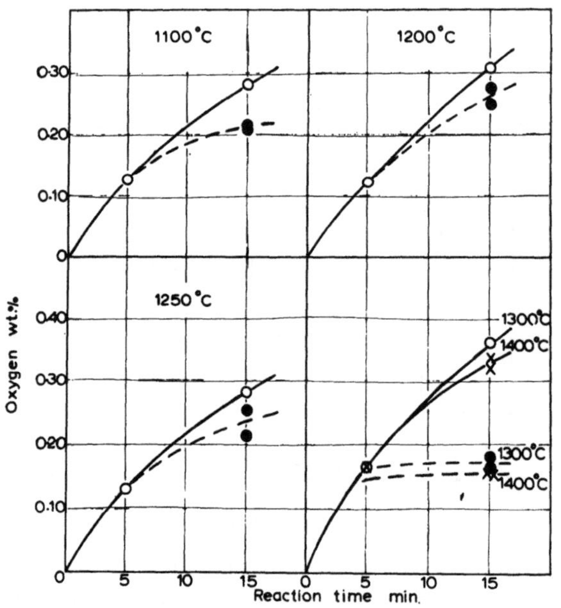

Fig. 10. Oxidation of cleaned drops under same gas conditions as for Fig. 9. with (● and X) and without (O) addition of silica, 5 min.after start of oxidation.

samples were treated with HF or not. But the reaction rate was reduced by some 75% during its first 10 minutes when silica powder was added.

Mass Transfer Experiments with Naphthalene

Naphthalene spheres were prepared by dipping copper spheres soldered to wires, into molten naphthalene. By successive dippings they were brought to the 4 mm and 5 mm diameters of the drops used in the experiments. They were then held in nitrogen, at the room temperature flow rates of the high temperature experiments, in order to establish the relationship between Sherwood and Reynolds Number for the geometrical conditions of the experiments. The results are considered later (Fig. 11).

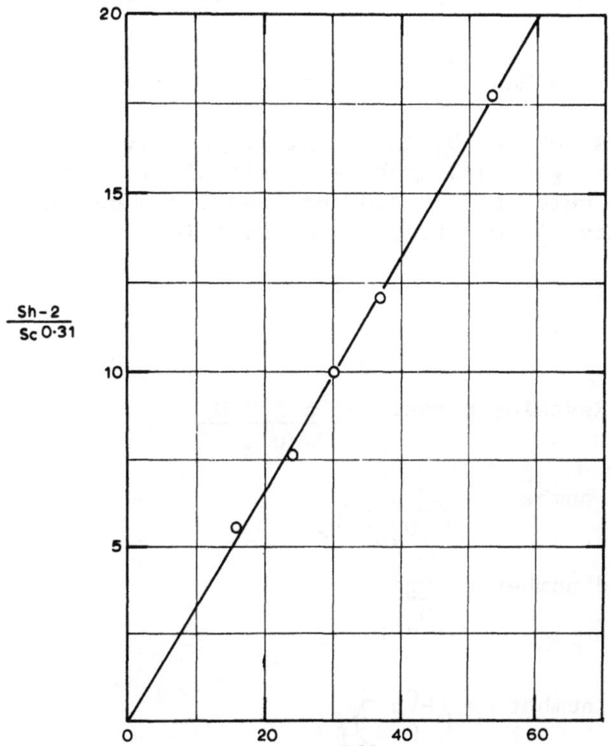

Fig. 11. Results obtained at 24°C for the evaporation of Napthalene from spheres 4 mm dia. into N_2 under the same geometrical conditions as for the levitation experiments.

DISCUSSION

Steady States or Equilibria

With cold gas impinging on hot droplets of metal, there is the possibility that the steady state partition of oxygen attained between gas or metal, may be different from the equilibrium partition obtained under isothermal conditions. The results obtained previously by Toop and Richardson [1] on the reactions of copper and nickel with CO_2 had indicated that the steady state concentrations were identical with the equilibrium concentrations. This earlier work was however influenced by the presence of silica films arising from the 2 ppm silicon in the metal used, so the question needs further checking. Recent studies with fully desiliconised drops, made by Mr. A. Forster in this laboratory, have now confirmed that the steady state values for ($p\ CO_2/p\ CO.\ \%O$) are the same as the equilibrium values. The results shown in Fig. 6 indicate that this is also the case for the reaction of copper with mixtures of oxygen and nitrogen.

Gas Mass Transfer Coefficients

From previous work by Distin et al [2] and by others [3], it has been shown that gas mass transfer coefficients about small heated drops of metal held in cold flowing gases, can be derived with reasonable accuracy (± 20%) from the correlation

$$Sh = 2 + 0.5(GrSc)^{0.5} + nRe^{0.62}Sc^{0.31} \qquad (1)$$

where,

Re = Sphere Reynolds number $= \dfrac{\rho_f V D}{\mu_f}$

Sc = Schmidt number $= \left(\dfrac{\mu}{\rho D_{AB}}\right)_f$

Sh = Sherwood number $= \dfrac{k_g d}{D_{AB_f}}$

Pr = Prandtl number $= \left(\dfrac{\mu C_p}{K}\right)_f$

Gr = mean Grashof number $= Gr_m + \left(\dfrac{Sc}{Pr}\right)^{0.5} \cdot Gr_H$

Gr_m = Grashof number for mass transfer $= \dfrac{gd^3\rho_f(\rho_s - \rho_b)}{\mu_f}$

Gr_H = Grashof number for heat transfer $= \dfrac{g d^3 \rho_f^2 \quad (T_s - T_b)}{\mu_f^2 \quad T_f}$

d = diameter of drop, cm.

D_{AB} = diffusion coefficient in a binary gas system, $cm^2 sec^{-1}$.

g = acceleration due to gravity, cm sec^{-2}.

ρ = gas density, gm cm^{-3}

μ = gas viscosity, gm cm^{-1} sec^{-1}.

Cp = gas heat capacity, cal gm^{-1} $^oK^{-1}$.

K = gas thermal conductivity, cal $sec^{-1} cm^{-1} {}^oK^{-1}$.

V = linear velocity of gas, cm sec^{-1}.

Subscripts, s,b,f, refer to gas properties calculated at the temperature of the specimen (Ts), of the bulk gas (Tb) and at the film temperature (T_f) respectively. T_f is equal to $\dfrac{T_s + T_b}{2}$.

In order to find n in Eq. 1, for the flow conditions and gas mixtures used in the current experiments, the naphthalene results were plotted as shown in Fig. 11. (There being no temperature gradients in the naphthalene case, the middle term on the right hand side of equation [1] disappears). The value found is 0.33.

In calculating mass transfer coefficients from Eq. 1, the viscosities and Prantl numbers for nitrogen were taken from Kays [4] and its interdiffusion coefficient with oxygen was calculated from the Hirschfelder Eq. [5]. The diffusivities for copper vapour in nitrogen were calculated in the manner proposed by Turkdogan [6].

The Reaction Between Copper and Oxygen

If mass transfer in the gas is rate controlling one may write

$$\frac{d\,n_O}{dt} = k_G \cdot A \cdot (C^b - C^s) \tag{2}$$

where C^s and C^b are oxygen concentrations in the gas at the surface of the drop and in the bulk i.e. at a distance from the drop.

In terms of partial pressures of oxygen one may then write

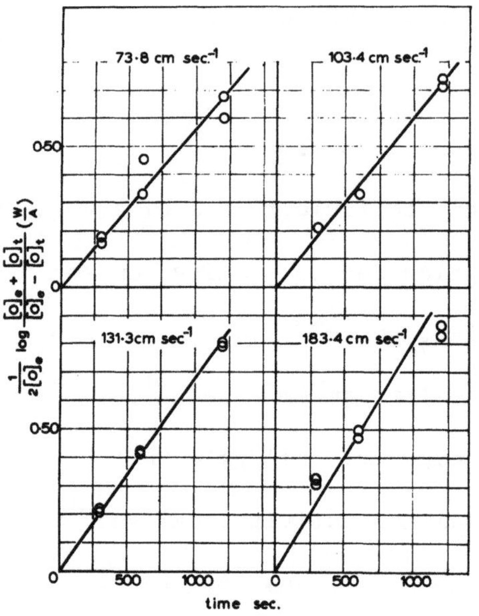

Fig. 12. Typical plots for deriving k_G for the reaction of copper with $O_2 + N_2$ mixtures at 1500°C.

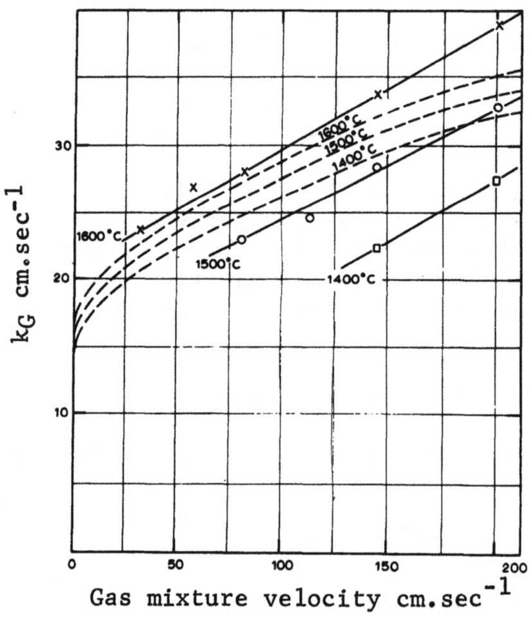

Fig. 13. Values of k_G derived for the reaction of copper with $O_2 + N_2$ mixtures shown as a function of the flow rate for various temperatures. Broken lines show values derived from equation (1).

$$\frac{dn_0}{dt} = \frac{k_G . A}{RT_f} \quad (p^b{}_{0_2} - p^s{}_{0_2}) \qquad (3)$$

and

$$\frac{d[\%0]}{dt} = \frac{3200}{W} \frac{k_G A}{RT_f} \quad (p^b{}_{0_2} - p^s{}_{0_2}) \qquad (4)$$

where [0] is in wt %.

Putting $p_{0_2} = (a[0])^2 /K$ and taking γ [0] as constant:

$$\frac{d\,[0]}{dt} = \frac{3200}{W} \frac{k_G A}{RT_f\, K} \quad ([0_e]^2 - [0_t]^2) \qquad (5)$$

where $[0_e]$ is the oxygen concentration which would be in equilibrium with the bulk gas and $[0_t]$ the oxygen concentration in the metal at time t. Integration then gives

$$\frac{1}{2[0_e]} \quad \ln \frac{[0_e] + [0_t]}{[0_e] - [0_t]} = \frac{3200\ k_G\ A}{WRT_f\ K} t \qquad (6)$$

Fig. 12 shows four typical plots for obtaining k_G by the use of Eq. 6. It can be seen that the results conform well with the linear relationship required by the equation. The values of k_G derived from the slopes are shown in Fig. 13, where the broken lines show values derived from Eq. 1. Mass transfer in the gas phase appears to be rate controlling at 1600°C and 1500°C and possibly at 1400°C. The disagreement at the lower temperature is not understood for it is improbable that chemical kinetics could become rate controlling for so simple a reaction at this temperature. It is equally improbable that mass transfer in the metal phase could have been rate controlling, as values of k_M in any way compatible with the rates, would have been about 10^{-4} cm sec^{-1}. Such values are improbably low. Distin et al [2] have, for example, reported k_M for carbon being removed from 2 gm drops of iron levitated in oxygen at about 1800°C, to be about 3 x 10^{-2} and the diffusion coefficients of carbon in iron at 1800°C and of oxygen in copper at 1400°C are about the same [7]. For the movement of oxygen into iron the value could be as low as 3 x 10^{-3} cm sec^{-1}, but this seems unlikely [2] unless very different degrees of interfacial turbulence are associated with different chemical reactions. Mr. A. Forster in this laboratory has shown that k_M for oxygen being removed from 0.5 gm drops of copper levitated in mixtures of hydrogen and nitrogen at 1200°C exceeds 9 x 10^{-2} cm sec^{-1}.

The Vaporisation of Copper

A nominal mass transfer coefficient, k_V, may be calculated for vaporisation, if it is assumed that the saturation vapour pressure of the metal is maintained at the metal-gas interface, and that the pressure is zero in the bulk of the gas. The values of k_V so obtained are plotted in Fig. 14, where the values calculated from Eq. 1 are also shown. The experimental values are some 3 times greater than those calculated. This finding is in line with the accepted view that evaporation from a hot body into a cold gas may be substantially enhanced by condensation in the steep temperature gradient in the gas near the interface [8]. This condensation makes the partial pressure gradient of the vapour steeper than it would otherwise be and so greatly increases the vaporisation rate and makes k_V much greater than the true k_G.

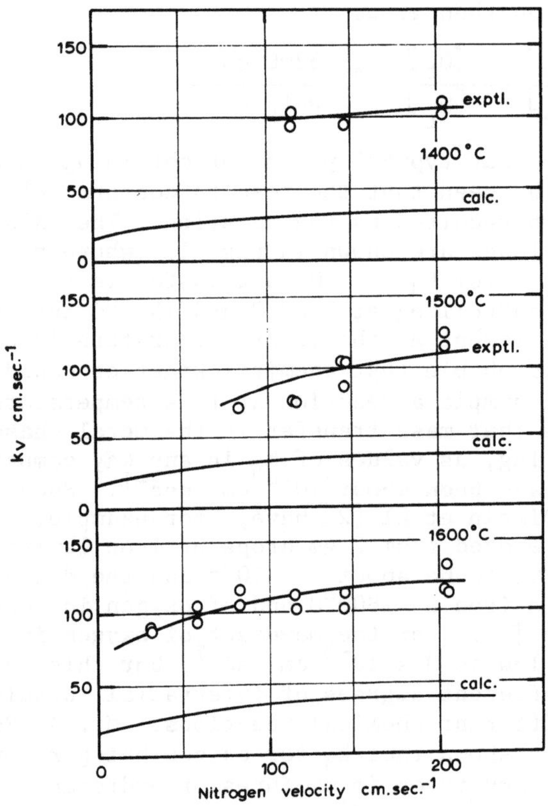

Fig. 14. Values of the nominal mass transfer coefficients for the vaporisation of copper from 0.25 cm dia. drops in nitrogen: calculated values from equation (1).

Role of Surface Films

The surface films of silica formed on specpure copper containing about 2 ppm of silicon may have been up to 70Å thick, and those produced by adding 20 ppm of silicon or one mgm of silica powder, were much thicker. It is to be expected that these thicker films should have taken longer to evaporate and so retarded the reaction with $O_2 + N_2$ at 1500° and 1600°C.

The differences between the ways in which the 2 ppm film behaves towards the vaporisation of copper (insignificant effect), and the oxidation by oxygen + nitrogen mixtures (substantial effect), may be explained as follows:

For the vaporisation of copper where the film intervenes between metal and gas, we may write the copper concentrations as follows:

<u>Copper</u>	Cu^i_{fm} <u>film</u> Cu^i_{fg}	Cu^i_g <u>gas</u> Cu^b_g

where the superscripts i and b stand for interface and bulk respectively, and the subscripts f,m and g stand for film, metal and gas.

Then for transport through the film we may write

$$\dot{n} = k_f^{Cu} (Cu^i_{fm} - Cu^i_{fg}). \tag{7}$$

If $Cu^i_g / Cu^i_{fg} = m'$, and we assume equilibrium at the interfaces,

$$\dot{n}_{Cu} = \frac{k_f^{Cu}}{m'} (m' \, Cu^i_{fm} - m' \, Cu^i_{fg}). \tag{8}$$

For transport in the gas we may write

$$\dot{n}_{Cu} = k_G^{Cu} (Cu^i_g - Cu^b_g) \tag{9}$$

$$= k_G^{Cu} (m'. \, Cu^i_{fg} - Cu^b_g) \tag{10}$$

So that, if $Cu^b_g = 0$

$$\dot{n} \, Cu \left(\frac{m'}{k_f^{Cu}} + \frac{1}{k_G^{Cu}}\right) = m' \, Cu^i_{fm} \tag{11}$$

This is equal to Cu_g^i for copper, free from any film.

As the film has been found to have little or no effect on vaporisation

$$\frac{m'}{k_f^{Cu}} << \frac{1}{k_G^{Cu}}$$

$$<< \frac{1}{95} \quad \text{at } 1400^\circ\text{C}.$$

The oxygen concentrations across the system may be written,

$$\text{Copper} \;\Big|\; O_{2_{fm}}^{i} \quad \text{film} \quad O_{2_{fg}}^{i} \;\Big|\; O_{2_g}^{i} \quad \text{gas} \quad O_{2_g}^{b}$$

If $O_{2_g}^{i} / O_{2_{fg}}^{i} = m''$, one may deduce that

$$\dot{n}_{O_2}\left(\frac{m''}{k_f^{O_2}} + \frac{1}{k_G^{O_2}}\right) = O_{2_g}^{b} - m'' \, O_{2_{fm}}^{i} \qquad (12)$$

$$= O_{2_g}^{b} - O_{2_g}^{e} \qquad (13)$$

where $O_{2_g}^{e}$ is the oxygen gas concentration which would be in equilibrium with the copper. At 1400°C the film has a marked effect on the oxygen + nitrogen reaction, so that

$$\frac{m''}{k_f^{O_2}} >> \frac{1}{k_G^{O_2}}$$

$$>> \frac{1}{22}$$

The conclusions may be summarised as,

$$\frac{m''}{k_f^{O_2}} >> \frac{1}{22} \quad \text{and,} \quad \frac{1}{95} >> \frac{m'}{k_f^{Cu}}$$

In a lime-alumina-silica slag at 1500°C, the solubility of copper is about 0.055 at.% when the copper vapour pressure is 0.43 mm [9]. Thus m' for a silicate film might be about 2×10^{-4}

and for silica glass nearer 3×10^{-4}. If $m'/k_f^{Cu} << 1/95$, then $k_f^{Cu} >> 3 \times 10^{-2}$ cm sec^{-1}. k_f^{Cu} may be put equal to D/x where D is the diffusion coefficient of copper in the film and x is its thickness - about 7×10^{-7} cm. Then D becomes $>> 2 \times 10^{-8}$ cm sec^{-1} and this is quite possible for fused silica [10]. It looks as though the solubility of oxygen molecules in the film would be much less than that of copper at the same gas pressures, so that $m' << m''$; also that the diffusion coefficient of the oxygen molecule would be less than that of the copper atom, so that $k_f^{Cu} > k_f^{O_2}$. Thus $m''/k_f^{O_2} >> m'/k_f^{Cu}$.

The oxygen might, conceivably, also be transferred by the copper which dissolved in the film, being oxidised at the gas-film interface and carrying oxygen back as oxide. Copper oxide cannot form a liquid slag with silica at temperatures between 1200° and 1500°C until the copper oxide activity is greater than about 0.5 [11]; thus copper oxide present at activities of the order of 0.05 (as in the experiments reported here) may have exceedingly high activity coefficients in the silica film. Oxide may therefore be present in the film at no higher concentrations than oxygen itself, and it may not diffuse any faster in view of the possible "locked-in" nature [12] of metal ions at low concentrations in liquid silica.

Spreading and Vaporisation of the Silica

It appears from the results shown in Fig. 9 that silica added as powder at high temperature does not spread readily over the copper drop below about 1250°C. It is reasonable that there should be a lower temperature limit for spreading, for this phenomenon is exhibited by organic substances which spread on water. In both cases the spreading temperatures are less than the melting points. It is to be noted that the tendency for the silica not to spread rapidly at 1250°C and below, is unaffected by the presence of oxygen in the drop before the silica powder is added.

When the silica film is produced by the oxidation of dissolved silicon, as in the case when elemental silicon is added to the metal before levitation or possibly in the case of the uncleaned metal containing 2 ppm of silicon, a coherent film is formed at all temperatures.

The fact that the silica film can apparently be removed by evaporation is in line with our knowledge of the vapour pressure of silica and the dissociation reaction [13],

$$2SiO_2 = 2SiO + O_2, \qquad (14)$$

for which $\Delta G^o = 385,700 - 120.8T$. From this expression, the SiO pressures† in equilibrium with oxygen pressures of 10^{-4}, and 10^{-6} and 10^{-8} atm were calculated and these are listed in Table I together with SiO_2 vapour pressures. Assuming p(SiO) and $p(SiO_2)$ are equal to zero at a distance from the drop, the flux of SiO_2 from the film can then be calculated from the equation

$$\frac{dn_{SiO}}{dt} = \frac{\cdot A}{RT_f} \left(k'_V p_{SiO} + k''_V p_{SiO_2}\right) \quad (15)$$

TABLE I

Calculated effects of oxygen pressure and temperature on times to evaporate SiO_2 films (2 ppm Si) for kV eqn

Drop Temp. °C	P_{O_2} atm.	P_{SiO} atm.	P_{SiO_2} atm.	Calculated time min.	Observed time min.
1400	10^{-4}	10^{-10}	8×10^{-10}	530	400-500
1500	"	$3x10^{-9}$	8×10^{-9}	49	7-8
1600	"	$5x10^{-8}$	7×10^{-8}	4	∿ 1
1400	10^{-6}	10^{-9}	8×10^{-10}	270	35-40
1500	"	$3x10^{-8}$	8×10^{-9}	14	2-3
1600	"	$5x10^{-7}$	7×10^{-8}	0.8	< 1
1400	10^{-8}	10^{-8}	8×10^{-10}	45	-
1500	"	$3x10^{-7}$	8×10^{-9}	2	-
1600	"	$5x10^{-6}$	7×10^{-8}	0.1	-

where k_V and k''_V are the mass transfer coefficients for the vaporisation of SiO and SiO_2. To a first approximation these may both be taken as equal to k_V for copper vapour (100 cm sec^{-1}), for the SiO will condense by recombination with oxygen and because of this k_V will be substantially greater than k_G estimated for SiO from Eq. 1. The difference is likely to be greater than for copper because of the greater heat of evaporation of both SiO and SiO_2. From the assumed values of k_V and Eq. 15 the times taken to evaporate all the

† accurate to about a factor of 2

silica formed from 2 ppm of silicon in a 0.25 gm drop, have been calculated for different conditions and these are listed in Table I.

These calculations are qualitatively consistent with observations recently made by Mr. R. L. Harvey in this laboratory. He has shown that the film can be removed by holding the copper drops at 1500°C in nitrogen containing 10^{-6} atm O_2 for 2-3 minutes at 1500°C and 35-40 at 1400°C. When $p(O_2)$ is raised to 10^{-4} atm, the times are about 7 minutes at 1500°C, 25 at 1450°C and 400-500 at 1400°C. The film actually formed may be thinner than calculated for 2 ppm, and the values assumed for k_V may be too low. The results in Figs. 5 and 6 show that the film is sufficiently removed not to impede the reaction with O_2, during the time the drop is being brought to 1500°C and held there for a few minutes in nitrogen containing less than 10^{-6} atm of oxygen.

Effects of Other Oxides

The destruction of the blocking effect of the silica film by additions of CaO or MgO powder is presumably caused by formation of silicates. These either coagulate into islands or form slags in which copper oxide is sufficiently soluble to permit rapid transport of oxygen across the film by oxidation of Cu^o and Cu^+ to Cu^+ and Cu^{2+} at the gas-film interface and reduction at the metal-film interface: as $CaO+SiO_2$ and $MgO+SiO_2$ do not give rise to any liquid phases below 1436°C and 1543°C respectively, the first explanation seems the more likely. Zirconia presumably does not react at all with the silica, and alumina either does not react or joins in the film as mullite or some other alumino-silicate. Both these oxides seem unable to spread on copper on their own.

REFERENCES

1. G. W. Toop and F. D. Richardson: *Advances in Extractive Metallurgy*, Inst. Min. and Met. London, 1968, p. 181.
2. P. A. Distin, G. D. Hallett and F. D. Richardson: J. Iron Steel Inst. London, 206, 821 (1968).
3. L. A. Baker, N. A. Warner and A. E. Jenkins: Trans. Met. Soc. AIME, 230, 1228 (1964); 239, 857 (1967).
4. W. M. Kays: *Convective Heat and Mass Transfer*, McGraw Hill, 1966.
5. J. O. Hirschfelder et al.: *Molecular Theory of Gases and Liquids*, Wiley and Sons, 1954, p. 539.
6. E. T. Turkdogan: *Steelmaking*, Chipman Conference June 1962, MIT Press 1965, p. 77.
7. G. R. Belton and M. G. Hocking: (University of Pennsylvania), Private Communication.

8. E. T. Turkdogan: Trans. Met. Soc. AIME, 230, 740 (1964).
9. H. W. Meyer and F. D. Richardson: Trans. Inst. Min. and Met. London, 71, 201 (1961-2).
10. R. H. Doremus: *Modern Aspects of the Vitreous State*, 2, Butterworth & Co. 1962.
11. E. M. Levin, C. R. Robbins and H. F. MCMurdie: *Phase Diagrams for Ceramists*, Amer. Ceram. Soc. 1964.
12. J. O'M. Bockris, J. W. Tomlinson and J. L. White: Trans. Farad. Soc., 52, 299 (1956).
13. L. Yang, C. L. McCabe and R. Miller: *Physical Chemistry of Steelmaking*, Chapman and Hall Ltd., 1958, p. 63.

DISCUSSION OF PAPER

S. E. Khalafalla. I am interested in the correlation you used to calculate the mass transfer coefficient, namely:

$$Sh = 2 + 0.5\,(Gr\ Sc)^{0.5} + n\ Re^{0.62}\ Sc^{0.31}$$

We use a similar equation given by Stewart and Lightfoot in which n=0.6, the power of Reynold number is 2/3 and the power of Schmidt number is 1/3. What value of n do you get from your results? Secondly, how certain were you in calculating the diffusivities and viscosities in binary gas mixtures? I believe you get large uncertainties when calculating for ternary gas mixtures.

F. D. Richardson. The value of n varies slightly with the geometry. We obtain a value of about 0.4 in this work. The values for the diffusivities were calculated by the various equations that have been proposed. There is some uncertainty here and I am always surprised that these things work out as well as they do. The methods worked very well in the case of the decarburization of drops of iron in flowing CO_2, both in Jenkins' work and our own.

D. E. Rosner. Is the quasi-steady configuration of the silica film itself modified by the shear stresses associated with the imposed flow? For example, is the imposed flow large enough to "open up" a portion of the upstream hemispheric surface of the droplet (thereby exposing pure Cu(ℓ)?

F. D. Richardson. We have no evidence of this but we only went up to gas velocities (cold) of about 150 cm/sec.

Y. A. Chang. Have you done similar types of studies by bubbling gases through a batch of molten copper?

F. D. Richardson. Dr. Diaz, when he was with us, did some experiments in which CO/CO_2 mixtures were bubbled through molten copper that had not been especially de-siliconized. He obtained relationships of the form expected for simple chemical kinetics and rates which are in reasonable agreement with the measured rates for the cleared drops. It would appear that the bubbles rise more quickly than the silica film can form.

THE REACTION OF LIQUID IRON AND ITS ALLOYS IN PURE OXYGEN

D. G. C. Robertson* and A. E. Jenkins**

*Department of Metallurgy, Imperial College, London

**School of Metallurgy, University of New South Wales
Sydney, Australia

INTRODUCTION

When liquid iron or its alloys are exposed to reaction with oxidizing gaseous atmospheres, the sequence of events appears to be closely related to the composition of the alloy, to the oxygen potential and to the flow characteristics of the ambient atmosphere. There have been numerous laboratory investigations covering a variety of such situations and these have been complemented in a few instances by reports of actual process studies, notably from oxygen steelmaking. The results reported in this paper refer to the behaviour in pure oxygen of pure iron and its alloys with carbon and silicon. Specimens of one gram mass were levitated in an electromagnetic system and subsequently reacted and simultaneously photographed with the aid of a high-speed cinematographic camera capable of operating up to a speed equal to 9,000 pictures per second. In many instances a temperature record of the event was obtained by means of a focussed two-colour pyrometer sighting downwards along with the camera onto the upper surface of the levitated specimen. Although a full description of the experimental technology is to be published elsewhere several factors of experimental significance should be mentioned at this point. The successful interpretation of the sequence of events, occurring often over a period of several milliseconds (msec), was achieved only after the successful development of a high speed cinematographic technique†. The camera adopted was a HITACHI Type 16K "HIMAC" No. 192 fitted with a 105 mm Nikkor lens mounted on a 50 mm extension piece giving a field of view equal to 18 mm diameter. A timing light in the camera left a streak of 1/200 second duration on the edge of the film every 1/100 second. The time interval between separate events could therefore be precise-

† A 10 minute film covering the topics outlined in this paper was presented at the Conference.

ly assessed. Frequently a conventional cine camera was utilized to measure the duration of a reaction and to assess such important parameters as the approximate time lapse leading up to a significant reaction change in the specimen or to the approximate estimation of image brightness changes due to a significant temperature increase or emissivity change. The change in temperature experienced by a reacting specimen was obviously an important experimental factor in this work. Two distinctly different methods were used. In the first instance a two-colour pyrometer, calibrated at the melting point of "Battelle" iron was fitted with an extension filter and further calibrated against a tungsten filament lamp to provide a reliable 1000°C range between 1600° to 2600°C. The continuous output from the instrument was presented on a strip recorder, the maximum speed of response being equivalent to 12°C per millisecond. The second method utilized the fact that if the film was exposed so that the logarithm of the image density fell within the range 0.8 to 1.5, then it was proportional to the alteration in image brightness. The image density in this work was not calibrated against objects of known brightness and no precautions were taken to utilize monochromatic radiation, which prevented the data being utilized to calculate absolute temperature of the specimen. However for surfaces of the same emissivity, an estimate could be made of the temperature increment suffered by the specimen.

The extreme rapidity of the reactions involved required the briefest of possible delays in changing the gas atmosphere surrounding the specimen. This was accomplished by the use of a revolving plate valve which permitted pure oxygen flowing at 18 standard litres per minute (s.l.m.) to replace the volume occupied by inert argon or helium in 35 msec. The oxygen flow rates of 5 and 18 s.l.m. used in this work were equivalent to approach velocities of 67 and 242 cm/sec at 20°C which should be compared with the velocity of a drop after 3 feet of free fall and the velocity in a Mach 0.5 jet, namely 420 cm/sec and 1.6×10^4 cm/sec.

Finally from initial experiments with a number of alloys exposed to oxygen it became clear that the reactions could be "blocked" by oxide films if the alloys were not cleaned immediately prior to exposure to oxygen. They were held for two minutes to a flow of pure hydrogen and were then degassed for a similar period in pure helium. Following this treatment, all alloys of the same composition behaved in an absolutely reproducible fashion.

PURE IRON-OXYGEN REACTION

Although the work of Turkdogan, Grievson and Darken [1] on the fuming of pure iron in oxygen has shown that the evaporation of iron may be enhanced as a consequence of the formation of FeO fume in the boundary layer, there has been surprisingly little work in oxygen atmospheres exceeding 0.1 atmosphere pressure (approx.) where the existence of a stable FeO film was proposed to account for the cessation of fuming. Kirschfeld [2], and Harrison and Yoffe [3] have studied the burning of pure iron wires in oxygen but their experimental conditions are somewhat unrelated to the present discussion.

In the present work, one gram levitated specimens of both Battelle iron and soft magnetic iron (major impurities C 0.02, Si 0.02, Mn 0.11, P 0.6, S 0.03, Ni 0.05 and Cr 0.08%) were reacted in pure oxygen flowing at 18 s.l.m., the reaction commencing with the iron at approximately 1600°C. During the first two or three milliseconds of reaction, an oxide phase spread rapidly as a thin veil across the surface producing a threefold increase in specimen brightness due to the emissivity change. The subsequent reaction to form a non-volatile liquid oxide phase could be followed quite clearly by the two-colour pyrometer and the high-speed cine camera, there being a total absence of fuming. Several successful runs were performed in which the specimen was quenched into a copper mould during the course of the reaction and a quantitative assessment made of the mass and chemical analysis of the oxide. The oxide contained 72% Fe. Because of the high temperature reached it is impossible to say whether this represented an oxygen-saturated condition but all the iron in the quenched slags was determined to be ferrous iron. The time of reaction could be measured to within 0.02 seconds and the remnant iron mass to within $\pm$ 0.01 g. This data enabled a calculation to be made of the rate of oxidation of pure iron under these conditions. The rate was equivalent to the oxidation of 0.46 g Fe/second in the initial stages of reaction which would have required an amount of oxygen equal to 7.6 s.l.m. This is well inside the supply rate of 18 s.l.m., and so the measured iron reaction rate could be taken as the maximum burning rate under these conditions. It is hardly surprising therefore that the strongly exothermic reaction resulted in the specimen temp. rising from 1600°C to approx. 2800°C in 0.4 seconds. At this point the two-colour pyrometer limit had been reached but the films clearly showed that the specimen reached a much higher temperature before reaction ceased due to the melt falling from the coil. A total reaction time of nearly 2 seconds was achieved with the Battelle iron which appeared to reach a temperature in excess of 3000°C, i.e. above the boiling point of iron. The drop did show some signs of becoming unstable as it approached this temperature but this was insufficient to shatter the drop. In the case of the soft magnetic iron, these very high temperatures were not achieved; the specimen shattered unpredictably due to the oxidation of the small traces of carbon. Fig. 1 presents the appearance of the iron during four stages of reaction.

OXIDATION OF IRON-CARBON ALLOYS

The presence of carbon in liquid iron produces a significant change in the oxidation behaviour of the metal. The reaction between dissolved carbon and oxygen may take place at the free surface between the condensed and gaseous phase or it may occur within the melt either homogeneously or heterogeneously. The liberation of the gaseous carbon monoxide as a surface decarburization process should not produce a bulk distortion of the melt or ebullition effect and the situation may be referred to as non-boiling. However should nucleation and growth of CO bubbles occur within the melt, then the resultant release of the gas produces a boiling situation as a consequence of which various side effects ensue. There have been several studies of the problem recently of which could be mentioned those by Baker, Warner and Jenkins [4], Swisher and Turkdogan [5], Baker and Ward [15], and Distin, Hallett and Richardson [6]. The behaviour depends ultimately on the balance between the rate at which oxygen is being supplied to the surface of the melt and that

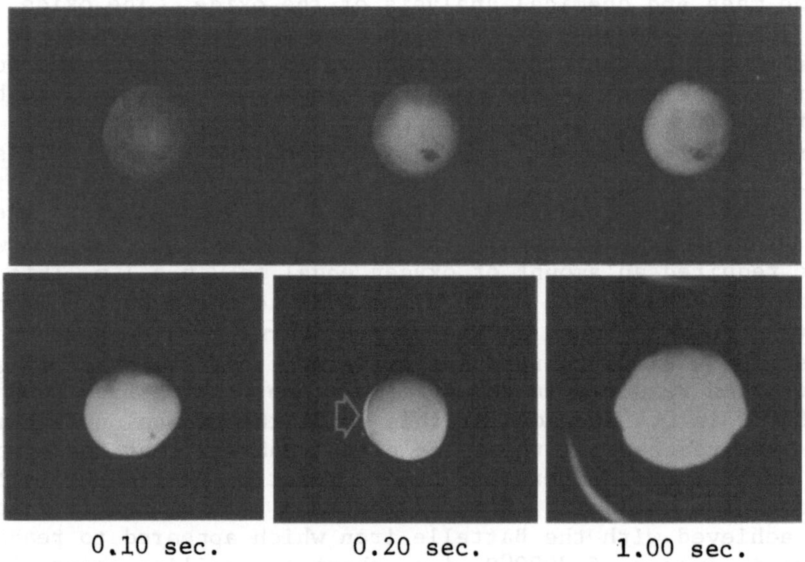

Fig. 1. Burning drop of high-purity iron. The top row of pictures shows the growth of the initial oxide layer at intervals of 1/1500 sec. The oxide hanging beneath the drop can be seen after 0.20 sec ⇨. After 1.00 sec. the drop was superheated, bubbling occurred and drop size increased.

at which carbon is diffusing to that surface. Specifically, the current study examines in detail the behaviour prior to, during and following the transition from quiescent decarburization to full boiling conditions.

Near carbon-saturated liquid iron specimens were reacted in pure oxygen under identical conditions to those previously described in this paper. The runs extended up to 6 seconds duration. Great care was taken in the preparation of extremely pure alloys by non-consumable arc melting, after it was established that a 5% C, 0.3% Si, 0.2% Al iron alloy behaved after a fashion to be described in the next section on Fe-C-Si ternary oxidation. The emphasis in this work was directed to establishing the various reaction sequences and processes rather than the rate of reaction but several runs were interrupted to provide carbon loss data which agreed well with the earlier results of Baker [4]. In general the specimens were reacted until the violence of the boiling caused them to hit the side of the reaction tube.

The commencement of reaction was not signalled on this occasion by the appearance of a veil of oxide spreading across the specimen, as occurs with pure iron. The temperature merely increased linearly with time ($165 \pm 8^{o}C$) as the carbon similarly decreased (7.28×10^{-3} gram C/sec) during the non-boiling period, the oxygen flow rate on this occasion being 5 s.l.m. This rate of temperature rise was approximately 7% of that observed with pure iron which is consistent with the observation noting the absence of iron oxide formation at this point. After nearly two seconds reaction a flame front surrounding the specimen became visible and at about this time the first signs of a carbon boil appeared. This event was manifested by bursting of small bubbles close to the surface, which led to the ejection at high speed of small particles from the liquid metal. The films clearly showed these particles igniting in the gas, leaving as they moved a trail of fine fume behind.

After about 3.5 seconds reaction time, the flame front was fully established around the specimen, Fig. 2, which showed very little sign of internal instability. At 4.0 seconds (approx. 1% C), However, the main specimen bulk was clearly distorted due to the onset of subsurface CO formation, although particle ejection, the flame front and hence surface decarburization were all still occurring. At 4.3 seconds, the flame front was still present and the main bulk of liquid metal was heavily distorted by a deep-seated boil which swelled the specimen considerably to the point where at 4.5 seconds the reaction ceased as the liquid bulk hit the tube walls. At no time during this typical sequence of events did the film show any indication of the formation of liquid iron oxide. The temperature rose as high as 2200°C.

It was obvious too that below the 1% C level, the ejection of small particles at high velocity had become much less frequent. At this carbon level, large bubbles grew from deep within the liquid and deformed the drop as they broke through the surface.

A feature of the reaction appeared to be the appearance of extremely vigorous turbulence and surface waves on the metal surface, even in the early stages of reaction when there was no boil. A close study of individual photographic frames showed that velocities associated with such phenomena reached a maximum of 200 cm/sec with the direction of movement frequently changing radically over distances equal to only 2 mm on the drop surface. This turbulence in a reacting drop was of far greater severity than that observed on a metal surface levitated in an inert atmosphere where a maximum velocity of 10-20 cm/sec can be produced by the electromagnetic field visible as small oxide particles move right around the drop, i.e. over much larger distances than the intense local eddies produced during oxidation. The consequences and possible causes of this intense turbulence are discussed in a later section .

THE OXIDATION OF IRON-SILICON ALLOYS

The oxidation of liquid Fe-Si alloys has been investigated by Filippov and Martynov [7], R. Baker [8] and Desforges and Charles [9], the first workers using an oxygen-argon jet and the second and third groups an oxygen tower, through which the alloy drops plunged in free fall. The indications from this work suggested that the alloys containing less than 3% Si reacted vigorously, whereas for silicon in excess of 6%, the reaction was suppressed due to the initial formation of a passive slag cover.

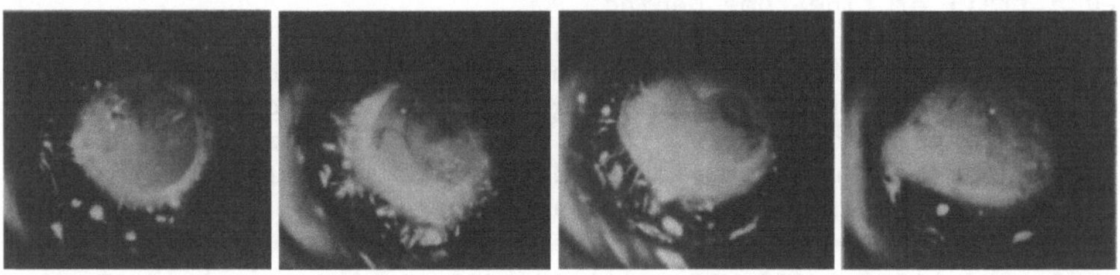

3.51 sec. 4.09 sec. 4.31 sec. 4.43 sec.

Fig. 2. Boil reaction in a pure iron-carbon alloy drop. Carbon content 1 w/o C and temperature 2000°C. Exposure time 1/10,000 sec. See text for details.

In this study, one gram specimens of the binary Fe-Si alloys were levitated in an inert atmosphere of argon or helium and brought to a temperature of 1660°C and then reacted in pure oxygen or gas mixtures containing 10% or 20% oxygen with argon as diluent. The reaction was interrupted after varying times by either simultaneously turning off the gas and casting into a copper mould or by replacing the active gas with an inert stream while levitated and subsequently casting out when the slag had collected as a pendant drop. This enabled an excellent separation to be made of metal and slag. In many instances the course of the reaction was followed by the usual pyrometer and cinematographic techniques.

The range of alloy compositions and oxygen potentials covered in this work is given in Fig. 3, which also classifies the reaction behaviour as "passive" or "burning". The passive terminology has

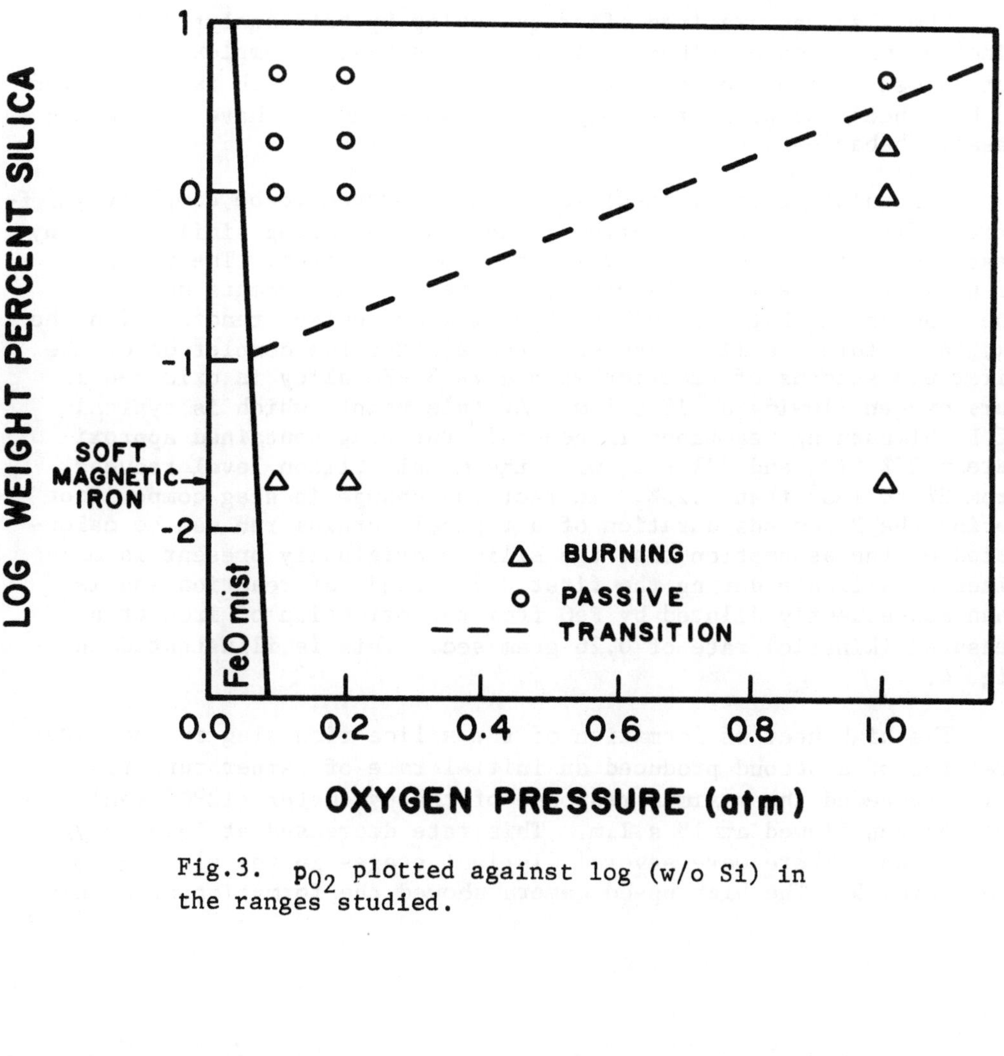

Fig.3. p_{O_2} plotted against log (w/o Si) in the ranges studied.

been used to describe the situation which arose when a liquid slag layer approaching 50μ in thickness formed during the first seconds of reaction at 1600°C and subsequently acted as a barrier to further rapid oxidation of the alloy. During burning reactions on the other hand, the slag did not so act and rapid oxidation ensued.

PASSIVE VERSUS BURNING BEHAVIOUR

The intermediate 1% and 2% Si alloys showed a transition from passive to burning stages as the oxygen potential increased. The passive specimens commenced to cool as the higher emissivity of the slag took effect. Being in the levitated state and beneath a liquid slag, it is scarcely surprising that the metal super-cooled to approximately 1340°C before solidifying.

The slag was examined after quenching by both analytical and electron microscope methods. It proved to have a complex structure but showed a fayalite layer at the metal surface which is considered to have been the barrier to oxygen transport and to have led to the passive behaviour.

The burning of the iron-silicon alloys proved to be completely different from that just described, the reaction being similar in many respects to that previously described for pure iron. The results of slag-metal chemical analyses and kinetic measurements support the conclusion that the bulk of the silicon content reacted with the oxygen to form a silica-rich slag phase after the completion of the first 0.3 seconds of reaction when a 2% Si-Fe alloy is oxidized in pure oxygen flowing at 18 s.l.m. At this point, which is typical of Fe-Si burning reactions in general, the slag contained approximately 27% SiO_2 and 73% FeO, with the metal silicon level reduced from 2% to less than 0.25%. In fact the change in slag composition during the 2 seconds duration of a typical burning run can be calculated on the assumption that the silicon originally present is oxidized to silicate during the first 0.3 seconds of reaction and is then subsequently diluted by FeO from the oxidation of iron at a measured (kinetic) rate of 0.26 gram/sec. This is illustrated in Fig. 4.

The high heat of formation of the silica-rich slag in the first fraction of a second produced an initial rate of temperature rise which exceeded the maximum response of the pyrometer (12°C/msec) when oxygen flowed at 18 s.l.m. This rate decreased at lower oxygen flow rates. There were several distinct stages in the slag formation, Fig. 5. The high speed camera showed the formation of a thin veil of oxide across the complete surface during the first few milliseconds which rapidly changed in appearance as brighter island regions of slag made their appearance and swept across the liquid surface, to collect beneath the drop. When the oxygen supply was cut off, the formation of these mobile islands ceased. The temperatures reached during a full 2 second reaction varied according to the oxygen flow conditions but were lower than with the pure iron specimens because the rate of slag formation was slower - iron being oxidised at 0.26 g/sec instead of 0.46 g/sec.

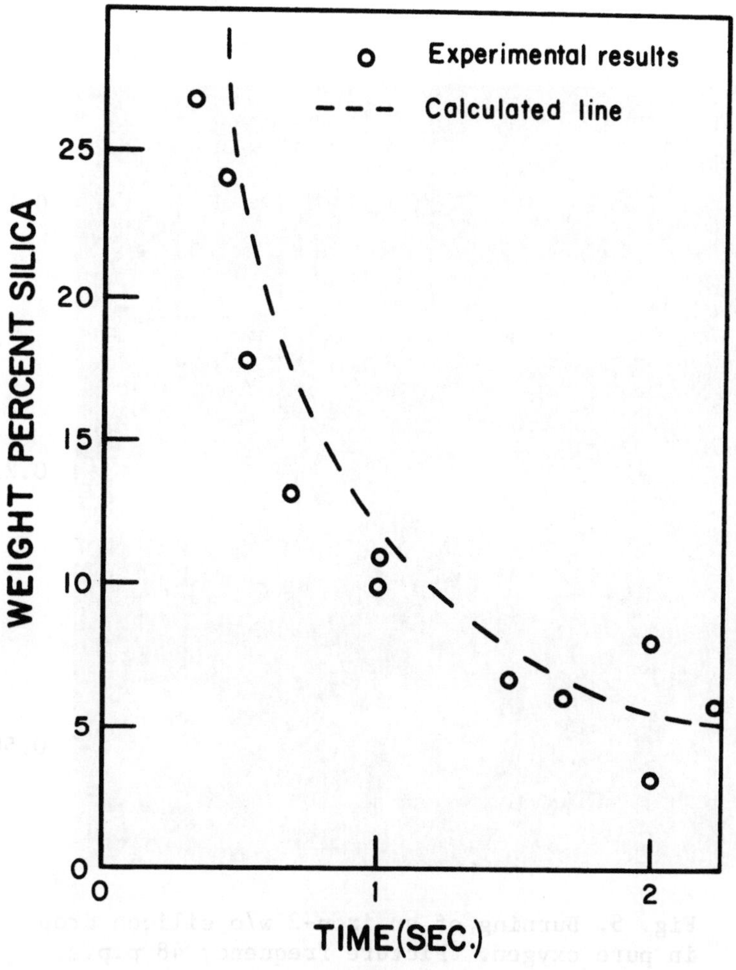

Fig. 4. Analysis of slag formed during Fe-2Si burning as a function of time. The line was calculated on the assumption that all the silicon originally present in the metal was oxidized to silica in slag forming at 0.337 g/sec, the approx. rate derived from the measured kinetics.

THE BEHAVIOUR OF IRON-CARBON-SILICON ALLOYS IN OXYGEN

In view of the completely different behaviour of the iron binary alloys with silicon and carbon, it is to be expected that the ternary alloys will behave in a rather complex fashion. There have been several studies in this field, notably by Sano et. al. [10], Morris, Riott and Illig [11] and R. Baker [8]. However in view of the detail involved, a comprehensive treatment of past and present work on this subject will be presented elsewhere with only the significant issues being now considered.

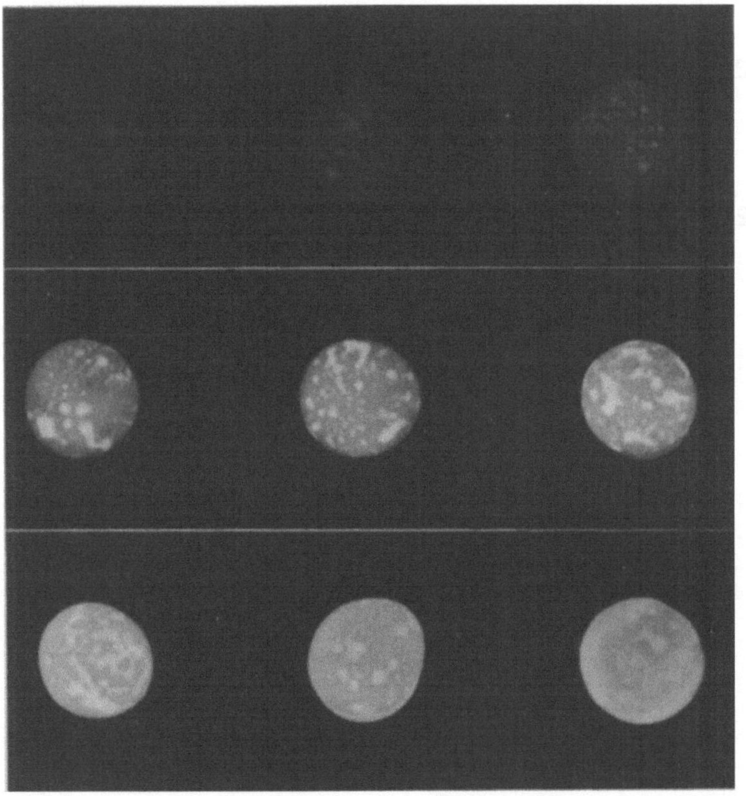

0.1 sec.

0.25 sec.

0.50 sec.

Fig. 5. Burning of an iron-2 w/o silicon drop in pure oxygen. Picture frequency 48 p.p.s. prints enlarged under the same conditions to illustrate the increase in brightness during burning.

0.1. sec.	appearance of FeO-rich islands.
0.25 sec.	islands slide beneath drop.
0.50 sec.	formation of islands ceases when oxygen is turned off.

The experimental technique was identical to that previously described and alloys used comprised a range covering 4% and 1% C and 2%, 1% and 0.1% Si. Specimens from a commercial blast furnace iron, major constituents 4.6% C, 0.88% Si and 0.24% Mn, were also used.

The sequence of events in the oxidation of the Fe-4C-Si ternary alloys as revealed by high speed photography followed initially that shown by the iron-silicon binary alloys suffering burning. A thin oxide film spread right across the metal surface during the first three milliseconds with a consequent sharp increase in specimen brightness. This was followed by the appearance after another 10 milliseconds of distinctly brighter mobile islands of oxide which grew and coalesced. At about 40 milliseconds, the camera showed the first signs of metal particles being ejected from regions beneath these slag islands which opened and closed as the ejection took place. After 90 milliseconds, the drop began to swell and the burning of ejected particles increased. Subsequently the reaction produced an alternate swelling and collapse of a massively deformed metal mass as gas was liberated from within the drop, Fig. 6. The slag layer foamed in the case of the higher carbon alloys. Reaction times of up to 2 seconds duration were achieved in some cases but the events could be interrupted quite often by the intensive boil causing the specimen to hit the walls of the tube. There was no evidence for the existence of a flame front around these alloys during oxidation. The intensive particle ejection process gave rise to the generation of large quantities of fume which prevented the reasonable use of the two-colour pyrometer. Several runs were interrupted after approximately 0.5 seconds and the specimens quenched into a copper block. The slags were extensively foamed. Slag layers were examined by the electron microprobe in steps of 2 microns across the 5 to 20 micron width. The compositions were very close to that of a silica-saturated $FeO-SiO_2$ slags at 1700-1900°C.

The Fe-1C-Si drops reacted in an identical manner to the Fe-4C-Si drops until the beginning of the carbon boil. The bursting of bubbles from the Fe-1C-Si drops was much more vigorous. Waves spread around the drop from the sites of bursting and the bubbles were not caught in the slag, so no slag foam was formed. The boil rapidly became extremely violent and after 0.16 seconds the drops disintegrated and hit the side of the tube.

DISCUSSION

In assessing the significance of the data obtained in this study of the fundamental steel-making reactions, the discussion on this occasion will have to be restricted to three issues relating to decarburization, slag formation and finally, fuming.

The Iron-Carbon Reaction

When liquid iron containing carbon reacts with pure oxygen, there can be no doubt that even small traces of aluminum or silicon can induce a slag forming reaction which alters the subsequent behaviour of the alloy. Whilst the lower limits to this effect have not been determined in our work, the presence of 0.3% Si and 0.2% Al in a 5% C alloy were sufficient to produce a slag cover through which subsequent decarburization had to proceed by ebullition rather than by the surface decarburization achievable with pure iron-carbon alloys. The use of crucibles to prepare ferrous alloys for these fundamental studies must therefore be avoided. The present work with high purity alloys demonstrated that a surface oxide is *not* formed initially even in pure oxygen when a near carbon-saturated alloy

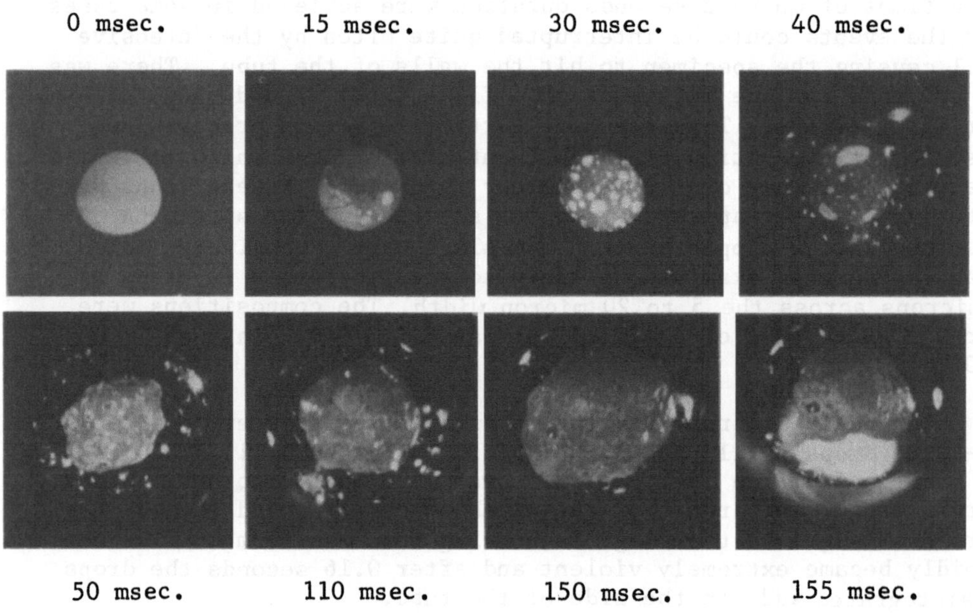

Fig. 6. Reaction of Fe-4C-1Si drop in pure oxygen. The sequence of reaction processes is described in the text. The swelling and collapse illustrated here after 150 msec. reaction occurred at approximately 50 msec. intervals. Exposure time 1/10,000 sec.

reacts in oxygen at 1600°C. A very rapid decarburization ensues at a rate which is in accord with the earlier results of L. A. Baker [4]. Oxygen gas access to the surface is restricted by the counter-diffusing carbon monoxide which is not being burnt to carbon dioxide (to any major extent) within the boundary layer during the initial decarburization period. This is supported by film evidence, the results of Distin [6] and Baker [4]. In addition if the surface decarburization did involve the strongly endothermic carbon dioxide reaction, the observed temperature rise of the specimen would require the existence from the very first instance of an intense flame front around the specimen. This was not observed. Detailed heat balance calculations will be given elsewhere.

One of the most significant features of the present work related to the insight now obtained into the transition from surface decarburization to a deep-seated carbon boil. It has long been recognized that the oxygen level during decarburization may rise to a significantly higher level than a postulated "equilibrium" amount but since the level achieved falls far short of that required for homogeneous carbon monoxide nucleation a concept involving adventitious nuclei has been invoked to support heterogeneous nucleation.

It is considered that the observations relating to the existence of intense surface turbulence in this work provide an as yet unexplored avenue to explain the sub-surface nucleation and growth of carbon monoxide bubbles which lead to the carbon boil. This turbulence, which has nothing whatsoever to do with the electromagnetic field, is of such intensity that if the eddy centres were assumed to constitute free vortices, then negative pressures of the order of 30,000 atmospheres could exist at 4 microns from their centers [12]. certainly they would have an extremely short lifetime of perhaps one millisecond or less prior to degenerating but a proportion could give rise to a new form of nucleation of carbon monoxide at solute levels approaching those observed.

This nucleation is neither heterogeneous nor homogeneous in the accepted sense and will be referred to as vortex nucleation. The vortices could also serve as funnels through which slag or gas could be entrained in the metal. Although space does not permit any extensive discussion as to the cause of this turbulence the phenomenon is frequently encountered at interfaces across which there exists a process involving rapid mass transfer. In this case the large effect which oxygen exerts on the surface tension of the liquid is likely to lead to a high degree of interfacial turbulence as a result of mass transfer.

Slag Formation

In discussing the initial formation of oxide layers in this

work it should be remembered that the oxygen content in the back-mixed oxygen with inert cooling gas mixture increased from zero to 100% over a time of the order of 70 milliseconds (dead volume from valve to specimen replaced twice). This time is relatively slow compared with the initial spread of oxide across the complete surface of specimens in five milliseconds. The oxygen content of the gas and at the metal surface should be considered as rising gradually.

Consider the clean surface of an alloy drop at 1600°C exposed to a gas containing a proportion of oxygen molecules. Table 1 gives the calculated dissolved oxygen levels in liquid alloys which would be in equilibrium with the phases noted. These values give an

TABLE I

Oxygen-Slag Equilibria, Temp. 1600°C

ALLOY Fe-2Si		ALLOY Fe-4C-2Si	
Slag	O%	Slag	O%
		CO (1 atm.)	0.0005
SiO_2	0.004	SiO_2	0.0006
$FeO.SiO_2$	0.020	$FeO.SiO_2$	0.035
$2FeO.SiO_2$	0.042	$2FeO.SiO_2$	0.085
FeO	0.226	FeO	0.36

indication of the overpotential of oxygen which must exist before the formation of the different phases. The initial silicate film which was formed on Fe-2Si and Fe-4C-Si alloy drops was clearly metastable with respect to the bulk metal but kinetic considerations, which included the rapid arrival of oxygen at the surface and the ease of nucleation of a liquid phase which wets the substrate, favoured its formation. In the case of the Fe-4C alloy drops it appeared that the over-potential did not rise high enough to form FeO before the carbon monoxide formation got under way. When the Fe-4C drops contained impurities like Si and Al then these formed mixed oxides at lower over-potential than FeO and led to coverage of the surface by an oxide layer before surface formation of carbon monoxide.

The silicate layers formed initially on Fe-Si and Fe-C-Si drops in pure oxygen showed identical but unexpected behaviour. Blobs of FeO-rich slag were nucleated, grew and coalesced and flowed round underneath the drop. In the case of Fe-Si burning these blobs were sites for the oxidation of iron, which was the sole reaction after

0.3 seconds when all the silicon had been oxidized, and the reason for the slower iron oxidation reaction on the Fe-2Si drops was that part of the surface covered by the slower reacting silicate film.

In the case of the carbon-bearing ternary alloys and the blast furnace metal the carbon-oxygen reaction got going within about 40 milliseconds from beneath the silicate slag islands and, as a direct consequence of the ejection of the gas product, *metal* droplets issued forth to ignite in the gas and form fume. The complete absence of this process from areas other than the FeO blobs and the fact that metal droplets were ejected and not slag suggested that the carbon-oxygen reaction was occurring just within the surface of the metal. The existence of severe turbulence as a result of rapid mass transfer was therefore inferred to exist beneath the blobs.

The concept of a critical supersaturation before CO nucleation was retained in this work and is a consequence of vortex nucleation theory. The fact that there was a delay in the start of the carbon monoxide boil reaction suggested that while silicon was being oxidized in the first 40 msec of reaction the dissolved oxygen concentration in the metal did not rise to a sufficiently high level to nucleate carbon monoxide.

After 0.16 seconds of reaction with the ternary Fe-4C-Si alloy drops no more slag was produced, but the existing slag was foamed by the issuing gas. This slag foaming process only occurred at higher carbon levels because the ejection of gas was not always sufficiently energetic for bubbles to break through the slag layer at high carbon levels and many bubbles were captured in the slag. At 1% C the bubbles were able to burst through the slag and no slag foam was formed. If the growth of a CO bubble once nucleated was restricted by the supply of oxygen, then this would be less of a problem at 1% C as against 4% C which in turn would explain the more energetic bubble release.

Fuming

The evidence provided by the high speed films and subsequent analysis of the collected fume strongly supported the view that the mechanism of fume formation during the boil of pure iron-carbon drops involved firstly the ejection of metal as a result of the boil. These particles ranged in size up to 100 micron diameter moving at speeds up to 1500 cm/sec in the case of very small droplets. There were also signs of some very large droplets around 1 mm diameter moving much more slowly. The droplets burned rapidly in the gas stream, the smaller ones leaving behind a trail of fine fume as they moved across the field of view. Fume from the oxidation of the ternary alloys was formed from the explosion of 100 micron diameter ejected particles which reacted vigorously in the oxygen gas and rose to very high temperature. Their explosion produced a very bright flame which could only have been caused by the burning of iron vapour. In appearance and composition this fume was identical to that reported by Morris [11] and Szekely [16] and also to that collected from inductrial furnaces.

REFERENCES

1. E. T. Turkdogan, P. Grieveson and L. S. Darken: J. Phys. Chem. 67, 1647 (1963).
2. L. Kirschfeld: Angew. Chem. 71, 663 (1959).
3. P. L. Harrison and A. D. Yoffe: Proc. Roy. Soc. 261A, 357 (1961).
4. L. A. Baker, N. A. Warner and A. E. Jenkins: Trans. AIME 239, 857 (1967).
5. J. H. Swisher and E. T. Turkdogan: Trans AIME 239, 602 (1967).
6. P. A. Distin, G. D. Hallett and F. D. Richardson: J.I.S.I. 206, 821 (1968).
7. S. I. Filippov and S. Z. Martynov: Izvest. Vuz. Chern. Met. 1, 5 (1961).
   H. Brutcher Transl. No. 5134.
8. R. Baker: J.I.S.I. 205, 637 (1967).
9. K. Ito and K. Sano: I.S.I. Japan Trans. 8, 165 (1968).
10. C. D. Desforges and J. A. Charles: British J. Appl. Phys. 15, 983 (1964).
11. J. P. Morris, J. P. Riott and E. G. Illig: J. Metals 18, 803 (1966).
12. R. B. Dean: J. Appl. Phys. 15, 446 (1944).
13. N. Sano and R. G. Ward: McMaster University, Hamilton, Ontario, Canada.
14. R. S. Kaplan and W. O. Philbrook: Paper presented at 98th AIME Annual Meeting, Washington, D. C. 1969.
15. L. A. Baker and R. G. Ward: J.I.S.I. 205, 714 (1967).
16. A. G. Szekely: Paper presented at the 96th AIME Annual Meeting, Los Angeles 1967.

THE INTERACTION OF CARBON MONOXIDE AND LIQUID IRON

T. B. King*, R. A. Karasev**, and P. Dastur***

*Massachusetts Institute of Technology, Cambridge, Mass.
**A. A. Baikov Institute for Ferrous Metallurgy, Moscow
***Republic Steel Corporation, Cleveland, Ohio

Absorption of carbon monoxide by and desorption of carbon monoxide from liquid iron were studied with a modified Sieverts technique. It is shown that the rate of absorption for all carbon contents, and the rate of desorption from alloys of medium carbon content, are compatible with rate control by transport of oxygen in the liquid. The rate of desorption from alloys of low carbon content is anomalously slow, and may best be explained on the basis that a slow surface reaction, on sites limited in availability by the strong adsorption of oxygen, controls the overall rate.

Most studies of the carbon-oxygen reaction in liquid iron have simulated conditions occurring in one or another phase of the steelmaking process. The mechanism of the reaction, at least under some experimental conditions, is fairly well understood and the appropriate rate-limiting factors have been recognized. Nevertheless, there are some features of the reaction, as it occurs during solidification, during the late stages of vacuum refining, and during flushing with a neutral gas, which are imperfectly understood.

While much of the earlier research has some bearing on the results of the present investigation, the work which comes closest in intent and execution is that of Parlee, Seagle and Schuhmann [1]. These authors measured the rate at which CO is absorbed by or evolved from liquid iron. A modified Sieverts' apparatus was used and the reaction was made to proceed by suddenly raising or lowering the pressure (though not sufficiently to induce bubble formation). However, the lowest carbon content of the iron was 0.15%. It is for carbon contents somewhat below this value that slow removal of carbon monoxide in vacuum degassing, or in flushing with argon, is found in practice.

The present work has been carried out under the simplest conditions, i.e. the total pressure over the melt remained at 1 atm. and there was only natural convection in the liquid iron.

EXPERIMENTAL

The experimental procedure consisted of melting, under argon, vacuum-refined master alloys, of various carbon contents and very low oxygen content. The furnace was essentially the hot volume of a Sieverts apparatus. Studies of the rate of absorption of CO by liquid iron were carried out by melting under argon, flushing out the argon with CO, then sealing the system and measuring the volume change, at constant pressure, as CO was absorbed. In desorption experiments, the metal was first brought to equilibrium with pure CO by the above method and the system was then flushed rapidly with argon, sealed, and the volume change due to CO evolution measured as a function of time, again at constant pressure.

The analyses of the master alloys are shown in Table 1*. It should be noted that the oxygen contents are uniformly low, exceeding 20 parts per million only for the 0.018% carbon alloy. The alloys were obtained as vacuum-refined, cast and forged bars. Specimens of appropriate size were cut from the bars for each experiment. The two gases used, carbon monoxide and argon, were dried by passing through anhydrone columns. Carbon monoxide was also prepurified, to remove CO_2, by passing through an ascarite column. Argon was prepurified by passing through a copper furnace.

Figure 1 shows the main features of the apparatus, consisting of a furnace (5) connected to a measuring burette, of 50 cm^3 capacity (3) through a three-way stopcock (4). The system pressure is maintained at 1 atm. by adjusting the level of fluid in the measuring burette with the 20 cm^3 capacity syringe (1) in which a tight-fitting plunger (2) operates. Argon or carbon monoxide is introduced into the furnace through a three-way stopcock (6). The gases may be flushed through the furnace and out of the system by opening both stopcocks. Butyl phthalate was used as a fluid in the burette and gas bubblers.

A more detailed view of the furnace is shown in Figure 2. The design is such as to minimize the hot volume which was (including the alumina tubes leading into the furnace) about 120 cm^3. The furnace chamber consists of a recrystallized alumina crucible (11), 40 mm. outside diameter, and 100 mm. in height. The resistance heating element (13) is platinum, 40% rhodium alloy wire, 1 mm. in diameter, arranged in a hairpin-type winding and maintained in position by threading the wire through double-bore alumina insulators. The crucible and winding are set in bubble alumina insulation (14) and the furnace shell (9) is stainless steel with transite endplates (8).

*See appendix.

The furnace is sealed by a water-cooled brass head (6) attached to the alumina crucible with a silicone rubber gasket (5). Gas is introduced into and removed from the furnace through recrystallized alumina tubes (10), 5 mm inside diameter and 25 cm. long, set into the furnace head through Wilson seals (3). Pyrex T pieces (2) are attached to the alumina tubes with tygon tubing (4) and viewing ports are made of pyrex flats (1) cemented to the tubes. The iron melt (15) is contained in a recrystallized alumina crucible (16) 31 mm outside diameter and 31 mm high. A special, high-fired grade of alumina was used so that reaction with the melt would be minimized. Temperatures are measured by a platinum-rhodium thermo-couple, inserted through the furnace head and almost touching the melt. Alumina radiation shields (7) minimize heat losses from the top.

During a given run the temperature could be maintained at $1580^{o} \pm 7^{o}C$ by manual adjustment of the power supply or by a simple, on-off controller.

In a typical experiment the weight of iron-carbon alloy varied between 50 and 80 gm. The alloy was melted in pure argon and, after the temperature and, hence, the volume of the system was stabilized the system was flushed with carbon monoxide for 2 to 5 minutes. Complete flushing of the system was facilitated by adjusting the

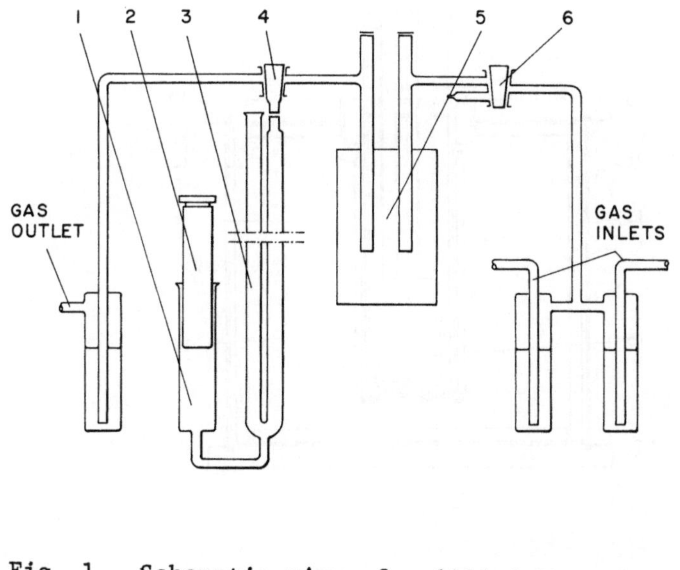

Fig. 1. Schematic view of modified Sieverts apparatus.

height of the outlet alumina tube so that short circuiting of the gas would not readily occur. Both inlet and outlet stopcocks were then closed, and readings of the gas volume taken at intervals of 1 min. or 30 sec. When there was no further change in the burette reading, the system was flushed again with pure argon, and readings of the gas volume during desorption of CO were taken. In most cases, the melt was then allowed to solidify and the amount of gas evolved during solidification was measured. As will be discussed later, this enables an accurate value to be obtained for the oxygen content. Since solidification results in the removal of almost all the oxygen dissolved in the liquid metal, absorption and desorption processes could be carried out repeatedly in the same run.

The reaction path during absorption and desorption of CO is shown in Figure 3, which is a plot of the equilibrium carbon-oxygen product for the temperature at which all experiments were carried out, 1580°C [2]. The lines with arrows show the path of reaction duringabsorption and desorption for various initial carbon contents.

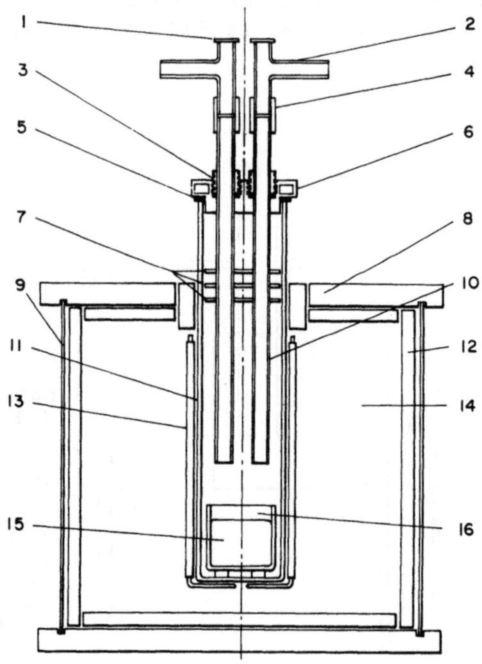

Fig. 2. Enlarged section of furnace.

As will be shown later, absorption resulted in equilibrium with carbon monoxide at 1580°C and 1 atm. pressure. Figure 3 also illustrates that the total volume of gas which can be absorbed or desorbed increases rapidly as the carbon content of the starting alloy decreases.

REACTION WITH THE CRUCIBLE

Since the highest carbon content of the iron was 0.3% and the temperature was only 1580°C, reaction with the high-fired alumina crucible was not expected to be a problem. This was confirmed by holding the melts initially under argon; the burette reading did not increase with time. Further confirmation was obtained from solidification experiments. After the melts were brought to equilibrium with CO at 1 atm pressure they were cooled to the freezing point and allowed to solidify. The volume of CO evolved during cooling and solidification was recorded (a correction must be applied for the temperature change of the hot volume). The solid specimens were then analyzed for carbon and oxygen. For carbon contents above about 0.04% the analyzed oxygen was always 10 parts per million or less. The amounts of carbon and oxygen removed could be calculated from the CO evolved, enabling the carbon and oxygen contents of the liquid to be estimated. The results of six such experiments are given in Table II. The last column shows the calculated carbon-oxygen products. Since these are close to the equilibrium value for 1580°C of 1.97×10^{-3} it may be concluded that equilibrium is reached during absorption, further, that essentially all

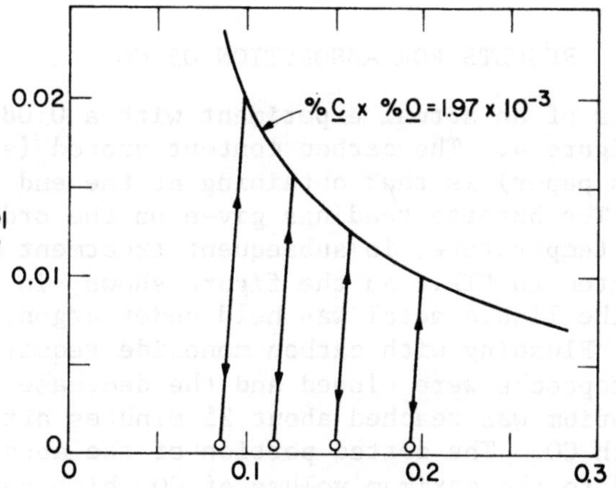

Fig. 3. Variation of % $\underline{C}$ and % $\underline{O}$ during absorption and desorption.

the oxygen in the liquid metal is removed as CO during solidification, and, finally, that reaction with the crucible is unimportant. The relatively small amount of CO_2 in the gas for the low carbon melts was not taken into account; as Table II shows, these also give somewhat lower figures for the carbon-oxygen product.

Table II

Solidification of Melts equilibrated with CO at 1 Atm. and 1580°C

Wt. of Melt(G) (gm)	%C(solid)	CO Evolved On Solidification (cm^3 STP)	Calculated $\%\underline{C} \times \%\underline{O}$ (in liquid)
85.5	0.19	11.7	2.13×10^{-3}
55.0	0.045	21.2	1.85×10^{-3}
52.5	0.162	7.5	1.95×10^{-3}
62.5	0.154	8.4	2.0×10^{-3}
51.7	0.035	22.0	1.85×10^{-3}
48.2	0.118	9.7	1.96×10^{-3}

RESULTS FOR ABSORPTION OF CO

The results of an actual experiment with a 0.084% carbon alloy are shown in Figure 4. The carbon content quoted (as is the case throughout this paper) is that obtaining at the end of the absorption process. The burette readings given on the ordinate of Figure 4 are for room temperature; in subsequent treatment of the data volumes are corrected to STP. As the figure shows, in the first ten minutes while the liquid metal was held under argon, there was no volume change. Flushing with carbon monoxide required 2 minutes and then the stopcocks were closed and the decrease in volume measured. Equilibrium was reached about 25 minutes after the start of flushing with CO. The dotted portion of the absorption curve is extrapolated to the maximum volume of CO which could be absorbed (according to the oxygen content of the metal at equilibrium). During the flushing period some CO is being absorbed but the CO pressure does not reach 1 atm. until flushing is completed.

The shape of the absorption curve suggests first-order behavior.

Intuitively one might expect this to be so since the likeliest rate-controlling step in the overall reaction is the transport of oxygen in the metal, with the surface reaction at equilibrium. This would hold true for all but the lowest carbon contents where transport of carbon would have to be taken into account. Since the temperature distribution in the furnace is such that thermal convection in the metal is inevitable, we may, as a simplification, consider transport of oxygen in the liquid metal to be equivalent to diffusion across a diffusion boundary layer, of thickness δ_O. While this may not be the most reasonable representation of the actual convection conditions, it will serve in the derivation of the appropriate rate equation.

Making the above assumptions, the rate at which CO is absorbed, in mol sec^{-1}, may be expressed as follows [3]:

$$\frac{dn_{CO}}{dt} = A \cdot \frac{D_O}{\delta_O} [C_O^* - C_O] = A \cdot \frac{D_C}{\delta_C} [C_C^* - C_C] \qquad (1)$$

In this equation, D_O and D_C are the diffusion coefficients of oxygen and carbon in liquid iron, δ_O and δ_C are the respective

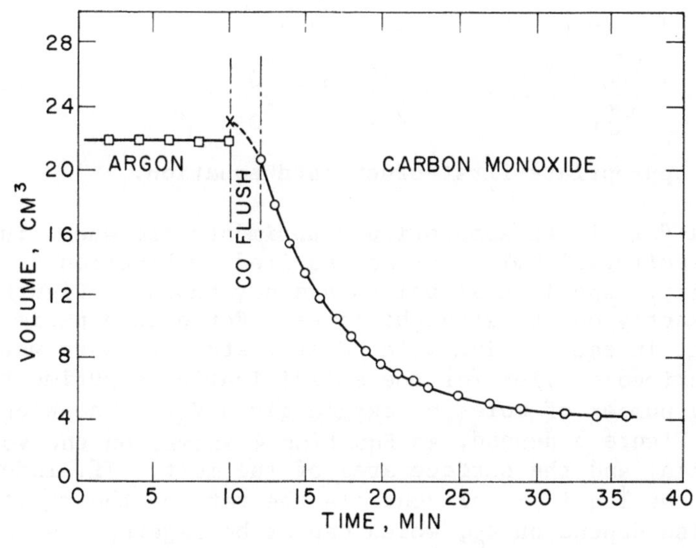

Fig. 4. Plot of direct experimental data. Run No. 5. %C (final) = 0.084. Temp. 1580°C.

diffusion boundary layer thicknesses, A is the area of the metal surface, C_O^* and C_C^* are the surface concentrations of carbon and oxygen, in mol. cm^{-3}, in equilibrium with CO, and C_O and C_C are the time-dependent, bulk concentrations.

Making the further assumption that $C_C >> C_O$ (not valid for very low carbon alloys) $C_C \approx C_C^*$ and C_O^* will be constant and calculable from the equilibrium constant for the reaction, CO (g) = $\underline{C} + \underline{O}$.

Since the alloys, at the start of absorption, contain essentially no oxygen, we may replace C_O by $n_{CO}\rho/G$, and C_O^* by $n_{CO}^*\rho/G$, where G is the weight and ρ the density of metal, n_{CO} the number of moles of CO absorbed during time, t, and n_{CO}^* the total number of moles of CO absorbed when equilibrium is reached.

Substituting into Equation 1, we have:

$$\frac{dn_{CO}}{dt} = \frac{A\cdot\rho}{G}\frac{D_O}{\delta_O}(n_{CO}^* - n_{CO}) \qquad (2)$$

Since, at constant pressure and temperature, the volume of CO absorbed is directly proportional to the number of moles:

$$\frac{dV_{CO}}{dt} = \frac{A\ \rho}{G}\cdot\frac{D_O}{\delta_O}(V_{CO}^* - V_{CO}) \qquad (3)$$

The integrated form of this equation is:

$$\log\left(\frac{V_{CO}^* - V_{CO}}{V_{CO}^*}\right) = -\frac{1}{2.3}\frac{A\cdot\rho}{G}\frac{D_O}{\delta_O}\cdot t \qquad (4)$$

which is the appropriate first-order rate equation.

The data for eleven absorption experiments are shown in Figure 5 on such a semilogarithmic plot as required by Equation 4. For the sake of clarity, experimental points are not shown in Figure 5, but these fall exactly on the straight lines. For each experiment, the oxygen content in equilibrium with CO at 1 atm. pressure was calculated as described earlier for the solidification experiments; the corresponding number of moles of oxygen gives V_{CO}^*. The slopes of the lines in Figure 5 depend, as Equation 4 shows, on the volume of the liquid metal and the surface area of the melt. If, indeed, diffusion in the liquid metal controls the rate of the reaction, the slope will also depend on δ_O, which cannot be regarded as constant from one experiment to another. For example, runs 10 and 11 were conducted after a new heating element had been installed in the furnace; the temperature distribution in the liquid metal is not expected to be the same as with the original furnace element.

The data of Figure 5 are collected in Table III in which the last column lists the calculated velocity constant, D_O/δ_O (according to Equation 4). The values of the velocity constant are not unreasonable for the particular experimental setup used. For example, using the value of 2.3×10^{-5} $cm^2 sec^{-1}$ for the diffusion coefficient of oxygen in liquid iron recently determined by McCarron and Belton [4], the diffusion boundary layer thicknesses vary from 1.1×10^{-2} to 4.1×10^{-3} cm.

However, a conclusion that the rate of CO absorption is indeed determined by diffusion in the liquid metal cannot be made unambiguously. A rate equation similar to Equation 4 can be derived for

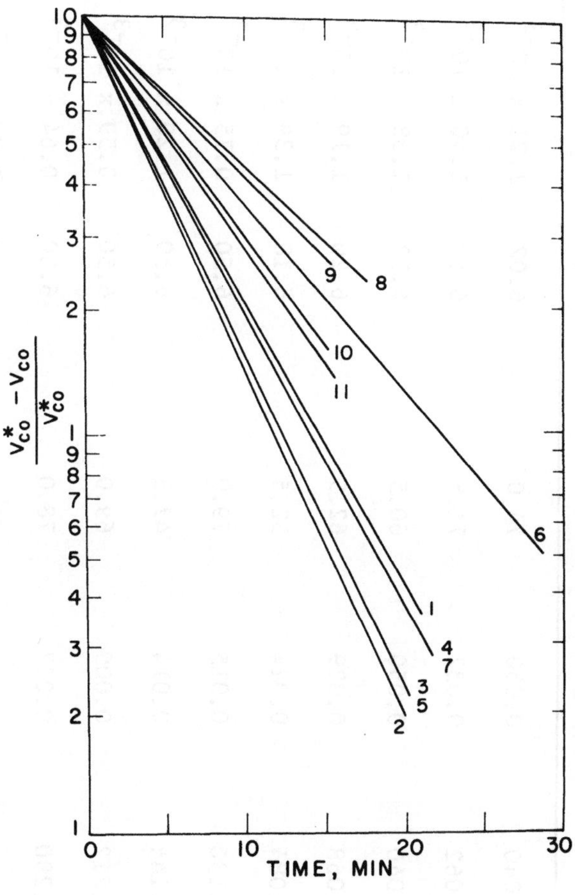

Fig. 5. $(V^*_{co} - V_{co})/V^*_{co}$ versus time (min.). Plot of data from experiments I through 11. Absorption of CO at 1580°C.

TABLE III

EXPERIMENTAL DATA FOR ABSORPTION OF CO AT 1580°C

Run No.	%C (final)	%O (equilibrium)	Metal weight (G) M (gm)	Area (A) (cm^2)	Slope (from Fig. 5) (sec^{-1})	D/δ ($cm\ sec^{-1}$)	V^*, cm^3 (STP)
1	0.040	0.050	74,0	6.02	1.21×10^{-3}	4.86×10^{-3}	51.2
2	0.062	0.032	71.5	6.50	1.40×10^{-3}	5.08×10^{-3}	31.9
3	0.040	0.050	60.5	6.72	1.38×10^{-3}	5.60×10^{-3}	42.0
4	0.068	0.029	62.5	6.20	1.19×10^{-3}	3.95×10^{-3}	25.5
5	0.084	0.024	52.5	6.18	1.39×10^{-3}	3.86×10^{-3}	17.3
6	0.130	0.015	79.0	6.50	0.75×10^{-3}	2.97×10^{-3}	16.8
7	0.144	0.014	47.5	6.50	1.20×10^{-3}	2.88×10^{-3}	9.12
8	0.212	0.009	69.0	6.50	0.59×10^{-3}	2.04×10^{-3}	9.0
9	0.290	0.007	78.0	6.50	0.64×10^{-3}	2.52×10^{-3}	7.4
10	0.090	0.022	70.0	6.50	0.88×10^{-3}	3.10×10^{-3}	21.5
11	0.210	0.009	70.0	6.50	0.93×10^{-3}	3.30×10^{-3}	8.8

the case that the surface reaction controls the overall rate.

We assume that adsorption of CO is rapid and the rate is determined by the reaction:

$$CO(ads) = \underline{C} + \underline{O} \tag{5}$$

The net rate of the absorption reaction is then given in mol.sec.$^{-1}$ cm^{-2}.

$$\frac{1}{A}\frac{dn_{CO}}{dt} = k_1 P_{CO} - k_2 \cdot C_C \cdot C_O \tag{6}$$

As in the derivation of Equation 4, we make the tentative assumption that C_C is substantially constant.

At equilibrium, $\frac{dn_{CO}}{dt} = 0$ and:

$$k_1 P_{CO} = k_2 C_C^* . C_O^* \simeq k_2 C_C C_O^*$$

Hence,

$$\frac{1}{A}\frac{dn_{CO}}{dt} = k_2 C_C \, (C_O^* - C_O) \tag{7}$$

which is the same form as Equation 1. We may therefore write, immediately, the integrated form:

$$\log \left(\frac{V_{CO}^* - V_{CO}}{V_{CO}}\right) = \frac{1}{2.3} \cdot \frac{A\rho}{G} \; k_2 C_C \cdot t \tag{8}$$

The velocity constants of Table III are plotted in Figure 6 as a function of the carbon content. For comparison the similarly derived rate constants of Parlee, Seagle and Schuhmann [1] are included. The latter are somewhat larger than those obtained in the present work, but it should be pointed out that the induction stirring in their experiments would result in smaller diffusion boundary layer thicknesses. Both sets of results show that the rate constant increases fairly rapidly as the carbon content decreases. If the reaction rate is controlled by diffusion of oxygen in the metal, it is possible that the lowering of the density at higher carbon contents decreases the convective flow due to thermal gradients. On the other hand, if the reaction rate is controlled by the surface reaction, the observed dependence on the carbon content is the opposite of that expected from Equation 8.

RESULTS OF DESORPTION OF CO

The first desorption experiment was inadvertently carried out on a low carbon alloy. Surprisingly, it was found that the rate of desorption was considerably slower than the rate of absorption. Further experiments showed that the rate of desorption was markedly dependent upon the carbon content. The results of the desorption experiments are presented in Figure 7 in the same manner as the results for the absorption experiments, but V^*_{CO} is now the maximum volume of CO which can be removed from the metal and V_{CO} is the amount desorbed in time, t. It can be readily shown that the same assumptions as regards transport control or determination of the overall rate of desorption by the rate of the surface reaction lead to equations exactly similar to Equation 4 and Equation 8, respectively. On this basis, the results of desorption experiments for alloys of initial carbon content varying from 0.05% to 0.21% are plotted in Figure 7. For comparison, the rate of absorption, for the 0.21% carbon alloy, is also included in Figure 7. Clearly, the rates of absorption and desorption for higher carbon alloys are similar, but, for low carbon alloys, the rate of desorption is much lower than the rate of absorption. As Figure 7 also shows, first-order behavior is no longer observed for very low carbon alloys, but the rate of desorption decreases much faster than would be expected. It is true that strict adherence to the first-order rate law cannot be expected

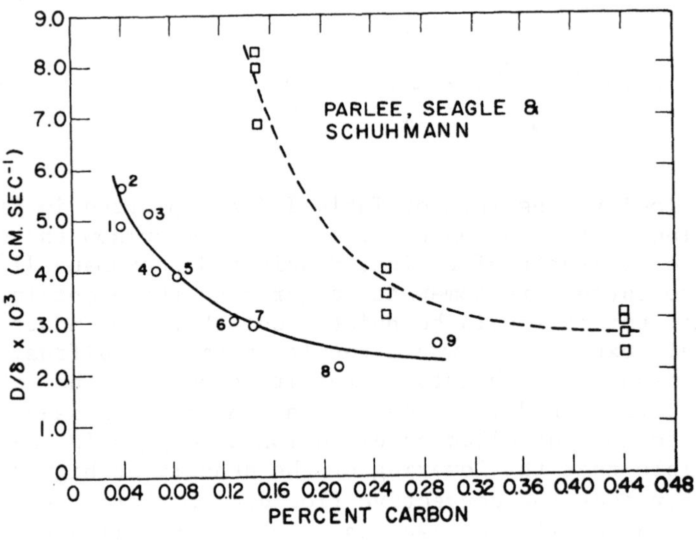

Fig. 6. Rate parameter D/δ from runs 1 through 9 versus %C (final). Absorption at 1580°C.

for the low carbon alloys, in which the change of the carbon content during absorption or desorption is appreciable, but this effect should be less apparent at short times. However, even the initial slope of the lines is much less for the low carbon alloys.

In at least one respect the desorption process differs from that of absorption. Desorption takes place into an argon atmosphere so that diffusion of CO in the gas phase is a necessary step in the overall reaction. The possible effect of the increasing concentration of CO in the bulk gas phase during reaction can be dismissed since in most cases (and particularly in low carbon runs) less than 10 cm^3 of gas were evolved into a total volume, including the measuring burette, of about 160 cm^3. The possibility that the reaction rate is limited by slow transport in the gas phase must, however be considered. To test for this effect, the apparatus was modified so that the furnace and measuring burette formed a closed loop. through which the gas could be circulated by a simple solenoid-activated pump (shown schematically in Figure 8). One of the alumina tubes entering the furnace was lowered to a position close to the metal surface so that the gas was continually swept over the surface. Such circulation had little effect on the rate of evolution of carbon monoxide, as can be seen from Figure 9, which shows the results for a 0.1% carbon alloy. The difference between the two curves is within the limits of reproducibility from one experiment

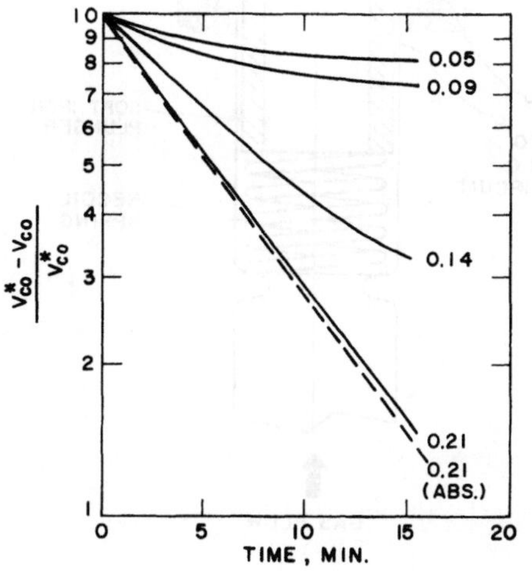

Fig. 7. $(V^*_{CO} - V_{CO})/V^*_{CO}$ versus time (min.) Semilogarithmic plot of desorption data at 1580°C. Carbon content as indicated.

to another. It may therefore be concluded that transport of CO in the gas phase does not limit the rate of desorption.

Effect of Surface-Active Solutes

The drastic variation of the rate of desorption with the carbon content of the alloy seems incompatible with the previous tentative assumption that the rate is limited by transport in the metal. It is, however, not incompatible with a slow surface reaction, the rate of which could be influenced by adsorption of impurities on the surface of the metal. The alloys of low carbon content are also those of high oxygen content and oxygen in iron is a particularly surface active solute [5]. Since sulphur is also surface active in iron, experiments in which sulphur was added to both low and high carbon alloys were conducted. The results of these experiments are

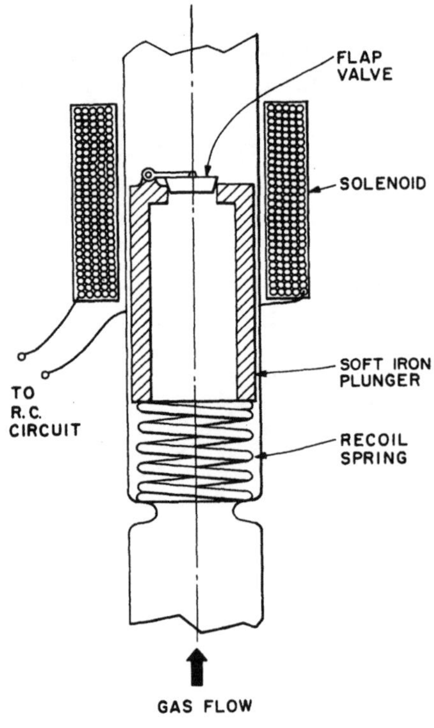

Fig. 8. Sketch of solenoid, circulating pump.

presented in Figure 10 as the percent complete reaction, 100 V_{CO}/V^*_{CO} versus time. The addition of sulphur does not retard the rate of reaction, though such retardation might have been expected in the 0.2% C alloy. If, indeed, surface adsorption affects the reaction rate, the effect would seem to be specific to oxygen.

Isotopic Exchange Reaction

The striking discrepancy between the rates of absorption and desorption for low carbon alloys was explored further in experiments in which, after equilibration with carbon monoxide, a mixture of carbon monoxide and argon (rather than pure argon) was used as the flushing gas and the rate of desorption of CO into the gas mixture was followed. The results, which are typical of a number of runs carried out on low carbon alloys, are presented in Figure 11. It is clear that admixture of carbon monoxide with the argon increases the rate of desorption. For a 0.14% alloy, the presence of 25% CO in the gas increased the desorption rate noticeably. For a 0.09%

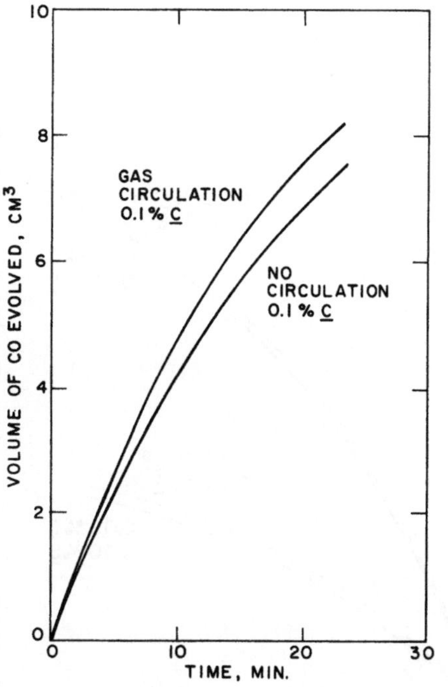

Fig. 9. Effect of gas circulation on rate of desorption of CO from 0.1%C alloy.

carbon alloy, 50% CO in the gas was necessary to increase the rate of desorption until it approached the rate of absorption (also shown in Figure 11). A further experiment on the 0.14% carbon alloy, in which desorption took place with an argon-10% hydrogen atmosphere is somewhat inconclusive. The rate of desorption is increased slightly but the result may be within the limits of reproducibility. Experiments on a very low carbon alloy (0.05%) were unsuccessful. The addition of as much as 75% CO to the argon had no effect on the rate of desorption.

To study further the influence of the composition of the gas phase on the rate of desorption, experiments on isotopic exchange between gas and melt were carried out. For this purpose the gas phase was enriched in CO containing an appreciable amount of ^{14}C isotope. The exchange reaction may be represented as follows:

$$^{30}CO(g) + {}^{12}\underline{C} = {}^{28}\underline{CO}(g) + {}^{14}\underline{C} \quad (9)$$

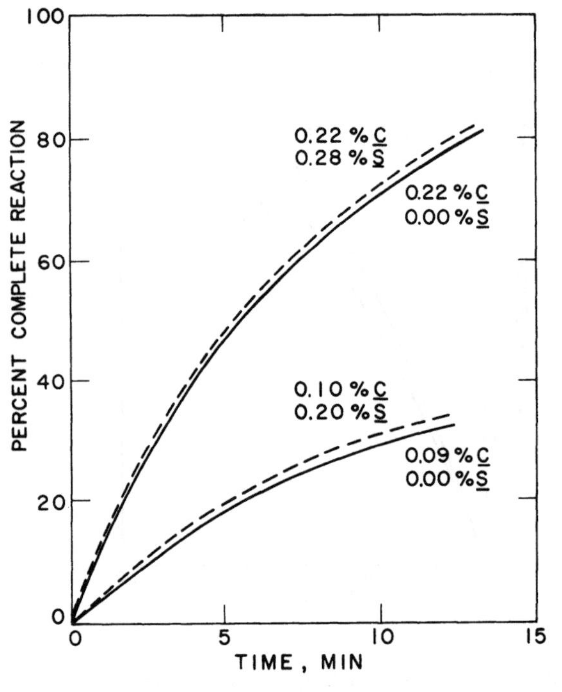

Fig. 10. Effect of sulphur on rate of desorption of CO from 0.22%$\underline{C}$ and 0.10% $\underline{C}$ alloys.

Thus the exchange reaction necessarily involves both absorption and desorption; if the latter is a slow process, as has been found for low carbon alloys, the exchange reaction should also be slow. There are, however, two advantages in studying the exchange reaction. The melt remains in chemical equilibrium with CO and there is no change in the carbon and oxygen content of the iron.

The experimental procedure consisted of melting the alloy under argon, flushing the system with ^{28}CO, as in other experiments, and allowing the gas-metal equilibrium to be reached. Some of gas was then flushed out with CO enriched in ^{30}CO, until a concentration suitable for monitoring by a counter was obtained. The amount of ^{30}CO added was such that, after exchange equilibrium was reached, the counting rate was negligible compared with the initial value, CPM^*. This permitted representation of the results in the form, $(CPM^*-CPM/CPM$ versus time, in a manner analogous to the previous representation in the form, $(V^*_{CO} - V_{CO})/V^*_{CO}$ versus t. Further modification of the apparatus was made so that a Baird-Atomic scintillation counter could be installed. The circulating pump was also necessary to maintain a uniform gas composition.

The results of experiments on alloys of four different carbon contents are shown in Figure 12. For comparison, the previous results for desorption into argon are included for two of the alloys (dashed lines) as are results of two experiments in which 0.30%S was added to the alloy. As in the desorption experiments, sulphur has no appreciable effect on the reaction rate. The exchange

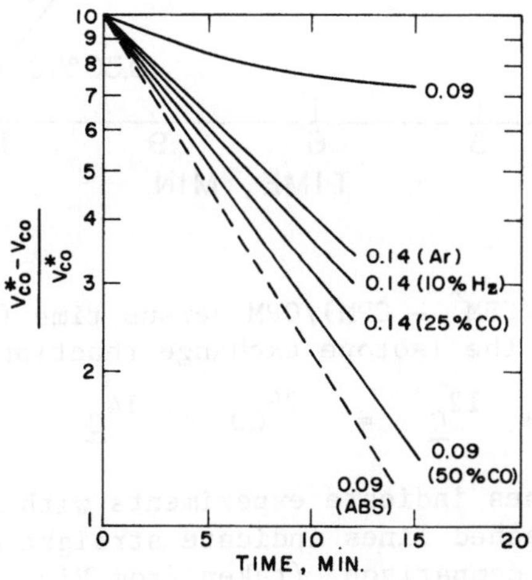

Fig. 11. $(V^*_{CO} - V_{CO})/V_{CO}$ versus time (min). Semilogarithmic plot of data on desorption into various mixtures of Ar and Co, N H_2. Dotted line indicates absorption run for comparison.

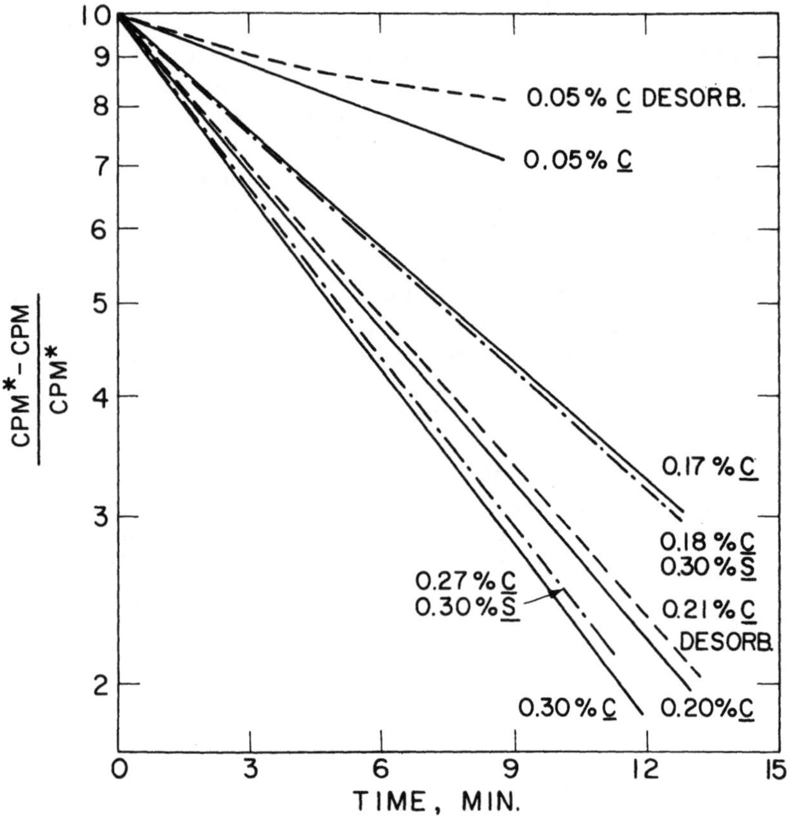

Fig. 12. $(CPM^* - CPM)/CPM$ versus time (min.) Results for the isotope exchange reaction

$$^{30}CO + {}^{12}\underline{C} = {}^{28}CO + {}^{14}\underline{C}$$

Dot-dash lines indicate experiments with added sulphur. Dashed lines indicate straight desorption results for comparison (taken from Fig. 7).

reaction is seen to be a first-order process but the rate constant decreases as the carbon content of the alloy decreases. For the 0.21%C alloy, the rates of the exchange and desorption reactions are similar. For the 0.05%C alloy, the rates of the exchange and desorption reactions are also similar in the early stages, but the rate falls off more rapidly for the desorption reaction.

DISCUSSION

While the data on absorption of CO by iron-carbon alloys are reasonably consistent with transport in the liquid alloy as the rate-determining step, the data on the reverse reaction, desorption of CO into argon are not. In particular, when the carbon content is low (less than 0.2%) and the oxygen content correspondingly high, the desorption reaction becomes very slow. The possibility of a slow surface reaction, influenced by the presence of strongly adsorbed oxygen, must therefore be considered. Eyring [6] has considered a number of such cases, particularly for gas-solid reactions. Pehlke and Elliott [7] have shown that the rates of absorption and desorption of nitrogen by liquid iron are strongly dependent on the oxygen content of the iron, and have explained their results in terms of a slow surface reaction when the surface is largely covered with adsorbed oxygen. Swisher and Turkdogan [8] have suggested that a slow surface reaction may be a general feature of gas-metal reactions at high temperatures when a strongly adsorbed solute is present, and have applied such reasoning to explain the rate of decarburization of iron alloys. In the same paper, these authors have also reconsidered the results of Parlee, Seagle and Schuhmann, Pehlke and Elliott, and a number of other authors. In view of this publication, the present discussion need not be extensive, although some particular features of the present work must be dealt with.

The measurements of Halden and Kingery [5], and, more recently, of Kozakevitch and Urbain [9], of the surface tension of iron-oxygen alloys, show that almost complete surface coverage, $\Gamma_0 = 10^{15}$ atom.cm^{-2}. where Γ_0 is the excess surface concentration calculated from the Gibbs equation is attained for bulk oxygen contents between 0.02 and0.03%.

In the present work, the discrepancy between absorption and desorption rates begins to be apparent for alloys containing 0.02%C and is very pronounced for an 0.05% C alloy. The corresponding equilibrium oxygen contents (those obtaining at the start of desorption) are 0.01% and 0.04%. According to Swisher and Turkdogan [8] the fraction of available sites covered by oxygen atoms, θ_0, is, respectively, 0.65 and 0.95. These figures are for binary oxygen alloys. The surface tension of ternary Fe-C-O alloys has not been measured but, since Kozakevitch and Urbain [10] have shown that carbon is not surface-active in iron, the possible effect of carbon on the ternary may be ignored.

During desorption (but not during absorption) the initial oxygen content of the low carbon alloys is high. We may, therefore, examine the possibility that the rate of desorption is determined, primarily, by the surface reaction, which is slow because suitable surface sites are blocked by adsorbed oxygen atoms. Following Eyring 6 , we treat such sites as reactants and write the desorption reaction as:

$$\underline{C} + \underline{O} + S = (CO\text{-}S)^{\dagger} \rightarrow CO\ (g) \qquad (10)$$

in which the slow step is the combination of carbon and oxygen at a suitable surface site, S, to form the activated complex, $(CO\text{-}S)^{\dagger}$, which then decomposes, forming CO. The processes of adsorption of reactants on and desorption of products from the surface are not considered to be rate determining.

The rate of reaction 10 is proportional to the concentration (assumed to be small) of the activated complex:

$$\vec{v} = \frac{kT}{h} C^{\dagger} \qquad (11)$$

where $\frac{kT}{h}$ is Eyring's universal rate constant. Equilibrium between the adsorbed reactants (and, hence, from the previous assumption, the reactants in solution) and the activated complex is presumed; we write, making the appropriate approximation:

$$K^{\dagger} = \frac{C^{\dagger}}{\%\underline{C}\cdot\%O\cdot C_S} \qquad (12)$$

and, thus:

$$\vec{v} = \frac{kT}{h} K^{\dagger}\ C_S\cdot\%\underline{C}\cdot\%\underline{O} \qquad (13)$$

Now, if L is the total number of surface sites and if oxygen is the only species adsorbed in significant quantity, $C_S \simeq L(1-\theta_O)$, and:

$$\vec{v} = \frac{kT}{h} K^{\dagger} \cdot L(1 - \theta_O)\cdot\%\underline{C}\cdot\%\underline{O} \qquad (14)$$

we need not consider the reverse reaction, which may be treated in a similar manner, since p_{CO} in the atmosphere above the melt is negligible during desorption. The net rate of the desorption reaction may therefore be written (in mole sec^{-1}) as:

$$\frac{dn_{CO}}{dt} = k(1 - \theta_O)\cdot\%\underline{C}\cdot\%\underline{O} \qquad (15)$$

where the constant, k, includes the terms in Equation (14) and the surface area of the metal.

If surface coverage by oxygen is high, $1-\theta_0$ is roughly proportional to $1/\%\underline{O}$ and Equation (15) therefore reduces to:

$$\frac{dn_{CO}}{dt} = k'\%\underline{C} \qquad (16)$$

An integrated form of Equation (16), in terms of V^*_{CO} and V_{CO} may be obtained, but it is not necessary to do this in order to test the applicability of the equation to the present data. According to Equation (16) the rate of gas evolution in desorption experiments should be proportional to the carbon content. From Fig. 7, we may obtain the slopes of the lines, i.e. $d \log (V^*_{CO} - V_{CO})/dt$ (where the lines show curvature, we may take the initial slope). It can be readily shown that dV_{CO}/dt at $t = 0$ is proportional to this slope multiplied by V^*_{CO} (which itself is proportional to the reciprocal of the initial carbon content $\%\underline{C}(i)$.) In Table IV, values of the slopes of the lines in Fig. 7 and, in the third column, the slopes divided by $\%\underline{C}(i)$ are collected. The latter are proportional to dV_{CO}/dt at $t=0$ and, according to Equation (16), should increase as the initial carbon content increases. In fact, the values are approximately constant.

TABLE IV

RATE CONSTANTS FOR DESORPTION RUNS

$\%\underline{C}$	$-\frac{d \log(V^*_{CO} - V_{CO})}{dt}$	$-\frac{1}{\%\underline{C}(i)}\frac{d \log(V^*_{CO} - V_{CO})}{dt}$
0.05	0.23×10^{-3}	4.6×10^{-3}
0.09	0.39×10^{-3}	4.3×10^{-3}
0.14	0.64×10^{-3}	4.5×10^{-3}
0.21	0.93×10^{-3}	4.4×10^{-3}

Clearly the surface reaction cannot, in reality, be quite so simple as Equation (16) suggest. Blocking of surface sites, suitable for the formation of the activated complex, ought to occur when sulphur is present, yet sulphur has been shown to have no effect on the desorption rate. In addition, the presence of CO in the atmosphere (and its consequent adsorption on the surface) facilitates the desorption reaction when the oxygen content of the metal is high. This effect suggests that the liquid may not, as far as adsorption is concerned, be in equilibrium with the bulk metal; one might, for example, expect the results of measurements of the surface tension of iron-oxygen alloys by the maximum bubble pressure method to depend upon the composition of the gas used. Nonetheless, the concept that the surface reaction determines the rate of desorption accords well with the experimental facts.

In the absorption experiments,even at low carbon contents, the initial oxygen content is very low and the gas atmosphere is pure CO. Both factors combin to maintain θ_0 low and, hence, the rate of absorption is fast, compatible with control of the rate by transport in the metal.

In the light of these experimental results and the interpretation suggested, it is possible that the observed slow rate of decarburization or deoxidation in vacuum refining [11] or argon flushing may have less to do with transport problem and crucible reaction (though the latter is undoubtedly a factor) than with a slow surface reaction caused by oxygen adsorption at low carbon contents.

ACKNOWLEDGEMENTS

The authors wish to thank the American Iron and Steel Institute for their generous support of this work which was carried out, as part of the graduate studies of two of the authors, in the chemical metallurgy laboratories of the Department of Metallurgy and Materials Science, Massachusetts Institute of Technology.

REFERENCES

1) N.A.Parlee, S. R. Seagle, and R. Schuhmann: Trans. T.M.S.-AIME, 212, 132.8 (1958).
2) J. F. Elliott, M. Gleiser, and V. Ramakrishna: *Thermochemistry for Steelmaking*, Addison-Wesley, II, 548 (1963).
3) See, for example,T. B. King in *Electric Furnace Steelmaking*, Interscience II, 342 (1963).
4) R. L. McCarron and G. R. Belton, Trans. TMS-AIME, 245, 1161-66 (1969).
5) F. A. Halden and W. D. Kingery: J. Phys. Chem 59, 557-50 (1955).
6) S.Glasstone,K. J.Laidler, and H. Eyring: *The Theory of Rate Processes*, McGraw-Hill, 369-89, 1941.
7) R. D. Pehlke and J. F.Elliott, Trans. TMS-AIME, 227, 844-55 (1963).
8) J. H. Swisher and E. T. Turkdogan, Trans. TMS-AIME 239, 602-10 (1967).
9) P. Kozakevitch and G. Urbain, Rev. de Met. 58, 517-34 (1961).
10) P. Kozakevitch and G. Urbain, Rev. de Met. 58, 931-47 (1961).
11) See, for example, A. E.Rathke and S. K. Tarby: Trans. TMS-A AIME 245, 119-26 (1969).

APPENDIX

TABLE I

ANALYSIS OF MASTER ALLOYS (Wt. Percent)

C	0.018	0.039	0.155	0.300
Cr	<0.01	<0.01	<0.003	<0.003
Mo	<0.01	<0.01	<0.006	<0.006
Si	<0.01	<0.01	0.01	<0.01
Al	<0.003	<0.003	<0.005	<0.005
Co	0.009	0.009	<0.01	<0.01
Cu	<0.005	<0.005	<0.003	<0.003
Mn	<0.01	—	<0.005	<0.005
Ni	<0.005	—	<0.005	<0.005
N	<0.0001	0.0002	0.0002	0.0002
O	0.0026	0.0011	0.0008	0.0002

DISCUSSION OF PAPER

W. L. Worrell: Your observed *increase* in the desorption of CO from liquid iron with an increase in CO (g) pressure appears to be inconsistent with your derived rate equation. Combining your rate equation for the forward reaction with that for the reverse reaction, one would calculate that the rate of desorption should *decrease* with increasing pressure of CO (g). Would you like to comment? Could you also comment on possible mechanisms for the observed increase in rate with increasing CO (g) pressure?

T. B. King: Dr. Worrell is correct in expecting, from the form of the rate equation, that the rate of desorption of CO from liquid iron should decrease with increasing pressure of CO in the gas phase. However, since exactly the opposite is observed in practice, we may postulate that the presence of CO in the gas phase has another effect which overrides this. In the derivation of the rate equation it is implicitly assumed that the surface of the liquid iron is in equilibrium with the bulk with respect to adsorption of oxygen. This may not be true. The presence of CO at an appreciable partial pressure in the gas phase may, in fact, alter the nature of the surface sites in such a way that desorption of CO from the metal is facilitated. I cannot really comment on a detailed mechanism to account for the observation but it perhaps illustrates the folly of an oversimplified view of the effect of adsorption on the rate of a surface reaction.

F. D. Richardson: In order to establish that gas flow rates had no effect on his results, Prof. King used a circulating pump and showed

The results were unaffected by flow: Could he comment on this?

T. B. King: Professor Richardson, I agree that experiments with a little circulating pump operated at various speeds are not conclusive. It suggests that the effect is quite small, but I think one has to take into account the fact that when we deliberately add carbon monoxide to the gas phase, for those alloys of low carbon content, the rate of desorption is increased. I would not expect this if I had to contend with transport in the gas as a problem.

L. S. Darken: Do you feel that the interpretation in terms of oxygen adsorption adequately explains the assymmetry of absorption and desorption?

T. B. King: Interpretation of results in terms of oxygen adsorption does not, I agree, adequately explain the asymmetry of adsorption and desorption. One should be careful to note, however, that, at the start of desorption, the oxygen content of the bulk metal is quite high and one would therefore expect the surface coverage by oxygen to be high. On the other hand, at the start of adsorption, the oxygen content of the metal is extremely low and one would not, therefore, expect the surface to be blocked by adsorbed oxygen. In all the absorption experiments, the initial condition of the metal is that it contains carbon but the oxygen content is very low. In addition, since the gas phase is pure CO in the absorption experiments, one would expect that the effect referred to in the reply to Dr. Worrell would prevent the reaction from becoming slower, even as the oxygen content of the metal builds up.

A STEADY-STATE ELECTROCHEMICAL CELL STUDY OF THE KINETICS OF THE REACTION OF WATER VAPOR WITH LIQUID Pb-Te ALLOYS

G. R. Belton and F. J. Harvey

School of Metallurgy and Materials Science

University of Pennsylvania

The effect of tellurium on the rate of absorption of oxygen by liquid lead from water vapor-hydrogen mixtures has been studied by means of a steady-state electrochemical cell technique. Evidence is presented that tellurium causes the interfacial reaction step to be rate controlling in the process. It is concluded from this work and from published studies of the surface activity of tellurium in liquid lead that almost full coverage of the surface by a complex occurs and that dissociation of dilute adsorbed species leads to the dissolution of oxygen.

Although the instantaneous rate of absorption of a gas into a liquid metal can sometimes be measured by gas volumetric techniques, as for example in the work of El Tayeb and Parlee [1] or of King et al. [2], studies of the rates of absorption of solutes from a *complex* gas phase have usually required the determination (by chemical analysis) of the solute content of the metal as a function of time. One of the present authors and McCarron [3], for example, have used this latter approach in studies of the diffusion of oxygen and sulfur in liquid iron. However, if the rate of absorption is governed by an interfacial reaction step, the dependence of the rate on the gas composition can become quite complex [4], and a large number of separate experiments might be needed to establish this dependency. Clearly, it would be desirable to have a method of continuously measuring the flux of the solute into the metal while varying the composition of the gas phase; in addition, the boundary conditions for diffusion of the solute should be well defined. Accordingly, a steady-state electrolyte technique has been developed in which the flux of oxygen through the bottom of a column of liquid metal is measured, the upper surface of the liquid metal being in contact with the gas phase of interest. In the present paper, the method is briefly described and results are presented of a study of the effect of small additions of tellurium on the interfacial reaction rate in the absorption of oxygen by liquid lead from water vapor-hydrogen-argon mixtures.

A more detailed description of the technique is presented elsewhere [5], where it is demonstrated that in the absence of tellurium the rate of absorption of oxygen is governed by diffusion in the liquid lead. The resulting diffusivity of oxygen was found to be in experimental agreement with the value reported by Arcella and Fitterer [6], who applied the solid electrolyte cell method of Rickert and Steiner [7], in a study of the diffusivity of oxygen in liquid lead.

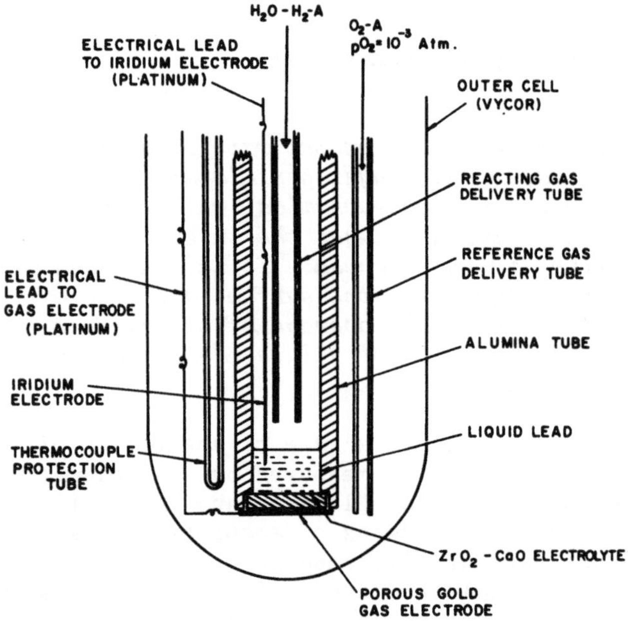

Fig. 1. Electrochemical Cell

EXPERIMENTAL

The design ofthe electrochemical cell is shown in Fig. 1. The short column of liquid lead (5-8 mm high) was contained within an alumina tube, approximately 9 mm I.D. A ZrO_2-CaO electrolyte pellet (Zircoa), about 3 mm thick, was cemented with high temperature alumina cement (Aremco) into a counterbore in the bottom of this alumina tube; the electrolyte thus forming the base of the lead column. A gas electrode, formed from a plátinum wire mesh, was attached to the outer face of the electrolyte pellet by means of a thin layer of gold paste (DuPont 5780).

The H_2O-H_2-Ar reaction gas mixtures were delivered to within about 5 mm of the upper surface of the liquid lead by a 4 mm bore vycor tube. An iridium wire, fastened to the delivery tube, made contact with the liquid lead; electrical leads to this wire and to the gas electrode were of platinum. The assembly was contained within a vycor cell through which an argon - 0.1% oxygen reference gas mixture was passed at about 500 ml/min.

The reaction gas was prepared in the usual manner by saturating H_2-Ar mixtures [8], particular care being taken, after each change in hydrogen pressure, to allow sufficient time (2-4 hrs.) for equilibration of the hydrogen content of the water in the saturator system. Very high purity, pre-analyzed gas mixtures were used and argon, for dilution purposes, was purified by passing through Ti-Zr sponge at 850°C.

The applied potential difference across the cell (gas electrode positive) was controlled by a constant voltage source, and the current passing through the cell was determined by measuring the voltage drop across a standard resistor (10Ω) incorporated in the circuit. This voltage drop was potentiometrically recorded. Provision was also made for measurement of the resistance of the cell by means of an A.C. conductivity bridge.

Procedure

A lead-tellurium alloy was charged to the cell and the whole assembly slowly raised, over about 12 hrs., to the experimental temperature. With both the reaction gas and the reference gas passing, the cell was left under an applied potential such that the apparent oxygen pressure at the lead/electrolyte interface was about 10^{-18} atm.; this corresponds to less than 0.5 ppm of oxygen in lead at 767°C, accordingto the data of Alcock and Belford [9]. The potential was calculated from the following equation:

$$V = IR + \frac{RT}{4F} \ln \left[P'(O_2)/P''(O_2)\right] \qquad (1)$$

where I is the current passing through the cell, R is the resistance of the cell, $P'(O_2)$ and $P''(O_2)$ are the oxygen pressures of the reference gas and lead/electrolyte interface respectively, and the other symbols have their usual meanings.

After several days, the resistance of the cell reached a constant value (100-400 Ω) and the gas electrode behaved in an apparently reversible manner - as indicated by a rapid response of the current to changes in the composition of the reference gas. The cell was then tested for "voltage independence" by raising the potential by about .02 volts and observing the return of the current to the original value in a period of 30-60 min. This was considered necessary since the measured value of the resistance, needed in equation (1), included an unknown contribution due to cell capacitance effects. This voltage independence check was carried out on two or three of the experiments at each tellurium content.

The current or flux was then followed as a function of gas composition, steady state values being obtained within three to four hours after each change. The effect of flow rate of the reaction gas was also studied at several gas compositions and it was found that a significant region existed where the flux was independent of flow-rate. This region usually extended from about 80 - 200 ml/min (STP). Above about 250 ml/min the flux increased in an erratic manner, presumably because of disturbance of the liquid surface; measurements were restricted therefore to this "plateau" region.

Additions of tellurium or Pb-Te alloy through the delivery tube permitted the use of some cells for more than one series of experiments, although it was not possible to avoid mechanically disturbing the cell during this procedure. All the alloys were subsequently analyzed for the tellurium content and it was found that within the limits of analytical error ($\pm$ 5% of Te content), no loss of tellurium had occurred.

RESULTS

Two main sequences of experiments were carried out at 767°C, this being the temperature at which the diffusion studies were made [5]. The results obtained in the first sequence of experiments are presented in Fig. 2 in the form of electrical flux versus pressure of water vapor. The gas compositions, noted in the figure, represent compositions of the *dry* argon-hydrogen mixtures before saturation with water vapor.

Two types of behavior appear to exist: at low tellurium contents the flux varies with both the pressure of water vapor and the pressure of hydrogen, whereas at higher tellurium contents the flux is a function of the pressure of water vapor only. At all three compositions however, the flux of oxygen is significantly decreased by the presence of tellurium. This is illustrated in Fig. 2(b) where the dotted lines represent the flux which would be observed in the absence of tellurium, i.e. under diffusion control (taken from the previous study [5]). It will also be noted that the lines drawn through the results show a positive flux at zero water pressure. This flux represents the "leak rate" of the cell for a given experiment. This same phenomenon was found in the diffusion studies [5], and probably represents leakage of oxygen through micro-cracks in the alumina cement.

The second main sequence of experiments yielded the results shown in Fig. 3. A somewhat wider range of hydrogen pressures was used, but, again, the same behavior was observed. The results shown in Fig. 3(b) were obtained by keeping the pressure of water vapor constant at a series of values and switching back and forth between different argon-hydrogen mixtures, the gas trains being arranged such that prolonged flushing could be carried out before each change-over. The 20 ppm H_2 gas mixture was produced by diluting two different standard gas mixtures and the results were found to be coincident. Unfortunately, the hydrogen pressures which were used with the 360 ppm Te alloy did not vary widely, but, within experimental error, the results appear to be independent of hydrogen pressure.

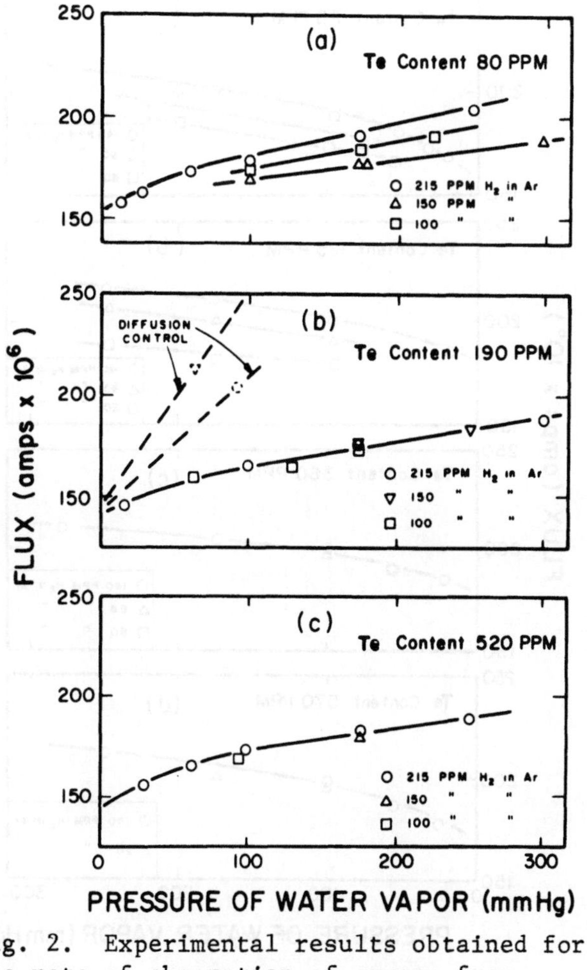

Fig. 2. Experimental results obtained for the rate of absorption of oxygen from H_2O-H_2-Ar mixtures. Temp. 767°C.

The results from a third series of experiments are shown in Fig. 4. In this case a wide range of hydrogen pressures was used for the 720 ppm Te alloy (Fig. 4(a)), and, within a reasonable experimental scatter, the results are independent of hydrogen partial pressure. In Fig. 4(b) the results are given for the alloy with the highest tellurium content to be investigated.

In view of the complexity of the behavior at low tellurium contents, the temperature dependence was investigated only for one high tellurium content alloy (725 ppm Te). These results are given in Fig. 5. The temperature range was somewhat restricted because: (a) at low temperatures the changes in flux became too small to be accurately determined (with the existing instrumentation), and (b) at temperatures above about 780°C a steady drift in the flux was observed.

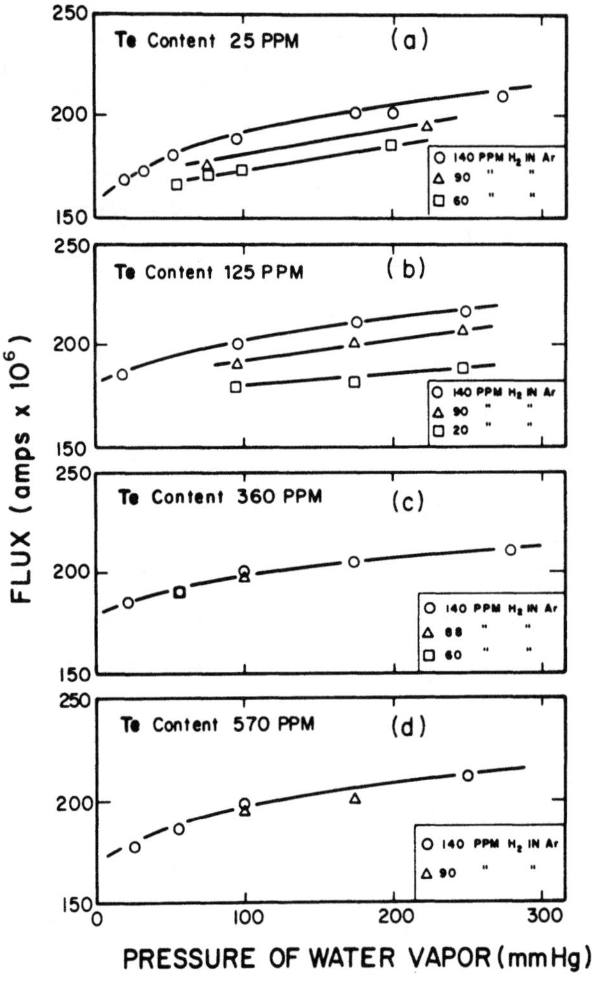

Fig. 3. Experimental results obtained in the second series of experiments. 767°C.

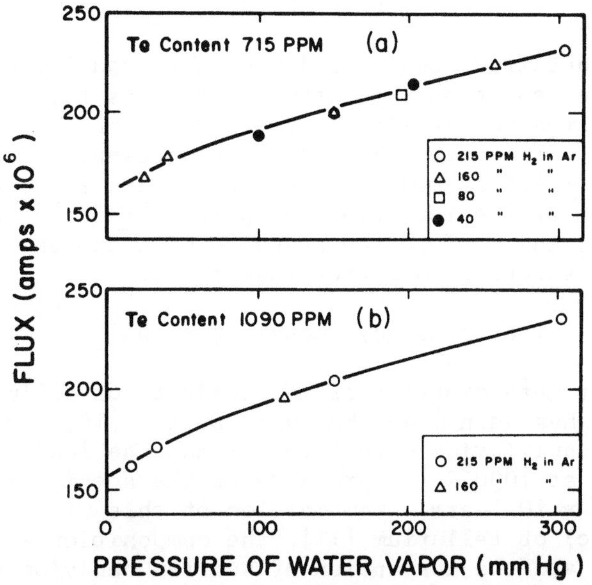

Fig. 4. Experimental results obtained in the third series of experiments. 767°C.

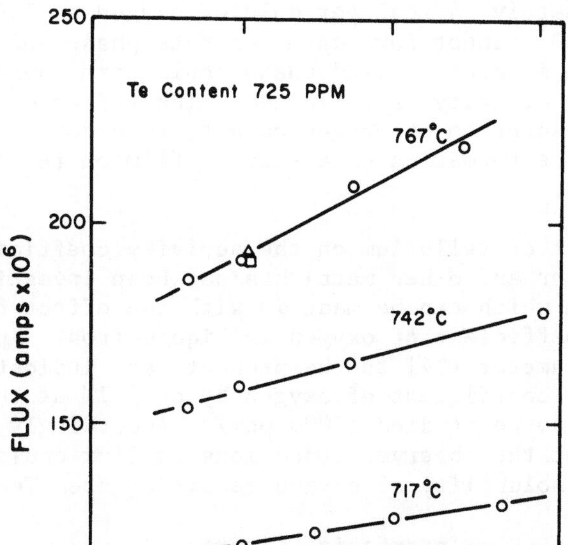

Fig. 5. Effect of temperature on the measured fluxes.

DISCUSSION

Since a detailed consideration of the behavior of the cell is included in the report on the diffusion studies [5], this discussion will be restricted to the effect of tellurium on the rate of absorption of oxygen. The uncertainties in the present results are considered to be: $\pm$ 4μa in measured current (principally due to drift in the calibration of the recording equipment), $\pm$ 3% in the pressure of water vapor, and $\pm$ 5% in the pressure of hydrogen. Temperatures are considered accurate to better than $\pm$ 2°C.

Thermochemical Considerations

No direct determination of the activity of tellurium in liquid lead solutions has been made; however, Porter [10] has investigated the vapors in equilibrium with PbTe (s) and the lead-rich solution. He showed that at 1000°K the pressure of the species Te had a maximum value of 5 x 10^{-6} atm. Combination of this with the vapor pressure (monomeric) of tellurium [11], the composition at the phase boundary [12], and the assumption of Henrian behavior yields the activity of tellurium at 1000 ppm in lead as approximately 2 x 10^{-4} (with respect to liquid tellurium).

The condensed oxide of tellurium, TeO_2, is less stable than PbO by approximately 24 kcal per mole of oxygen at 1000°K [13]; accordingly, TeO_2 cannot form as a separate phase and it is most unlikely that a separate mixed phase could form, particularly in view of the low activity of tellurium. The effect of tellurium on the rates of absorption of oxygen cannot, therefore, be reasonably attributed to the formation of a skin or film on the surface of the liquid lead.

The effect of tellurium on the activity coefficient of oxygen in liquid lead or any other metal has not been investigated. The only comparison which can be made is with the effect of sulfur on the activity coefficient of oxygen in liquid iron. Applying this interaction parameter [14] to the present case indicates a reduction in the activity coefficient of oxygen by only 1% at the highest tellurium content to be studied (1090 ppm). Accordingly, it is extremely unlikely that the observed reductions in flux could be caused by changes in the solubility of oxygen resulting from Te-O interactions.

Interfacial Kinetics

In view of the foregoing, it would appear that the reduction in flux must be attributed to the competitive adsorption of Tellurium or a tellurium-containing species at the liquid lead surface, leading to a reduction in the surface available for adsorption and dissociation of an oxygen containing species.

Two studies have been made of the surface activity of tellurium in liquid lead. Bradhurst and Buchanan [15] reported tellurium to have a negligible effect on the surface tension of lead at 750°C. However, since these workers melted lead containing pieces of PbTe under 10^{-9} atm, it is probable that the tellurium was completely lost by vaporization, the vapor pressure of PbTe being 2×10^{-4} atm. at 750°C [10]. D. J. Harvey [16] has studied the surface tension between 0.098 and 4.23 wt % Te by the maximum bubble pressure technique. He found a significant reduction in surface tension, and interpolated results from his paper are shown in Fig. 6. The line shown drawn through the data yields, by the Gibbs Adsorption isotherm, an excess surface coverage of 7×10^{-11} moles/cm^2. This corresponds to less than 4% of close-placked monolayer coverage by the largest ion, Te^{--} [17]. Accordingly, unless Harvey's work is in serious error, the reduction in the flux of oxygen observed in the present work cannot be attributed to a simple coverage of the surface by tellurium.

Since coverage of the surface by tellurium atoms or ions is insufficient to explain the phenomena, a surface complex must be proposed. Evidence that active surface complexes do exist in liquid metal systems is already available: a marked lowering of the surface tension of liquid iron occurs in the presence of the pairs of solutes chromium and carbon [18, 19], and tungsten and carbon [19]; whereas the presence of these solutes singly has little or no effect on the surface tension [17].

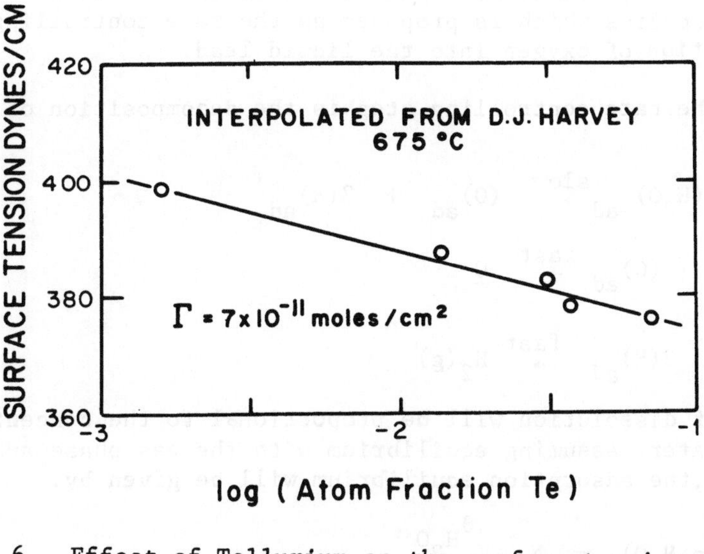

Fig. 6. Effect of Tellurium on the surface tension of lead. Taken from D. J.Harvey [16].

The gaseous species TeO has a large bond enthalpy of -76 ($\pm$ 10) kcal at 1000°K [13, 11]; accordingly, this species appears to be a reasonable choice for the adsorbed complex. The activity of adsorbed TeO, if the surface phase is assumed to be in equilibrium with the gas phase and dissolved tellurium, will be given by:

$$a\,(TeO) = K \frac{p(H_2O) \cdot a\,(Te)}{p(H_2)} \tag{2}$$

where K is an equilibrium constant. It is also reasonable to expect the complex to occupy two surface sites; accordingly, if ideal Langmuir adsorption is assumed, the activity of adsorbed TeO will also be given by the expression:

$$a\,(TeO) = \Phi \frac{\theta_{TeO}}{(\theta_v)^2} \tag{3}$$

where θ_{TeO} is the fraction of surface sites covered by the complex, θ_v is the fraction of vacant sites, and Φ is a proportionality constant. If $\theta_{TeO} \approx 1$ (almost full coverage) the fraction of vacant sites may be obtained by combining Eq. (2) and (3):

$$\theta_v = \left[\frac{\Phi}{K} \frac{p(H_2)}{p(H_2O) \cdot a\,(Te)}\right]^{1/2} \tag{4}$$

Several minor species may be competively adsorbed on the surface, e.g. TeH, H_2O, TeOH, O, H etc., and it is the dissociation of the minor species which is proposed as the rate controlling step in the absorption of oxygen into the liquid lead.

If the rate controlling step is the decomposition of water, i.e.

$$\begin{aligned} (H_2O)_{ad} &\xrightarrow{slow} (O)_{ad} + 2(H)_{ad} \\ (O)_{ad} &\xrightarrow{fast} \underline{O} \\ 2(H)_{ad} &\xrightarrow{fast} H_2(g) \end{aligned} \tag{5}$$

the rate of dissolution will be proportional to the concentration of adsorbed water. Assuming equilibrium with the gas phase and ideal adsorption,the adsorption equilibrium will be given by:

$$p(H_2O) = \Phi_1 \frac{\theta_{H_2O}}{\theta_v} \tag{6}$$

Rearrangement of this equation and substitution for θ_v from Eq. 4 leads to the following equation for the forward flux of oxygen per unit area:

$$\frac{\dot{n}_o}{A} = K' \left[\frac{p(H_2O)\ p(H_2)}{a(Te)} \right]^{1/2} \qquad (7)$$

where the apparent rate constant K' is a combination of the true rate constant, the equilibrium constant K, and the proportionality constants Φ and Φ_1.

It should be noted that following the formalism of the activated complex theory will have no effect on the form of this equation as long as the same form of adsorption isotherm is assumed. The fractional coverage by the activated complex will have the same dependence as the fractional coverage by adsorbed water, since the activity of the activated complex would be considered proportional to the activity of $(H_2O)_{ad}$ which is proportional to $p(H_2O)$.

The dependence of the flux per unit area on $[p(H_2O) \cdot p(H_2)]^{1/2}$, required by Eq.7, is shown to be in reasonable agreement with the results from the experiments at *low* tellurium contents in Fig. 7. The agreement shown in Fig. 7(c) is particularly striking since these experiments involved the widest range of hydrogen pressures.

The slopes shown in Fig. 7 decrease with increasing tellurium content but with a lower power dependence than is required by Eq. 7. The decrease supports the essential model, i.e. coverage by a complex containing tellurium and dissociation of a simpler adsorbed species, but the lower power dependence possibly indicates a more complicated surface complex. A TeO-PbO complex would, for example, yield almost exact agreement.

The results from the experiments with higher tellurium contents are not in accord with the foregoing dependency (Eq. 7), and it is necessary to assume a change in mechanism to account for the apparent independence of the flux on hydrogen pressure. Dissociation of the adsorbed species TeOH as the major rate controlling step gives the required dependence. Keeping the same basic assumptions as before, i.e. equilibrium of the surface phase with the gas and the tellurium, the activity of the adsorbed TeOH will be given by:

$$a\ (TeOH) = \frac{K_1 p(H_2O) \cdot a\ (Te)}{p(H_2)^{1/2}} \qquad (8)$$

The activity will also be given by:

$$a\ (TeOH) = \Phi_2 \frac{\theta_{TeOH}}{\theta_v} \qquad (9)$$

if ideal adsorption is again assumed. Equating these two expressions, substituting for θ_v from Eq. 4, and setting the rate proportional to the concentration of adsorbed TeOH yields the following equation for the forward flux of oxygen per unit area:

$$\frac{\dot{n}_O}{A} = K'' \ [p(H_2O) \cdot a\ (TeO]^{1/2} \qquad (10)$$

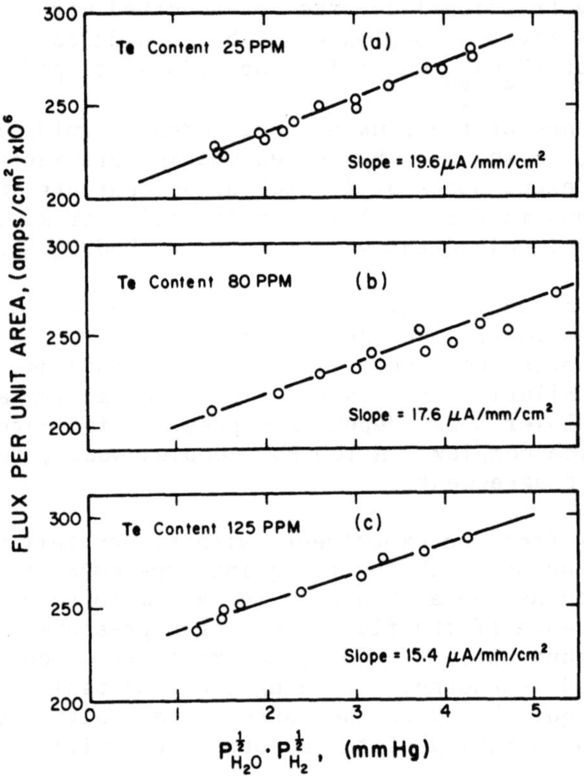

Fig. 7. Dependence of flux of oxygen on gas composition at 767°C for alloys of low tellurium content. Test of equation 7.

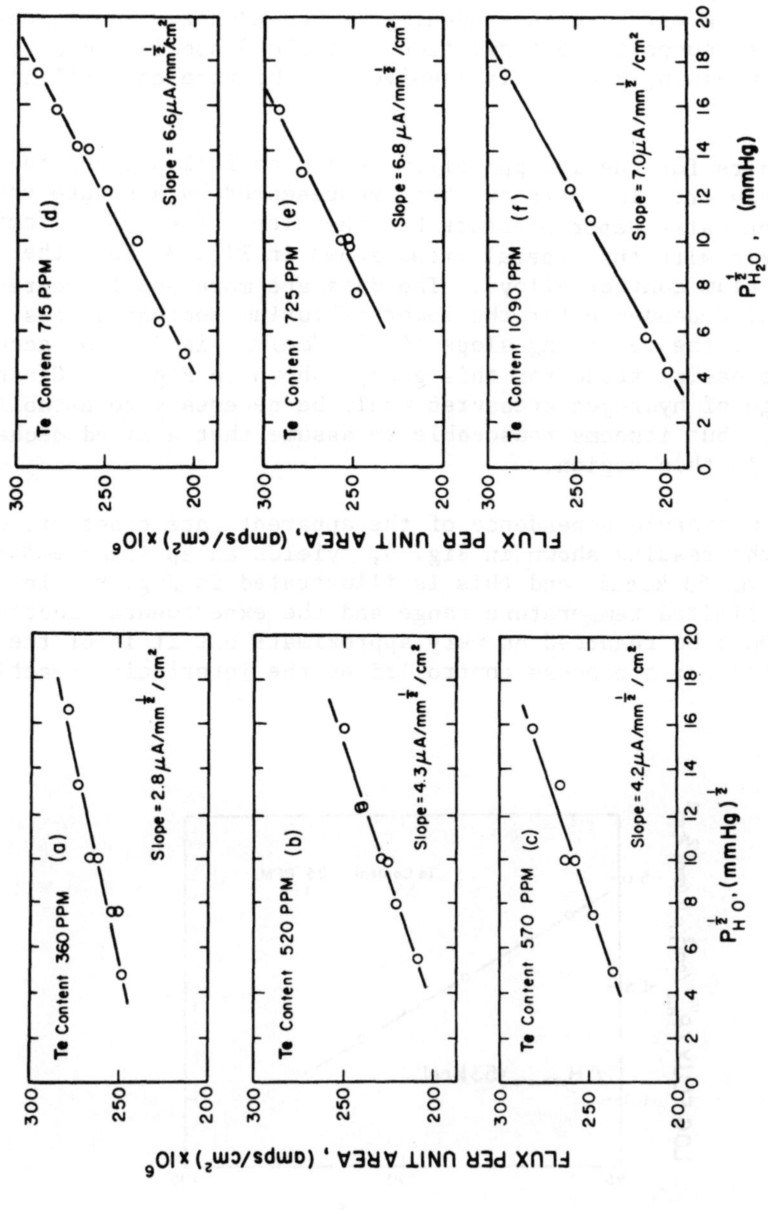

Fig. 8. Dependence of flux of oxygen on gas composition at 767°C for alloys of high tellurium content. Test of equation 10.

In Fig. 8, the required $[p(H_2O)]^{1/2}$ dependence is seen to be in good agreement with the results for alloys containing at least 360 ppm Te. According to Eq. 10, the rate should be proportional to the square root of the activity of tellurium. The slopes shown in Fig. 8 increase with the tellurium content but the experimental scatter is too great to give a meaningful dependence. The increase does, however, support the suggestion that the dissociation of a tellurium-containing species is involved in the rate controlling step.

The results for the 190 ppm alloy appear to fall between the two types of behavior. The data are best represented by a square root dependence on water vapor pressure but the slope of $4\mu A/mm^{1/2}/cm^2$ does not agree with the general trend shown in Fig. 8 for the higher tellurium content alloys. The data are more poorly represented by the dependence for the lower tellurium content alloys ($\pm 6\mu A/cm^2$) but the resulting slope of $13\mu A/mm/cm^2$ is in good accord with the decreasing trend for this group, shown in Fig. 7. Clearly, a wider range of hydrogen pressures would be necessary to establish the behavior, but itseems reasonable to assume that a mixed mechanism occurs in this region.

The temperature dependence of the apparent rate constant, derived from the results shown in Fig. 5, yields an apparent activation energy of 53 k.cal and this is illustrated in Fig. 9. In view of the limited temperature range and the experimental scatter, this value must be regarded as very approximate but it is of the order expected for a process controlled by the interfacial reaction step.

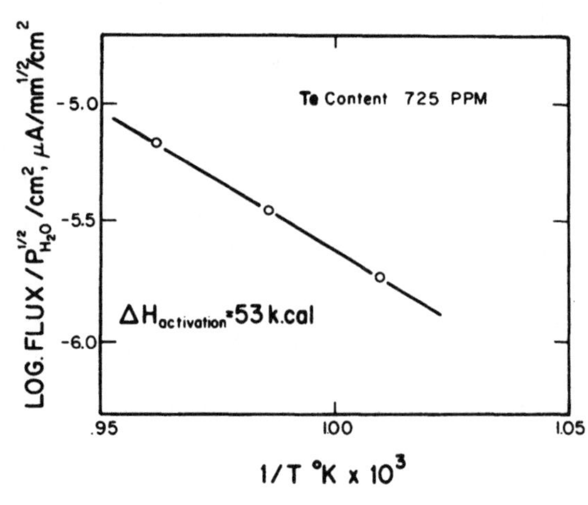

Fig. 9. Apparent activation energy for the absorption reaction at high tellurium content.

Conclusions

Because of the many assumptions involved in the foregoing treatment and the lack of corroborating surface studies, the suggested mechanisms can only be regarded as tentative. Other combinations of complexes and dissociating species could probably be suggested which would give satisfactory agreement with the experimental results, particularly if non-ideal adsorption behavior were to be assumed. The present study does, however, demonstrate that interfacial reaction rate control in the absorption of a solute from the gas phase into a liquid metal may be caused by an added impurity which, of itself, has insufficient surface activity to cover the surface.

The results on the present system are consistent with almost complete coverage of the surface by a complex formed between the impurity (tellurium) and a component of the gas phase with dissociation of minor species in the chemisorbed layer leading to the absorption of oxygen by the liquid lead. At low tellurium contents this minor species appears to be a simple component of the gas phase, e.g. H_2O; whereas at higher tellurium contents, dissociation of a tellurium-containing species appears to predominate.

ACKNOWLEDGMENTS

The study is, in part, a contribution of the Laboratory for Research on the Structure of Matter which is supported by the Advanced Research Projects Agency, Office of the Secretary of Defense. The authors wish to thank the Selenium-Tellurium Development Association for their additional support of this work through a grant-in-aid.

REFERENCES

1. N. M. El Tayeb and N. A. D. Parlee: Trans. Met. Soc. AIME, 239, 1345 (1967).
2. T. B. King, R. Karasev, and P. Dastur: This volume.
3. R. L. McCarron and G. R. Belton: Trans. Met. Soc. AIME, 245, 1161 (1969).
4. J. H. Swisher and E. T. Turkdogan: Trans. Met. Soc. AIME, 239, 602 (1967).
5. G. R. Belton and F. J. Harvey: Intended for publication in Metallurgical Transactions.
6. F. G. Arcella and G. R. Fitterer: 98th Annual Meeting Met. Soc. AIME, Washington, D. C., 1969.
7. H. Rickert and R. Steiner: Z. Phys. Chem. (Frankfurt), 49. 127 (1966).

8. G. R. Belton and F. D. Richardson: Trans. Faraday Soc., 58, 1562 (1962).
9. C. B. Alcock and T. N. Belford: Trans. Faraday Soc., 60, 822 (1964).
10. R. Porter: J. Chem. Phys., 34, 583 (1961).
11. D. R. Stull and G. C. Sinke: *Thermodynamic Properties of the Elements,* Amer. Chem. Soc., Washington, D. C., 1956.
12. M. Hansen, *Constitution of Binary Alloys,* McGraw-Hill Book Co, Inc., N. Y. 1958.
13. J. P. Coughlin: U. S. Bureau of Mines Bulletin 542, U.S. Government Printing Office, Washington, D. C., 1954.
14. J. F. Elliott, M. Gleiser, and M. Ramakrishna: *Thermochemistry for Steelmaking,* Vol.11, Addison-Wesley Publishing Co., Inc. Reading, Mass., 1963.
15. D. H. Bradhurst and A. S. Buchanan: J. Phys. Chem., 63, 1486 (1959).
16. D. J. Harvey: Trans. Met. Soc. AIME, 221, 266 (1961).
17. P. Kosakevitch, Plenary lecture, symposium held by Society of Chemical Industry, London, 20-22 Sept. 1967.
18. T. J. Whalen, S. M. Kaufman, and M. Humenik, Jr., Trans. ASM 55, 778 (1962).
19. P. Kosakevitch and G. Urbain: Mem. Sci. Rev. Met., 58, 931 (1961).

DISCUSSION OF PAPER

T.B.King: Does the mechanism assume that the surface is not simply in equilibrium with solutes, in other words does the gas composition affect the type of adsorbed species?

G. R.Belton: The experimental results strongly indicate that the surface is in equilibrium with one dissolved solute (Te), and the gas phase,and that a surface complex is formed.

W. L. Worrell: Would you like to comment on your assumption that the TeO complex occupies two surface sites and that the TeOH complex occupies only one surface site.

G. R. Belton: Unfortunately, we do not have independent surface studies which could answer this question with certainty. Assuming such surface complexes exist, the assumed occupancies are the ones which give best agreement with the experimental results. Since a TeOH molecule would be expected to be much more strongly polar than a TeO molecule, it might be expected that the OH group would point away from the metal surface. We would not expect such a tendancy for TeO.

THE IMPORTANCE OF MACROSCOPIC TRANSPORT PHENOMENA IN GAS/SOLID REACTIONS

A. W. D. Hills

Department of Metallurgy, Imperial College

London, S.W. 7, England

Previous workers have described the mechanisms of those gas solid reactions which result in a porous solid product in terms of one of three reaction models: the chemical control model, the transport control model, or the mixed chemical and transport control model. Decisions as to which model applies to any one reaction have been made by observing the pattern of reaction behaviour that it exhibits. However, this paper shows that the differences between the patterns of reaction behaviour predicted by the three models are too small for unequivocal decisions to be made in this way. An improved experimental strategy is proposed in which all the parameters involved in the transport control model are measured independently, and used to predict the rate of the reaction on the assumption that it is transport controlled. The other reaction models should only be considered if measured reactions rates are found to be significantly slower than these predictions. If this is found to be the case, the measured transport parameters can be used to calculate additional parameters involved in the other reaction models from the measured reaction rates.

Since this strategy involves a considerable amount of experimental effort,an alternate empirical strategy is also suggested so that the behaviour of industrial gas/solid reactors can be predicted without the mechanism of the reaction being fully understood.

REACTIONS CONSIDERED

There are many industrial processes in which gases and solids react together to form a new solid. The new solid usually forms on the outside of the original solid particles so that reactants must cross it either by gaseous diffusion or by solid state diffusion while the reaction is continuing. Solid state diffusion is an extremely slow process and it is not surprising that the majority of economically important reactions of this type give rise to a porous solid product easily permeated by the reactant and product gases. The present article is concerned with reactions of this type.

MECHANISM OF GAS/SOLID REACTIONS

Several different physical and chemical processes occur during these reactions. The reactant species in the gas mixture swept past the reacting particle must be transferred across the gas phase boundary layer surrounding the particle and then diffuse on through the product solid to be adsorbed onto the reaction surface. The gas must then dissociate to allow the chemical reaction to proceed. The chemical reaction itself occurs by the growth of one solid lattice from another; a process which involves the movement of dislocations and the diffusion of ions and lattice vacancies, once sufficient nuclei have been formed for the new lattice. When the product is porous-the case considered here - a marked molar volume change will be associated with the reaction so that large mechanical stresses will be set up at the interface causing the product solid to crack away, repeatedly exposing new surface.

Product gas formed by association after the chemical reaction has occurred will then desorb from the interface, diffuse through the product solid and the gas phase boundary layer surrounding the particle to be swept away in the bulk gas stream. In addition the heat of the reaction must be brought to the reaction surface - by radiation and convection to the outer surface of the particle and onto the reaction front by conduction.

REACTION MODELS

It is virtually impossible to develope a complete conceptual picture of the precise way in which all these processes influence the rate at which any one gas solid reaction occurs. Conceptual models are thus evolved for these reactions by selecting the processes which occur least readily, and hence control the rate at which all the other processes take place. Three important classes of model have been used to describe the mechanisms of the reactions which have been investigated by various workers: the chemical control model [1, 2, 3], the transport control model [4, 5], and the mixed chemical and transport control model [6, 7, 8].

In the chemical control model the 'chemical' processes at the reaction front are assumed to occur least readily. Relative to these, then, the mass transfer and diffusion processes take place so easily that the composition of the gas in the pores in the product solid at the reaction front is virtually equal to the bulk gas composition. The reaction rate is thus controlled by the rate of the chemical step at the reaction front.

The transport control model, on the other hand, assumes that it is the diffusion and mass transfer processes that occur least readily. The chemical processes at the reaction front, then, occur so easily that the gas mixture in the pores at the front is in equilibrium with the reactant and product solids. The rate of the reaction in this case is controlled by the rate at which the reactant and product gases are transported between the reaction front and the bulk gas phase.

The mixed control model postulates that there is little to choose between the ease with which the transport processes and the chemical processes occur, so that they both affect the rate of the reaction.

The heat of the reaction must be transferred to the reaction front so that a temperature difference will be set up between the reaction front and the environment of the reacting particle. Thus the reaction models described above should be modified to take account of this temperature difference, although there is some divergence of opinion as to the importance of doing this.

INDUSTRIAL IMPORTANCE OF REACTION MODELS

There are two ways in which the model that we evolve for any particular gas/solid reaction is important in an industrial context. In the first place, it determines the qualitative way in which we think about the reaction and, secondly, it allows us to base quantitative predictions about the rate of the reaction in certain circumstances on observations made under entirely different circumstances.

The qualitative influence of the reaction model can be seen, for example, in the influence that it would exert on decisions to be made concerning the ideal type of industrial reactor in which to carry out a reaction. It would, moreover, determine how we would expect the performance of existing plant to be influenced by changes made in operating variables. It will also play an important role in the evolution of simple process control models to be used in automating industrial gas/solid reactors.

The quantitative influence of the reaction model is to be seen in the detailed predictions about the performance of proposed plant that we would base on the results of experiments carried out in the laboratory or on small pilot plants. Indeed the whole problem of scale-up of gas/solid reactors depends upon an adequate methematical model for the reaction. Such a model could also be used to predict the maximum possible efficiency with which an industrial gas/solid process could be carried out, and in the development of complex control models.

EVOLUTION OF REACTION MODELS

There is no a-priori way of deciding which reaction model best describes the mechanism of any one gas/solid reaction. Kinetic experiments on gas/solid reactions thus have two functions. They must provide evidence as to which is the best reaction model and they must then provide numerical values of the parameters involved in that model. It had been thought that both these functions could be fulfilled by measuring the rates at which single particles react under closely controlled experimental conditions. This however is not so, and the reasons for this are discussed below.

PRESENT EXPERIMENTAL STRATEGY

Standard kinetic experiments of this type of gas/solid reaction are conducted by hanging a single particle about 1 cm in size in a high temperature furnace through which gas of known composition is passed. The rate of the reaction is followed by measuring the weight of the sample, or by sectioning partially reacted samples. The temperature of the furnace is normally kept constant, although it can be controlled to follow a predetermined path. The temperature of the reacting sample is sometimes measured by embedding a thermocouple into it, although this is done all too seldom.

The way in which the results of these experiments are analysed depends upon the reaction model which is considered to apply, but it is only by analysing the experimental results that information can be obtained to show which reaction model is the most applicable. Thus a circular argument can develop, and some previous workers have been led to incorrect conclusions as to the most suitable model for the reaction. This is because qualitative information has to be used to select the reaction model; the quantitative information - the measured reaction rate, etc. - being reserved to determine values of the parameters involved in the chosen reaction model. This is a dangerous experimental strategy since insufficient differences exist between the reaction behaviour predicted by the different reaction models.

Chemical Control or Transport Control

The similarity in the reaction behaviour predicted by these two models has been discussed previously [8, 9, 10], and only a brief review will be presented here.

Let us consider a general gas/solid reaction described by an equation of the form:*

$$\text{Solid I} + \beta_I(\text{Gas I}) \rightarrow \omega_{II}(\text{Solid II}) + \beta_{II}(\text{Gas II})$$

where β_I, β_{II} and ω_{II} are stoichiometric factors.

The simple chemical control model for this reaction would represent its overall rate as a balance between a forward reaction:

$$\dot{n}''_F = K_F \left\{ \frac{[p_I]_G}{R\theta_R} \right\}^{\beta_I} \quad (1)$$

and a back reaction:

$$\dot{n}''_B = K_B \left\{ \frac{[p_{II}]_G}{R\theta_R} \right\}^{\beta_{II}} \quad (2)$$

where the suffix G is used because the gas composition at the reaction front is assumed to be the same as that in the bulk gas phase.

When this bulk gas phase is in equilibrium with the two solid phases these two reaction rates balance since no reaction can occur. Consequently the overall reaction rate under non-equilibrium conditions is given by:

$$\dot{n}'' = \frac{\dot{n}}{A_R} = \frac{K_C}{(R\theta_R)^{\beta_I}} \left\{ [p_I]_G^{\beta_I} - \frac{[p_{II}]_G^{\beta_{II}}}{[K_{eq}]_R} \right\} \quad (3)$$

where K_C is equal to K_F and is known as the chemical reaction rate constant.

Many workers [1, 3, 11, 12, 13, 14] have established that the reaction surface is defined adequately when the porosity of the original solid particle approaches zero. Thus the value of A_R can be considered known under these conditions, and the value of K_C can be determined by measuring the rate at which dense particles react at known temperature with a gas of known composition. If no other measurements are to be made, the value of K_C determined in this way cannot tell us whether the chemical model is the most suitable model

* Symbols and variables are defined and tabulated at the end of the paper.

with which to describe the mechanism of the reaction. We can only find this out by comparing the reaction behaviour predicted by the various reaction models with the behaviour observed experimentally. The reaction behaviour can be described in terms of either (a) the shape of curves showing how the reaction rate, or certain functions of the unreacted mass, vary with time, or (b) the effect of gas velocity, gas composition, reaction temperature, particle size and total gas pressure on the rate of the reaction.

The chemical control model predicts a fairly definite pattern of behaviour in these respects. For example, the equation of continuity applied to reactions in dense pellets shows that the velocity of the reaction front is related to the rate of the reaction by the equation:

$$\dot{n} = - A_R \, \rho_I \, \overline{V} \tag{4}$$

Comparison between Eq. (4) and Eq. (3) shows that the chemical control model predicts that the reaction front would move at a constant velocity throughout the reaction. The position of the reaction front can be related to the unreacted mass of the pellet, so that this prediction implies that a certain function of the unreacted mass would vary linearly with time. In spheres, for example, where the reaction front remains spherical, the radius of the reaction front is related to the fractional unreacted mass by the equation:

$$r/_{r_o} = (m^*)^{1/3} = \left(\frac{m_\tau - m_\infty}{m_o - m_\infty} \right)^{1/3} \tag{5}$$

Thus the chemical control model would predict that $(m^*)^{1/3}$ varies linearly with time.

It would also predict that the time taken for a pellet of any shape to react completely would be proportional to its size, and that its rate of reaction would vary with the composition of the gases passed over it as predicted by Eq. (3). The flow rate of these gases would only affect the reaction if it failed to maintain a gas mixture of constant composition around the reacting particle. The effect of reaction temperature and total gas pressure on a chemical controlled reaction is somewhat less well defined. The rate of a chemical controlled reaction would be expected to follow an Arrhenius type relationship with temperature, and to have a fairly large apparent activation energy. The rate of the reaction would also be expected to increase with total gas pressure although a limit would soon be reached whence any further pressure increase would not affect the reaction rate.

The transport control model predicts a very much less well defined pattern of reaction behaviour than the chemical control model. largely because it contains more reaction parameters. Moreover, this pattern of behaviour frequently includes the precise behaviour predicted by the chemical control model.*

For example, the general equations derived in Appendix 1 for a transport controlled reaction show that the fractional reaction front radius in a dense spherical particle has a value $\lambda^* = r/r_o$ after a dimensionless time period given by:

$$\tau^* = \int_{\lambda^*}^{1} \frac{\lambda^* - \lambda^{*2}(1 - Bi_M)}{Y} \, d\lambda^* \tag{6}$$

where Y is the root of the equation:

$$\frac{\left\{\frac{[p_{II}]_G}{P_G} + \frac{Y}{\xi}\right\}^{\beta_{II}}}{[K_{eq}]_G \left\{\frac{P_G}{P_o}\right\}^{(\beta_I - \beta_{II})}} - \left\{\frac{[p_I]_G}{P_G} - P - \frac{Y}{\xi}\right\}^{\beta_I} \exp\left\{\frac{6YH^{*2} f(\lambda^*)}{\beta_{II} k^* - YH^* f(\lambda^*)}\right\} = 0 \tag{7}$$

and where:

$$f(\lambda^*) = \frac{\frac{1}{\lambda^*} - 1 + Bi_H}{\frac{1}{\lambda^*} - 1 + Bi_M} \tag{8}$$

Bi_M in Eqs. (6) and (8) is the ratio between the mass transfer resistance outside the sphere and the diffusion resistance within the product solid, and Bi_H in Eq. (8) is the ratio of the corresponding resistances for heat transfer. These variables are defined by:

$$Bi_M = \frac{D}{r_o \alpha} \; ; \quad Bi_H = \frac{k}{r_o h} \tag{9}$$

where D and k are, respectively, the diffusion coefficient of the product gas in the product layer and the thermal conductivity of the layer, and α and h are the mass and heat transfer coeficients to the surface of the sphere.

---

* In those cases where this is not so, the effect of gas velocity and the effect of particle size, insignificant variations would be expected under the experimental conditions normally used [9, 10] .

Eq. (8) shows that $f(\lambda^*) = 1$ throughout the reaction if $Bi_M = Bi_H$. Under these conditions, or if $H^*/k^* << 1$, Eq. (7) shows that Y can only vary during the reaction, if the gas composition, furnace pressure or furnace temperature are varied. If these parameters are kept constant, Eq. (6) can be integrated analytically to give:

$$3\lambda^{*2} - 2\lambda^{*3} - 2Bi_M(1-\lambda^{*3}) = 1 - 6Y\tau^* \qquad (10)$$

Since τ^* is proportional to time, Eq. (10) shows that the value of Bi_M controls the shape of curves relating the radius of the reaction front within a spherical particle to the reaction time. A series of such curves have been presented previously [10] and they show that the reaction front radius varies linearly with time during the first 94% of a reaction when $Bi_M = 0.33$. Even when Bi_M varies within the range $0.15 < Bi_M < 0.6$ the variation from linearity is extremely small for most of the reaction. A similar set of curves are presented in Fig. 1, calculated from the equations presented in Appendix 1 for cylindrical samples, and these curves show the same tendency, except that the straightest curve is for $Bi_M = 0.5$.

Thus a transport controlled reaction can exhibit the same experimental behaviour as a chemically controlled reaction. Even more significant is the fact that the value of Bi_M would be expected to fall in the range $0.15 < Bi_M < 0.6$ [10], so that we should expect a transport controlled reaction to behave in exactly the same way as a chemically controlled reaction if Y is constant.

General equations cannot be written down if Y is not constant, since Eq. (6) must then be evaluated by numerical integration. Y will vary during a reaction if there is a significant heat effect which causes the temperature of the reaction front to vary during the reaction, or if the diffusion coefficients of the reactant and product gases are not equal.* Very few such situations have been dealt with theoretically in the existing literature. The decomposition of calcium carbonate has been treated and it has been shown that the reaction front moves at a constant rate during the first 90% of the reaction, even though its temperature can change some 80°C [15] during this time. More general equations are developed in Appendix 1, and they have been used to obtain the curves shown in Figs. (2) and (3) for the four typical sets of reaction conditions set out below:

Case 1: A reduction reaction producing a coarsely divided metal product layer of moderate porosity, through which the reactant and product gases diffuse by molecular diffusion. There is a moderate heat of reaction - about

* This can come about if Knudsen diffusion is important in the product solid. The influence of Knudsen diffusion is discussed later, and in Appendices 1 and 2.

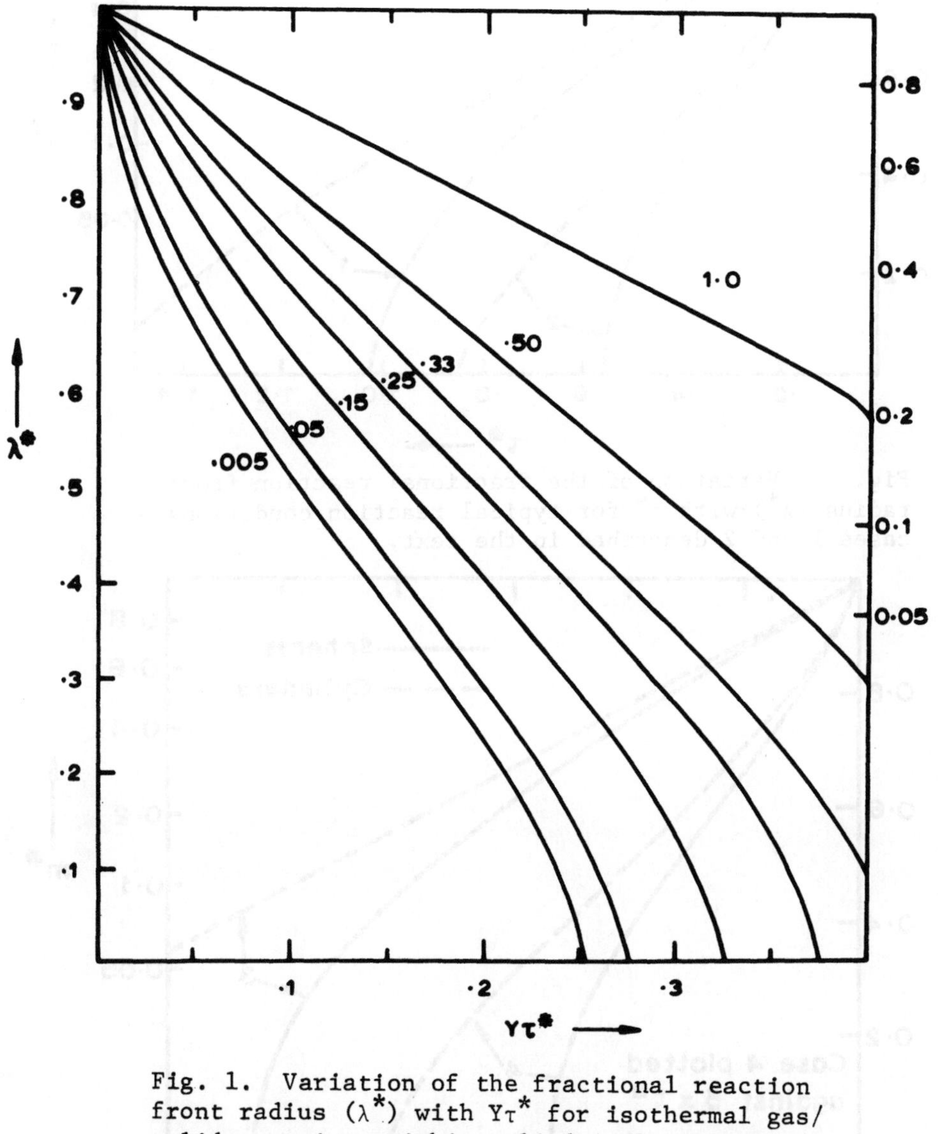

Fig. 1. Variation of the fractional reaction front radius (λ^*) with $Y\tau^*$ for isothermal gas/solid reactions within cylindrical compacts, plotted with different parametric values of Bi_M.

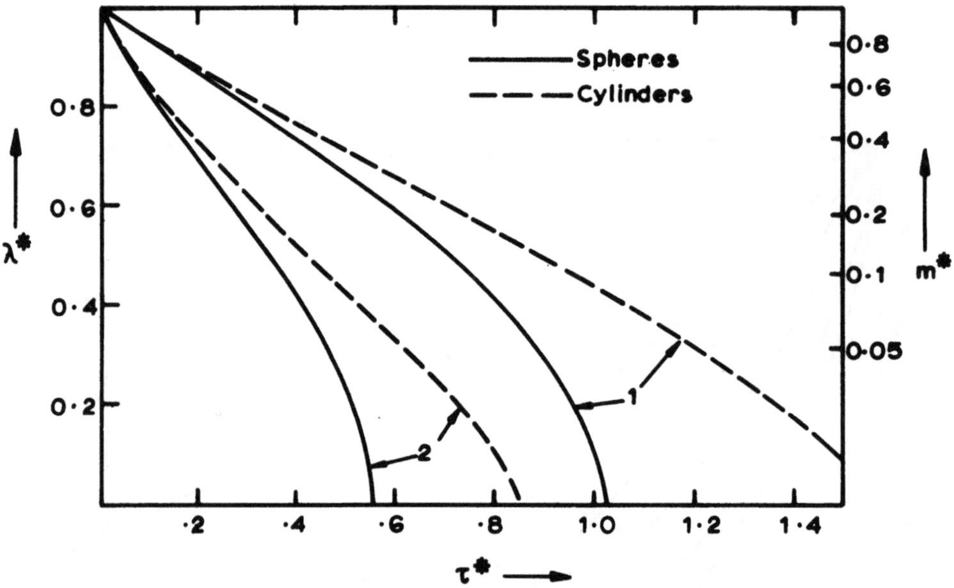

Fig. 2. Variation of the fractional reaction front radius (λ^*) with τ^* for typical reaction conditions - cases 1 and 2 described in the text.

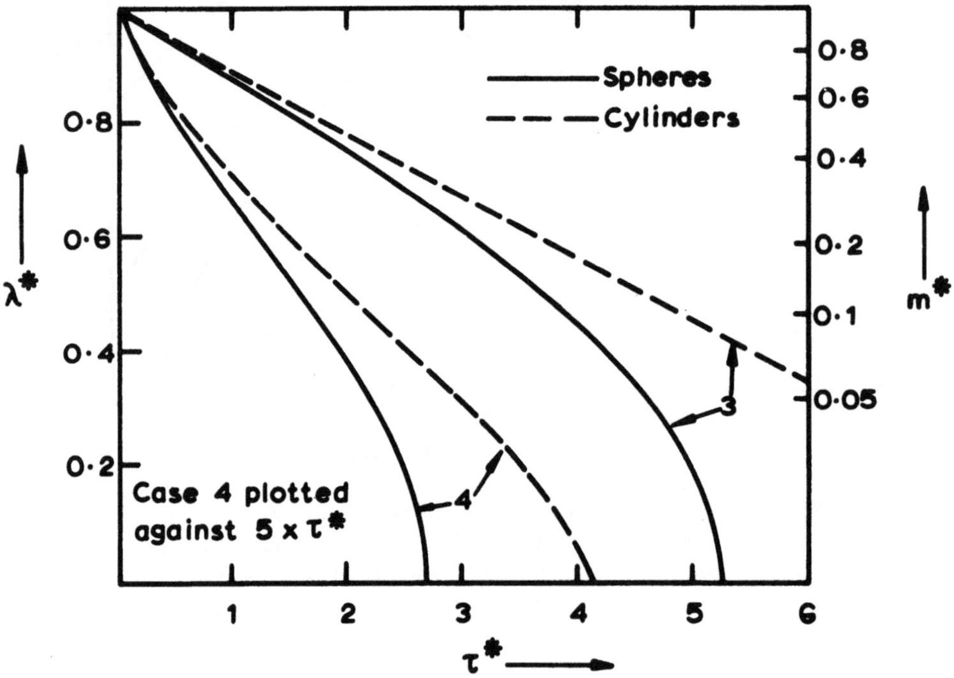

Fig. 3. Variation of the fractional reaction front radius (λ^*) with τ^* for typical reaction conditions - cases 3 and 4 described in the text.

4Kcals/g-mole, and the reaction takes place at about 800°C, in pure hydrogen.

Case 2: This case is the same as case 1 except that the product metal layer is considerably more finely divided, so that Knudsen diffusion becomes important. Thus the gases diffuse through the solid by a mixed diffusion mechanism.

Case 3: The reduction of a metal chloride by hydrogen to produce a very open metal product layer through which the gases diffuse by molecular diffusion. The reaction is strongly endothermic - about 34 K cals/g-mole and is assumed to take place at 600°C in pure hydrogen.

Case 4: A strongly endothermic equi-molar reaction in which an extremely finely divided non-metallic product is formed, through which the gases diffuse by Knudsen diffusion. The reaction is considered to take place in pure reactant gas.

The relevant values of the reaction parameters used to obtain the curves set out in Figs. (2) and (3) are listed in Table I.

Table I

Values for the Reaction Parameters for the Curves in Figures 2 and 3

Case	Bi_M	Bi_H	H*	K*	K_G	β_I	β_{II}	D*
1	0.5	15.0	2.0	400	0.54	1	1	1
2	0.136	15.0	2.0	400	0.54	1	1	2.04
3	0.5	1.1	20	50	$6x10^{-3}$	1	2	1
4	0.04	2	16	1000	0.6	1	1	2

Once again, the curves show that the reaction front moves at an almost constant rate during the first 90% of all these reactions. Thus it must be concluded that there is no significant difference between the reaction rate curves predicted by the chemical and transport control models and that these curves cannot be used to distinguish between them.

Nor can any significant differences be found between other aspects of reaction behaviour predicted by the two models. Varying the composition of the gases passed over the reacting particle changes the value of Y and would change the rate of a transport controlled reaction in a manner almost identical to that predicted by

Eq. (3) [10]. If the flow rate of the gases is sufficient to maintain the gas composition constant around the sample, any further increase would have a negligible effect on the reaction rate - at least over the range of gas flow rates that can be realised in laboratory experiments [8, 9, 10], so there is no difference in this aspect of reaction behaviour. Temperature affects the rate of a transport controlled reaction in a complicated way. The equilibrium coefficient will change with temperature as will the various transport parameters - diffusion coefficients, thermal conductivities, a and transfer coefficients. Thus the rate of transport controlled reaction can go up very rapidly with temperature, and would not behave in a way markedly different from a chemically controlled reaction [10]. Although a reaction controlled by molecular diffusion would be relatively insensitive to changes in total pressure, this would not be the case for a reaction in which Knudsen diffusion in the solid product is important. This point is discussed in the next section, and it is shown such a reaction would behave in precisely the same way as a chemically controlled reaction.

Thus we must conclude that the type of behaviour exhibited by a reaction of the type considered here cannot be used to show whether its mechanism is best described in terms of the chemical control model or the transport model.

Transport Control or Mixed Control

Recently, certain workers have measured mass transfer coefficients to reacting particles [7], or calculated them from standard correlations available in the literature [8, 16]. They have then used these values and measured reaction rates to show that the composition of the gas at the surface of the reacting particles could not be the same as the gas passed over the samples. This quantitative observation showed that the reactions were not chemically controlled, and these workers have postulated the mixed control model described earlier.

Although the mixed control model is more general than the chemical or transport control models, and therefore more intellectually satisfying, the evidence so far advanced in its favour cannot be considered as conclusive. This evidence, once again, depends upon specific items of reaction behaviour: (a) the shape of curves showing how certain functions of the fractional mass vary with time and, (b) the variation of the reaction rate with changes in total pressure. In both cases, the type of behaviour exhibited by the mixed control model could also be exhibited by the transport control model.

The mixed control model assumes that the reactant and product gases are not in equilibrium at the reaction front. Instead their partial pressures are related as in Eq. (3), except that the values $[p_I]_R$ and $[p_{II}]_R$ are substituted for $[p_I]_G$ and $[p_{II}]_G$. This equation then replaces the equilibrium relationship used in the transport control model as shown in Appendix 1. In most applications of the mixed control model, an equi-molar reaction has been involved, and it has been assumed that the reaction front is at the same temperature as the surroundings and that the reactant and product gases diffuse through the product solid by molecular diffusion.

Thus $\beta_I = \beta_{II}$; $[K_{eq}]_R = [K_{eq}]_G$ and $[p_I^*]_R = 1 - [p_{II}^*]_R$

Substituting Eq. (A-2-6) into the modified form of Eq. (3) gives:

$$\frac{\dot{n}\, R\theta_G}{A_R K_C p_G} = 1 - Y - [p_{II}^*]_G - \frac{Y - [p_{II}^*]_G}{[K_{eq}]_G} \qquad (11)$$

Substituting from Eqs. (A-2-3), (A-2-4) and (A-2-5) gives:

$$\frac{b}{K_C} + \frac{A_R}{SD} = \left\{[p_I^*]_G + \frac{[p_{II}^*]_G}{[K_{eq}]_G}\right\}\frac{b p_G A_R}{\dot{n}\, R\theta_G} - \frac{A_R}{A_0{}^{\alpha}} \qquad (12)$$

$$\text{where } b = \frac{[K_{eq}]_G}{1 + [K_{eq}]_G} \qquad (13)$$

Evidence in support of the mixed control model has been obtained by showing that plots of experimental results follow the form indicated by Eq. (12) or equations of a similar nature obtained in more complicated situations. Since A_R and S are related to the fractional unreacted mass of a dense particle, this comparison can be made using Eq. (12) in the form shown above or in its integrated form. For simplicity we will consider the equation in the form given above. This shows that the mixed control model predicts that a plot of $([p_I]_G + [p_{II}]_G/[K_{eq}]_G)\,(bp_G A_R/\dot{n}\, R\theta_G) - A_R/A_0$ against A_R/S should be a straight line of slope 1/D and intercept b/K_C.

Experimental results have been shown to give linear plots of this type, and these have been used as confirmation of the mixed control model, and as a means of measuring the values of D and K_C. If no other measurements other than the reaction rate and the mass transfer coefficient are made, the values obtained cannot be taken to confirm the mixed control model; this can only be done by the shape of the experimental plot.

Further evidence in support of the mixed control model has been provided by investigating how the rate of the reaction varies with total pressure. Since molecular diffusion coefficients in gases are inversely proportional to pressure, the product Dp_G will not vary with pressure, if the reactant and product gases diffuse through the porous solid by molecular diffusion. Because the thermal conductivity of a gas is also relatively insensitive to moderate changes in pressure, the equations developed in Appendix 1 show that the rate of a transport controlled reaction involving molecular diffusion would not vary with total pressure. Experimental results have shown that reaction rates increase rapidly with total pressure at pressures around one atmosphere. They then level off until they are almost constant at pressures higher than 20 atm.. This is exactly the type of behaviour that would be predicted by the mixed control model, and has been taken to confirm its validity.

Other aspects of reaction behaviour exhibited by the mixed control model will obviously lie between the predictions of the chemical control model and the transport control model. As the previous section has shown, the differences between these latter two models are too small for any distinction to be made between them, so we cannot expect any distinction between the mixed control model and either of the other models. Moreover, it is shown in the following two sections that the two specific aspects of reaction behaviour discussed above can also be included in the pattern of behaviour predicted by the transport control model. Even the observation that a certain reaction exhibits these aspects cannot, therefore, be taken to prove that the mixed control model applies.

Influence of Heat Transfer

Very few gas solid reactions have a zero heat of reaction. In general, then, the reaction front in a reacting particle will not be at the temperature of its surroundings, and this can affect the analysis of experimental results. It is surprising how seldom the actual temperature of a reacting particle has been measured, in spite of the ease with which this can be done. Even in the reduction of iron oxide, where the heat of the reaction is relatively small, the reaction front can be some 20°C below the temperature of the reaction furnace [23]. The largest temperature differences always occur at the beginning of the reaction and always slow down the reaction rate. This effect is sufficient to reduce the initial curvature in the curves of λ^* against time that would otherwise be predicted by a mass transfer and diffusion model, in which $Bi_M < 0.15$.

Radiation plays a significant part in the transfer of heat to the surface of a reacting particle. Even in pure hydrogen at 800°C, for example, the convective heat transfer coefficient to a sphere 1 cm in diameter is only 128 Watt/m^2 °K whereas the radiation coefficient could be as high as 280 Watt/m^2 °K. This latter figure cannot be calculated very accurately because it will depend upon the emissivity of the product solid, and upon the extent to which the particle 'sees' cold surfaces. Thus the heat transfer coefficient to a reacting particle is, to some extent, a function of furnace construction and this explains slight differences in reaction behaviour that have been observed by different workers, especially in the case of strongly endothermic reactions such as the decomposition of calcium carbonate [4].

Experimentally measured rates of a transport-controlled reaction could appear to follow Eq. (12) if the influence of heat transfer were to be overlooked. This is only likely to happen if the heat of the reaction is not very large i.e. when $H^*/k^* << 1$. Thus the exponential term in equation A-1-24 in Appendix 1 can be expanded as a Taylor series, and second order terms neglected. For an equi-molar reaction involving molecular diffusion $\beta_I = \beta_{II}$ and $P = 0$, so that the equation becomes:

$$\frac{[p^*_{II}]_G + Y}{[k_{eq}]_G} = \left\{[p^*_I]_G - Y\right\}\left\{1 - \frac{YH^{*2}f(\lambda^*)}{k^* - YH^*f(\lambda^*)}\right\} \qquad (14)$$

Using the definition of Y, this equation becomes:

$$\frac{bH^{*2}f(\lambda^*)\left\{[p^*_I]_G - Y\right\}}{k^* - YH^*f(\lambda^*)}\left\{\frac{A_R}{SD} + \frac{A_R}{\alpha A_0}\right\} + \frac{A_R}{SD}$$

$$= \left\{[p^*_I]_G - \frac{[p^*_{II}]_G}{[k_{eq}]_G}\right\}\frac{bp_GA_R}{n\,R\theta_G} - \frac{A_R}{A_0\alpha} \qquad (15)$$

The first term on the left hand side of this equation corresponds to the term containing K_C in Eq. (12). The term varies quite considerably during a typical reaction, but its variation with A_R/SD is approximately linear for most of the reaction. A plot of the right hand side of this equation against A_R/SD for a reaction controlled by heat and mass transfer would still give a straight line and the reaction would therefore appear to follow the predictions of the mixed control model. Thus heat transfer must be rigorously excluded from the reaction mechanism before linear plots of this nature can be taken to verify mixed control reaction models. The importance of heat transfer can be simply assessed by measuring the actual temperature of the sample while the reaction is taking place.

Influence of Knudsen Diffusion

Knudsen diffusion plays a much greater part in the transport of gases through porous product solid than is generally realised. Carbon dioxide liberated during the decomposition of Calcium Carbonate [4, 15], for example, diffuses through the porous lime almost entirely by Knudsen diffusion. The measurements made by Nabi & Lu [16] on the hydrogen reduction of hematite to magnetite suggest that Knudsen diffusion is also of over-riding importance in this system. Bogdandy has suggested that Knudsen diffusion plays an important part in the diffusion of hydrogen/water vapour mixtures through porous iron formed during the reduction of hematite in hydrogen [18], although this suggestion appears to have found little favour with recent workers. However, the calculations outlined in Appendix 4, based on pore size measurements made by Warner [7] and diffusion measurements made by McKewan and Olsson [20], confirm Bogdandy's suggestion.

Knudsen diffusion occurs when gas molecules in the pores of the product solid collide much more frequently with the walls of the pores than with one another, i.e. when their mean free path is much greater than the width of the pores. Molecular diffusion occurs when collisions between gas molecules and the walls are relatively rare, but a large range of pore sizes exists in which neither mechanism dominates. Both diffusion mechanisms occur together, the resulting process being called mixed diffusion. During Knudsen diffusion the reactant and product gases diffuse independently of one another, their diffusion coefficients are not equal, and the total pressure at the reaction front is different from that in the gas phase surrounding the particle. During molecular diffusion the two diffusion coefficients are equal and the total pressure within the pores in the solid is equal to that in the gas phase surrounding the reacting particle. Mixed diffusion is similar to pure Knudsen diffusion except that the effects are not so great.

Since the Knudsen diffusion coefficient of a gas species is inversely proportional to the square route of its molecular weight, large effects can occur during reduction reactions involving hydrogen. Some doubt must therefore fall on reaction models that have been based on the assumption that diffusion coefficients of hydrogen and water vapour are equal in these systems, unless definite experimental evidence exists that Knudsen or mixed diffusion does not occur.

A heat and mass transfer reaction model involving mixed diffusion would predict that the reaction rate would vary with total pressure in the same manner as the chemical control, or mixed chemical and transport control models. The equations derived in Appendix 1 can be used to illustrate this, using data relevant to the reduction of hematite in pure hydrogen. The data determined in Appendix 4 shows that the effective diffusion coefficient of water vapour in porous iron at 800°C under a pressure of one atmosphere is:

$$D_{800,p_o} = \left\{\frac{1}{2.94} + \frac{1}{0.92}\right\}^{-1} \text{cm}^2/\text{s} \qquad (16)$$

Since the molecular diffusion coefficient in a gas is inversely proportional to pressure, the effective coefficient at any other pressure is:

$$D_{800,p_G} = \left\{\frac{p_G}{2.94\ p_o} + \frac{1}{0.92}\right\}^{-1} \text{cm}^2/\text{s} \qquad (17)$$

and that for hydrogen is:

$$D_{H_2,800,p_G} = \left\{\frac{p_G}{2.94\ p_o} + \frac{1}{2.81}\right\}^{-1} \text{cm}^2/\text{s} \qquad (18)$$

As p_G increases, the first term in each of these expressions becomes more and more important, and the second term eventually ceases to make a sensible contribution to the values of the effective diffusion coefficients. When this occurs, the diffusion process within the porous iron is virtually pure molecular diffusion, the effective diffusion coefficient is inversely proportional to pressure, and the rate of the transport controlled reaction would be unaffected by further changes in the total pressure.

The equations developed in Appendix 1 show that the reaction front would have moved half way towards the centre of a spherical particle during a heat and mass controlled reaction, after a dimensionless time period:

$$\tau^*_{0.5} = \int_{0.5}^{1} \frac{\lambda^* - \lambda^{*2}(1-Bi_M)}{Y}\, d\lambda^* \qquad (19)$$

Thus the mean reaction front velocity during the first 94% of such a reaction is given by:

$$\overline{V}_{0.5} = \frac{DP_G}{2R\theta_G r_o \rho_I \beta_{II}} \left[\int_{0.5}^{1} \frac{\lambda^* - \lambda^{*2}(1-Bi_M)}{Y}\, d\lambda^*\right]^{-1} \qquad (20)$$

and the mean velocity at any pressure p_G is related to the velocity under one atmosphere by the equation:

$$\frac{\left[\bar{V}_{0.5}\right]_{P_G}}{\bar{V}_{0.5\ P_o}} = \frac{D_{P_G} P_G \left[\int_{0.5}^{1} \frac{\lambda^* - \lambda^{*2}(1-Bi_M)}{Y} d\lambda^*\right]_{P_o}}{D_{P_o} P_o \left[\int_{0.5}^{1} \frac{\lambda^* - \lambda^{*2}(1-Bi_M)}{Y} d\lambda^*\right]_{P_G}} \quad (21)$$

This expression is plotted in Fig. 4 using the data set out in Table II for the reduction of hematite to iron in pure hydrogen, and the diffusion coefficient expressions set out in Eqs. [17] and [18]. Reaction rates measured by Warner [7] and McKewan [19] are also plotted on the diagram.

It can be seen that the theoretical curve reproduces the experimentally observed variation quite closely, although the experimental results have previously been taken to justify a chemical, or a mixed chemical and transport control model [7, 8]. Fig. 4 shows that they can equally well be taken to justify a transport control model.

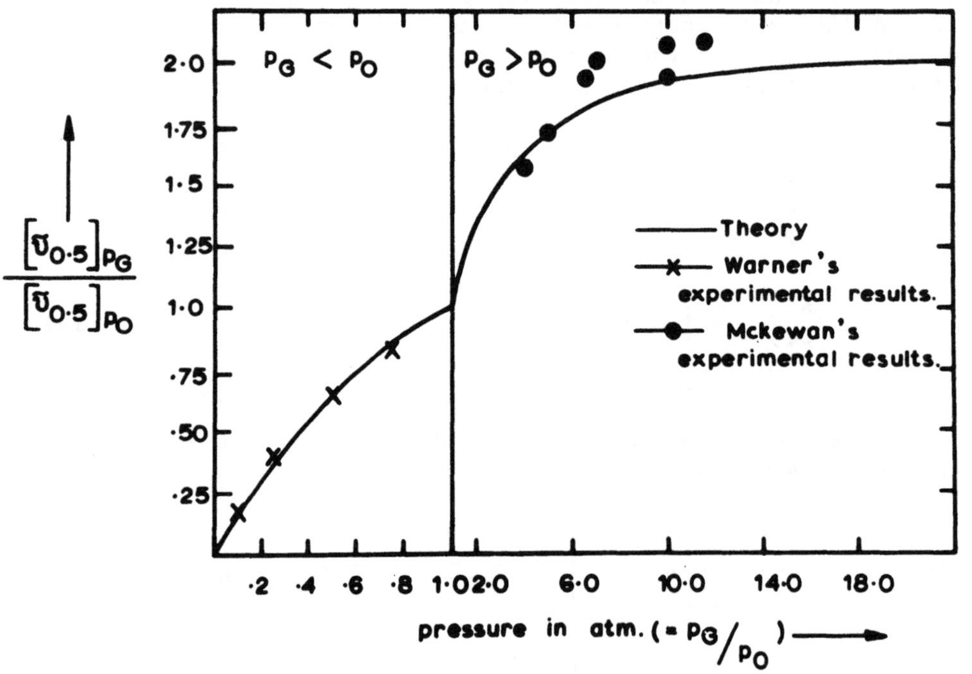

Fig. 4. Variation of the mean reaction front velocity with the total gas pressure during the reduction of hematite spheres in pure hydrogen at 800°C.

AN IMPROVED EXPERIMENTAL STRATEGY

The previous section has shown that qualitative observations of the reaction behaviour of dense single particles cannot be used to determine which reaction model best describes the mechanism of any one gas solid reaction. Insufficient differences exist between the patterns of reaction behaviour predicted by the various models even when the reacting particles are dense spheres in which the reaction surface is closely defined and precise mathematical expressions are available for its area and for the shape factor of the diffusion path from the surface of particle. The situation is even less clear when the reaction of porous compacts is considered since the reaction can take place over an extensive zone within the compact. The area of this zone cannot be specified, nor can any value be given to the shape factor of the diffusion path from the surface of the particle to the reaction zone.

Thus a new experimental strategy must be evolved if unequivocal decisions are to be made concerning the control mechanisms of gas/solid reactions. A clue to the nature of this strategy is provided by the one clear decision that was described in the previous section. The quantitative argument used by Warner [7] based on measured mass transfer coefficients showed that the chemical control model could not be applied to the reduction of hematite in hydrogen. The extension of this principle to include the diffusion coefficients of the reactant and product gases within the product solid would allow similar clear decisions to be made between the transport control model and the mixed transport and chemical control model. These decisions could be made by comparing reaction rates predicted from the transport control model and the measured diffusion coefficients, with the results of actual kinetic experiments. Only if the measured reaction rates were found to be significantly slower than the predicted rates should the mixed chemical and transport control model be considered. In this case, the measured diffusion coefficients could be used to determine the actual partial pressures of the reactant and product gases at the reaction front during the reaction and thus allow the true value of the chemical rate constant to be determined from the observed reaction rates. In either case, it is obvious that the transfer coefficients to the reacting particle must be known, and that mathematical expressions must be available to show how A_R and S vary with the extent of the reaction. It is thus possible to draw up the experimental strategy described below, and to be certain that it will allow the most suitable reaction model to be clearly defined.

1. Kinetic experiments should be carried out using dense spherical samples and then samples of increasing porosity, under conditions such that the mass and heat transfer coefficients to their surfaces are known, particular attention being paid to the construction of radiation

shields so that the heat transfer coefficient may be known with some accuracy.

2. The actual temperature of the reacting sample should be measured during the reaction by embedding a thermocouple into the sample.

3. Partially reacted samples should be sectioned and examined under the microscope, particular attention being paid to the nature of the reaction zone, and to any signs of sintering in the product layer. Measurements should be made of the average pore size in the porous solid, or of its specific surface area.

4. Diffusion-coefficient measurements should be made in the porous product solid, produced from material identical to that used in the kinetic experiments. These measurements should ideally involve the reactant and product gases although this is not absolutely necessary as long as a gas pair with a large difference in molecular weights is used - He/Ar is very suitable. If possible the diffusion coefficients of the two gases should be measured separately and measurements should be made at different total gas pressures.

5. The emissivity of the product solid should be measured under reaction conditions, or the heat transfer coefficient from a reacted particle should be measured in the reaction furnace.

6. If the slightest doubt exists about the thermodynamic data for the reaction it should be redetermined using material identical to that used in the kinetic experiments.

The results of these experiments will allow a quantitative prediction to be made of the rate of the reaction assuming it to be transport controlled, this prediction being based on data that is completely independent of the kinetic experiments. A comparison between this prediction and the reaction rate actually measured will allow the correct reaction model to be chosen.

The procedure outlined above has been carried out for the decomposition of calcium carbonate [4, 15]. The diffusion coefficient of carbon dioxide has been measured in lime produced by decomposing calcium carbonate compacts of different porosity. These values have been used with equations similar to those derived in Appendix 1 to predict the rate at which these compacts would decompose. The theoretically predicted rates are compared with experimentally measured decomposition rates in Fig. 5, and it can be seen that the two rates

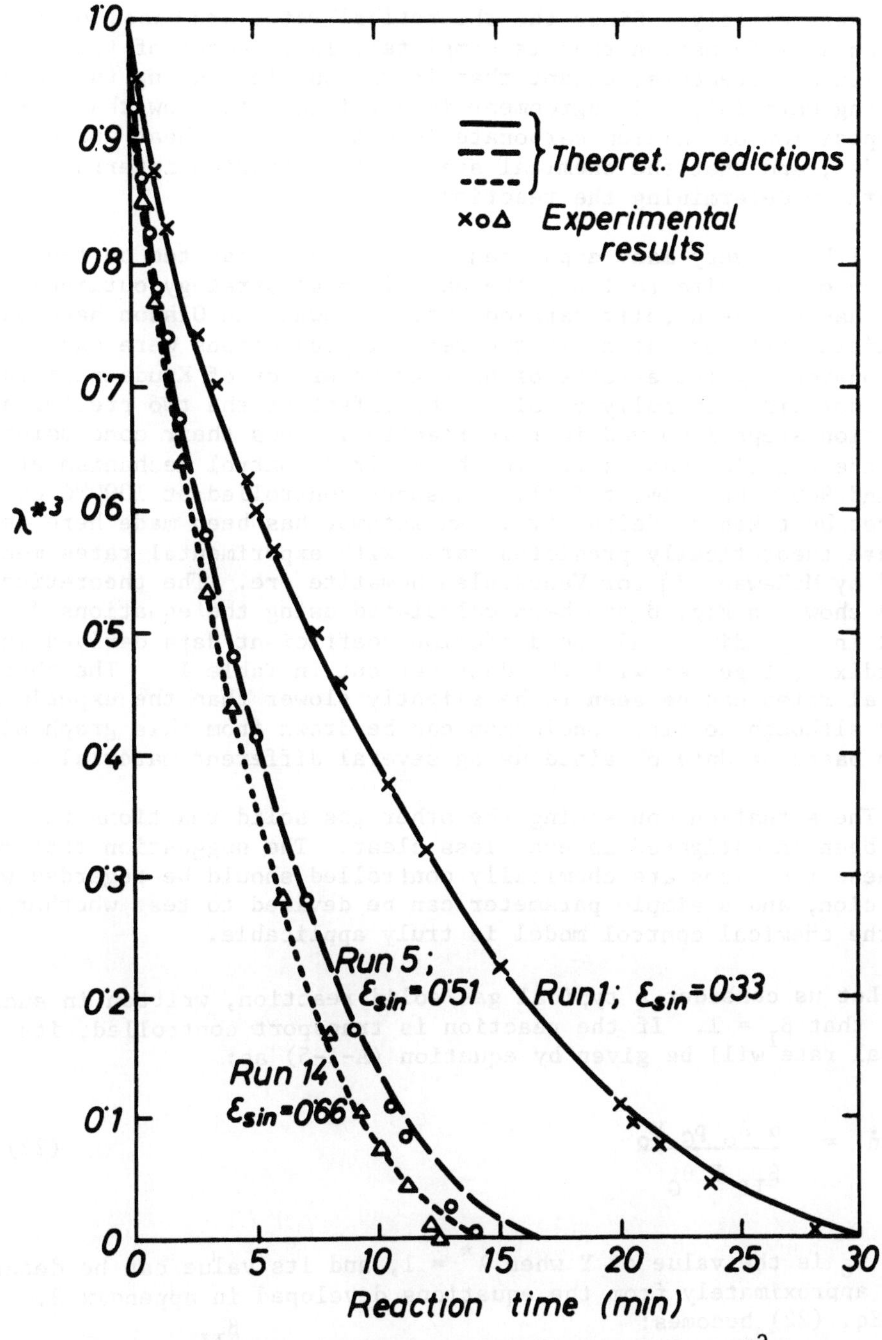

Fig. 5. Variation of the fractional mass ($m^* = \lambda^{*3}$) during the decomposition of calcium carbonate compacts of different initial porosity at 900°C in pure air. Theoretical results predicted using heat and mass transfer model and measured CO_2 diffusion coefficients[17].

agree very closely. Since the theoretical rates are based on experimental information that is completely independent of the decomposition measurements, except that it was obtained using the same starting material, this agreement is sufficient to show that the decomposition of calcium carbonate is controlled by heat and mass transfer, and that the chemical step at the reaction interface plays no part in determining the reaction rate.

Although very many experiments have been conducted on the reduction of hematite to iron, the experimental strategy outlined above has not been fully carried out. McKewan and Olsson have made an initial attempt but their theoretical predictions were made without considering the effects of heat transfer, or of Knudsen diffusion, nor did they fully consider the effect of the two preliminary reduction steps involved in this reaction. Thus their conclusion that the reaction was controlled by a mixed control mechanism at 800 and 900°C but almost fully transport controlled at 1000°C cannot yet be taken as definitive. An attempt has been made here to compare theoretically predicted rates with experimental rates measured by McKewan [3] for Venezuelan hematite ore. The theoretical curve shown in Fig. 6 has been calculated using the equations derived in Appendix 1 and the diffusion coefficient data derived in Appendix 4, together with the data set out in Table II. The theoretical rates can be seen to be slightly slower than the experimental rates although no firm conclusion can be drawn from this graph since it is based on data obtained using several different materials.

The situation concerning the other gas solid reactions that have been investigated is even less clear. The suggestion that many of these reactions are chemically controlled should be regarded with suspicion, and a simple parameter can be devised to test whether or not the chemical control model is truly applicable.

Let us consider a typical gas/solid reaction, written in such a way that $\beta_I = 1$. If the reaction is transport controlled, its initial rate will be given by equation (A-1-5) as:

$$\dot{n} = \frac{\alpha A_o p_G Y_o}{\beta_{II} R \theta_G} \tag{22}$$

where Y_o is the value of Y when $\lambda^* = 1$, and its value can be determined approximately from the equations developed in appendix 3. Thus Eq. (22) becomes:

$$\dot{n} = \frac{2\alpha A_o}{\beta_{II} R\theta_G \left(\Omega + \sqrt{\Omega^2 - 4\Gamma\Lambda}\right)} \quad [p_I]_G - \frac{[p_{II}]_G^{\beta_{II}}}{[k_{eq}]_G} \tag{23}$$

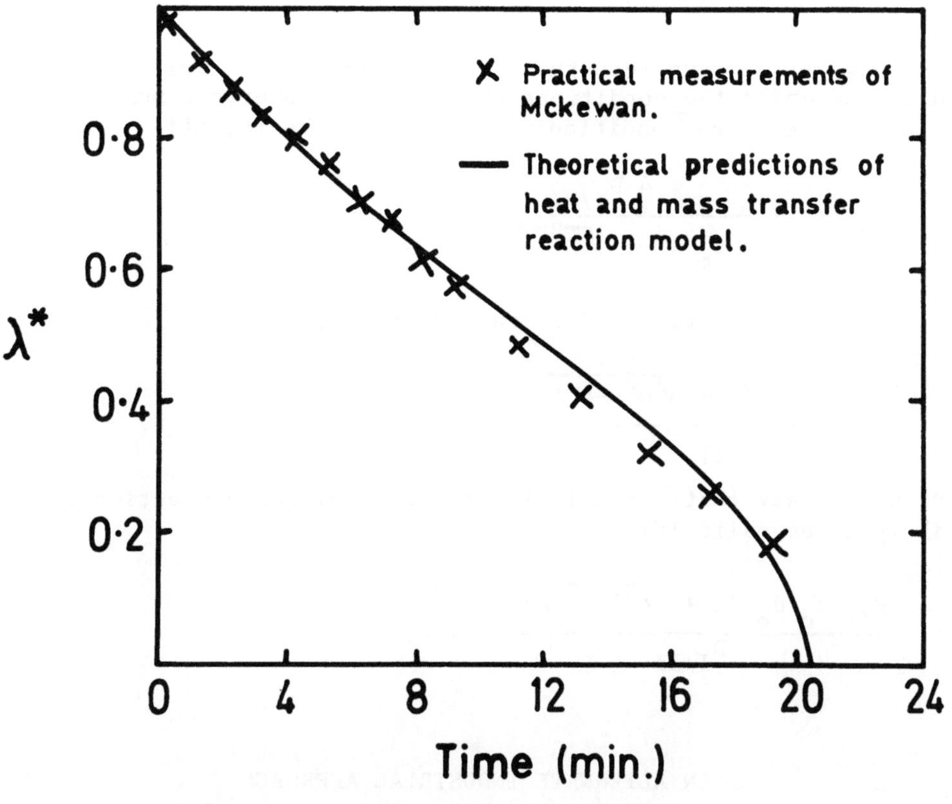

Fig. 6. Variation of the fractional reaction front radius (λ^*) with time during the reduction of hematite in pure hydrogen at 800°C. Theoretical results predicted using the heat and mass transfer model and the diffusion coefficient data calculated in appendix 4 compared with experimental results obtained by McKewan[3].

If the reaction is chemically controlled its initial rate will be very much smaller than the rate predicted by this equation. Eq. (3) shows that this is equivalent to saying that the reaction will only be chemically controlled if:

$$k_c \ll \frac{2\alpha}{\beta_{II}\ (\Omega + \sqrt{\Omega^2 - 4\Lambda\Gamma})}$$

However, we have seen that kinetic experiments are normally conducted under the conditions such that the Sherwood number is about 2. The above condition will therefore be met if:

$$k_c \ll \frac{4\ D_{I/II}}{\beta_{II}\ d_o\ (\Omega + \sqrt{\Omega^2 - 4\Gamma\Lambda})}$$

We can thus define a chemical control modulus

$$\frac{k_c\ \beta_{II}\ d_o\ (\Omega + \sqrt{\Omega^2 - 4\Gamma\Lambda})}{D_{I/II}}$$

and we can say that a chemical control model for a reaction is only likely to be valid if:

$$\frac{\beta_{II}\ k_c\ d_o\ (\Omega + \sqrt{\Omega^2 - 4\Gamma\Lambda})}{D_{I/II}} \ll 4.$$

AN ALTERNATE INDUSTRIAL APPROACH

We have seen that simple kinetic experiments in which the rate of a gas/solid reaction is measured under specified conditions do not in themselves provide enough information for us to decide which model best describes the mechanism of the reaction. Several additional experiments must be carried out, and it is possible that there will not always be sufficient time or money to enable this to be done. Thus it is important to ask whether the quantitative information contained in the results of simple kinetic experiments can be used without the reaction mechanism being fully understood. In a previous paper [24] the author reported the results of a "theoretical experiment" that was designed to answer this question.

Table II

Parameters Involved in the Heat and Mass Transfer Model for the Reduction of Hematite

Variable	Value	Units	Source
Diffusion coefficient of H_2O in porous iron (D)	0.70	cm^2/sec	App. 4
Diffusion coefficient of H_2 in porous iron (D_{H_2})	1.44	"	"
Molar density of Venezuelan ore Pellets (ρ_I)	$2.62x10^{-2}$	$g\text{-mole}/cm^3$	McKewan [3]
Radius of pellets λ_o	0.5	cm	"
Bulk gas pressure (p_G)	101.3	kN/m^2	--
Thermal conductivity of porous iron (k)	20.0	$Watt/m^oK$	0.55 x (the value for pure iron [25].
Heat of Reaction (ΔH)*	75.5	kJ/g-mole	
Mass transfer coefficient (α)	16.6	cm/s	(Sh=2)
Convective heat transfer coefficient (h_c)	98.0	$Watt/m^{2o}K$	(Nv=2)
Radiative heat transfer coefficient (h_r)	45.6	$Watt/m^{2o}K$	Calculated as outlined by Hills [4] taking emissivity of iron as 0.16 [25].
$\beta_I = \beta_{II}$	3	—	—
Fractional pressure of H_2 in bulk gas phase $\left(\left[p^*_{H_2}\right]_G\right)$	1	—	McKewan [3]
Equilibrium constant at Gas temp. (k_G)	.814	—	Ref. [26]

* The reduction of hematite in pure hydrogen at 811°C is complicated because there are, effectively, two values of ΔH. The overall reaction is $Fe_2O_3 + 3H_2 \rightarrow 2Fe + 3H_2O$, for which $\Delta H = 57.4$ kJ/g-mole [26], and it is this reaction which determines the heat requirement. However, the equilibrium conditions at the interface are controlled by the reaction $2Fe_{0.95}O + H_2 \rightarrow 1.9Fe + H_2O$, for which $\Delta H = 16.6$ kJ/g-mole [26]. Thus the value of H^{*2} which appears in Eq. (A-1-24) is given by:

The decomposition of calcium carbonate is controlled by heat and mass transfer. However, many authors have suggested that the chemical control model applies to this reaction, and have determined values of an apparent chemical rate constant from the results of their kinetic experiments. Single particle reaction rates predicted using these values agree with experimental results almost as well as rates predicted from the heat and mass transfer model. In the theoretical experiment referred to above, the author used the two models to make two predictions of the behaviour of a packed bed reactor containing calcium carbonate spheres. A substantial difference was found to exist between the two predictions, although they were both based on the results of the same set of single particle experiments. These differences did not appear to be serious where a single chemical reaction was concerned but could result in a large error if two or more reactions were occurring together, especially since the most marked differences were to be found in predicted temperatures of the solid particles in the bed.

The differences between the two predictions arise because the chemical control model drastically under-estimates the initial reactivity of solid particles in the bed. Since it assumes that the chemical rate constant is not affected by the environment of the reacting particles, it predicts that their reactivity in the packed bed reactor is the same as in the single particle experiment. However, the mass transfer coefficient to a solid particle in a packed bed can be up to 100 times greater than to the same particle isolated in a laboratory furnace. Thus calcium carbonate particles would start to react much more readily in the packed bed than they would do in a single particle experiment so that the behaviour of the packed bed predicted by the chemical control model is bound to be in error.

This theoretical experiment showed that some understanding of the mechanism of a gas/solid reaction is essential if we are to predict the behaviour of a packed bed, or indeed of any other industrial gas/solid reactor, from the results of simple kinetic

---

$$H^{*2} = \frac{57.4 \times (3\times2\times16.6)}{(R\theta_G)^2} \quad kJ^2/(g\text{-mole})^2$$

which corresponds to a mean value of ΔH = 75.5 kJ/g-mole. The use of this value in Eq. (A-1-24) produces very little error in the divisor in the exponential term since k^* is such a large number, in this case.

---

experiments. We must at least know the part played by gas-phase mass transfer in controlling the rate of the reaction. We can then assume that the rate controlling processes that occur within the solid particles are unaffected by the change between their environment in the single sphere experiment and in the industrial reactor, provided that the particles are not crushed or abraded in the reactor. As far as a quantitative prediction of the behaviour of the reactor is concerned, the actual nature of these processes is unimportant; we are only concerned with their quantitative effect on the rate of the reaction. This can be determined if we carry out kinetic experiments under known gas phase mass transfer conditions, and measure the actual temperature of the sample during the reaction. In the case of a spherical particle, for example, the reaction rate can be represented by an equation of the form:

$$\beta_{II}\dot{n} = \frac{[p'_{II}]_R - [p_{II}]_G}{\dfrac{R\theta_G}{4\pi r_o}\left\{\dfrac{1}{\alpha r_o} + f(m^*,\theta_S, p_G)\right\}} \tag{24}$$

where $p'_{II\ R}$ is a measure of the partial pressure of the product gas at the reaction front, and $f(m^*, \theta_S, p_G)$ represents the reaction resistance that is due to the rate controlling processes that occur within the particle - whatever they may be.

Provided we can calculate $[p'_{II}]_R$, then, the value of $f(m^*,\theta_S p_G)$ can be determined at any time during the reaction from the measured reaction rate,the measured particle temperature, and the total pressure in the bulk gas phase. Carrying out kinetic experiments over the entire range of solid temperatures and total gas pressures that would apply in the industrial reactor should yield sufficient values of $f(m^*,\theta_S,p_G)$ for an accurate prediction to be made of the reactor's behaviour. Since, in any case, a computor would be involved in such a prediction, these values could be stored within the computor as a three-entry table, or in the form of an empirical function determined from the experimental results by multiple regression analysis.

A small problem arises over the value of $[p'_{II}]_R$. Since we do not know what the rate controlling processes within the solid are, we cannot calculate the true value of the partial pressure of the product gas at the reaction front within the pores of the product solid. We can however, calculate $[p'_{II}]_R$ on the assumption that it is the partial pressure of the product gas at some point within the product solid where the reactant and product gases are in equilibrium and where the total gas pressure is equal to the total gas pressure in the bulk gas phase. No error will result from these assumptions since they are correct at the start of the reaction when

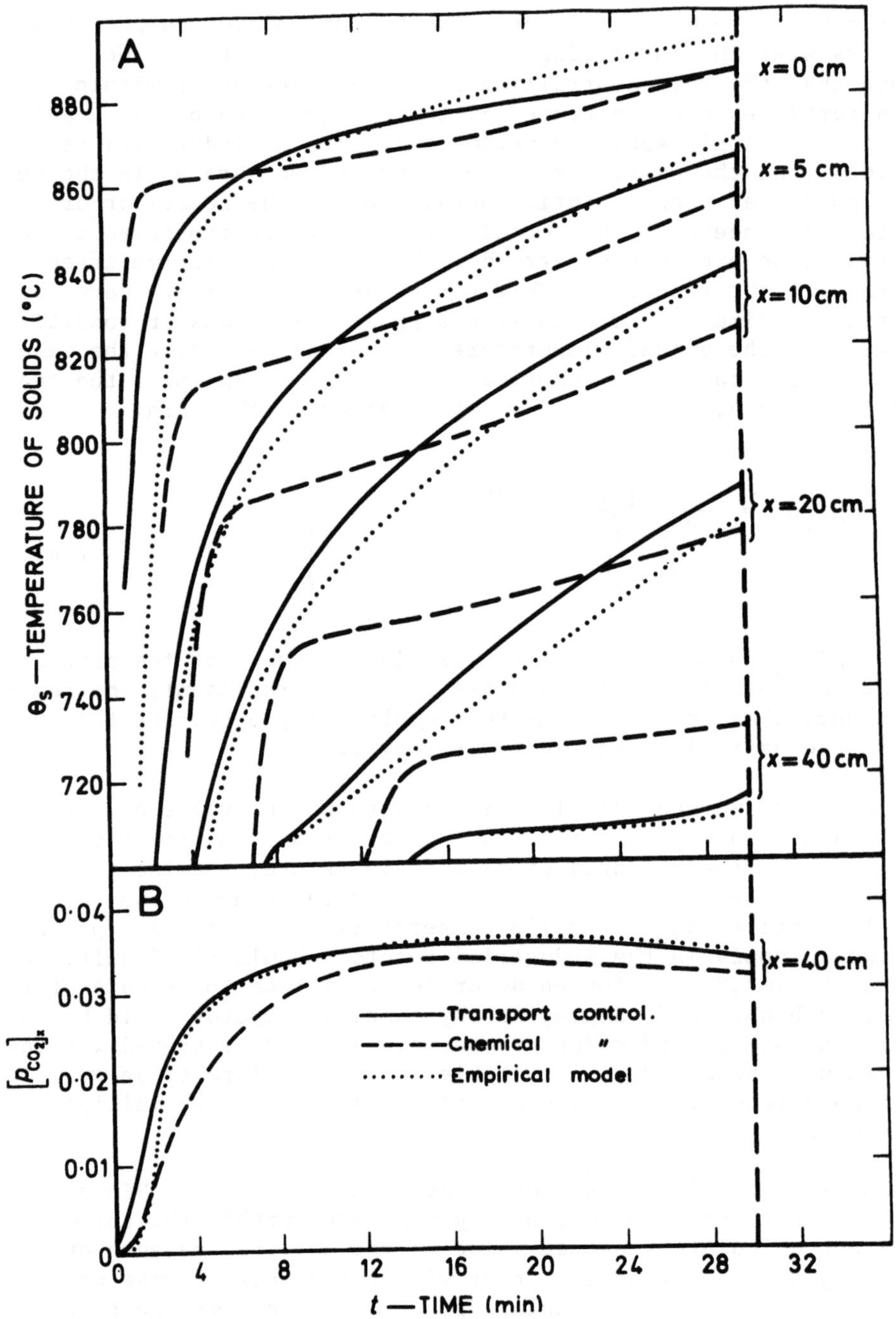

Fig. 7. Behaviour of a packed bed reactor containing calcium carbonate spheres, air being passed through at 900°C. The packed bed equations are derived, and described in reference 24.

$f(m^*,\theta_S,p_G)$ is equal to zero. At any subsequent time during the reaction the calculated value of $f(m^*,\theta_S,p_G)$ will automatically compensate for the assumptions, should they be invalid.

An experimental strategy of this nature could greatly reduce the amount of work involved in predicting the performance of an industrial gas/solid reactor from the results of single particle experiments. Its accuracy can be tested by carrying out a further theoretical experiment of the type referred to earlier. Reaction rates and sample temperature determined in a previous investigation [4] during the decomposition of calcium carbonate under known gas phase mass transfer conditions can be used to determine relevant values of $f(m^*,\theta_S,p_G)$. Since the investigation was carried out at atmospheres pressure, we must restrict consideration to situations in which $p_G = 1$ atm. Under these conditions, we find that the experimental values of the solid state resistance to the reaction can be correlated by an equation of the form:

$$f(m^*,\theta_S) = \left\{\frac{2085(^{o}C)}{\theta_S - 268(^{o}C)} - 1.53\right\}\left\{\frac{1}{m^*} - 1\right\}^{0.827} \text{ s/cm}^2 \quad (25)$$

Packed bed reaction rates have been calculated using this expression and Eq. (24) for the conditions set out in Table III. They are compared in Fig. 7 with predictions based on the same experimental results, but using the correct heat and mass transfer control model, and the incorrect chemical control model. The computations have been carried out using the numerical procedure outlined in Ref.24.

It can be seen that empirical model agrees very closely with the heat and mass transfer (transport) model, and certainly very much better than the chemical control model. An even better agreement could be achieved between the empirical model and the transport control model if more terms were to be taken in the expression given in Eq. (25).

This industrial strategy is inferior to the more detailed strategy outlined in the previous section. Its only advantage lies in the reduction of experimental effort that it offers in cases where a quantitative prediction is required for the performance of a particular type of reactor. It could be argued that an incorrect reaction model could also be used as long as it took proper account of gas phase mass transfer. However, the empirical rate equation has the advantage that it offers no more information than can be obtained from simple kinetic experiments - that is it tells us what the reaction rate will be at a certain temperature and total gas pressure under specified gas phase mass transfer conditions. Since

Table III

Parameters Involved in Predicting Reaction Rates in a Packed Bed of Calcium Carbonate Spheres

Variable	Value	Units	Reference
Radius of spheres	0.57	cms	
Initial mass of spheres	0.0179	gm-moles	
Initial solid temperature	20	^oC	
Specific heat* of lime	54.7	J/g-mole ^oK	[27]
Specific heat* of calcium carbonate	125.6	"	[27]
Thermal conductivity of porous lime	1.0	W/m ^oK	[4]
Number of spheres in unit volume of bed	0.645	cm^{-3}	
Specific molar velocity of air through bed	$3.32x10^{-3}$	gm-mole/sec cm	
Inlet temperature of air	901	^oC	
Total pressure in bed	101.3	kN/m^2	
Height of bed	0.4	m	
Specific* heat of CO_2	42	J/g-mole ^oK	[27]
Specific* heat of air	33.6	"	[27]
Mass transfer coefficient to sphere in bed	32.0	cm sec	calculated from the total gas flow through the bed at a mean temperature of 832°C using the equation of Rowe & Claxton [28], and the property values given below
Heat transfer coefficient to a sphere in the bed	209	W/m^2 ^oK	(as above)
Viscosity of air at 832°C	$4.48x10^{-4}$	Ns/m^2	[25]

Table III (continued)

Thermal Cond. of air at 832°C	7.27×10^{-2}	W/ms	[25]
Diffusion coefficient of CO_2 in air at 832°C	1.19	$cm^2 sec^{-1}$	[4]
Diffusion coefficient of CO_2 in porous lime†	$D/\theta_s = a + b(\theta_s - 1088)$ where $a = 7.55 \times 10^{-5}$ $cm^2/s\ ^oK$; $b = 0.015 \times 10^{-5}$ $cm^2/s\ ^oK$		[24]
Apparent chemical rate constant$^\Delta$	$13.9 \exp\left(\frac{c}{R\theta_G}\right)$ $g\text{-mole}/cm^2 s$ where $c = -1.40 \times 10^5$ $J/g\text{-mole}\ ^oK$		[24]

* Between 600°C and 900°C, the most significant range of temperature variation, these specific heats vary by about 5%.

† Used only in the prediction based on transport control.

Δ Used only in prediction based on the chemical control model.

it makes no attempt to provide information about the nature of the rate controlling processes within the solid, it cannot exert a false influence on the qualitative way in which we think about the reaction.

In situations where the qualitative way in which we think about the reaction is important, in decisions concerning the most suitable type of reactor or about the effect of possible pretreatment processes for example, we must know which of the reaction models is correct. As we have seen, this information can only be provided by the improved experimental strategy outlined in the previous section.

ACKNOWLEDGEMENTS

It gives the author very great pleasure to acknowledge the many stimulating discussions that he has had with professor A. V. Bradshaw and with Dr. A. Paulin and Mr. F. Campbell about this paper, and about gas/solid reactions in general, and to give special thanks to Dr. B. Partridge of A.E.R.E., Harwell for checking some of the mathematics presented in Appendix 1.

APPENDIX 1

General Equation for a Topochemical Reaction Controlled by Heat and Mass Transfer

Consider a gas/solid reaction described in terms of the equation:

$$\text{Solid I} + \beta_{I}(\text{Gas I}) \rightarrow \beta_{II}(\text{Solid II}) + \beta_{II}(\text{Gas II})$$

If we assume that heat and mass flows away from the reaction interface are positive, we can write:

$$\dot{n}_{II} = {}_{II}\dot{n} \; ; \; \dot{q} = -\Delta H \; \dot{n} \qquad \text{(A-1-1)}$$

where $\dot{n}$ is the rate at which solid I reacts.

Appendix 2 shows that the rate at which the product gas is transferred from the reaction interface can be described in terms of Eq.(A-1-2):

$$\dot{n}_{II} = \frac{\left[p^*_{II}\right]_R - \left[p^*_{II}\right]_G}{\dfrac{R\theta_G}{\xi D p_G}\left\{\dfrac{1}{S} + \dfrac{D}{A_0\alpha}\right\}} \qquad \text{(A-1-2)}$$

where S is the shape factor of the diffusion path through the product solid between the reaction interface and the surface of the particle. The exact meanings of the variables ξ and D depend upon the diffusion mechanism, as described in Appendix 2.

It is convenient if Eq. (A-1-2) is rearranged in terms of the following dimensionless groups:

$$S^* = \frac{S\lambda_0}{A_0} \qquad \text{(A-1-3)}$$

$$Bi_M = \frac{D}{\lambda_0\alpha} \qquad \text{(A-1-4)}$$

FOOTNOTE: The reaction rate is expressed in terms of the flux of the product gas so that decomposition reactions may be included in the general treatment.

$$Y = \frac{\dot{n}_{II}\, R\theta_G \lambda_O}{DA_O p_G} \left\{ \frac{1}{S^*} + Bi_M \right\} \qquad \text{(A-1-5)}$$

so that it becomes:

$$\left[p^*_{II}\right]_R - \left[p^*_{II}\right]_G = \frac{Y}{\xi} \qquad \text{(A-1-6)}$$

where λ_O is a characteristic linear dimension of the original reacting particle, whose precise definition depends upon the shape of the original particle.

Appendix 2 shows that under certain diffusion conditions the sum of the pressures of the reactant and product gases at the reaction front will be different from their sum in the gas phase around the reacting particle. Using the values of P derived in Appendix 2 we have:

$$\left[p^*_{I}\right]_R = 1 - \left[p^*_{II}\right]_R - P \qquad \text{(A-1-7)}$$

If the chemical step at the reaction interface does not affect the reaction rate, the pressures of the two gases must obey the equilibrium relation of the reaction:

$$\left\{\frac{\left[p_{II}\right]_R}{p_O}\right\}^{\beta_{II}} = \left[K_{eq}\right]_R \left\{\frac{\left[p_{I}\right]_R}{p_O}\right\}^{\beta_I} \qquad \text{(A-1-8)}$$

where p_O is the pressure at which the fugacities of the gases are taken as unity - normally 1 Atm. (101.325 kN/m^2).

Eq. (A-1-8) can be rearranged in the form:

$$\left[p^*_{II}\right]_R^{\beta_{II}} = K'_R \left[p^*_{I}\right]_R^{\beta_I} \qquad \text{(A-1-9)}$$

where K'_R is given by:

$$K'_R = \left[K_{eq}\right]_R \left\{\frac{p_G}{p_O}\right\}^{(\beta_I - \beta_{II})} \qquad \text{(A-1-10)}$$

Thus substitution from Eqs. (A-1-6) and (A-1-7) gives:

$$\left\{\left[p^*_{II}\right]_G + \frac{Y}{\xi}\right\}^{\beta_{II}} = K'_R \left\{\left[p^*_{I}\right]_G - P - \frac{Y}{\xi}\right\}^{\beta_I} \qquad \text{(A-1-11)}$$

The temperature at the reaction interface will be given by:

$$-\Delta H \dot{n} = \frac{\theta_R - \theta_G}{\frac{1}{k_S} + \frac{1}{A_0 h}} \qquad \text{(A-1-12)}$$

This equation can be rearranged in the form:

$$\theta_R - \theta_G = - \frac{\Delta H \dot{n}}{k \lambda_0} \left\{ \frac{1}{S^*} + Bi_H \right\} \qquad \text{(A-1-13)}$$

where

$$Bi_H = \frac{k}{\lambda_0 h} \qquad \text{(A-1-14)}$$

Eq. (A-1-13) can be further rearranged in the form:

$$\theta_R^* - 1 = - \frac{Y\ H^*}{\beta_{II} k^*} \left\{ \frac{\frac{1}{S^*} + Bi_H}{\frac{1}{S^*} + Bi_M} \right\} \qquad \text{(A-1-15)}$$

where

$$\theta_R^* = \frac{\theta_R}{\theta_G} \qquad \text{(A-1-16)}$$

$$H^* = \frac{\Delta H}{R\ \theta_G} \qquad \text{(A-1-17)}$$

and

$$k^* = \frac{\theta_G k}{D p_G} \qquad \text{(A-1-18)}$$

The progress of the reaction can be described in terms of a characteristic parameter of the reaction front, λ, or in terms of its dimensionless form $\lambda^* = \lambda/\lambda_0$. The term in curly brackets in Eq. (A-1-15) will be a function of λ^*, so that the equation can be written in the form:

$$\theta_R^* - 1 = - \frac{Y\ H^*}{\beta_{II} k^*} f(\lambda^*) \qquad \text{(A-1-19)}$$

The equilibrium constant at the interface will be a function of the temperature there, the function normally being represented in the form:

$$\left[K_{eq}\right]_R = \exp\left\{\frac{-\Delta H}{R\theta_G} + \frac{\Delta S}{R}\right\} \tag{A-1-20}$$

$$= \exp\left\{\frac{-\Delta H}{R\theta_G} + \frac{\Delta S}{R} + \frac{\Delta H}{R\theta_G} - \frac{\Delta H}{R\theta_R}\right\} \tag{A-1-21}$$

$$= \left[K_{eq}\right]_G \exp\left\{H^*\left(1 - \frac{1}{\theta_R^*}\right)\right\} \tag{A-1-22}$$

Substituting from Eqs. (A-1-10) and (A-1-19) gives:

$$K_R' = K_G' \exp\left\{\frac{-Y\ H^{*2}\ f(\lambda^*)}{\beta_{II}k^* - Y\ H^*\ f(\lambda^*)}\right\} \tag{A-1-23}$$

Substituting from Eq. (A-1-11) gives:

$$\frac{1}{K_G^{\dagger}}\left\{\left[p_{II}^*\right]_G + \frac{Y}{\xi}\right\}^{\beta_{II}} - \left\{p_{I\ G}^* - P - \frac{Y}{\xi}\right\}^{\beta_I} \exp\left\{\frac{-Y\ H^{*2}f(\lambda^*)}{\beta_{II}k^* - Y\ H^*\ f(\lambda^*)}\right\} = 0 \tag{A-1-24}$$

which is one equation linking Y and λ^*. They are also linked by the conservation equation at the interface:

$$\dot{n} = \frac{\dot{n}_{II}}{\beta_{II}} = - A_R\rho_1\ \overline{V} \tag{A-1-25}$$

where $\overline{V}$ is the average velocity of the reaction interface at any time. It is equal to the rate at which λ changes, so that Eq. (A-1-25) gives:

$$\frac{\dot{n}_{II}}{\beta_{II}} = - A_R\rho_1\lambda_o\frac{d\lambda^*}{d\tau} \tag{A-1-26}$$

Substituting from the definition for Y, Eq. (A-1-5), gives:

$$\frac{A_R}{A_0}\left\{\frac{1}{S^*} + Bi_M\right\}\frac{d\lambda^*}{d\tau} = -\frac{Dp_G}{R\theta_G\lambda_0^2\rho_I\beta_{II}}\, Y \qquad \text{(A-1-27)}$$

It is convenient to define a dimensionless time variable:

$$\tau^* = \frac{Dp_G\tau}{R\theta_G\lambda_0^2\rho_I\beta_{II}} \qquad \text{(A-1-28)}$$

so that Eq. (A-1-27) becomes:

$$\frac{A_R}{A_0}\left\{\frac{1}{S^*} + Bi_M\right\}\frac{d\lambda^*}{d\tau^*} = -Y \qquad \text{(A-1-29)}$$

which can be integrated to give:

$$\int_1^{\lambda^*} \frac{\frac{A_R}{A_0}\left\{\frac{1}{S^*} + Bi_M\right\}}{Y}\, d\lambda^* = \tau^* \qquad \text{(A-1-30)}$$

Before this equation can be evaluated, expressions must be available to show how A_R/A_0 and S^* vary with λ^*. These expressions can only be derived with any certainty if the reacting particles are spheres when the reaction interface remains spherical in shape. Thus:

$$\lambda = r;\quad S = \frac{4\pi r\, r_0}{r_0 - r};\quad A_0 = 4\pi r_0^2;\quad \frac{r}{r_0} = \lambda^* \qquad \text{(A-1-31)}$$

so that Eq. (A-1-30) becomes:

$$\int_1^{\lambda^*} \frac{\lambda^* - \lambda^{*2}(1 - Bi_M)}{Y}\, d\lambda^* = \tau^* \qquad \text{(A-1-32)}$$

This equation can only be integrated analytically if Y is not a function of λ^*. Eq. (A-1-24) shows that this occurs for a reaction involving molecular diffusion when H^* is zero, or when $Bi_M = Bi_H$. In both cases the reaction takes place isothermally; In the first case the temperature of the reaction interface is equal to that of furnace; in the second case $f(\lambda^*) = 1$ throughout the reaction, and the interface temperature is given by Eq. (A-1-19).*

* In this case, there will be a short period of time at the start of the reaction during which the interface temperature stabilizes itself. Hills [4] has shown that, even in the strongly endothermic decomposition of calcium carbonate, his time period has a negligible effect on the rate of the reaction.

In both these cases the integration of Eq. (A-1-32) gives:

$$3\lambda^{*2} - 2\lambda^{*3} - 2Bi_M(1 - \lambda^{*3}) = 1 - 6Y\tau^* \qquad \text{(A-1-33)}$$

The value of Y to be substituted into Eq. (A-1-33) must be determined by solving Eq. (A-1-24) using an iterative technique, such as the Newton-Raphson method. Fig. A-1 shows the function f(Y) plotted against Y for three typical sets of conditions. It can be seen that the function is well behaved and that no difficulties should develop in its numerical solution. An approximate value of Y is required at the start of the numerical integration procedure, however, and this can be obtained by the method outlined in Appendix 3.

When $Bi_M \neq Bi_H$ or for a reaction involving Knudsen or mixed diffusion, $f(\lambda^*)$ or $\bar{P}$ vary throughout the reaction so that Y will not be constant. Thus Eq. (A-1-32) must be solved by a numerical quadrature procedure such as Simpson's rule, Eq. (A-1-24) being solved for Y each time the integrand is evaluated. Plots of λ^* against τ^* obtained in this way are discussed in the main section of the paper.

Equations (A-1-30), (A-1-31), and (A-1-32) are valid for spheres but many workers have used cylindrical samples in their experiments. This is unfortunate since the shape of the reaction interface changes during the reaction so that A_R/A_O, and S^* cannot be evaluated as simple functions of λ^*. However, if the cylinder is sufficiently long for a central length L to be considered free of end effects, the reaction interface will remain cylindrical in this region, and we can write:

$$\lambda = r; \quad S = \frac{2\pi L}{Ln\left\{\frac{r_O}{r}\right\}}; \quad A_o = 2nLr_o; \quad \lambda^* = \frac{r}{r_o} \qquad \text{(A-1-34)}$$

so that Eq. (A-1-30) becomes:

$$-\int_1^{\lambda^*} \frac{\lambda^* \; Bi_M - Ln(\lambda^*)}{Y} \, d\lambda^* = \tau^* \qquad \text{(A-1-35)}$$

This equation can only be integrated analytically for the conditions under which Eq. (A-1-32) can be integrated. When these conditions are met, we obtain:

$$(1 - \lambda^{*2})(1 + 2Bi_M) + 2\lambda^{*2}Ln(\lambda^*) = 4\tau^* Y \qquad \text{(A-1-36)}$$

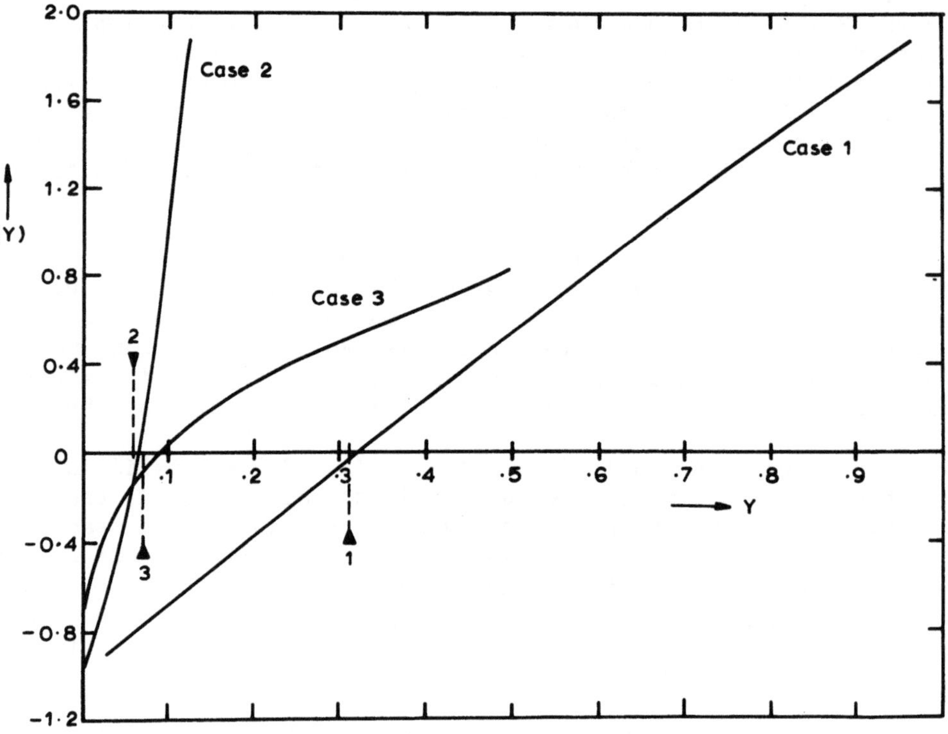

Fig. A-1. The function f(Y) appearing in equation A-1-24 plotted against Y for the for the three typical sets of reaction conditions described below. Values of the approximate conditions described below. Values of the approximate roots determined as described in Appendix 3 are indicated by the vertical dashed lines in the figure.

Case	$f(\lambda^*)$	H^*	k^*	K'_G	β_I	β_{II}	D^*
1	20	2	400	0.5	1	1	2
2	2	20	50	0.006	1	2	1
3	50	16	1000	0.6	1	1	1

APPENDIX 2

Derivation of Mass Transfer Equations for Different Diffusion Mechanisms

Generalized reaction rate equations for transport controlled gas/solid reactions are derived in Appendix 1, starting with an equation for the mass transfer rate of the product gas. This equation takes on different forms, represented by different values of D, ξ and P, depending on the type of diffusion mechanism that applies in the porous solid. The different forms are derived in this appendix for a series of reactions taking place in the absence of an inert dilutant gas.

The mechanism whereby a gas diffuses through a porous solid depends upon the sizes of the pores relative to the mean free path of the gas. When the pore size is very much greater than the mean free path, most of the collisions experienced by gas molecules are with one another, and the diffusion mechanism is the same as it is in a free gas-molecular diffusion. Knudsen diffusion takes place when the pore sizes are much smaller than the mean free path so that the gas molecules collide much more frequently with the walls than with one another. Under these conditions different gases in a mixture diffuse independently of one another.

In general, both diffusion mechanisms will operate within a porous solid, and the diffusion flux can be calculated from the following equation presented by Scott and Dullien [21] for a binary gas mixture:

$$\dot{n}''_{II} = - \frac{\nabla p_{II}}{R\theta_S \left\{ \dfrac{1 - (1 + \dot{n}''_I/\dot{n}''_{II})p_{II}/p_T}{D_{P,I/II}} + \dfrac{1}{D_{K,II}} \right\}} \qquad \text{(A-2-1)}$$

where $D_{P,I/II}$ is the binary molecular diffusion coefficient in the porous solid. It is related to the binary molecular diffusion coefficient in the free gas by:

$$D_{P,I/II} = \frac{D_{I/II}\gamma}{\tau} \qquad \text{(A-2-2)}$$

where γ is the porosity of the solid, and τ is a tortuosity factor that can be regarded as the length of the actual diffusion path through the interconnecting pores of the solid, divided by the direct path across the solid.

$D_{K,II}$ is the Knudsen diffusion coefficient of gas II in the porous solid. In their review of diffusion in porous solids Satterfield and

Sherwood [22] quote the following formula for the Knudsen diffusion coefficient:

$$D_{K,II} = \frac{9700\, r_e \gamma}{\tau} \sqrt{\frac{\theta_S}{M_{II}}} \qquad \text{(A-2-3)}$$

where r_e is the mean radius of the pores. It can be determined from micrographs of the solid, or from s_p, the specified surface area per unit mass of the solid, and ρ_p, the apparent density of the solid:

$$r_e = \frac{\gamma}{s_p \rho_p} \qquad \text{(A-2-4)}$$

A general treatment for the integration of Eq. (A-2-1) is extremely difficult, and we will restrict consideration to certain limiting situations.

Molecular Diffusion in the Absence of a Dilutant

When the pores in the solid are large, $D_{K,II} >> D_{P,I/II}$, and Eq. (A-2-1) shows that molecular diffusion is the dominant diffusion mechanism, since the equation becomes:

$$\dot{n}''_{II} = - \frac{p_G D_{P,I/II} \nabla p_I^*}{R\theta_G \left\{1 - (1 + \dot{n}_I/\dot{n}_{II}) p_I^*\right\}} \qquad \text{(A-2-5)*}$$

During a chemical reaction the diffusion process within the porous product solid effectively takes place under steady state conditions, so that Eq. (A-2-5) may be integrated directly. Moreover, the fluxes of the two gases are related by the stoichiometry of the reaction, so that this integration yields:

$$\dot{n}_{II} = \frac{\beta_{II} \,\mathrm{Ln}\left\{\dfrac{\beta_{II} - (\beta_{II} - \beta_I)\left[p_{II}^*\right]_S}{\beta_{II} - (\beta_{II} - \beta_I)\left[p_{II}^*\right]_R}\right\}}{\dfrac{(\beta_{II} - \beta_I) R\theta_G}{S p_G D_{P,I/II}}} \qquad \text{(A-2-6)}$$

---

* When the temperature of the product solid is not constant, the temperature and diffusion coefficients appearing in this equation should be mean values in the solid. However, the diffusion coefficient does not change very rapidly with temperature, and the variation of its ratio with temperature is very small indeed. Thus very little error is introduced when this equation is used under non-isothermal conditions.

where S is the shape factor of the diffusion path between the surface of the particle and the reaction front; $\left[p_{II}^*\right]_S$, the fractional pressure at the surface of the particle, is unknown. However it can be determined from the equation for mass transfer between the surface and the bulk gas phase:

$$\dot{n}_{II} = \frac{\beta_{II} \mathrm{Ln}\left\{\dfrac{\beta_{II} - (\beta_{II} - \beta_I)\left[p_{II}^*\right]_G}{\beta_{II} - (\beta_{II} - \beta_I)\left[p_{II}^*\right]_S}\right\}}{\dfrac{(\beta_{II} - \beta_I)R\theta_G}{A_0 \alpha p_G}} \qquad \text{(A-2-7)}$$

Eliminating $\left[p_{II}^*\right]_S$ from Eqs. (A-2-6) and (A-2-7) gives:

$$\dot{n}_{II} = \frac{\beta_{II} \mathrm{Ln}\left\{\dfrac{\beta_{II} - (\beta_{II} - \beta_I)\left[p_{II}^*\right]_G}{\beta_{II} - (\beta_{II} - \beta_I)\left[p_{II}^*\right]_R}\right\}}{\dfrac{(\beta_{II} - \beta_I)R\theta_G}{p_G D_{P,I/II}}\left\{\dfrac{1}{S} + \dfrac{D_{P,I/II}}{A_0 \alpha}\right\}} \qquad \text{(A-2-8)}$$

which shows that the value of ξ to be used in Eq. (A-1-2) during binary molecular diffusion is given by:

$$\xi = \frac{\beta_{II} \mathrm{Ln}\left\{\dfrac{\beta_{II} - (\beta_{II} - \beta_I)\left[p_{II}^*\right]_G}{\beta_{II} - (\beta_{II} - \beta_I)\left[p_{II}^*\right]_R}\right\}}{(\beta_{II} - \beta_I)\left\{\left[p_{II}^*\right]_R - \left[p_{II}^*\right]_G\right\}} \qquad \text{(A-2-9)*}$$

and that 'D' in the equation represents $D_{p,I/II}$

Equation (A-2-9) is indeterminate for an equi-molar reaction in which $\beta_I = \beta_{II}$: However, the logarithm may be expanded as an infinite series whence:

$$\lim_{\beta_I \to \beta_{II}} (\xi) = 1 \qquad \text{(A-2-10)}$$

---

* ξ is a generalization of the log-mean pressure concept used when only one species in a mixture is diffusing. Under such conditions, ξ would be equal to the reciprocal of the log-mean pressure, and it will be termed the modified reciprocal log-mean pressure.

Even when $\beta_I \neq \beta_{II}$ the value of ξ is not often far from unity, and changes only slowly with the values of $[p^*_{II}]_R$ and $[p^*_{II}]_G$. The value of ξ can be calculated from the thermodynamics of the reaction, since $[p^*_I]_R = 1 - [p^*_{II}]_R$, and from the composition of the gas surrounding the reacting particle. $[p^*_{II}]_R$ varies during a reaction in which the temperature of the reaction front changes. Strictly speaking, ξ should be calculated whenever Eq. (A-1-24) is used to determine Y during the reaction, but it changes so slowly with $[p^*_{II}]_R$ that a single calculation is sufficient for the accuracies normally required. If this is not so, ξ and Y must be calculated at each point using an iterative procedure.

Since there is no variation in the total pressure across the product layer when the gases diffuse through it by molecular diffusion, $[p_I]_R + [p_{II}]_R = p_G$, and P in Eq. (A-1-7) is thus given by:

$$P = 0 \tag{A-2-11}$$

Knudsen Diffusion

When the pores in the solid are very small, $D_{K,II} << D_{P,I/II}$ and Knudsen diffusion is the dominant mechanism, since (A-2-1) becomes:

$$\dot{n}''_{II} = - \frac{D_{K,II}}{R\theta_S} \nabla p_{II} \tag{A-2-12}$$

which can be integrated to give:

$$\dot{n}_{II} = \frac{S\, D_{K,II}}{R\theta_S} \left\{ [p^*_{II}]_R - [p^*_{II}]_S \right\} \tag{A-2-13}$$

Once again, the value of $[p^*_{II}]_S$ that appears in this equation should be eliminated using Eq. (A-2-7), but this would result in some cumbersome algebra because a log mean pressure term incorporating $[p^*_{II}]_S$ would be involved. However, convective mass transfer in the gas plays a relatively minor role in determining the reaction rate when Knudsen diffusion is involved in the solid. We therefore introduce very little error if we take a linear approximation for Eq. (A-2-7), and write:

$$\dot{n}_{II} = \frac{[p_{II}]_R - [p_{II}]_G}{\dfrac{R\theta_G}{D_{K,II}} \left\{ \dfrac{1}{S} + \dfrac{D_{K,II}}{A_0 \alpha} \right\}} \tag{A-2-14}$$

We can see that the value of ξ to be used in Eq. (A-1-2) when Knudsen diffusion is involved is:

$$\xi = 1 \tag{A-2-15}$$

and 'D' in the equation represents $D_{K,II}$.

Eq. (A-2-14) can be rearranged in terms of the variable Y used in Appendix 1:

$$\left[p^*_{II}\right]_R - \left[p^*_{II}\right]_G = Y \tag{A-2-16}$$

However:

$$\left[p^*_I\right]_G - \left[p^*_I\right]_R = \frac{\dot{n}_I \; R\theta_G\lambda_o}{D_I p_G A_o}\left\{\frac{1}{S^*} + \left[Bi_M\right]_I\right\} \tag{A-2-17}$$

$$= \frac{\beta_{II}}{\beta_I \; D^*}\left\{\frac{\frac{1}{S^*} + \left[Bi_M\right]_I}{\frac{1}{S^*} + Bi_M}\right\} Y \tag{A-2-18}$$

Let us define:

$$g(\lambda^*) = \frac{\frac{1}{S^*} + \left[Bi_M\right]_I}{\frac{1}{S^*} + Bi_M} = \frac{\frac{1}{S^*} + D^* \; Bi_M}{\frac{1}{S^*} + Bi_M} \tag{A-2-19}$$

$$\text{where} \quad D^* = \frac{D_I}{D} \tag{A-2-20}$$

Subtracting Eq. (A-2-17) from Eq. (A-2-16) gives:

$$\left[p^*_{II}\right]_R + \left[p^*_I\right]_R = \left[p^*_{II}\right]_G + Y + \left[p^*_I\right]_G - \frac{\beta_{II} \; g(\lambda^*)}{\beta_I \; D^*} Y \tag{A-2-21}$$

But $\left[p^*_{II}\right]_G + \left[p^*_I\right]_G = 1$, so that P in Eq. (A-1-7) is given by:

$$P = \left\{\frac{\beta_{II} \; g(\lambda^*)}{\beta_I \; D^*} - 1\right\} Y \tag{A-2-22}$$

Mixed Knudsen and Molecular Diffusion During an Equi-Molar Reaction

During an equi-molar reaction, $\dot{n}'' = -\dot{n}''_{II}$ so that Eq. (A-2-1) becomes:

$$\dot{n}''_{II} = \frac{\nabla p_{II}}{R\theta_s \left\{ \frac{1}{D_{P,I,II}} + \frac{1}{D_{K,II}} \right\}} \qquad \text{(A-2-23)}$$

This case is mathematically identical to the case treated previously except that no error is involved in describing convective mass transfer outside the particle in terms of a linear equation, since the modified reciprocal log mean pressure term, ξ, is always unity when equi-molar mass transfer is involved.

Thus we can see that D in Eq. (A-1-2), for these conditions, is given by:

$$D = \left\{ \frac{1}{D_{P,I/II}} + \frac{1}{D_{K,II}} \right\}^{-1} \qquad \text{(A-2-24}$$

and D_I to be used in the definition of D^* is given by:

$$D_I = \left\{ \frac{1}{D_{P,I/II}} + \frac{1}{D_{K,II}} \right\}^{-1} \qquad \text{(A-2-25}$$

The values of ξ and P are then the same as those given in Eqs. (A-2-25) and (A-2-22).

APPENDIX 3

Calculation of Approximate Roots of the Reaction Rate Equation

Equation (A-1-24) must be solved by an iterative technique such as the Newton-Raphson method, for which a starting value is needed. This starting value can be determined by expanding Eq. (A-1-22) in a Taylor series and taking only the first two terms:

$$\left[K_{eq}\right]R = \left[K_{eq}\right] G \left\{ 1 + H^*(\theta^*_R - 1) \right\} \qquad \text{(A-3-1)}$$

Substituting from Eqs. (A-1-19) and (A-1-10) gives:

$$K'_R = K'_G \left\{ 1 - \frac{Y\, H^{*2}\, f(\lambda^*)}{\beta_{II} k^*} \right\} \qquad \text{(A-3-2)}$$

Substituting this approximate value for K_R into Eq. (A-1-11) gives:

$$\frac{1}{K'_G}\left\{\left[P^*_{II}\right]_G + \frac{Y}{\xi}\right\}^{\beta_{II}} = \left\{1 - \frac{Y\,H^{*2}f(\lambda^*)}{\beta_{II}k^*}\right\}\left\{\left[P^*_I\right]_G - P - \frac{Y}{\xi}\right\}^{\beta_I} \quad \text{(A-3-3)}$$

For many technically important situations, this equation can be rearranged in the form of a quadratic equation:

$$\Lambda Y^2 - \Omega Y + \Gamma = 0 \quad \text{(A-3-4)}$$

where the constant term Γ represents the thermodynamic driving force for the reaction:

$$\Gamma = \left[P^*_I\right]_G^{\beta_I} - \frac{\left[P^*_{II}\right]_G^{\beta_{II}}}{K'_G} \quad \text{(A-3-5)}$$

Eq. (A-3-4) can be rearranged in the form:

$$Y = \frac{\Omega \pm \sqrt{\Omega^2 - 4\Gamma\Lambda}}{2} \quad \text{(A-3-6)}$$

No reaction occurs under equilibrium conditions, so that Y = 0 when $\left[P^*_I\right]_G^{\beta_I} = \left[P^*_{II}\right]_G^{\beta_{II}}/K'_G$. Thus we can see that it is the negative sign that is valid in Eq. (A-3-6). Since the product of the roots of a quadratic equation is equal to the constant terms, Eq. (A-3-6) is more conveniently represented in the form:

$$Y = \frac{2\Gamma}{\Omega + \sqrt{\Omega^2 - 4\Gamma\Lambda}} \quad \text{(A-3-7)}$$

Values of Ω, Γ and Λ that are relevant to different conditions are listed below:

<u>Case 1:</u> $\beta_{II} = \beta_I = 1$. Molecular diffusion with no dilutant gas, under these conditions, P = 0 and $\xi = 1$.

Rearranging Eq. (A-3-3) in the form of Eq. (A-3-4), gives:

$$\Lambda_1 = \frac{H^{*2}f(\lambda^*)}{k^*} \quad \text{(A-3-8)}$$

$$\Omega_1 = 1 + \frac{1}{K_G} + \left[P^*_I\right]_G \frac{H^{*2}f(\lambda^*)}{k^*} \quad \text{(A-3-9)}$$

$$\Gamma_1 = \left[P^*_I\right]_G - \frac{\left[P^*_{II}\right]_G}{K_G} \quad \text{(A-3-10)}$$

<u>Case 2:</u> $\beta_{II} = 2$, $\beta_I = 1$. Molecular diffusion with no dilutant gas. Under these conditions, P = 0 but ξ is finite. Thus we have:

$$\Lambda_2 = \frac{H^{*2}f(\lambda^*)}{2\xi k^*} - \frac{1}{K_G\xi^2} \quad \text{(A-3-11)}$$

$$\Omega_2 = \frac{1}{\xi} + \frac{2\left[p_{II}^*\right]_G}{K_G\xi} + \left[p_I^*\right]_G \frac{H^{*2} f(\lambda^*)}{2\ k^*} \qquad \text{(A-3-12)}$$

$$\Gamma_2 = \left[p_I^*\right]_G - \frac{\left[p_{II}^*\right]^2_G}{K_G} \qquad \text{(A-3-13)}$$

Case 3: $\beta_I = 1 = \beta_{II}$. Knudsen or mixed diffusion.

Appendix 2 shows that, under these conditions:

$$P = \left\{\frac{g(\lambda^*)}{D^*} - 1\right\} Y \qquad \text{(A-3-14)}$$

but that ξ is equal to one.

$$\Lambda_3 = \frac{H^2 f(\lambda^*)\ g(\lambda^*)}{K^*\ D^*} \qquad \text{(A-3-15)}$$

$$\Omega_3 = \frac{g(\lambda^*)}{D^*} + \frac{H^{*2} f(\lambda^*)}{K^*} \left[p_I^*\right]_G + \frac{1}{K_G} \qquad \text{(A-3-16)}$$

$$\Gamma_3 = \left[p_I^*\right]_G - \frac{\left[p_{II}^*\right]_G}{K_G} \qquad \text{(A-3-17)}$$

APPENDIX 4

Diffusion in Porous Iron

Warner [7] has reported that pore sizes in porous iron produced by reducing hematite in hydrogen at 800°C are in the range 0.4-1.0 μm. Substituting a mean pore radius of 0.35 μm into Eq. (A-2-3) gives:

$$D_{K,H_2} = 7.8 \frac{\gamma}{\tau}\ \text{cm}^2\ \text{sec}^{-1}, \quad D_{K,H_2O} = 2.6 \frac{\gamma}{\tau}\ \text{cm}^2\ \text{sec}^{-1}$$

Using literature values for the binary free gas diffusion coefficient in an H_2/H_2O mixture at 800°C and at one atmosphere we have:

$$D_{p,H_2/H_2O} = 8.3 \frac{\gamma}{\tau} \text{ cm}^2\text{sec}^{-1}$$

Since this value is of the same order as the Knudsen diffusion coefficient values calculated above, diffusion in porous iron at 800°C under a pressure of one atmosphere must be mixed diffusion, neither Knudsen nor molecular diffusion dominating.

This conclusion is born out by an observation made by McKewan during his measurements of diffusion coefficients in porous iron [20]. He measured the rate at which granules of wüstite were reduced in a container sealed with a porous iron plug, formed by reducing hematite, and held in a furnace through which mixtures of hydrogen and water vapour were passed. He observed that a slight positive pressure was developed within the container showing that hydrogen could diffuse into the container more easily than water vapour could diffuse out of it. This could only arise if Knudsen diffusion made an appreciable contribution to the diffusion process within the porous iron plug.

McKewan computed diffusion coefficients from the difference in hydrogen partial pressure between the furnace and the container, so that his values are actually $(1/D_{p,H_2/H_2O} + 1/D_{K,H_2})^{-1}$. At 800°C, he reports a value of 1.46 cm^2/s when the porosity of the iron plug was 0.55. Thus the tortuosity of the iron is given by:

$$1.46 = \left\{ \frac{\tau}{8.3x0.55} + \frac{\tau}{7.9x0.55} \right\}^{-1} \qquad \text{(A-4-1)}$$

which gives $\tau = 1.55$.

This is a more reasonable value than the value of 3.17 calculated by McKewan on the assumption that H_2/H_2O mixtures diffused through porous iron by molecular diffusion.

Using the tortuosity factor calculated above we find that the following values apply in iron of 55% porosity when it is at 800°C under a pressure of one atmosphere:

$$D_{K,H_2} = 2.81 \text{ cm}^2/\text{s} \; ; \; D_{K,H_2O} = 0.92 \text{ cm}^2/\text{s} \; ; \; D_{P,H_2O} = 2.94 \text{ cm}^2/\text{s}.$$

REFERENCES

1. C. N.Satterfield and F. Feakes: A.I.Ch.E.J. 5, 115 (1959).
2. T. R. Ingraham and P. Marier: Can. J. Chem. Engng 41, 170 (1963).
3. W. M.McKewan: Trans. Met. Soc. A.I.M.E. 218, 2 (1960).

4. A. W. D. Hills: Chem. Eng. Sci. 23, 297 (1968).
5. E. Kawasaki, J. Sanscrainte and T. J. Walsh: A.I.Ch.E. 8, 48 (1962).
6. B. B. L. Seth and H. U. Ross: Can. Met. Quart. 5, 315 (1966).
7. N. A.Warner: Trans. Met. Soc. A.I.M.E. 230, 163 (1964).
8. R. H.Spitzer, F. S. Manning and W. O. Philbrook: Trans Met. Soc. A.I.M.E. 236, 726 (1966).
9. N. A. Warner: Proc. Aust. Inst. Min. & Met. 210, 31 (1964).
10. A. W. D. Hills: *Heat & Mass Transfer in Process Metallurgy*, A. W. D. Hills (Ed.), 1967, p. 39 (London: Inst. Min. Met.).
11. G. Bitsianes and T. L. Joseph: Trans. Met. Soc. A.I.M.E., 200, 150 (1954).
12. C. C. Furnas: Ind. Eng. Chem. 23, 534 (1931).
13. W. D.Spenser and B.J. Topley: J. Chem. Soc., 2633 (1929).
14. N. A. Warner and T. R. Ingraham: Can. J. Chem. Eng. 40, 263 (1962).
15. A. Paulin: Ph.D. Thesis, University of London, 1968.
16. G. Nabi and W-K. Lu: Trans. Met. Soc. A.I.M.E. 242, 2471 (1968).
17. A. W. D.Hills and A. Paulin: to be published
18. L. V. Bogdandy, H. P. Schultz, I. N. Stranski and B. Wurzner: Ber Bunsenges 67, 958 (1963).
19. W. M.McKewan: *Reactivity of Solids*, G. M. Schwab, ed., Elsevier Pub. Co., (Amsterdam) 1965.
20. W. M. McKewan and R. G. Olsson, Trans. Met. Soc. A.I.M.E. 236, 1518 (1966).
21. D. S. Scott and F. A. Dullien: A.I.Ch.E.J. 8, 113 (1962).
22. C. N. Satterfield and T. K. Sherwood: *Role of Diffusion in Catalysis*, Addison-Wesley, 1963, p. 17.
23. P. K. Strangeway, S. Toppie and H. U. Ross: Can. Met. Quart. 5, 211 (1966).
24. A. W.D.Hills: *The Engineering of Gas/Solid Reactions*, J. R. Pirie (Hon. Ed.). (London: Inst. Chem. Eng.) p. 37, 1969.
25. E. R.G. Eckert and R. M. Drake, Jr.: *Heat and Mass Transfer*, 1959, (New York, McGraw Hill).
26. J.F.Elliot and M. Gleiser: *Thermochemistry for Steelmaking*, 1960,(Massachusetts, Addison-Wesley).
27. O.Kubaschewski and E. LL. Evans, *Metallurgical Thermochemistry*, 1956 (Oxford, Pergamon Press Ltd.).
28. P. N. Rowe and K. T. Claxton: Trans. Inst. Chem. Eng. 43, T321 (1965).

List of Symbols

(Variables are to be measured using S. I. (Système International d'Unités) units, as specified.)

A Surface area (cm^2)
D Diffusion coefficient (cm^2/s)

ΔH	Heat of reaction (J/g-mole)
h	Heat transfer coefficient (W/m^2 ^{o}K)
K_C	Chemical rate constant (cm/s)
K_{eq}	Equilibrium coefficient
k	Thermal conductivity (W/m ^{o}K)
L	Length of a cylinder (cm)
M	Molecular weight (g/g-mole)
m	Mass of a reacting particle (m_τ - at time τ, m_o initial, m_∞ - final) (g)
$\dot{n}$	Reaction rate (g-mole/s)
p	Pressure (kN/m^2)
p_o	Standard pressure (=101.325 kN/m^2)
$\dot{q}$	Heat flow rate (W)
R	Gas constant (J/g-mole ^{o}K)
r	Radius of a reacting pellet (cm)
r_e	Pore radius (cm)
ΔS	Entropy of reaction (J/g-mole ^{o}K)
S	Shape factor for diffusion or heat transfer (cm)
s_p	Specific surface of porous material (cm^2/g)
V	Velocity of reaction front (cm/s)
α	Mass transfer coefficient (cm/s)
β	Stoichiometric factor
γ	Porosity of product solid
Γ	Coefficient in quadratic equation (Appendix 3)
θ	Temperature (^{o}K)
Λ	Coefficient in quadratic equation (Appendix 3)
λ	Parameter describing position of reaction interface (cm)
ρ_I	Molar density of original solid (g-mole/cm^3)
ρ_P	Apparent mass density of solid product (g/cm^3)
τ	Tortuosity factor
ω	Stoichiometric factor
Ω	Coefficient in quadratic equation (Appendix 3)

Defined Variables

(Defined in equations listed in brackets)

Bi_H	Ratio between heat transfer resistances within product solid and outside particle (9 and A-1-4)
Bi_M	Ratio between mass transfer resistances within product solid and outside particle (9 and A-1-4)
D*	Ratio between diffusion coefficients of reactant and product gases (A-2-20)
H*	Dimensionless heat of reaction (A-1-17)
K'	Modified equilibrium constant (A-1-10)
k*	Modified thermal conductivity of product solid (A-1-18)
m*	Fractional mass of reacting particle (5)
p*	Dimensionless pressure of gas {= P/P_G}

P	Fractional increase of pressure at reaction front over pressure in gas phase (A-2-11 and A-2-22)
S^*	Dimensionless shape factor (A-1-3)
Y	Dimensionless reaction rate parameter (A-1-5)
ξ	Modified log mean pressure term (A-2-9 and A-2-15)
θ^*	Dimensionless temperature (A-1-16)
λ^*	Dimensionless reaction front parameter (A-1-31 and A-1-34)
τ^*	Dimensionless time (A-1-28)

Sufficies

B	Back reaction
F	Forward reaction
G	Bulk gas phase
I	Gas I
II	Gas II
I/II	For gas mixture I and II
0	At the start of the reaction
P,I/II	For the gas mixture in the pores of the product solid
K,I	Relating to Knudsen diffusion for gas I
R	Reaction front
S	In the solid
T	Total pressure in porous material

DISCUSSION OF PAPER

S. E. Khalafalla: Your picture is very correct, but also very gloomy in that it leaves no hope of measuring the intrinsic chemical kinetics of the surface reaction. I believe we can overcome most of the difficulties by measuring initial reaction rates. In addition to that, calculation of the mass transfer coefficient becomes easy to manage because we don't deal with diffusion through a solid reaction product. The intrinsic reaction rate coefficient k_i adds up to the mass transfer rate coefficient k_m in an Ohm's law fashion for conductances in series to give the apparent reaction rate coefficient, i.e. $1/k_a = 1/k_i + 1/k_m$. We can measure k_a, calculate k_m and hence estimate k_i.

A. W. D. Hills: I cannot agree because a) one cannot measure with any accuracy the initial rate of a reaction since this entails drawing a tangent at zero time, necessitating an assumed curve; and b) if heat transfer is involved there is a highly unstable state in the initial stages, when the reaction front is dropping greatly in temperature.

T. B. Lindemer: Your paper is well taken. In our own work presented earlier the classical models used were not sufficient to determine the rate controlling mechanisms. Auxillary information from reactions that had different geometrical configurations were needed before the rate-controlling mechanisms could be deduced. Again referring to the models normally used, Ishida and Wen showed that for certain cases the activation energies determined from the improper use of the models could be in error by a factor of two.

A. W. D. Hills: I would agree with your comments.

J. B. Wagner: Would you please tell us the nature of the experiment on the decomposition of $CaCO_3$ for which you determined an activation energy of some forty kcal/mole for your reported gas transport reaction.

A. W. D. Hills: We measured the reaction rate per unit area as a function of temperature as did all the workers who postulated reaction rate control. The experiments were carried out in air.

N. J. Themelis: Dr. Hills mentioned that all too often erroneous conclusions were drawn on the rate controlling step of gas-solids reactions on the basis of the dependence of the rate of reaction on the particle diameter. It can be readily shown that in the case of chemical rate control the rate of reaction is inversely proportional to the particle diameter, while for diffusion through the reduced shell the dependence is on the square of the particle diameter.

A few years ago, we correlated (1) the results of about fifteen different investigations on the reduction of iron oxides by hydrogen by means of the McKewan type of equation, which assumes control at the reaction interface between the reduced outer shell and the unreduced core of the particle.

The results are shown in Figure 1 and encompass particle diameters ranging from 40 microns to three centimeters (30,000 microns). Although there is considerable scatter in the experimental data, the dependence of the reaction rate (y-axis of Fig. 1) on the inverse of the particle diameter (x-axis) is unmistakable.

It must be noted that the data shown in Figure 1 were obtained under experimental conditions of adequate hydrogen mass transfer to the particle so that the reaction was not controlled by diffusion through the boundary layer.

On the basis of these data, it may be concluded that in the reduction of iron oxides, the interface reaction plays a predominant

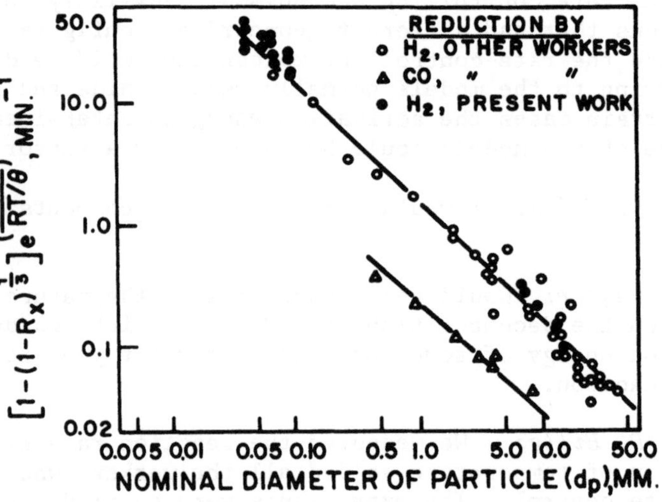

Fig. 1. Correlation of reduction rate of iron oxide particles to particle diameter.

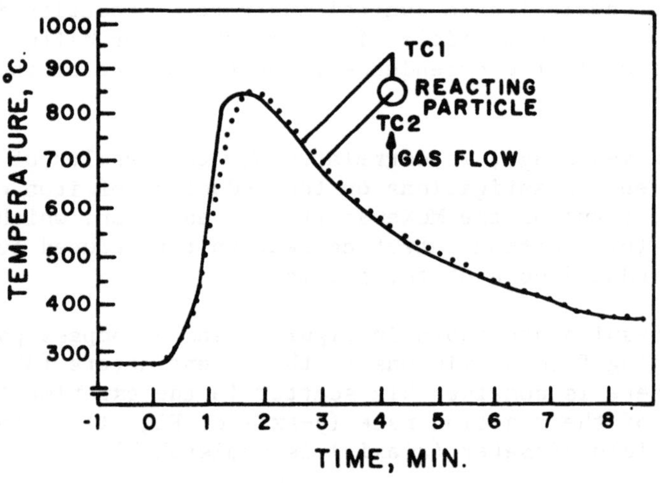

Fig. 2. Temperature on the surface and at the center of a reacting cupric oxide particle (d_p = 0.610 in., Reynolds number - 94).

role in comparison to diffusion through the reduced layer. Never-the less, with artificially made dense hematite particles, "pockets" of unreduced wustite may be left behind the advancing reaction interface, as shown by the works of Gray and Henderson (2), and Warner(3).

Coming to the importance of temperature effects in the study of gas-solids reactions, the writer agrees with Dr. Hills' conclusions. Figure 2(4) shows the variation in temperature in the case of a cupric oxide particle (0.610-in. dia.) reduced in a stream of hydrogen. In this case, the reaction is highly exothermic and the particle temperature rose from about 300° to 900°C in a few minutes. The temperature difference which was measured between the center of the particle and its surface is shown in Figure 2.

In the case of cupric oxide, the reaction is completed in about seven minutes. However, for an iron oxide sphere of the same size reduced by hydrogen, the reaction would extend over at least 40 minutes. The heat of CuO reduction is -19,300 cal./g. mole (298°K) while for Fe_2O_3 reduction by hydrogen is +25,106 cal./g. mole (298°K). Therefore, the decrease of about 20°C quoted by Dr. Hills as measured in iron oxide reduction appears to be reasonable in comparison to the data shown in Figure 2 for cuprix oxide reduction.

REFERENCES

1. Themelis, N. J. and Gauvin, W. H., Trans. Met. Soc. AIME <u>227</u>, 290 (1963).
2. Gray, N. B. and Henderson, J., Trans. Met. Soc. AIME <u>236</u>, 1213 (1966).
3. Warner, N. A., Trans. Met. Soc. AIME, <u>230</u>, 163 (1964).
4. Themelis, N. J. and Yannopoulos, J. C., Trans. Met. Soc. AIME <u>236</u>,414 (1966).

KINETIC STUDIES AT TEMPERATURES NEAR 4000°K USING THE FALLING SPECIMEN TECHNIQUE: THE REACTION $Zr(\ell)+O_2(g)\rightarrow ZrO_2(\ell)$

Lloyd S. Nelson and Herman S. Levine

Sandia Laboratories

Albuquerque, New Mexico 87115

Submillimeter samples of reactive materials have been studied at the temperatures near 4000°K by a free fall technique. This technique eliminates containment problems, yet permits accurate measurements to be made at temperatures where severe experimental difficulties are normally encountered.

The laser-initiated combustion of molten Zr in gaseous O_2 is used here as an example of the free fall technique. The reaction was studied as it occurred by photographic, spectroscopic and pyrometric techniques, and the specimens were studied by gravimetric techniques at room temperature, either after being quenched during the reaction in liquid argon or after oxidizing completely.

The droplets retain metal as they burn near their boiling temperature in pure oxygen near atmospheric pressure and lose metal when burning temperatures are lowered by reducing the oxygen concentration. This is explained by the return of vaporized oxide via condensation and coalescence from a spherical layer of oxide fog that surrounds the parent droplets during combustion.

There is growing need for and interest in understanding the behavior of materials at very high temperatures. Aside from its fundamental importance, this information has great practical value in such fields as propellants, nuclear fuels and reentry science. Our experimental capabilities, unfortunately, have not kept up with the need for high temperature studies. We describe here a falling specimen method for high temperature studies that avoids many of the previous experimental problems. Its use will be illustrated by describing our investigations of the reaction of molten zirconium with gaseous oxygen in which temperatures reach and occasionally exceed 4000°K.

The high temperature world in the laboratory is very uneven, as shown in Fig. 1. The left side of the figure shows the temperatures that can be achieved reproducibly in the gas phase -- namely in the range extending to 10^6-10^7 °K [1]. On the right side of Fig. 1, however, the maximum temperatures normally achieved in careful laboratory studies on solids and liquids seldom exceed 3000°K, even though at atmospheric pressure condensed phases exist at temperatures as high as 6000°K (the normal boiling point of tungsten), and even higher at increased pressures.

In both the gaseous and condensed phases, the upper temperature limit appears to have been set by the failure and/or reactivity of the container in which the experiment is performed. In both cases, a good solution is to avoid containment; that is, to keep the hot sample from touching the container while retaining the ability to make meaningful measurements on a well-defined system.

A popular technique for studying uncontained condensed phases is levitation of the sample in a high frequency field [2, 3, 4, 5]. This method provides a sample of known dimensions which can be examined at close range with optical measuring devices such as cameras, optical pyrometers and spectrographs. The temperature can be controlled fairly well, and flow conditions around the sample can be

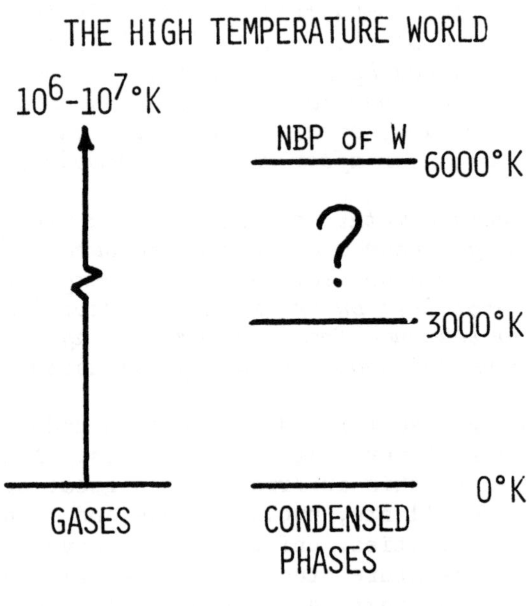

Fig. 1. The high temperature world. Falling specimen technique permits investigations in the region designated by the question mark.

selected over wide ranges. However there are several limitations inherent in this method, such as restrictions on sample diameter and density, the inability to control heating rates independently of the levitating forces, rapid electromagnetic stirring of the sample while levitated and the requirement that the sample be an electrical conductor at all times. (Besides ruling out studies of poor conductors, this feature limits the ability to perform a chemical reaction on a levitated sample that will yield a non-conducting product.)

We have developed a simpler technique for studying uncontained, unsupported samples at high temperatures. The method involves study of hot specimens in free fall, with no serious limitations from size, density, electrical properties or the extent of chemical reaction that the sample undergoes. Moreover, the sample is isolated from vibration and other mechanical effects. Simple instrumentation allows us to photograph the sample and record its temperature and emission spectrum as it falls.

We have used the falling sample technique primarily to learn how molten metals burn in oxidizing atmospheres.* In this communication, we describe a series of experiments performed to determine the kinetics and mechanism of the combustion of zirconium droplets. Although much new has been learned about reaction, it has also been demonstrated that the reaction is considerably more complex than had been anticipated on the basis of the concepts that prevailed before the experiments were started.

EXPERIMENTAL

In the falling sample technique, it is usually desirable to keep the sample size compatible with a given room height. We have found empirically that most high temperature chemical reactions may be performed on spherical samples with diameters less than one millimeter in a laboratory with ordinary ceiling height. This provides reaction times of 0.5 - 0.7 seconds.

In the experiments to be described here, droplets of zirconium were formed by heating squares of zone refined foil with an intense pulse of light from either a helical flash lamp (energies up to $\sim$ 25 thermal J cm^{-2} per milli-second pulse) [13] or a focused Nd-glass laser (energies up to $\sim$ 500 thermal J cm^{-2} per millisecond pulse) [14]. In both instances, the foil was melted and ignited in an atmosphere of high purity oxygen at pressures up to 625 Torr, or

* The falling specimen technique is currently being used by other groups to study metal combustion [5,6,7,8,9,10]. Also, high temperature techniques which involve both levitation and free fall of metallic specimens are being used in metal combustion [5] and calorimetric studies [11] and in the preparation of supercooled alloys by the splat technique [12].

Fig. 2. Time-exposed (streak) photograph of laser-ignited zirconium droplet that burned under standard conditions (initial dia. = 525 μm; pure oxygen at 625 Torr). Chamber is 10 cm wide. Arrow indicates onset of solidification (spearpoint).

in oxygen diluted with a rare gas, with the mixture at a total pressure of 625 Torr. After irradiation, a brightly luminous droplet resulted which burned as it fell downward through a windowed oxidation chamber 7.5 cm across and up to 3 meters high.

The behavior of the droplets was studied by their self-luminosity during the oxidation (still and high speed photography, photomultiplier recording, spectroscopy and both brightness and ratio pyrometry) and by conventional analyses (microscopy, x-ray diffraction and weighing) of the particles, either after the oxidation had been interrupted by a liquid argon quench [15], or after the reaction had gone to completion in a long oxidation chamber [16].

RESULTS

The experimental results obtained for the reaction

$$Zr(\ell) + O_2(g) \rightarrow ZrO_2(\ell)$$

are presented below. For this system the melting and normal boiling temperatures are 2130°K and about 4500°K, respectively, for Zr and 3000°K and about 4500°K, respectively for ZrO_2.

A particular set of experimental parameters, chosen for convenience and used extensively in previous reports [15, 16, 17, 18], has been termed standard conditions (S.C.). These conditions apply to a 525 μm initial droplet diameter of zirconium that reacts in pure oxygen at 625 ± 5 Torr (prevailing local atmospheric pressure).

Combustion in Oxygen at Atmospheric Pressure

The luminous history of a laser-ignited droplet of zirconium as it burns in pure oxygen under standard conditions is illustrated in Fig. 2, which shows a time-exposed photograph of the integrated light emitted during the combustion reaction.

The upper portion of the trace in Fig. 2 is heavily broadened due to overexposure, indicating a very intense emission of light. Below this, the trace shows a steady decrease in luminosity until an abrupt intensification occurs, which is indicated by the arrow. Below this, the luminosity decreases gradually until it is extinguished altogether. The abrupt intensification at the arrow has been shown to coincide with the sudden solidification of the droplet after a period of supercooling [18]. The abrupt intensification is referred to as the "spearpoint". By the time the droplet solidified, its composition was essentially stoichiometric ZrO_2.

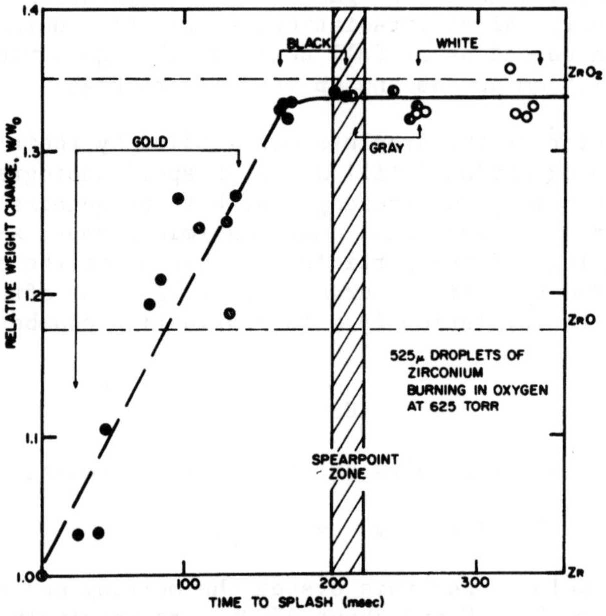

Fig. 3. Plot of relative weight change of zirconium droplets after various combustion times (times to splash in liquid argon). Arrows indicate surface color of droplets quenched during time intervals shown. "Spear-point zone" is time interval in which droplets solidify.

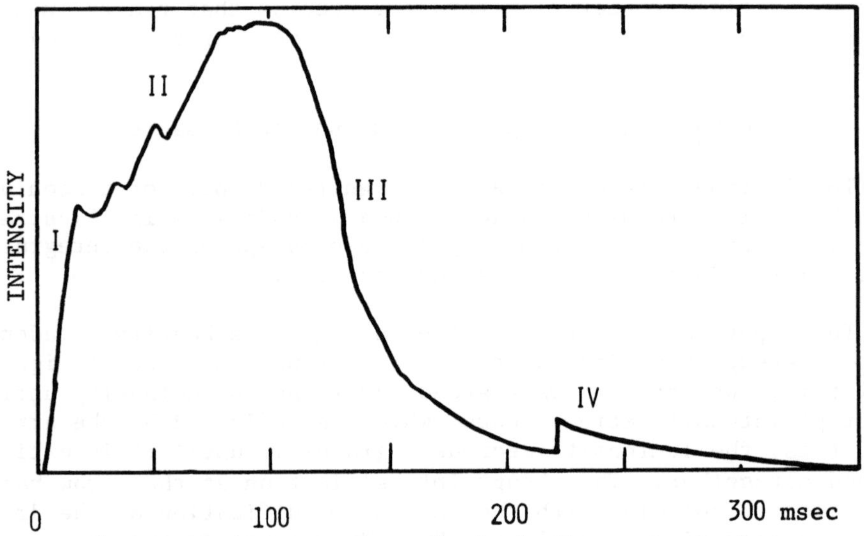

Fig. 4. Luminous emission of single laser-ignited zirconium droplet (initial dia. = 525 μm) as it burned in pure oxygen (pressure = 625 Torr). Four regimes of rate-limitation are indicated (see text).

The external colors of the droplets that were quenched in liquid argon at various stages in the oxidation indicate the progress of the reaction. The surface colors change progressively from the original silvery gray of the metal through gold, black and gray and ultimately to white. The gray colorations occurs at a composition of $ZrO_{1.96}$ [19], whereas the white coloration represents stoichiometric ZrO_2 [19]. These color changes are indicated in Fig. 3 on a plot of the relative weights of the droplets, W/W_o, vs. reaction time (see references 15 and 17). As illustrated in Fig. 3, the droplets increase in weight smoothly from an initial value of unity to a final value of 1.35. The gray color and the spearpoint zone approximately coincide (at S.C.). The subsequent change of surface color to white indicates complete reaction to stoichiometric ZrO_2. It is interesting to note that if the reaction occurs in ultrapure oxygen at 625 Torr, the zirconium initially present is retained in the droplet to within about 3% (see references 15, 16 and 17). This experimentally observed conservation of metal by the burning droplet will be a major point of discussion here.

As the weight of zirconium droplet increases toward the stoichiometric value, the luminosity of the droplet changes in a characteristic and reproducible manner. A typical photo-multiplier record for a droplet burning under standard conditions is shown in Fig. 4. A comparison of Figs. 3 and 4 shows that the weight of the droplet increases throughout almost the entire period of luminous emission, even though the luminosity rises and falls dramatically during the same period. Note that the luminosity increases rapidly at first (stage I), then increases at a somewhat lower rate until a maximum is reached (stage II), then decreases steadily to a low value (stage III), then increases abruptly at the spearpoint and finally decreases again until extinction (stage IV).

The reaction of $Zr(\ell)$ with $O_2(g)$ has been described [17] in terms of a sequence of rate-limiting processes as follows: Stage I represents a period of rapid heating in which molten zirconium reacts with gaseous oxygen with little or no diffusional limitation in the gas phase. During stage II, a slower heating period occurs, due to diffusional limitations set up by a cloud of metal-bearing vapors emitted by the droplet. Stage III represents a cooling period in which the reaction is governed by diffusion of oxygen within the liquid droplet. Finally, stage IV is essentially a cooling period which is controlled mainly by radiant and convective heat loss after the droplet solidifies.

The maximum temperature at which the droplet burns has been estimated to be $\sim 4300°K$ by ratio pyrometry [17]. This measurement is consistent with similar measurements by Bracco and Weisberger [20] and Nijland and Schröder [21] for the peak temperatures in combustion flashbulbs filled with zirconium and oxygen. There is similar agreement with calculations of maximum burning temperatures for zirconium-oxygen flames made by Grosse and Conway [22].

It is surprising to note that the zirconium droplets burn in pure oxygen at 625 Torr within a few hundred degrees of the normal boiling point of both metal and oxide (and even closer to the boiling point at 625 Torr) with hardly any loss of metal!

Combustion at Reduced Oxygen Concentrations

When zirconium droplets initially 525 μm in diameter were burned in oxygen diluted with either helium or argon, with the total pressure in the chamber held at 625 Torr, the burning temperatures decreased as the dilution increased, as shown in Fig. 5. Note that the maximum burning temperature decreases smoothly from about 4300°K in pure oxygen to about 3200°K in a 20% O_2/ He mixture. The trend when argon is the diluent appears similar to that in helium. Although it has not yet been measured, it seems reasonable to assume that an analogous decrease will occur in undiluted oxygen as the pressure is reduced, particularly because the maximum luminous emission from the droplets decreases in roughly the same manner as when the oxygen concentration is reduced by rare gas dilution.

The degree of conservation of zirconium changes in a most unusual fashion when the oxygen pressure is reduced. Although one might expect, a priori, that the small loss of metal that occurs at the highest burning temperatures would disappear at lower burning temperatures, this expected result is not the case. The ultimate weight ratio, $U=W_f/W_o$ = ultimate weight/initial weight, of zirconium droplets burned in pure oxygen shows that the loss of metal actually increases as the gas pressure is reduced (see Fig. 6). (Independent studies [17] have shown that the ultimate composition of droplets burned in oxygen is in the range $ZrO_{1.96}$ to $ZrO_{2.00}$ regardless of pressure (for the oxygen pressures studied here).) The losses increase slowly at first as pressure is reduced from 625 to ∿ 200 Torr, then become wildly erratic between about 200 and about 75 Torr, and finally level off at 50-60% loss of metal at oxygen pressure below about 75 Torr. These unexpected results show that *the loss of zirconium increases as the burning temperature decreases!*

These conservation phenomena are best explained by an ultimate return of the zirconium-bearing vapors to the burning droplet via a fog layer and follower droplet mechanism, as reported recently by Marshall, Nelson and Richardson [23] and Marshall and Pellett [6]. (See the photograph of the phenomenon in Fig. 7.) Products that vaporize from the burning zirconium droplet apparently condense at an approximately spherical front that envelops the falling droplet. The result is a swirling layer of liquid oxide droplets which are swept upward in the gas flow and coalesce to form a secondary liquid oxide droplet that follows in the null region just above the parent

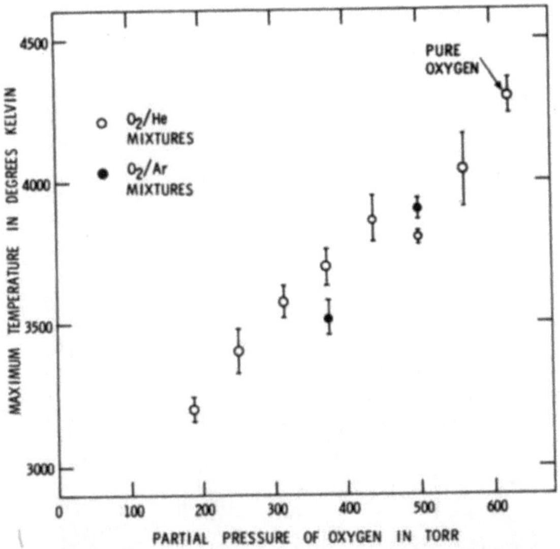

Fig. 5. Maximum burning temperatures for 525 μm initial diameter zirconium droplets during combustion in oxygen/argon and oxygen/helium mixtures at a total pressure of 625 Torr. See text for assumptions involved

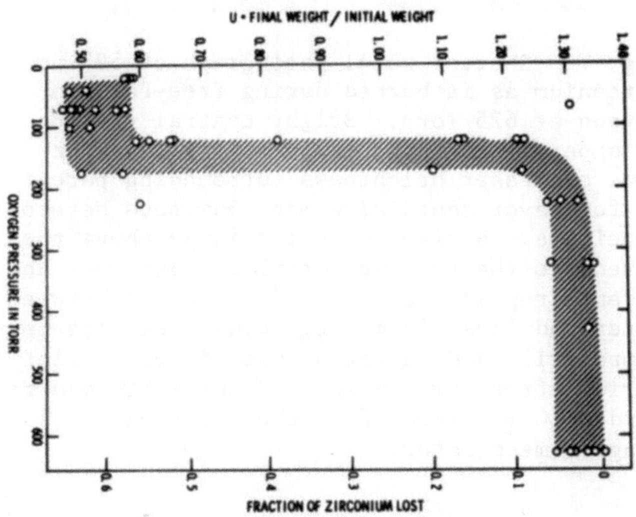

Fig. 6. Fractional weight changes of 525 μm initial diameter zirconium droplets burned to completion in pure oxygen at various pressures.

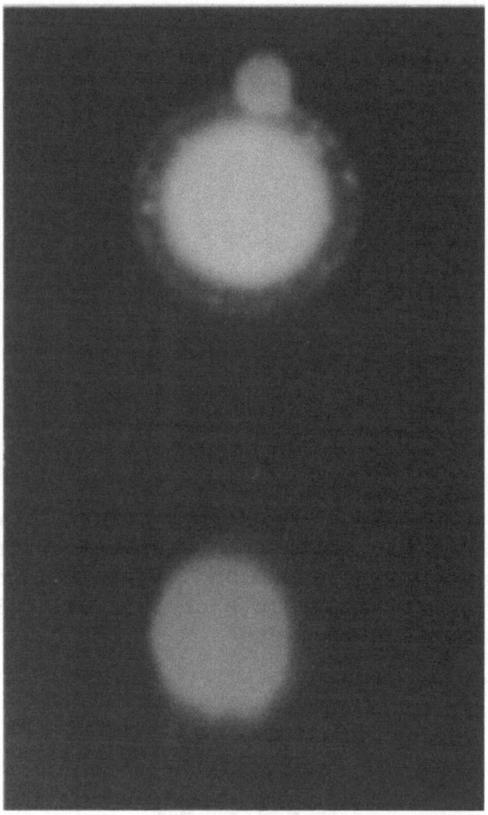

Fig. 7. Shutter-wheel photograph of 525 μm zirconium as it burned during free-fall in oxygen at 625 Torr. Bright central portion of upper image is parent droplet. Annular zone of lesser brightness surrounding parent is fog layer containing many luminous heterogeneities. Smaller circular image above the parent is the follower droplet. Follower and parent droplets touched and coalesced between upper and lower image; luminous zone disappeared temporarily and parent undulated for a brief period after lower image. Upper image was recorded 40 ± 3 msec after the laser pulse; lower image, 3 msec later.

droplet. Periodically, this follower droplet becomes large enough to touch the parent and coalesce with it. The repetitive coalescence of parent and follower droplets gives rise to an overall *reflux action* that returns the vaporized metal.

The characteristics of the fog layer and the ultimate fate of the follower droplet at different oxygen pressures apparently determine the degree of zirconium conservation by the system. This affects the nature of the luminosity curves. Representative luminosity curves for replicate experiments are shown in Fig. 8 for the combustion of zirconium droplets of 525 μm initial diameter in pure oxygen at 125 Torr. This pressure was chosen because it lies in the region of erratic conservation of metal shown in Fig. 6.

As illustrated in Fig. 8, at relatively low oxygen pressures, the center portion of the oscilloscope records seems to divide into two parts -- a plateau on which a hump may be superimposed. We have

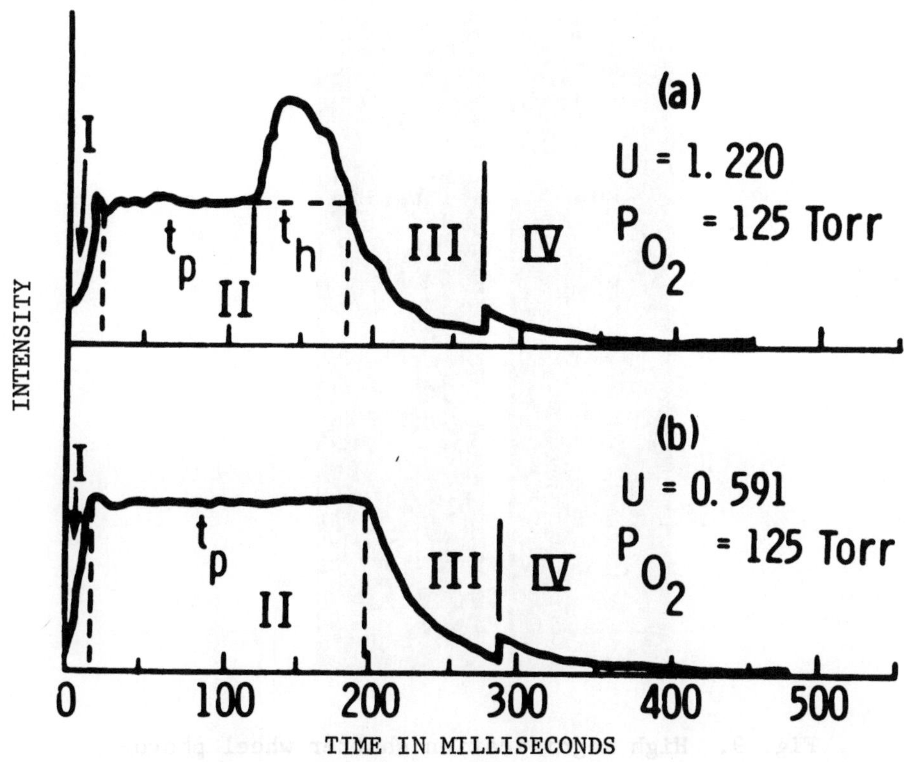

Fig. 8. Representative photomultiplier records of light emitted at 8300 ± 50 Å by 525 μm initial diameter zirconium droplets as they burned in pure oxygen. Traces correlate with photographic images in Figure 9.

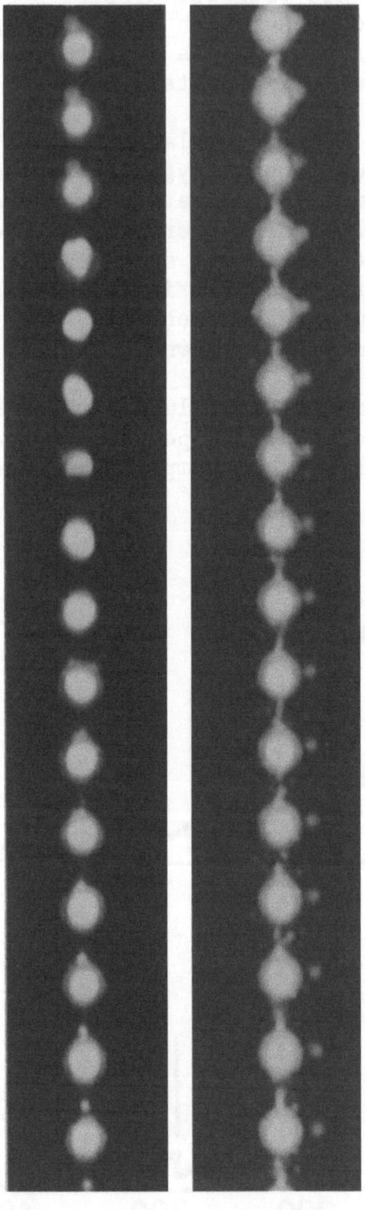

Fig. 9. High magnification shutter wheel photographs of 525 μm initial diameter zirconium droplets burning in free fall through pure pxygen at 175 Torr. Images are spaced 1.4 msec apart, (a) coalescence of parent follower drop, (b) follower drop veered off to one side without coalescing.

noted [17] that the duration of the hump in these traces is quite variable in replicate experiments. We also noted that metal tends to be retained when the humps are long and to be lost when the humps are short or absent. This has been shown quantitatively as a direct correlation between the conservation of metal and the ratio of hump duration to the sum of the durations of hump and plateau (i.e., the greater this ratio, the more metal is retained during the combustion).

It can also be seen in Fig. 8 that the fate of the metal may be decided quite late in the burning reaction. Thus, in this region of erratic weight gain behavior, combustions apparently proceed with identical luminous emissions for about half the reaction time, and then either retain much of the metal if the hump forms (upper trace in Fig. 8) or lose a lot if the hump fails to form. Apparently almost half of the zirconium originally in the droplet may linger close by and may or may not coalesce with the parent [10].

This behavior at intermediate oxygen pressures has been found to be caused by the formation of a follower droplet that may have a diameter as large as 80% of the parent (see note added in proof to ref. 17). This follower droplet may collide with the parent, giving rise to the hump, or it may veer off to one side of the parent, leaving only a plateau on the luminosity trace. This can be seen in Fig. 9, where in (a), the large follower drop collides with the parent, while in a replicate experiment shown in (b), the follower passes off to the right of the parent without coalescence.

DISCUSSION

An explanation of the effect of reduced oxygen concentration on conservation of zirconium is proposed as follows: At high oxygen concentrations, the burning temperature is high, products vaporize easily, and condensation occurs freely forming the fog layer close to the parent droplet. In this highly concentrated layer, fog droplets apparently coalesce with high probability and ultimately form the follower droplet. Since the fog layer is close to the parent, the follower never grows very large before it touches and unites with the parent droplet. Thus at high burning temperatures, the reflux action very effectively returns essentially all of the zirconium that may be vaporized in some form during the combustion.

At reduced oxygen pressures, the temperature of burning is lowered and the condensation front moves outward from the droplet and becomes more diffuse. With a lower vaporization rate and a reduced probability of coalescence in the fog layer, the follower droplet forms more slowly and also grows larger before it unites with the parent (because of the remote position of the fog layer). Often, the follower droplet may miss the parent altogether. Thus the vaporized products may be lost not only by a reduced probability of coalescence in the fog layer, but also by a reduced likelihood of collision between parent and follower. It is this hit-or-miss behavior of the follower droplet that is thought to give rise to the wild scatter of ultimate weight ratios at oxygen pressures in the region 75-200 Torr (see Fig. 6).

At very low oxygen pressures (below 75 Torr), we suggest that the burning temperatures may be so low that the vaporized products will condense as solid particles -- as smoke -- rather than as liquid droplets. Then coalescence into a follower droplet would not be likely to occur, and a large loss of metal would be observed.

It is interesting to compare the overall suppression of vaporization loss of metal oxide by the fog layer and follower droplet mechanism discussed here with the enhancement of vaporization by analogous fog layers in other high temperature systems [24, 25, 26, 27].

ACKNOWLEDGMENTS

It is a pleasure to acknowledge the experimental assistance of Norman L. Richardson. We also are grateful to D. E. Rosner and S. C. Kurzius of AeroChem Research Laboratories, Inc. and R. L. Marshall of NASA/Langley Research Center for valuable suggestions. The work was supported by the U. S. Atomic Energy Commission.

REFERENCES

1. A. F. Haught and D. H. Polk: Phys. Fluids 9, 2047 (1966).
2. E. C. Okress, D. M. Wroughton, G. Comenetz, P. H. Brace and J. C. R. Kelly: J. Appl. Phys. 23, 545 (1952).
3. A. E. Jenkins, B. Harris and L. Baker: Met. Soc. Conf. (K. A. Gschneider, Jr., M. T. Hepworth, N. A. D. Parlee, editors, Gordon and Breach, Science Publishers, New York, N. Y.) 22, 23 (1964).
4. W. A. Peifer: J. Metals 17, 487 (1965).
5. L. A. Baker and R. G. Ward: J. Iron Steel Inst. 205, 714 (1967).
6. R. L. Marshall and G. L. Pellett: Preprints, Spring Meeting, Western States Section, The Combustion Institute, China Lake, California, April 1969.
7. R. L. Marshall: M. S. Thesis, Virginia Polytechnic Institute, Blacksburg, Virginia (1968).
8. J. L. Prentice and L. S. Nelson: J. Electrochem Soc. 115, 809 (1968).
9. R. P. Wilson, Jr.: Preprints, Spring Meeting, Western States Section, The Combustion Institute, China Lake, California, April 1969.

10. R. Baker: J. Iron Steel Inst. 205, 637 (1967).
11. J. L. Margrave: Abstracts, Calorimetry Conf., Midland, Mich., August, 1968; see also Chem. Eng. News, October 28, 1968.
12. R. H. Willens and E. Buehler: Trans. Met. Soc. AIME 236, 171 (1966).
13. L. S. Nelson and N. L. Richardson: J. Phys. Chem 68, 1268 (1964).
14. L. S. Nelson, N. L. Richardson and J. L. Prentice: Rev. Sci. Instr. 39, 744 (1968).
15. L. S. Nelson: Eleventh Symposium (International) on Combustion, Berkeley, California, August 1966, (Publ. 1967), The Combustion Institute, Pittsburgh, Pennsylvania, p. 409.
16. L. S. Nelson and H. S. Levine: High Temp. Sci 1, 163 (1969).
17. L. S. Nelson, D. E. Rosner, S. C. Kurzius and H. S. Levine: Twelfth Symposium (International) on Combustion, Poitiers, France, July 1968 (Publ. 1969), The Combustion Institute, Pittsburgh, Pennsylvania, p. 59.
18. L. S. Nelson: Nature 207, 741 (1965); 210, 410 (1966).
19. R. J.Ackermann and R. J. Thorn: Progress in Ceramic Sci. 1, 39 (1961).
20. D. J. Bracco and S. Weisberger: Appl. Opt. 5, 1275 (1966).
21. L. M. Nijland and J. Schröder: Philips Res. Repts. 21, 304 (1966).
22. A. V. Grosse and J. B. Conway: Ind. Eng. Chem. 50, 663 (1958).
23. R. L. Marshall, L. S. Nelson and N. L. Richardson: Combust. Flame 13, 216 (1969).
24. E. T. Turkdogan, P. Grieveson and L. S. Darken: J. Phys. Chem. 67, 1647 (1963).
25. E. T. Turkdogan: Trans. Met. Soc. AIME 230, 740 (1964).
26. D. E. Rosner: Int. J. Heat Mass Transfer 10, 1267 (1967).
27. D. E. Rosner and M. Epstein: Trans. Met. Soc. AIME 242, 2355 (1968).

GENERAL DISCUSSION

In opening the general discussion, *C. B. Alcock* noted that many of the conference papers had emphasized chemical-reaction control at an interface. This represented a change in emphasis since the 1950's when local equilibrium at interfaces was generally presumed for high temperature reactions. He pointed to the necessity of establishing clear criteria which would enable one to differentiate between a diffusion-controlled reaction and one which is controlled by a chemical-reaction at the interface. Experimentally we are faced with the problem of eliminating the influence of mass transport in flowing gas mixtures.

J. Szekely commented that the conference had provided excellent opportunity for the process or fluid mechanics people to converse with the chemical-reaction people. A strong plea was made for investigators to use fluid mechanics whenever possible, to describe mass transfer phenomena at interfaces.

C. Wagner suggested that the phrase *Divide et Impera* is most appropriate for high-temperature kinetic studies. In many cases one can separately determine the diffusion rate and the chemical-reaction rate. For example, in a study of the dissolution of hydrogen in solid palladium, it was possible to measure separately the rate of the chemical reaction on the surface and the rate of diffusion of hydrogen in the palladium. It was possible to conclude that the reaction was not diffusion controlled.

L. S. Darken emphasized that chemical reactions are highly specific. One should consider only individual cases and not attempt to develop a broad approach which would describe everything.

S. E. Khalafalla suggested the use of inhibitors or promoters to determine if a reaction is diffusion or chemical-reaction controlled. As an example, he mentioned his results for the reduction of wüstite (FeO) by carbon monoxide. It was observed that trace amounts of boron or sulfur inhibited the rate. It is difficult to understand how a mass-transport controlled reaction could be affected by trace amounts of elements. Thus these results indicated chemical-reaction control.

T. B. King asked if one would expect chemical-reaction control only when there is adsorption of impurities or reactants at the interface. In reference to Szekely's earlier comments, he noted that the most convenient experimental design is often one which is most difficult to describe in terms of heat and mass transfer. We should devise new experimental designs and techniques in which there would be a much better control of heat and mass transfer conditions.

G. R. Belton cautioned against making definite conclusions from the agreement of experimental data with a particular mass-transfer model. As an example, he mentioned Y. K. Rao's results, which were presented at the 1968 AIME annual meeting. The measured rates of reduction of graphite-hematite (Fe_2O_3) mixtures agreed quite well with a fairly well-established mass transfer model. However, Rao was able to demonstrate conclusively that the reduction was controlled by either a gas reaction or a surface reaction by adding alkaline metal promoters in the manner described previously by Khalafalla. Regarding the problem of gas-transport, Belton suggested that one could often make intelligent deductions from the variation of rate with gas composition, since the interdiffusivities in some gas mixtures are fairly well known.

A. W. D. Hills indicated that he was in agreement with most of the previous comments, but pointed out that heat transfer in many high-temperature reactions may be rate limiting - this possibility should always be borne in mind.

E. T. Turkdogan stated that many investigators are and have been fully aware of the complications arising from dual control. However, he felt that progress had been made, particularly when care had been taken to simplify the system. One should work with the limiting case whenever possible. There are many cases where one can eliminate mass-transport control in order to study the surface reaction. Conversely, one can often eliminate experimentally the slow surface reaction in order to study diffusivities. He also mentioned the difficulty of predicting gas diffusivity in porous media. To know accurately the pore diffusivity, one must measure it. As an example, he referred to his results for the oxidation of carbon.

C. B. Alcock pointed out that sintering might cause changes in pore structure during an experiment.

J. B. Wagner, Jr. raised the problem of interpreting diffusion in solid compounds with large deviations from stoichiometry. There may be large interactions between the various defects; thus one should first ascertain if the theory is applicable to such compounds with wide deviations from stoichiometry. He also emphasized the importance of including as much tabular data as possible in the published results.

In reference to some of the earlier comments, *C. E. Birchenall* suggested that one of the most potent ways to determine the rate-controlling process is to identify the site of the driving forces. For example, the determination of the compositional gradients in solids may be just as useful as determining the relative mobilities of the species. Considering the role of defects in chemical reactions involving solids, he stated that experience indicates that defects in solids are essential for attaining equilibrium and that they tend to accelerate the rates of reactions.

E. Gulbranson hoped that investigators would give careful consideration to the thermodynamics or thermochemistry of a system before starting kinetic studies. For example, unsuspected volatile species could often influence the kinetics of high-temperature reactions.

J. N. Ong endorsed Gulbransen's statements and suggested that there ought to be very definite thermodynamic criteria by which one could determine if a reaction is heat or mass transfer controlled. He also mentioned that the higher the temperature, the more carefully we have to consider each possibility of mass-transfer or chemical-reaction control.

C. B. Alcock, acting as chairman, questioned how adsorption on the surface could control the rate of a surface reaction. He questioned the concept of self-poisoning and what it means.

W. L. Worrell stated that self-poisoning is a confusing and misleading term. It misleads one to think incorrectly that it is purely a surface coverage or blockage which slows the rate of a surface reaction. As an example, he considered the reaction of $H_2S(g)$ with FeS. From the available data, one is reasonably confident that the rate-controlling step in the surface reaction is the dissociation of adsorbed H_2S molecules on the FeS surface. One now asks how could the adsorption of other molecules on the surface affect the rate of dissociation of adsorbed H_2S. An answer is obtained following the activated-complex approach. If the surface is almost fully covered with adsorbed sulfur ($\theta s \rightarrow 1$), one obtains a rate equation for the dissociation of H_2S on FeS which fits the experimental data. In using this approach, no information is required regarding the defect structure of the surface or of the solid. The derived rate equation could be applicable to any solid or liquid surface almost fully covered with sulfur. Another approach, which is discussed in our conference paper, yields a similar rate equation. In the second approach, it is assumed that two electrons are necessary for the dissociation of an adsorbed H_2S molecule on an FeS surface. Thus the dissociation rate would depend on the defect equilibria in FeS. At this time it is impossible to state which approach is correct.

L. S. Darken felt that he was responsible for the term self-poisoning, but he usually tries to avoid using this highly subjective term. However one can consider a poisoning effect to exist in any experimental region in which the surface-reaction rate decreases with increasing activity of a particular species. From a broader viewpoint, one must consider all the irregularities on a solid surface (terraces,ledges, kinks, etc). Thus it is surprising that the surface coverage approach works at all. If one asks which sites are covered, he would probably guess that all sites are filled, surface adsorption experiments indicating a monolayer coverage. Any large departure from ideal adsorption behaviour can be accommodated by using an activity coefficient. This activity coefficient may be quite different from the normal one used in solution thermodynamics; it may involve changes in the nature of the surface or of the relative amounts of adsorption on the different types of sites. However, from the small amount of available evidence, it does seem reasonable to use the concept of coverage of surface sites.

T. B. King commented that the nature of the activated complex in some simple homogeneous reactions is understood. However, he questioned our understanding of the nature of an activated complex on a surface. He noted that the concept of coverage of the surface sites by adsorption of some poison, does not explain his results presented at this conference. One needs to know more about the nature of the activated complex which yields the products.

C. Wagner suggested that we restrict the concept of poisoning to a specific situation of the type which was described recently in a paper by Bechtoldt. Bechtoldt investigated the reaction of hydrogen with diatomic sulfur vapor catalysed by platinum at about 300°C. The partial pressure of H_2 was about 10^{-3} torr, and the sulfur partial pressure was much lower, possibly between 10^{-6} and 10^{-5} torr. The initial reaction rate was observed to increase with increasing partial pressure of S_2. However, the reaction rate decreased with a further increase in p_{S_2}. This decrease can properly be called poisoning. These observations can be interpreted to indicate that the adsorption of hydrogen molecules is essential for the reaction and that the reaction rate is dependent on the number of free sites for hydrogen adsorption. The number of free sites decreases as the sulfur pressure is increased, and accordingly the reaction rate decreases.

S. Wagner suggested that a surface should be considered as a separate phase. When the surface plays an essential role in a chemical reaction, there should be other investigations on the surface itself, for example, surface-tension measurements.

T. B. Lindemer mentioned that one should not assume an infinite interface when working with small discrete particles or powders.

One should not only consider the mechanisms at the particular interface, but should look carefully at the rest of the particle where something entirely different could be occurring, which could invalidate the analysis for the entire reaction.

Considering gas-solid reactions, *N. J. Themelis* stated that the mass-transfer coefficient in gas mixtures can often be calculated at the critical Reynold's number. The calculated mass-transfer rate can then be compared with the measured reaction rate. The critical Reynold's number can be established from correlations based upon several high-temperature studies. These studies, which include levitation-melting research and the work of Schwerdtfeger and Turkdogan at U. S. Steel, have shown that correlations developed for the evaporation of droplets, etc., are reasonably accurate. If one cannot define the system completely using fluid dynamics, then one should use a semi-empirical correlation.

J. Szekely suggested that the use of spherical particles in a reduction experiment is not helpful. The gradually shrinking reaction interface can cover up a lot of effects. More work should be done with linear systems. For example, Y-K. Lu at McMaster has been using cylinders in which the walls are blocked off. He also stressed the importance of knowing how to design the structure of solid pellets or sinter to give optimal reduction characteristics in large scale industrial furnaces. Does one want a lot of defects? Should one dope it, etc? It also would be useful to know the heat-transfer characteristics in order to locate the optimal temperature range for reduction. At too high a temperature, one might sinter.

A. W. D. Hills noted that while linear geometry is very helpful for mass-transfer analyses it is poor for heat transfer. The shape factor for thermal conduction is different from that for diffusion, which results in a distortion of the interface and makes the results difficult to interpret. Commenting on reaction models, he stated that one reason for evolving a reaction model is for intellectual satisfaction. One wishes to understand the universe in which one lives. However, another important use of reaction models is to make technologically accurate predictions. He cautioned that there can be large differences between a laboratory and an industrial environment which can lead to erroneous predictions for reaction rates in industrial reactors.

At this point *C. B. Alcock* invited comments on reactions involving liquids.

F.D.Richardson suggested one way in which interfacial turbulence on a liquid metal surface might complicate the kinetics. Consider the transfer of oxygen from a gas phase to a stirred liquid. If the metal surface is reasonably saturated with oxygen, there are no substantial differences in surface tension along the surface and no interfacial turbulence. However, if one is removing oxygen with H_2(g) or CO(g) one could have uneven rates of arrival of H_2(g) or CO(g) over the interface, and hence uneven rates of removal of oxygen. These could conceivably lead to strong local differences in surface tension and hence to interfacial turbulence. Interfacial turbulence can be important even in strongly stirred systems and generally arises under conditions where the driving forces are high, that is when there are high concentration differences across the boundary regions. One example of this can be found in the work of Subramanian et al. [J.I.S.I. 206, 1968, 576-583] on the oxidation of indium dissolved in mercury by ferric iron in aqueous solution at room temperature. It was studied in a system strongly agitated by bubbles and the process was found to be transport controlled. This control lies either in the metal phase or in the aqueous phase depending on the relative concentrations of indium and ferric ions. However, at high ferric ion concentrations the aqueous phase mass transfer coefficient increases by a factor of three over its constant value at lower concentrations. This increase is associated with the onset of interfacial turbulence which has now been observed by Schlieren photography. In all experiments, therefore, involving transfer across a liquid-gas interface or a liquid-liquid interface, one must consider the possibility of interfacial turbulence occurring and varying with concentration. Otherwise one might reach erroneous conclusions about other kinetic factors and the chemical effects of adsorbed species.

E. T. Turkdogan mentioned some of his observations of interfacial turbulence on a mercury surface. If one removes the lid from a container of mercury containing as little as 10-100 ppm sodium, one observes sodium oxide being formed in a number of places on the surface. There is also a tremendous surface motion which continues until there is complete coverage of the surface with sodium oxide.

G. R. Belton commented that there is no substantial experimental evidence for self-poisoning in liquid systems. With CO_2/CO mixtures in liquid lead, he finds interfacial control. In his H_2O/H_2 studies with liquid copper and liquid lead, diffusion control is observed. Thus there does not seem to be any self-poisoning by oxygen. A strong plea was made for pertinent physico-chemical measurements in addition to mass-transfer studies. More surface-tension data are needed for ternary and quaternary liquid systems. One should be hesitant about making definite conclusions from empirical rate data unless there are independent physico-chemical measurements which support one's deductions.

C. B. Alcock closed the general discussion session by thanking the chairmen and contributors for a most interesting and stimulating conference.

INDEX

REGISTRANTS

Dr. C. B. Alcock, University of Toronto, Toronto, Ontario.
Dr. Terrence J. Allen, The University of Michigan, Ann Arbor, Michigan.
Dr. G. R. Belton, University of Pennsylvania, Philadelphia, Pennsylvania.
Dr. Robert Benz, Los Alamos Scientific Laboratory, Los Alamos, New Mexico.
Dr. Joan B. Berkowitz, Arthur D. Little, Inc., Cambridge, Mass.
Dr. C. E. Birchenall, University of Delaware, Newark, Delaware.
Dr. Gary D. Blue, Battelle Mem. Institute, Ohio State University, Columbus, Ohio.
Dr. Charles Brookes, University of Waterloo, Waterloo, Ontario.
Dr. John W. Burlingame, Frankford Arsenal, Philadelphia, Pennsylvania.
Dr. Y. Austin Chang, University of Wisconsin, Milwaukee, Wisconsin.
Mr. C. D. Cassler, University of Pittsburgh, Pittsburgh, Pennsylvania.
Dr. P. E. Childs, Northwestern University, Evanston, Illinois.
Dr. John M. Chilton, Bell Telephone Laboratories, Allentown, Pennsylvania.
Dr. Horn D. Chu, Selas Corporation of America, Maple Shade, New Jersey.
Mr. Charles K. Clarke, Lehigh University, Bethlehem, Pennsylvania.
Dr. Pedro Danús, Chilean Copper Corporation, Santiago, Chile.
Dr. L. S. Darken, United States Steel Corporation, Monroeville, Pennsylvania.
Dr. Carlos M. Diaz, University of Chile, Santiago, Chile.
Dr. John F. Elliott, Massachusetts Institute of Technology, Cambridge, Massachusetts.
Dr. Toshihiko Emi, The University of Michigan, Ann Arbor, Michigan.
Dr. Nicholas G. Eror, Sprague Electric Company, North Adams, Mass.
Dr. H. Eyring, The University of Utah, Salt Lake City, Utah.
Dr. L. Eyring, Arizona State University, Tempe, Arizona.
Dr. Donald I. Finch, Leeds and Northrup Company, North Wales, Pennsylvania.
Dr. G. R. Fitterer, University of Pittsburgh, Pittsburgh, Penna.
Dr. Karl Foster, Westinghouse Electric Corporation, Pittsburgh, Pennsylvania.

Dr. Richard J. Fruehan, United States Steel Corporation, Monroeville, Pennsylvania.
Dr. Charles T. Fujii, Naval Research Laboratory, Washington, D. C.
Dr. D.R. Gaskell, University of Pennsylvania, Philadelphia, Pennsylvania.
Dr. Paul W. Gilles, University of Kansas, Lawrence, Kansas.
Dr. Robert K. Gould, Aerochem Research Laboratory, Inc. Princeton, New Jersey.
Mr. Ronald C. Gower, Lehigh University, Bethlehem, Pennsylvania.
Dr. Donald E. Grimes, Inland Steel Company, East Chicago, Indiana.
Dr.E. A. Gulbransen, Westinghouse Research Laboratories, Churchill Borough, Pittsburgh, Pennsylvania.
Dr. Helmut G. Hadrys, Bethlehem Steel Corporation, Bethlehem, Pennsylvania.
Dr. John P. Hager, Colorado School of Mines, Golden, Colorado.
Dr. Francis J. Harvey, Westinghouse Research Laboratories, Churchill Borough, Pittsburgh, Pennsylvania.
Dr. H. Hatwell, Union Carbide Corporation, Indianapolis, Indiana.
Dr. James M. Henderson, American Smelting and Refining Company, South Plainfield, New Jersey.
Dr. A. W. D. Hills, Imperial College of Science and Technology, London S. W. 7, England.
Dr. Charles E. Houck, Bethlehem Steel Corporation, Bethlehem, Pennsylvania.
Dr. Thomas S. Jones, Allegheny Ludlum Steel Company, Natrona Heights, Pennsylvania.
Dr. Andrew S. Jordon, Bell Telephone Laboratories, Murray Hill, New Jersey.
Mr. Howard I. Kaplan, University of Pennsylvania, Philadelphia, Pennsylvania.
Dr. Donald G. Kesterke, U. S. Bureau of Mines, Sparks, Nevada.
Dr. Sanaa E. Khalafalla, U. S. Bureau of Mines, Twin Cities, Minnesota.
Dr. T. B. King, Massachusetts Institute of Technology, Cambridge, Massachusetts.
Dr. Gerrit J. W. Kor, U. S. Steel Corporation, Monroeville, Pennsylvania.
Dr. Dwight A. Kraai, Allegheny Ludlum Steel Company, Brackenridge, Pennsylvania.
Dr. Avinash D. Kulkarni, Chase Brass and Copper Company, Cleveland, Ohio.
Dr. Carlos Landolt, University of Chile, Santiago, Chile.
Dr. Leon W. Laub, Northwestern University, Evanston, Illinois.
Dr. Regis J. Leonard, Bethlehem Steel Corporation, Bethlehem, Pennsylvania.
Dr. H. S. Levine, Sandia Laboratories, Albuquerque, New Mexico.
Dr. T. B. Lindemer, Oak Ridge National Laboratory, Oak Ridge, Tennessee.
Dr. W-K. Ku, McMaster University, Hamilton, Ontario.

Dr. Donald R. MacRae, Bethlehem Steel Corporation, Bethlehem, Pennsylvania.
Dr. Russell L. McCarron, International Nickel Company, Incorporated, Sterling Forest, Suffern, New York.
Dr. E. Scott McCormick, Purdue University, W. Lafayette, Indiana.
Dr. Clifford W. McCoy, Heppenstall Company, Pittsburgh, Pennsylvania.
Dr. William M. McKewan, U. S. Steel Corporation, Monroeville, Pennsylvania.
Dr. John D. McKinley, National Bureau of Standards, Washington, D. C..
Dr. Alex McLean, Jones and Laughlin Steel Corporation, Pittsburgh, Pennsylvania.
Dr. William R. Morcom, Westinghouse Electric Corporation, Bloomfield, New Jersey.
Dr. L. S. Nelson, Sandia Laboratories, Albuquerque, New Mexico.
Mr. John D. Oakey, Lehigh University, Bethlehem, Pennsylvania.
Dr. John N. Ong, University of Wisconsin, Milwaukee, Wisconsin.
Dr. Joseph A. Pask, Lawrence Radiation Laboratory, University of California, Berkeley, California.
Dr. John W. Patterson, Iowa State University, Ames, Iowa.
Dr. J. P. Pemsler, Kennecott Copper Corporation, Lexington, Mass.
Dr. Frederick S. Pettit, Pratt and Whitney Aircraft, Division of United Aircraft Corporation, Middletown, Connecticut.
Dr. Trikur A. Ramanarayanan, Ohio State University, Columbus, Ohio.
Dr. Krishna Y. Rao, Columbia University, New York, New York.
Dr. Richard H. Rein, Union Carbide Corporation, Tarrytown, New York.
Dr. Graham D. Renshaw, Pennsylvania State University, University Park, Pennsylvania.
Dr. Peter C.Rhee, 1275 Elgin Street, Burlington, Ontario.
Professor F. D. Richardson, Imperial College of Science and Technology, London S. W. 7, England.
Dr. Victor S. Robinson, III, New Jersey Zinc Company, Palmeton, Pennsylvania.
Dr. Colin Roscoe, Pennsylvania State University, University Park, Pennsylvania.
Dr. G. M. Rosenblatt, Pennsylvania State University, University Park, Pennsylvania.
Dr. D. E. Rosner, Yale University, New Haven, Connecticut.
Dr. H. Juergen Schoennagel, Mobil R & D Corporation, Princeton, New Jersey.
Dr. Ram A. Sharma, Argonne National Laboratory, Argonne, Illinois. .
Dr. John S. Sheasby, University of Western Ontario, London, Ontario.
Mr. David W. Short, The Ohio State University, Columbus, Ohio.
Dr. George Simkovich,The Pennsylvania State University, University Park, Pennsylvania.
Dr. Subhash C. Singhal, Chromalloy American Corporation, San Antonio, Texas.
Dr. B. C. H. Steele, Imperial College of Science and Technology, London S. W. 7, England.
Dr. Robert C. Svedberg, Westinghouse Astronuclear Laboratory, Pittsburgh, Pennsylvania.

Dr. James H. Swisher, Bell Telephone Laboratories, Incorporated, Murray Hill, New Jersey.
Dr. Julian Szekeley, State University of New York, Buffalo, New York.
Dr. Richard L. Tallman, Westinghouse Research Laboratories, Pittsburgh, Pennsylvania.
Mr. Edwin S. Tankins, Naval Air Development Center, Philadelphia, Pennsylvania.
Dr. Peter P. Tarassoff, Noranda Research Centre, Pointe Claire, Quebec.
Dr. T. A. Taylor, Union Carbide Corporation, Indianapolis, Indiana.
Dr. Nickolas J. Themelis, Noranda Research Center, Pointe Claire, Quebec.
Mr. Marshall K. Thomas, Naval Air Development Center, Warminster, Pennsylvania.
Dr. Johann G. Tschinkel, Pratt and Whitney Aircraft, Division of United Aircraft Corporation, Middletown, Connecticut.
Dr. R. C. Tucker, Union Carbide Corporation, Indianapolis, Indiana.
Dr. E. T. Turkdogan, United States Steel Corporation, Monroeville, Pennsylvania.
Professor C. Wagner, Max-Planck-Institut Für Physikalische Chemie, Göttingen, Germany.
Dr. J. B. Wagner, Jr., Northwestern University, Evanston, Illinois.
Dr. Sigurd Wagner, Ohio State University, Columbus, Ohio.
Dr. Patrick N. Walsh, Union Carbide Corporation, Indianapolis, Indiana.
Dr. Charles J. Warde,Westinghouse Research Laboratories, Churchill Burrough, Pittsburgh, Pennsylvania.
Dr. Thomas E. Weyand, University of Missouri, Rolla, Missouri.
Dr. Stirling G. Whiteway, National Research Council, Halifax, Nova Scotia.
Dr. Frank E. Woolley, Corning Glass Works, Corning, New York.
Dr. W. L. Worrell, University of Pennsylvania, Philadelphia, Pennsylvania.
Dr. John Yannopoulos, Newmont Exploration Ltd., Danbury, Connecticut.
Dr. Edward G. Zubler, General Electric Company, Cleveland, Ohio.
Dr. Alan D.Zunkel, St. Joseph Lead Company, Monaca, Pennsylvania.
Dr. J. Drowart, Université Libre de Bruxelles, Brussels, Belgium
Dr. P. Entner, University of Pennsylvania, Philadelphia, Pennsylvania.

MIX
Papier aus verantwortungsvollen Quellen
Paper from responsible sources
FSC® C105338

If you have any concerns about our products,
you can contact us on
ProductSafety@springernature.com

In case Publisher is established outside the EU,
the EU authorized representative is:
Springer Nature Customer Service Center GmbH
Europaplatz 3, 69115 Heidelberg, Germany

Printed by Libri Plureos GmbH
in Hamburg, Germany